Surveying Practice

Third Edition

SURVEYING PRACTICE

The Fundamentals of Surveying

Philip Kissam, C.E.

Professor Emeritus of Civil Engineering, Princeton University

Author of: Surveying, Surveying for Civil Engineers

Gregg Division

McGraw-Hill Book Company

New York   St. Louis   Dallas   San Francisco   Auckland   Bogotá
Dusseldorf   Johannesburg   London   Madrid   Mexico   Montreal
New Delhi   Panama   Paris   São Paulo   Singapore   Sydney
Tokyo   Toronto

Library of Congress Cataloging in Publication Data

Kissam, Philip, (date)
    Surveying practice.

    Includes index.
    1. Surveying. I. Title.
TA545.K55 1978   526.9   77-9952
ISBN 0-07-034901-0

Surveying Practice, Third Edition

1 2 3 4 5 6 7 8 9 0 K P K P 7 8 3 2 1 0 9 8 7

*The editors for this book were William K. Fallon and George
McCloskey, the designer was Roberta Rezk, the art supervisor
was George T. Resch, and the production supervisor was
Regina R. Malone. It was set in Modern 8A by Monotype
Composition Company, Inc.*
Printed and bound by Kingsport Press, Inc.

Contents

v

Tables

Preface

Surveying Practice, Third Edition, is designated for use in introductory surveying courses in junior colleges, technical institutes, engineering technology schools, and on-the-job training programs, as well as for self-instruction. Like its two very successful former editions, the book prepares the student to perform the duties of any member of a surveying party, including chief-of-party, both in the field and in the office. The book also serves as a firm foundation for future studies in surveying and helps develop the competencies necessary to obtain a position that is recognized as one of the best starting points in many types of industrial and engineering organizations.

This third edition has been improved by the inclusion of new material and changes recommended by users of the previous editions. Significant additions include metric conversion tables, new material on laser-beam surveying, a chapter on drawing maps and keeping records, and a more complete treatment of conducting property surveys. Moreover, both text and illustrations have been updated to reflect modern practices and attitudes.

The book is written clearly and simply, and nearly every principle is explained with the help of an illustration. End-of-chapter activities consist of problems and suggested field exercises. Problems are given in pairs; the problems in each pair are similar except in detail. Answers to the even-numbered problems are given at the end of the book, and answers to odd-numbered problems are given in a separate solutions manual available from the publisher.

A review of plane trigonometry and logarithms is covered in chapter 14.

The book assumes a very elementary knowledge of algebra and the ability to use the A, B, C, and D scales of a slide rule. Since most surveying computations are now made on desk and pocket calculators, however, machines of this type should be available to students and instruction should be given in their use. It is also recommended that copies of "Eight-place Table of Trigonometric Functions for Every Sexagesimal Second of the Quadrant" by J. Peters, published by Edwards Brothers, Inc., Ann Arbor,

Michigan, be available for computation. This table gives the values of the natural sines, cosines, tangents, and cotangents to eight places for each second of arc. The student would then be able to fit into office routine with very little difficulty.

If it is desired to study celestial observations, an ephemeris for the year is required. "The Solar Ephemeris and Instrument Manual," published by Keuffel and Esser Co., Morristown, New Jersey 07960, contains instructions for observations and computation and all the data required. It is sent free on request. The procedures for celestial observations and computations in this manual are written by the author and have therefore been omitted in this book.

Philip Kissam

Acknowledgments

The author wishes to acknowledge the contributions of the following organizations that were kind enough to supply illustrations: C. L. Berger & Sons, Inc.; U.S. Coast & Geodetic Survey; W. & L. E. Gurley, Wild Heerbrugg Instruments, Inc.; Thorpe-Smith, Inc.; and, particularly, Keuffel & Esser Co., which supplied so many.

Thanks are also extended to Selwyn Lewis studio for the photographs in Figs. 3-14, 3-15, 3-17 through 3-20, 4-19 through 4-25, and 4-29.

Philip Kissam

UNITS OF MEASURE
LENGTH

U. S. System				*Metric System*	
1 mile (mi)	=	5280	feet	1 kilometer (km)	= 1000 meters
1 chain (ch)	=	66	feet	1 meter (m)	= 1000 millimeters
1 rod (rd)	=	16.5	feet	1 millimeter (mm)	= 1000 microns
1 yard (yd)	=	3	feet	1 micron (μ)	= 1000 millimicrons
1 foot (ft)	=	12	inches (in)	1 millimicron (mμ)	= 1000 millionth microns ($\mu\mu$)
1 nautical mile	=	6076.1155−	feet	1 meter	= 10 decimeters
1 fathom (fm)	=	6	feet	1 decimeter (dm)	= 10 centimeters (cm)
				1 millimicron	= 10 angstroms (A)

Conversion 1959-Foot System and Metric System

1 kilometer	=	0.62137119+	miles	1 mile	=	1.609344	kilometers
1 meter	=	3.2808399−	feet	1 foot	=	0.3048	meters
1 meter	=	39.370079−	inches	1 inch	=	25.4	millimeters

AREA and VOLUME

1959-Foot System			*Metric System*	
1 sq. mile	=	640 acres	1 sq. kilometer	= 100 hectares (ha)
1 acre (A)	=	10 sq. ch.	1 hectare	= 100 ares
1 acrew	=	43560 sq. ft.	1 are	= 100 sq. meters

Conversion 1959-Foot System and Metric System

1 hectare	= 2.4710538+	acres	1 acre	=	0.40468564+	hectares
1 cu. meter	= 1.30795+	cu. yards	1 cu. yard	=	0.764555−	cu. meters
1 cu. cm.	= 0.0610237+	cu. in.	1 cu. inch	=	16.3870+	cu. cm.

Note. In 1959, the Foot System was redefined by agreement among officials of the nations where it is used, as follows: 1 yard = 0.9144 International Meter exactly. This reduced the lengths of units of the existing United States Foot System approximately 2 parts in 1,000,000. The then existing United States system was defined as follows: 39.37 inches = 1 International Meter and the foot in that system is now called the American Survey Foot. The American Survey Foot is still used by the U. S. Coast and Geodetic Survey and therefore applies to all the horizontal and vertical control nets in the United States. This exception is essential, as all data in feet published by that Bureau are the result of conversion from International Meters according to the definition 39.37 inches = 1 International Meter.

The meter is now defined in terms of light waves which can be produced under precisely defined conditions so that its length can be reproduced wherever and whenever required.

(Keuffel & Esser Co., Morristown, N. J.)

Part 1

BASIC SURVEYING OPERATIONS

1

Introduction

1-1. Definition of Surveying. Surveying is the art of making relatively large precise measurements with a maximum of accuracy and with a minimum expenditure of time and labor.

1-2. Basis of Surveying. Surveying is based on method and on the two chief instruments employed, the transit and the level. By adroit use of the method and skillful use of the instruments, almost any measurement problem can be solved and the work facilitated. Conversely, it is difficult to solve any problem of relatively large measurement with reasonable facility without resorting to surveying methods and surveying instruments.

1-3. Importance of Surveying. The present-day development in technology is both expanding the need for surveying in its customary applications and introducing many new fields in which surveying plays an essential part. The demand for high accuracy in property surveys, the new regulations for land subdivisions, the construction of high-speed highways and modern interchange facilities, the increasing use of steel

3

and prefabricated housing, the development of aerial mapping—all are increasing the need for accurate surveys. The newer developments that would be impossible without accurate surveys include testing equipment like rocket tracks, accelerators for atomic research, and cinetheodolite installations, as well as control for both position and direction of rockets, intercontinental missiles, and spacecraft (Fig. 1-1).

It is clear that an ever-increasing number of people must be available to carry out the fundamental surveying operations on which these great new developments depend.

1-4. Uses of Surveying.   Surveying is used for two specific purposes. The first is to make maps, charts, and profiles; to measure land boundaries; and often to determine precise sizes, shapes, and locations. Survey-

Fig. 1-1. Without surveying procedures, no self-propelled missile could be built to the accuracy necessary for its operation; its guiding devices could not be accurately installed; its launching equipment could not be constructed; it could not be placed in position nor oriented on the pad, nor could its flight be measured for test or control. Moreover, its launching position, the position of its target, and the gravitational forces which affect its flight would be a matter of conjecture. Surveying is an integral part of every project of importance that requires actual construction. Ewing Galloway (*Keuffel & Esser Co.*)

Fig. 1-2. Practically every line recorded on this photograph was laid out with a transit, a steel tape, a level instrument, and a rod—the primary equipment of the surveyor. The roads, streets, houses, drainage facilities, property lines, highways and highway interchanges—all were planned on maps created by surveying, and their positions marked on the ground by survey operations. (*Keuffel & Esser Co.*)

ing is thus a means of **measuring the relative positions of existing objects.** In this capacity, it serves as the only means of providing the information required for planning or designing all but the very smallest projects, as well as a means of checking how closely the finished work conforms to the original plan. The process is often called the **preliminary survey.**

The second purpose is to lay out, or mark, the desired positions and elevations of objects to be built or placed as directed by a completed plan, or to mark the boundaries of property either according to the findings of the land surveyor, according to a court decision, or as directed by a subdivision plan. In this capacity surveying comprises the first step in any actual building process or boundary location and is often called **location surveying** or **construction surveying.** As nearly every detail of a large project must be laid out by surveying methods, this type of surveying continues throughout the building process and entails by far the largest proportion of surveying operations carried out today (Fig. 1-2).

1-5. Surveys Must Be Correct. All surveys **must be free from mistakes.** A mistake in either the preliminary or the location survey may result in large expenditures for altering or removing and rebuilding finished construction. Since everyone makes a mistake occasionally, three rules are necessary which **must be followed** in all surveying operations. They are:

1. Record all field data carefully in a field book at the moment they are determined. The recorder must never allow the field party to give him data faster than he can record it. The record must be in a standard form and clearly written. The record should be erased[1] and changed if the data or the record is found to be incorrect immediately after it is recorded. Otherwise it must never be altered. If later on, or in the office, a mistake is discovered, the record should be crossed out so that it remains legible, and the new value should be circled or written in some distinctive color.

   If the data taken in the field are copied for one reason or another, the original record must not be destroyed. It follows that it is **unpardonable to lose a field book.** In the office, field books are usually kept in a safe. Also, if a field book is lost before the data are used, the survey must be repeated.

2. All data must be checked at the time they are recorded, either by adopting a method which automatically checks them or by having two people take the same reading independently.

3. The survey must be arranged so that the results can be checked by office computation.

It is absolutely essential that these rules be followed, even if they may seem to be ridiculous. In the descriptions of the procedures that follow, note that these rules are always carried out.

1-6. Accuracy versus Economy.   No measurement can be perfect. The more nearly perfect it is, the more highly **accurate** it is said to be. Almost any desired accuracy can be attained in surveying. More comprehensive methods or more precise instruments will give more accurate results. However, since higher accuracy requires more time in the field, usually more costly equipment, and more extensive computation, it is always more expensive. The degree of accuracy necessary for a survey depends on the purpose of the survey and therefore, the accuracy desired varies widely. Thus, to keep the costs at a minimum, the instruments and methods to be used must be chosen so that the desired accuracy is attained but too high an accuracy is avoided.

---

[1] Most surveyors feel that **any** erasure in a field book is bad practice. They recommend that incorrect entries found immediately after they are recorded should be neatly lined out and the correct data entered.

2

The Surveying Method

2-1. The Elements of Surveying. The surveying method is based on certain elements that facilitate operations and increase the accuracy of the work. These elements are described in the following paragraphs.

Gravity as a Reference Direction. The direction of gravity is used as a reference for all measurements. By **vertical** is meant the direction of gravity, and by **horizontal** is meant the direction perpendicular to gravity. Since the earth's surface is not a plane,[1] the direction of gravity is different at every position on the earth's surface.

The effect of the differences in the direction of gravity on horizontal measurements is so slight within a radius of 12 miles or so that it is almost impossible to measure it. For all small surveys, therefore, the curvature of the earth is neglected in the use of horizontal measurements. Plane geometry and plane trigonometry are used for the computation of results, and such a survey is said to be a **plane survey.** It is the only

[1] It is nearly an oblate spheroid, i.e., the solid generated by an ellipse rotated on its minor axis.

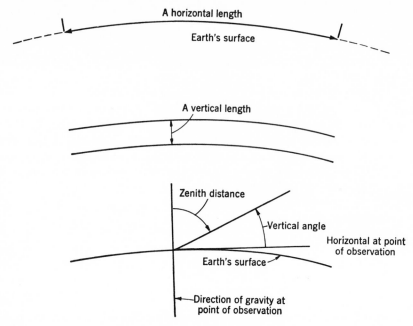

Fig. 2-1. Surveying measurements.

type of survey covered in this text. When spherical trigonometry or elliptical formulas are used, the survey is called a **geodetic survey.**

Measurements Made. Measurements are made of only four types of dimensions. They are (1) **horizontal lengths,** (2) **vertical lengths,** (3) **horizontal angles,** and (4) **vertical angles.**

Horizontal Length. A length measured horizontally throughout, that does not change in horizontal direction, is called a horizontal length or **distance** (Fig. 2-1). Sometimes a distance is measured on a slope and immediately reduced to the horizontal equivalent.

Vertical Length. A vertical length is measured along the direction of gravity and is equivalent to a difference in height (Fig. 2-1).

Horizontal Angle. A horizontal angle is an angle measured in a plane that is horizontal at the point of measurement. When a horizontal angle is measured between points that do not lie in this plane, it is measured between the perpendiculars extended to this plane from these points (Fig. 2-2).

Vertical Angle. A vertical angle is sometimes called the **altitude angle, angle of elevation,** or **site angle.** The vertical angle of a point is measured in a plane that is vertical at the point of observation and contains the point. Vertical angles are always measured up or down from the horizontal. Those measured upward are called **plus,** and those measured downward are called **minus** (Fig. 2-2). Sometimes the complement of the vertical angle is measured. This is the angle from the vertical above the point of the observation, i.e., the **zenith,** down to the point. Such an angle is called a **zenith distance** or **zenith angle** (Fig. 2-1).

2-2. Practical Concept.   The combined effect of these elements results in a very simple principle. All survey work is carried out as though the earth were flat. Of course, uphill and downhill slopes are taken care of; so a correct statement is this: All measurements are made as if the force of gravity were everywhere parallel to itself and as if underneath the irregular ground surface there existed a flat, horizontal reference plane.

There are only two exceptions to this as follows:

1. When a very long sight must be taken in leveling, the curvature of the earth is taken into account as described in Sec. 9-22.
2. When the survey covers a very large **area,** the field work is not changed, but the curvature of the earth is taken into account in the office computations.

2-3. Measurement of Horizontal Lengths.   Horizontal lengths are usually measured with steel tapes, generally graduated in hundredths of a foot, seldom in inches (see Fig. 2-3).

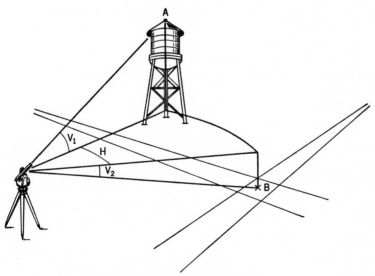

H is the horizontal angle between A and B
V_1 is the plus vertical angle from the transit to A
V_2 is the minus vertical angle from the transit to B

Fig. 2-2. Horizontal and vertical angles.

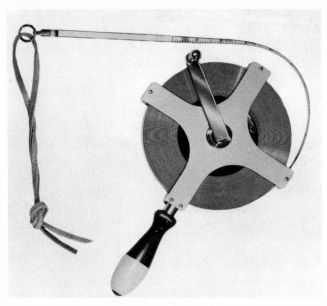

Fig. 2-3. Surveying tapes are graduated in feet and decimals of a foot. (*Keuffel & Esser Co.*)

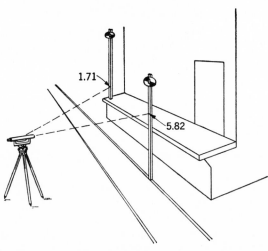

Fig. 2-4. Measuring a difference in height between a rail and a platform. Difference is 5.82 − 1.71 = 4.11.

Fig. 2-5. An engineer's transit. (*Keuffel & Esser Co.*)

2-4. Measurement of Vertical Lengths. Vertical lengths are usually measured with wooden rods held vertically and graduated in hundredths of a foot. The level instrument or its equivalent is used to observe the rods. A level consists of a telescopic line of sight, which can be made horizontal by an attached sensitive spirit level. The instrument can be turned in various directions around a stationary vertical axis. The differences in the readings on the rods are the differences in height of the points upon which the rods are placed (see Fig. 2-4).

2-5. Measurement of Horizontal and Vertical Angles. Horizontal and vertical angles are usually measured with a transit. A transit

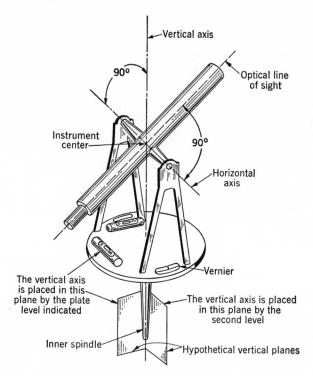

Fig. 2-6. Transit essentials. Schematic diagram of an alidade, which is the upper part of a transit.

consists essentially of an optical line of sight, which is perpendicular to and supported on a horizontal (or elevation) axis. The horizontal axis is perpendicular to a vertical (or azimuth) axis about which it can rotate. Spirit levels are used to make the vertical axis coincide with the direction of gravity. Graduated circles with verniers are used to read the angles (Figs. 2-5 and 2-6).

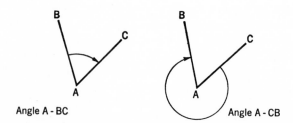

Fig. 2-7. Method of designating angles. Clockwise measurement is always assumed. The designation shows which of the two angles at *A* is actually measured.

Designation of Horizontal Angles. Horizontal angles are designated according to a system that differs slightly from the method used in geometry. The system is introduced so that it is possible to designate exactly which angle is measured. Figure 2-7 illustrates the surveying method of designating horizontal angles.

2-6. Measurement of Relative Horizontal Position. The relative horizontal positions of points are usually determined by **traverses** or by **triangulation.** A traverse consists of the measurement of a series of horizontal lengths called **courses** and the horizontal angles between these courses. Triangulation consists of the measurement of the angles of a series of connected triangles. At least one side of one triangle is measured. In both cases the final results are computed by trigonometry.

The results of a horizontal survey are best expressed by rectangular coordinates. One of the courses, or sides, is given a direction with respect to north, by measurement or assumption, and the directions of the other lines are computed from the measured angles. The direction used for north thus fixes the orientation of the coordinate system with respect to the survey courses (see Fig. 2-8).

Direction. The directions of the sides, or courses, are expressed either by **azimuths** or by **bearings.** An azimuth is ordinarily a **clockwise** horizontal angle from a reference direction, usually north. South is usually used for geodetic surveys that cover great areas. A bearing is the angle from the north or south, **whichever is nearest,** with the added designation of east or west, whichever applies. A bearing can never be

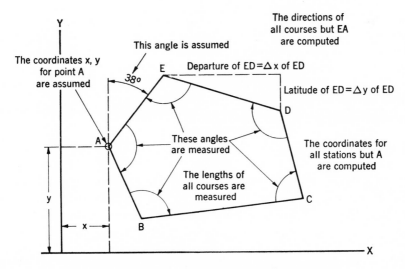

Fig. 2-8. Method of establishing a coordinate system.

greater than 90°. For example, the following four directions are expressed
by these three methods (see Fig. 2-9):

Azimuth$_N$		Azimuth$_S$		Bearing
120°	=	300°	=	S60°E
200°	=	20°	=	S20°W
290°	=	110°	=	N70°W
30°	=	210°	=	N30°E

The opposite direction to the one stated is often called a **back direction.** The back direction of a line can be found by adding $\pm 180°$ to the
forward direction. When bearings are used, this results in merely changing
both letters. For example, the back bearing of S27°10′E is N27°10′W, etc.

Horizontal Position.   The coordinates used are called **north** and **east**
or y and x. North, or y, ordinates are measured northerly from an east-
west line, or X axis. East, or x, abscissas are measured easterly from a
north-south line, or Y axis. To establish them, the coordinates of one of
the angle points are arbitrarily chosen. The coordinates of the other points
are computed by trigonometry. The coordinates of the starting point
should be so chosen that there will be no minus coordinates, i.e., the whole
survey will lie in the northeast quadrant. Every line has a Δy called

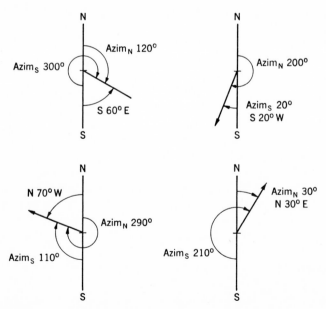

Fig. 2-9. Methods of expressing direction. Equivalent azimuths and bearings.

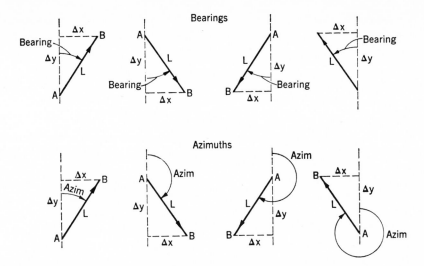

Fig. 2-10. Computation of latitudes and departures. In every case,

$$\Delta y/L = \cos \text{ direction} \qquad \Delta y = L \cos \text{ direction}$$
$$\Delta x/L = \sin \text{ direction} \qquad \Delta x = L \sin \text{ direction}$$

where Δy = latitude of AB
$\quad\;\; \Delta x$ = departure of AB
$\quad\;\;\; L$ = length of AB

a **latitude** and a Δx called a **departure** (see Fig. 2-8). The latitude of a line is equal to the length of the line multiplied by the cosine of its direction. The departure of a line is equal to its length multiplied by the sine of its direction (see Fig. 2-10). A marked point showing horizontal position is called a **station,** usually abbreviated to sta.

2-7. Measurement of Vertical Position. The relative vertical positions of points are determined by a series of level observations. Since the line of sight of the level is horizontal at each observation, the reference surface is made up of very short horizontal lines, or very nearly a curved surface everywhere perpendicular to gravity.

It is easier to refer the results of a level survey to a **standard datum.** Often, mean sea level is used. The vertical heights above the standard datum are called **elevations.** Sometimes the standard datum surface is called a datum **plane,** even though the surface is curved. A marked point of known elevation is called a **benchmark,** usually abbreviated to B.M. (see Fig. 2-11).

2-8. Solution of a Complete Surveying Problem. The method employed to attack a problem is **always the same** and applies to all engineering projects. It is often called the **engineering approach.** It can be best explained by an illustration. Consider for this illustration the process of constructing a hydroelectric power plant. There are five steps in the method, as follows: (1) reconnaissance, (2) preliminary survey, (3) map, (4) plan, and (5) location survey.

1. **Reconnaissance.** The general site is chosen by a careful study of existing small-scale maps and aerial photographs and by a thorough study of the actual ground, augmented by rough surveys.
2. **Preliminary survey.** A survey is made of the chosen site. This survey covers the watersheds, possible reservoir sites, dam sites, and sites for the buildings. It is called a preliminary survey and consists of determining the relative positions of existing topographic features and existing construction. Any survey which measures that which already exists is called a preliminary survey. The preliminary survey includes **control surveys** and **topographic surveys.**
   a. **Control survey.** A relatively few points called stations are permanently marked by monuments. They are arranged so that they can be easily surveyed, and their horizontal positions are determined by relatively precise triangulation and traverse (see Fig. 2-12). The elevations of the same or other permanent points, called benchmarks, are determined by relatively precise leveling. These positions and elevations provide an accurate framework upon which less accurate surveys can be based, without the accumulation of accidental errors or the high cost of making all measurements precise.
   b. **Topographic survey.** The topographic features are connected to the control surveys by measurements of comparatively low precision. Woven tapes, hand levels, plane tables, transit stadia, and photogrammetry may be used.
3. **Map.** Maps and profiles are drawn, giving all the required data.

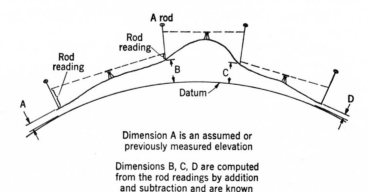

Dimension A is an assumed or
previously measured elevation

Dimensions B, C, D are computed
from the rod readings by addition
and subtraction and are known
as elevations

Fig. 2-11. Principle of vertical position, or elevation.

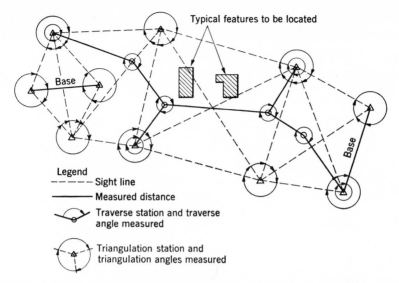

Fig. 2-12. A horizontal control system, showing a triangulation system with traverses measured from it.

4. **Design or plan.** By use of the map data, the construction plans are completed, and upon them are stated **location dimensions** to be measured from topographic features or control points. Vertical heights are usually given by elevations. Horizontal positions are given in the best practice by coordinates (see Fig. 2-13).

5. **Location survey.** The plans are executed first by marking on the ground the positions of the construction planned, according to the location dimensions. This is called **staking out.** It is accomplished by measuring from the control points or topographic features. This is the reverse process of the preliminary survey, requires different techniques, and is called a location survey.

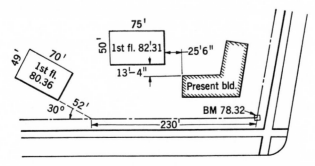

Fig. 2-13. Typical location dimensions found on plans.

2-9. Errors. It has been shown that surveying consists of making many measurements, and since no measurements can be perfect, it is obvious that measurements always contain errors. There are three types of errors: **blunders, systematic errors,** and **accidental errors.**

Blunders. A blunder is a mistake of some kind that causes **a wrong value to be recorded.** It is a **human** error. It may be caused by a mistake in reading the tape, the leveling rod, or the transit or by careless recording so that the number is not clearly written, is placed in the wrong position in the field book, is incorrectly labeled, or is not the same as the number read. It may be a mistake in counting, as would be the case in counting the number of tape lengths in a length measurement. It may be caused by using a wrong point in measuring an angle or a distance or by omitting a vital piece of information such as the fact that a certain measurement was made on a steep slope instead of horizontally.

Blunders can always be discovered, unless they are negligible, if the three rules given in Sec. 1-5 are always followed, except when two blunders happen to cancel each other. This very seldom occurs, as very few blunders can get into the record when every value is checked in the field as described in the rules. However, a type of blunder that is not discovered is staking out the wrong lot on a certain block or even on the wrong street, misreading a number on a plan, or counting one tape length too few in both very nearly parallel sides of a traverse. It follows that a surveyor must be continuously alert to avoid this type of blunder.

Certain kinds of mistakes are not blunders. These are mistakes due to lack of judgment or lack of knowledge. Mistakes of this kind can be avoided only by a thorough understanding of the principles of surveying.

Systematic Errors. Systematic errors are caused by imperfections in the equipment or the way it is used. They are **mechanical** errors. Under the same conditions, they always have the same size and sign. Since no equipment can be perfect, systematic errors always exist to a greater or lesser extent. Although they often introduce serious errors in the results, as they usually have no tendency to cancel, it is often difficult to be aware of their existence. In general, the surveyor must carefully consider the possible causes of systematic errors and take means to eliminate their effects on the results.

For example, suppose that a steel tape is the correct length at 68°F and that a survey had been made at some other temperature. The surveyor must realize that steel changes its size with temperature, must determine whether or not the error that results is large enough to be important, and, if it is important, he must correct all the measurements accordingly. See Appendix B.

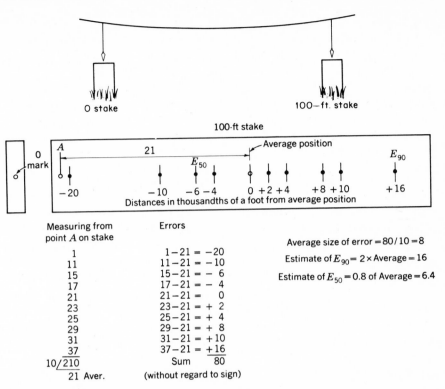

Measuring from point A on stake	Errors
1	$1 - 21 = -20$
11	$11 - 21 = -10$
15	$15 - 21 = -\ 6$
17	$17 - 21 = -\ 4$
21	$21 - 21 =\ \ \ 0$
23	$23 - 21 = +\ 2$
25	$25 - 21 = +\ 4$
29	$29 - 21 = +\ 8$
31	$31 - 21 = +10$
37	$37 - 21 = +16$
$10\overline{)210}$	Sum $\overline{80}$
21 Aver.	(without regard to sign)

Average size of error $= 80/10 = 8$

Estimate of $E_{90} = 2 \times$ Average $= 16$

Estimate of $E_{50} = 0.8$ of Average $= 6.4$

Fig. 2-14. Typical results of making 10 measurements of 100 ft with a 100-ft tape and plumb bobs. Scale greatly exaggerated. Maximum error, 0.02 ft. Measurement made from a point 100 ft to the left.

Accidental Errors—Theory.[1] An accidental error is the difference between a true quantity and a measurement of that quantity that is free from blunders or systematic errors. Accidental errors always occur in every measurement. They are the small **unavoidable** errors in observation that the observer cannot detect with the equipment and methods being used. Greater skill, more precise equipment, and better methods will reduce the size and over-all effects of accidental errors.

Consider the operation of placing two stakes, with pencil marks to show exact points, 100 ft apart in a field covered with high grass. A 100-ft steel tape is held level, and plumb bobs are held with their cords at the zero and 100-ft marks. If this operation were repeated a number of times,

[1] The theory of accidental errors should be studied only as far as desired at this time and finished later when the student is more familiar with surveying procedures.

there would be a series of marks on the 100-ft stake at slightly different distances from the mark on the zero stake, chiefly due to the endless swinging of the plumb bobs. The **average** position of these marks would be very close to the true distance between the zero mark and the 100-ft mark on the tape. The pencil marks would be closely grouped near the average position and farther apart on both sides of the average, somewhat as shown in Fig. 2-14. Obviously there would be some plus errors and some minus errors. Each of these errors is an accidental error. In the long run, the arrangement of the marks follows the laws of chance.

In Fig. 2-14 all values are given in thousandths of a foot and shown to an exaggerated scale. Ten measurements were made. The dots show where the marks came that were meant to be 100 ft from the zero point off to the left. The average position of the marks was determined by measuring from an arbitrary point A on the stake, averaging the values, and laying out this average distance (21) from A as shown. The distance of each mark from the average position was computed and called the error of that mark. These values are shown in the figure.

The closer the group of marks comes to the average, the greater the **precision** of the method of measurement. Two ways of indicating the precision of a measurement are used in this chapter, viz.:

1. The 90 per cent error (written E_{90}), which is just equal to, or greater than, 90 per cent of the errors
2. The 50 per cent error (written E_{50}) and known as the **probable error,** which is just equal to, or greater than, 50 per cent of the errors

In Fig. 2-14, $E_{90} = \pm 16$ because it is the ninth largest error, and $E_{50} = \pm 6$ because it is the fifth largest error.

Each is given a $\pm$ sign, as it is equally possible that it could have been plus or minus.

Both these values also can be approximately estimated from the average size of the errors, without regard to sign of the errors. In this case the average is 8.

$$\text{Estimate of } E_{90} = 2 \times \text{average} = \pm 16$$
$$\text{Estimate of } E_{50} = 0.8 \times \text{average} = \pm 6.4 \tag{2-1}$$

If this experiment were repeated a number of times, somewhat the same pattern of marks would occur in each experiment. If the E_{90} or the E_{50} values were averaged, the result would indicate a fairly accurate estimate of the 90 per cent error or the 50 per cent error of this method of measuring 100 ft.

How Accidental Errors Add Up. Assume that the 90 per cent error for measuring 100 ft turned out to be ±0.010 ft; what would be the 90 per cent error for measuring 1600 ft (16 tape lengths)? There would be 16 measurements of 100 ft, each with a 90 per cent error of ±0.010 ft. Since some of the errors would be plus and some minus, they would tend to cancel each other, but it would be very unlikely that they would completely cancel, so there would be a certain remaining error at 1600 ft.

It has been proved mathematically and shown by many tests that, in the long run, accidental errors of the same kind tend to accumulate in proportion to the square root of the number of observations of which they are composed. The equation can be written

$$\frac{E \text{ for } a}{\sqrt{a}} = \frac{E \text{ for } b}{\sqrt{b}} \tag{2-2}$$

where E = an error

a = number of observations in one determination

b = number of observations in another determination

(It is often assumed that the number of observations is proportional to the length of a survey.)

When b is the error of one observation and E_n is the error of n observations, Eq. (2-2) can be written

$$E_n = E_b \sqrt{n}$$

See also Appendix A.

In this case there are 16 observations; therefore

$$E_{90} \text{ for } 1600' = E_{90} \text{ for } 100' \times \frac{\sqrt{16}}{\sqrt{1}}$$
$$= 0.010 \sqrt{16}$$
$$= \pm 0.040'$$

Similarly, if E_{90} for 1600 ft were known, to compute the 90 per cent error for 100 ft,

$$E_{90} \text{ for } 100' = E_{90} \text{ for } 1600' \times \frac{\sqrt{1}}{\sqrt{16}}$$

Accidental Errors—Application. These principles make it possible to rate survey instruments, methods, procedures, and complete surveys by a per cent error computed from surveys previously performed. The most useful per cent error is the 90 per cent error for the result of a given number of observations. With this information, proper choice of instruments, methods, and procedures can be made when future surveys are planned. The following paragraphs describe how this is handled.

Accuracy and Precision. Since no measurement is perfect, the results obtained are qualified by some measure of **accuracy.** Usually this is estimated by the difference between two or more independent measurements. This difference is called the **error of closure,** or simply **closure.** When the accuracy is to be increased, greater **precision** must be used in the instruments, the methods, and the observations. Precision therefore can be defined as the degree of perfection **used** in the instruments, methods, and observations. Accuracy is the degree of perfection **obtained.**

Since, as has been stated, blunders and systematic errors can and must be eliminated from a survey, the degree of accuracy of a survey depends on the size of the accidental errors.

High precision is costly but necessary for high accuracy. The chief art of surveying is to obtain the data required, with the degree of accuracy desired, at the lowest cost.

2-10. Measures of Accuracy of Horizontal Measurement. The **degree** of accuracy of horizontal measurement is usually expressed as a ratio of the error of closure to the total distance measured; thus

$$H = \frac{C}{D} \tag{2-3}$$

where H = degree of accuracy
C = error of closure
D = distance measured

For example, if the error of closure is 0.18 ft made in measuring 577.80 ft, the degree of accuracy is computed as follows:

$$\frac{0.18}{577.80} = 1:3210$$

Ordinary measurement with a steel tape gives an accuracy of about 1:3000.

Note that the degree of accuracy depends on the assumption that the errors increase in proportion to the distance measured. Actually they increase only as the **square root of the distance** measured, which is a much smaller rate. Therefore, when the *same procedure* is used, the **degree** of accuracy of a long survey will be better than that of a short survey. In the example above, if the survey were four times as long, the estimated error of closure would be

$$0.18 \sqrt{4} = 0.36$$

and the **measure** of accuracy would be

$$\frac{0.36}{4 \times 577.80} = 1:6420$$

thus presumably twice as accurate.

2-11. Measures of Accuracy of a Level Survey. The accuracy of a level survey is expressed by V,

$$V = \frac{C}{\sqrt{M}} \tag{2-4}$$

where C = closure, ft

M = horizontal distance over which levels were run, miles

If, for example, in leveling 9 miles, the error of closure is 0.12 ft,

$$V = \frac{0.12}{\sqrt{9}} = 0.040$$

If the distance is expressed in feet, the formula becomes

$$V = C \sqrt{\frac{5280}{F}} \tag{2-4a}$$

where F = distance, ft

These equations are based on the assumption that the number of observations is proportional to the distance run. With this assumption, the error of a level survey should usually increase with the square root of the distance run [see Eq. (2-2)]. Since C is divided by the square root of the distance run to find V, the same precision will give the same value of V for any distance run.

2-12. Standards of Accuracy. Certain orders of accuracy have been generally accepted in the United States as standards. The most useful are given in Table 2-1.[1] The values are given in limits of error of closure.

Table 2-1

Order	Triangulation and traverse, H	Leveling, V, ft
First	1/25,000	0.012
Second	1/10,000	0.025
Third	1/5,000	0.050

2-13. To Rate Survey Instruments, Operations, Etc. It has been found that a survey should be planned so that 90 per cent of the work will be acceptable, as it is less expensive to rerun 10 per cent of the work than to attempt to reach perfection throughout. First the 90 per cent error should be determined from past experience. Two examples of this computation are given here.

Traverse. The most convenient value to use is the 90 per cent error for

[1] More complete information is given in Table V in the back of the book.

1000 ft. First, compute this value for each survey, as follows. Given: length of survey, 8500 ft; closure, 0.92 ft.

By Eq. (2-2):

$$\text{Computed error for 1000 ft} = 0.92 \frac{1}{\sqrt{8.5}}$$
$$= \pm 0.32'$$

Assume that the average computed error for 1000 ft for several surveys was 0.30 ft. Then, from Eq. (2-1),

$$E_{90} \text{ for } 1000' = 2 \times 0.30 = 0.60'$$

Level Surveys. Level surveys can be rated directly by their V's. If the average V for several surveys was 0.010, by Eq. (2-1)

$$E_{90} = 2 \times 0.010 = 0.020'$$

2-14. Choice of Survey Procedure. Usually the desired values for H and V are either specified or established by experience. Often merely the order of accuracy is stated. The surveyor should choose equipment and methods that have a rated 90 per cent error closely equal to the requirement.

Example 1. Required: a traverse about 10,000 ft long, to close with third-order accuracy, $H = 1:5000$. The maximum error of closure must be

$$\frac{1}{5000} \times 10,000 = 2'$$

Accordingly, the survey procedure chosen must have a 90 per cent error of 2 ft for 10,000. What is the corresponding rating for 1000 ft?

From Eq. (2-2)

$$\frac{E_{90} \text{ for } 1000}{\sqrt{1000}} = \frac{2}{\sqrt{10,000}}$$
$$E_{90} \text{ for } 1000 = \frac{2}{\sqrt{10}}$$
$$= 0.63'$$

A survey procedure so rated would be chosen.

Example 2. A level survey of any length is required to close with second-order accuracy ($V = 0.025$). A survey procedure rated at $E_{90} = 0.025$ would be chosen.

2-15. Summary of Error Theory. From the foregoing the following statements are evident:

1. Blunders can, and must, be eliminated.

2. Systematic errors may cause very large errors in the results, they can be recognized only by an analysis of the principles inherent in the equipment, and they must be eliminated by computation or field procedure.
3. Accidental errors are always present and control the measure of accuracy (the *H* and *V*) of the survey. They can be reduced at a higher cost by better field equipment and longer field procedures.
4. Accidental errors of the same kind accumulate in proportion to the square root of the number of observations in which they are found. This rule makes it possible to rate past surveys and select survey procedures for desired degrees of accuracy.

Problems

2-1. Express the following directions by two other means:

a. N20°10′E	*b.* A_N 130°30′	*c.* A_S 320°20′
d. N10°30′W	*e.* A_S 90°50′	*f.* S20°30′E
g. A_N 30°10′	*h.* A_S 40°20′	*i.* A_N 310°50′
j. A_S 210°20′	*k.* S40°10′W	*l.* A_N 250°40′

2-2. Express the following directions by two other means. Use three columns for the three methods.

a. N30°40′E	*b.* A_S 120°10′	*c.* A_N 350°40′
d. N40°20′W	*e.* A_N 10°30′	*f.* S0°30′E
g. A_N 90°20′	*h.* A_N 250°00′	*i.* S20°40′W
j. A_N 150°30′	*k.* A_S 160°10′	*l.* A_N 130°30′

2-3. Compute the back directions for the values given in Prob. 2-1.
2-4. Compute the back directions for the values given in Prob. 2-2.
2-5, 2-6. Express the following vertical angles by zenith distances.

	2-5	**2-6**
a.	20°10′	− 10°40′
b.	− 6 20	40 30
c.	60 40	0 10
d.	− 7 10	− 4 50

2-7 to 2-12. See Fig. 2-15. From a single position *O*, vertical angles *A* and *B* were measured to the tops of two flagpoles *A′* and *B′*. The distances from O to the flagpoles were found to be a and b, as shown. Find the difference in the elevations of the tops of the two flagpoles to the nearest 0.01 ft according to the data given below. If A' is above B', call the difference plus, and vice versa.

Fig. 2-15. Illustration for Probs. 2-7 through 2-12.

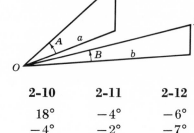

	2-7	2-8	2-9	2-10	2-11	2-12
A	20°	10°	16°	18°	−4°	−6°
B	10°	15°	−6°	−4°	−2°	−7°
a	100′	200′	300′	100′	200′	200′
b	200′	100′	200′	300′	100′	100′

2-13. Indicate by A, S, or B whether the following produce accidental errors, systematic errors, or blunders:

a. Swinging plumb bob

b. Reading 9 for 6

c. Repaired tape

d. No reading glass for transit

e. Aiming at the wrong point

f. Poor light

g. Recopying field data

h. Not aiming transit carefully

i. Transit not level

j. Failure to focus

2-14. Indicate by A, S, or B whether the following produce accidental errors, systematic errors, or blunders:

a. Level rod inaccurately graduated

b. Taking too long a sight with a level instrument

c. Carelessness in accurately centering the bubble of the spirit level in a level instrument when running a series of level observations

d. Level instrument out of adjustment so that the line of sight is not horizontal when the bubble is centered

e. Failure to check a reading

f. Failure to correct for temperature when measurements are made with a steel tape on a very hot day

g. Failure to hold the rod on the correct point

h. Leveling when "heat waves" make the rod apparently move rapidly up and down *i.* Using the end of the tape for measurement when the zero graduation is elsewhere

j. Poor eyesight

2-15, 2-16. Determine the accuracies of the following, and name the order of accuracy.

2-15.		Hor. Meas.		Levels	
	Error, ft	*Dist., ft*		*Error, ft*	*Dist., miles*
a.	10.00	23,361	*g.* 0.027		10
b.	0.50	3,005	*h.* 0.035		2
c.	1.27	14,000	*i.* 0.016		1
d.	0.09	1,002	*j.* 0.016		0.5
e.	1.00	25,000	*k.* 0.117		8
f.	0.84	8,400	*l.* 0.164		2
2-16.					
a.	8.00	30,560	*g.* 0.017		2
b.	0.07	2,000	*h.* 0.028		4
c.	1.32	8,460	*i.* 0.075		6
d.	0.13	1,709	*j.* 0.037		8
e.	1.00	17,543	*k.* 0.109		10
f.	0.72	1,800	*l.* 0.015		12

UNITS OF MEASURE
LENGTH

U. S. System

1 mile (mi)	=	5280	feet
1 chain (ch)	=	66	feet
1 rod (rd)	=	16.5	feet
1 yard (yd)	=	3	feet
1 foot (ft)	=	12	inches (in)

1 nautical mile	=	6076.1155−	feet
1 fathom (fm)	=	6	feet

Metric System

1 kilometer (km)	=	1000 meters
1 meter (m)	=	1000 millimeters
1 millimeter (mm)	=	1000 microns
1 micron (μ)	=	1000 millimicrons
1 millimicron (mμ)	=	1000 millionth microns ($\mu\mu$)

1 meter	=	10 decimeters
1 decimeter (dm)	=	10 centimeters (cm)
1 millimicron	=	10 angstroms (A)

Conversion 1959-Foot System and Metric System

1 kilometer	=	0.62137119+	miles	1 mile	=	1.609344 kilometers
1 meter	=	3.2808399−	feet	1 foot	=	0.3048 meters
1 meter	=	39.370079−	inches	1 inch	=	25.4 millimeters

AREA and VOLUME

1959-Foot System

1 sq. mile	=	640 acres
1 acre (A)	=	10 sq. ch.
1 acre	=	43560 sq. ft.

Metric System

1 sq. kilometer	=	100 hectares (ha)
1 hectare	=	100 ares
1 are	=	100 sq. meters

Conversion 1959-Foot System and Metric System

1 hectare	=	2.4710538+ acres		1 acre	=	0.40468564+ hectares
1 cu. meter	=	1.30795+ cu. yards		1 cu. yard	=	0.764555− cu. meters
1 cu. cm.	=	0.0610237+ cu. in.		1 cu. inch	=	16.3870+ cu. cm.

Note. In 1959, the Foot System was redefined by agreement among officials of the nations where it is used, as follows: 1 yard = 0.9144 International Meter exactly. This reduced the lengths of units of the existing United States Foot System approximately 2 parts in 1,000,000. The then existing United States system was defined as follows: 39.37 inches = 1 International Meter and the foot in that system is now called the American Survey Foot. The American, Survey Foot is still used by the U. S. Coast and Geodetic Survey and therefore applies to all the horizontal and vertical control nets in the United States. This exception is essential, as all data in feet published by that Bureau are the result of conversion from International Meters according to the definition 39.37 inches = 1 International Meter.

All values are exact except where followed by a + or − sign.

See also Appendix D.

3

Surveying Tapes and Taping

3-1. Steel Tapes. Except for determinations of low precision, horizontal lengths are measured by steel tapes (Fig. 3-1). In use, they are supported, preferably, either throughout their entire length or at regular intervals·(Fig. 3-2).

Whenever possible, a spring-balance handle should be attached to the forward end of the tape, even when only a low degree of accuracy is required. A spring-balance handle indicates the value of the pull applied. This ensures an accurate tension and speeds up the work by steadying the pull. When the accuracy required is greater than 1:3000, in addition to the spring-balance handle a thermometer must be used to measure the tape temperature. This is attached to the tape near one end with adhesive tape. The bulb should be in contact with the steel (Fig. 3-3).

3-2. Two Methods of Taping. When taping is used only to measure the distance between two points, a certain procedure is employed. When taping is used to mark certain positions on a given line, a more difficult procedure is necessary. There are many construction operations which

Fig. 3-1. A steel tape in a convenient reel (*Keuffel & Esser Co.*) and typical tape markings.

require marking positions on a given line, or "on line," as it is called; so the second type is encountered more often than the first.

As most distances are measured or laid out with a 100-ft or 30-m tape, their use is described here in detail.

3-3. Procedure for Taping a Distance between Two Points. Usually a 100-ft tape, graduated to hundredths of a foot, should be used. The inch is not generally used in surveying, for it introduces difficulties in computations and increases the opportunity for blunders. In the following description, a distance is to be measured from A to B. The zero mark on the tape is kept to the rear, although some surveyors prefer to keep the tape reversed. The position of the hands, etc., mentioned refers to right-handed persons.

A signal, usually a range pole (see Fig. 3-4), is set at station B. The head tapeman unreels the tape by walking toward B with the reel while the rear tapeman holds the zero end at A. He or she must always hold the zero of the tape **exactly** over the point, using a plumb bob when necessary, even when only a preliminary measurement is made. If he or she does not, the head tapeman will waste time clearing a place for the forward mark or often actually marking the point when the rear tapeman is holding the tape incorrectly. Frequently the head tapeman will raise the tape to clear obstacles to straighten it. The rear tapeman should raise the tape with him or her and still attempt to keep the zero mark as nearly as possible over the mark. When the head tapeman reaches the end of the tape, he or she removes the reel and attaches a handle or a thong at the end of the tape. The rear tapeman, sighting the signal at B, directs the head tapeman by voice until the head end of the tape is on line. He or she should name the direction and estimate length of movement, thus, "west, two-tenths," etc. The head tapeman pulls the tape straight and makes a rough measurement while the rear tapeman checks the alignment. The rear tapeman should keep his or her eyes above the mark, and the head tapeman should keep on one side of the tape so that the rear tapeman can see the target at B during this process. An error of 6 in. in alignment will not affect the results, and therefore subsequent to this rough measurement no further attention is given to alignment. The head tapeman prepares a place to mark the distance where the rough measurement fell. In grass, a small spot is rubbed clear

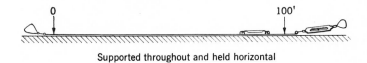

Supported throughout and held horizontal

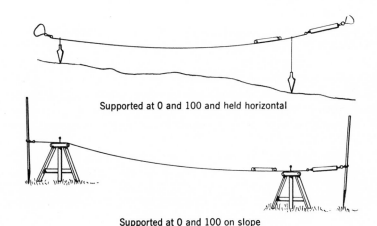

Supported at 0 and 100 and held horizontal

Supported at 0 and 100 on slope

The stretchers used facilitate the work

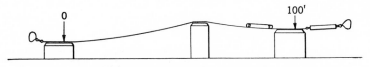

Supported at 0, 60, 100 over obstacles

Fig. 3-2. Methods of supporting a tape.

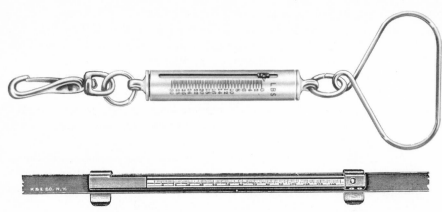

Fig. 3-3. A spring-balance handle and a tape thermometer. (*Keuffel & Esser Co.*)

of vegetation; on a pavement, she or he may mark a small spot with yellow
keel (see Fig. 3-5).

Next, the lengths of the plumb-bob cords are adjusted so that the bobs
will just swing clear when the tape is in position. The tape should be horizontal in the judgment of the head tapeman and should be as near the
ground as possible. If accuracy is desired, the tape should hang free of
all support. With the handles of the tape in their right hands, the tapemen should face the tape (their left sides toward each other). The plumb-
bob cord is held on the far side of the tape, bent over the tape, and held
on the proper graduation with the thumb of the left hand (Fig. 3-6). In
measuring, the tape is moved up and down slightly, tapping the point
of the bob on the mark to dampen the swing. The stance must be steady.
When the tape is high, the feet should be planted well apart along the line
of the tape; when the tape is low, one knee must be placed on the ground.

The head tapeman applies the tension gradually until the spring-
balance handle reads the correct tension, usually 20 lb. If no spring balance
is used, she or he must estimate the tension. When the head tapeman
applies the tension, the rear plumb bob may be pulled a short distance

Fig. 3-4. A range pole.

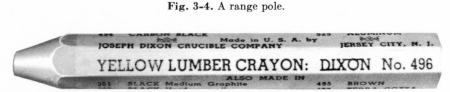

Fig. 3-5. Keel, often called lumber crayon. (*Keuffel & Esser Co.*)

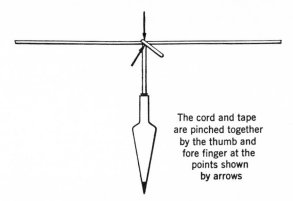

The cord and tape
are pinched together
by the thumb and
fore finger at the
points shown
by arrows

Fig. 3-6. Holding the plumb-bob cord on the tape.

off the point. The rear tapeman must pull the tape back at once with a smooth notion. He or she must try to hold the tape stationary and, as soon as it is in the correct position, call "mark." He or she should continue to call "mark" while it is in correct position and stop calling if it goes off. He or she stops calling "mark" when the head tapeman relieves the tension. When the tape becomes steady and the bob is still, the head tapeman lowers the tape so that the bob rests on its point. If the ground is soft, the hole the point makes is sufficient for the moment. He or she releases the tape and places a tack or a nail in the hole. Often a piece of cloth or marking tape is placed under the marker. Usually a chain pin (see Fig. 3-7) is placed in the ground near the tack so that the rear tapeman can find the tack and also so that it may act as a tally. If he or she is working on a pavement or other hard surface, the head tapeman holds the bob nearly upright, and thus the point remains where it marks the correct position. This is done by regulating the cord in the left hand. He or she then releases the tape, reaches the bob with the right hand, and marks the position of the point (Fig. 3-8). Usually this is done by making a scratch with the point

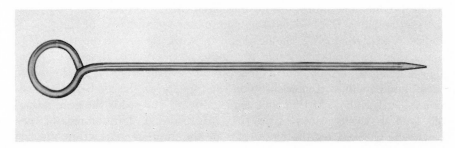

Fig. 3-7. Chain pin. (*Keuffel & Esser Co.*)

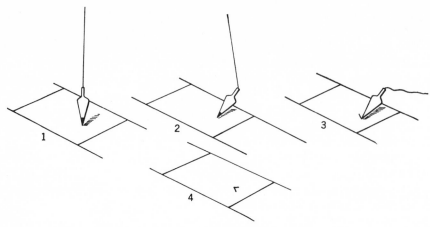

Fig. 3-8. Steps in marking a point on masonry.

from the position it occupies. The beginning of the scratch is the mark.
A second scratch is made from the mark, at right angles to the first,
forming a V. He or she then writes the number of tape lengths (stations)
on the pavement with keel. The rear tapeman calls the number of the
station **he or she occupies,** and the head tapeman calls the number
he or she is marking.

It is very difficult and very important to keep track of the number of
tape lengths. An error in counting tape lengths is one of the chief sources
of error in surveying. In addition to the two methods suggested here,
other systems may be worked out by the tapemen.

Frequently the measurement is repeated for a check after it is marked,
if either tapeman believes this to be desirable.

The head tapeman moves forward toward *B*, dragging the tape. The
rear tapeman recovers the chain pin. He or she *must not* pick up the rear
end of the tape. When the zero mark comes to the station, the rear tapeman
calls "chain" or the number of the station. The head tapeman stops and
can usually line himself or herself in within a foot or so by looking back
along the line. The procedure for measurement is then repeated.

Upon reaching *B*, the head tapeman either reels in part of the tape or
walks on past *B* carrying the head end forward. He or she returns to *B* to
make the measurement. While plumbing as previously described, he or
she slides the plumb-bob cord along the tape until the bob is on the mark;
then, holding the cord in position on the tape, he or she reads the graduations silently. The rear tapeman comes forward and reads the graduations
out loud. If the readings agree, they are recorded. The number of tape
lengths is checked by the chain pins or the number marked on the last
station.

When the tape is used with the 100-ft mark to the rear, the head tapeman holds the zero mark at *B* while the rear tapeman takes the reading and holds it while the head tapeman moves back to check.

To return the tape to the reel, the head tapeman first removes the handle or thong from the 100-ft end of the tape and passes the end into the reel between the two reel spacers opposite the handle where it will not come in contact with the spacers while it is being reeled in. He or she engages the end ring in the spindle so that the graduated side of the tape is up when the reel crank handle is on the right, facing the tape. He or she then reels in the tape and locks the reel. With the tape in this position, it can be used conveniently to measure less-than-tape-length distances.

3-4. Tapes with Graduations Back of Zero. Often a 100-ft tape has graduations extending 0.99 ft back of the zero mark, and thus outside the 100-ft length. These graduations are numbered backward; (see Fig. 3-9). If the end of these graduations is used instead of the zero mark, the distance 100.99 ft is measured. Accordingly, the two tapemen must make sure which kind of tape they are using and exactly where the zero mark is located. This kind of tape is usually graduated only at intervals of 1 ft. It is therefore cheaper than a fully graduated tape. In use, the graduated end and the zero mark are kept forward with the 100-ft mark at the rear. In laying out 100-ft intervals, the 100-ft mark and the zero mark are used. In measuring distances of less than 100 ft, for example, 64.32 ft, the head tapeman stops when point *B* is reached and holds the zero mark at that point. The rear tapeman finds that the previous point marked on the ground comes between the 64-ft mark and the 65-ft mark. Choosing the smaller, she or he calls "holding 64" and holds the 64-ft mark over the ground mark. The head tapeman reads the value of the backward graduation, i.e., 0.32 ft at point *B*. See Fig. 3-9.

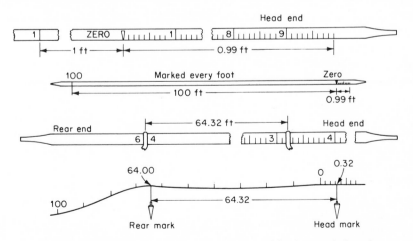

Fig. 3-9. A 100-ft tape with graduations outside the 100-ft length.

This type of tape has two advantages. It is cheaper, and at the end of a measurement the head tapeman does not have to either walk beyond the end point or reel in the tape before measuring.

It has three disadvantages. It is confusing to use, both because the foot graduations are numbered backward, which is unnatural, and because the end graduations are numbered in the opposite direction. Also, it is almost impossible to check the reading without tangling the tape; one of the tapemen must carry the tape mark she or he used to the other tapeman if two people are to read the tape independently (see Sec. 3-3). Finally, sooner or later, the head tapeman will use the end of the graduations by mistake instead of the zero mark.

3-5. Breaking Tape.   When at any time the slope is so great that the entire tape cannot be used, a process called **breaking tape** is employed. After carrying the tape out to the full distance, the head tapeman returns to the point where the tape can be held level. He or she selects a certain foot mark, which he or she announces to the rear tapeman. When the mark is set at this distance, the rear tapeman comes forward, taking the tape from the head tapeman so that there is no opportunity of using the wrong tape graduation, and then uses this graduation as though it were zero. No chain pin is set. Sometimes the **plus** (distance from the last station) is marked on the pavement (Fig. 3-10).

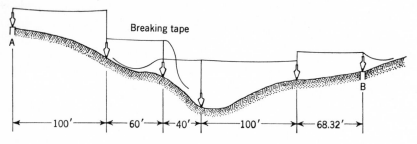

Fig. 3-10. Operations in taping.

3-6. Precautions to Avoid Damaging the Tape.   Although most steel tapes used for surveying will withstand a direct tension of 80 lb or more, it is very easy to break them by misuse. When a tape is allowed to lie on the ground, unless it is kept extended so that there is no slack, it has a tendency to form small loops like that shown in Fig. 3-11. When tension is later applied, the loop becomes smaller until either it jumps out straight or the tape breaks, as shown. If a tuft of grass or any object is caught by the loop, the tape almost always breaks or at least develops a permanent kink.

To avoid this, the tape must be handled so that no slack can occur.

For measurements of less than a full tape length, the tape should be kept on the reel. It should be reeled out to the necessary length and reeled in as soon as possible. The head tapeman, who handles the reel, must reel in any slack that might occur between the two tapemen while the tape is being handled. For measurements greater than the tape length, when the tape is off the reel, the tape should be kept fully extended in a straight line along the direction of measurement. It may be allowed to lie on the ground in this position, but, when it is to be moved, it must be dragged from one end only. If it is necessary to raise the tape off the ground, the two tapemen must lift the tape simultaneously and keep it in tension between them. Except for this operation, the rear tapeman must not touch the tape while it is being moved. If she or he picks up the rear end and moves forward faster than the head tapeman, she or he will form a U in the tape, which usually causes a loop to form when the tape is pulled taut.

When the end of the measurement is reached, where a less-than-tape-length measurement is required, the head tapeman must not pull in the tape hand over hand. This creates a pile of tape on the ground. This is safe only on a smooth surface. Instead, he or she must do one of three things:

1. Carry the end of the tape beyond the point, lay it on the ground, and walk back
2. Reel in the tape the requisite amount or
3. Take in the tape, forming figure-eight loops hanging from his or her hand

Each length of tape must be laid in his or her hand flat on the previous section and never allowed to change. Later, to extend the tape, he or she must lay it out carefully, as he or she walks forward, by releasing one loop at a time. This third method requires care and practice and should not be attempted until after considerable practice over a smooth floor where there is little danger.

If possible, no vehicle should be allowed to run over the tape. If the

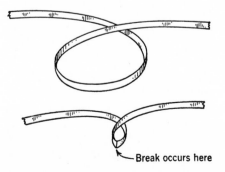

Fig. 3-11. How a loop breaks the tape.

Break occurs here

tape is across a smoothly paved street, a pneumatic tire can pass over the tape without damaging it if the tape is held flat and tightly pressed against the street surface by the two tapemen.

When a tape is wet, it should be carefully cleaned and oiled as soon as possible.

In general, it is well to remember that a tape is easily damaged but, with care and thought, damage seldom occurs.

3-7. Accuracy Obtained with Plumb Bobs.   Even experienced tapemen have difficulty in preventing the tape and the bobs from moving during measuring. The error from this source is between 1:5000 and 1:10,000. Variations in tension of 5 lb introduce an error of 1:10,000 with a tape of average cross section 100 ft long and supported at the ends. Temperature may introduce an error up to 1:5000, so that, all in all, this type of measurement has an accuracy seldom better than 1:2500. Using spring-balance handles will improve the accuracy to 1:3000; with temperature correction as well, an accuracy of 1:5000 can be reached.

The great advantage in this type of taping is that the tape is held level. The effect of slope increases nearly as the square of the difference in elevation of the ends; thus, when the tape is nearly level, the error can be disregarded. This eliminates the necessity of measuring the slope.

3-8. Procedure for Taping to Set Marks at Certain Distances on a Given Line.   This is known as **setting marks for line and distance.**   When a series of marks are set at measured distances, a standard system of naming the distances is almost always used. A zero position is established usually at the beginning of the survey or at the beginning of the line to be marked out. This zero point is called 0 + 00 or 0 + 0. Each 100 ft is called a **station.** The number of each station point is the part of the name in front of the plus sign. For example, the mark at 3 + 00 is station 3. A point 350 ft from the zero point is called 3 + 50; at 462.78 ft from zero it is called 4 + 62.78. The +50 and +62.78 are called pluses. The point 462.78 is said to have a plus of 62.78 from 4 or merely to have a plus of 4 + 62.78. Frequently, the enumeration is carried continuously throughout the whole survey. See Fig. 3-12.

The usual mark is a stake with a surveyor's tack driven in the top of the stake to mark the exact position. About 1 ft to the right of the stake is a "guard," usually a builder's wooden plaster lath or a 4-ft stripling with a piece of cloth or marking tape held in a split at the top. The station and plus are marked on the side of the guard facing so that a person walking forward (in the direction of increasing pluses) will see the plus on the guard and find the stake about a foot to the left of the guard.

When the stake cannot be driven because there is a root or a stone too near the ground surface, the earth is cleared off and a tack is placed in the root or a cross mark is chiseled in the stone. A point in masonry is

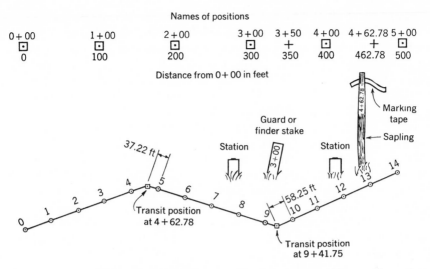

Fig. 3-12. Enumeration of stations and pluses.

marked with a chiseled cross, and a point in a hard-surfaced road with a roofing nail with the washer in place or with a special nail for the purpose. Each of these is circled, and the plus marked with paint or keel.

A transit is used to give line. It is set up over one end of the line and aimed at the other end. The tape is handled very much as when measuring between two points (Sec. 3-3).

Usually the measurement starts at the transit. The rear tapeman holds the zero end of the tape near the transit while the head tapeman carries all the equipment forward, holding the reel so that the tape unwinds. When the proper distance is reached, the head tapeman stops, and the rear tapeman places himself or herself under the transit, being careful to avoid touching the tripod legs. He or she holds the zero mark of the tape on the tack if conditions permit the tape to be level in this position. He or she can steady the tape against the stake but must take the full tension with his or her hand.

If the tape must be raised above the stake, the rear tapeman loosens the plumb-bob cord on the transit until about 8 in. slack has been taken in. He or she places the bob on the ground and holds the cord taut by pressing it against the tack with one hand. With the other hand he or she controls the tape so that the zero mark is at the cord (see Figs. 3-13, 3-14, and 3-15).

The head tapeman bends the plumb-bob cord over the tape at the proper graduation, holding it in position by squeezing the cord and tape together with one hand (Figs. 3-13 and 3-15). With the other hand he or she applies the tension, holding the tape at the proper height to keep it level. When the plumb bob is steady, he or she calls "line for stake." The transitman directs line by signal or voice, giving the compass direction and the amount of the movement, thus, "north two-tenths," "south five-hundredths," etc. When the cord is brought nearly on line, she or he signals or calls "good for stake."

At this call, the head tapeman releases the plumb bob so that it drops vertically, marking the ground slightly with its point. The longest dimension of the top of the stake is kept in the direction of measurement, and the stake is driven at the mark to a depth of 2 or 3 in. The position of the stake is then checked. The head tapeman calls "distance," stretches the tape, and checks the distance. He or she then calls "line for stake" and holds the bob for the transitman, moving it as directed. The stake is driven according to the results of this check, the transitman watching it go down as long as it is visible. He or she will call "keep it south" or or "south one-tenth" as the need arises.

Driving a Stake.   It takes considerable skill to drive a stake so that

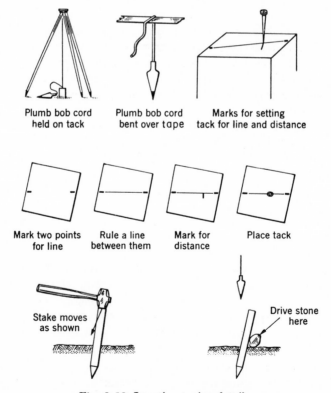

Fig. 3-13. Location taping details.

Fig. 3-14. Holding the plumb-bob cord taut against the tack.

the top remains in position. Frequently the head tapeman makes a second check when the stake is partly driven home. The top of the stake invariably moves toward the person driving it. Slight corrections can therefore be made by driving it from the position toward which the stake should move (see Fig. 3-13). When greater corrections are necessary, the ground should be pounded beside the stake. Still greater corrections can be made by driving stones into the ground beside it. Tapping the side of the stake to align it merely loosens the stake and sometimes breaks it.

When the stake is driven well into the ground and found to be out of position, the only recourse is to drive another stake beside it. If it is withdrawn, it will follow the old hole when redriven.

The stake must be driven until it is firm and usually with the top not more than a few inches above the ground.

Setting a Tack. A pencil is placed on the top of the stake, held slanting away from the transit or, preferably, balanced on its point. The pencil point is lined in and a pencil mark made on line.

If the transitman cannot see the pencil, he or she signals or calls "raise it" and a plumb bob is substituted. The head tapeman should hold the cord as close as possible to the bob without interfering with the transitman's view. The swing of the bob can be damped by tapping the point against the top of the stake.

When she or he is ready, the head tapeman calls "line for tack"; when the plumb bob (or other signal) has been brought precisely in line by directions from the transitman, the latter calls "good for tack." When satisfied, the head tapeman drops the bob to the stake by dropping one hand about $\frac{1}{2}$ in. Then, while holding the cord and bob in this position with one hand, he or she reaches the bob with the other hand and marks the point by making a hole in the stake with the point of the bob. If there is any doubt in his or her mind of the accuracy of the mark, the head tapeman calls for a check.

Frequently two marks are made for line near the edges of the top of the stake toward and away from the instrument, and a pencil line is ruled between them.

When conditions permit, the tape is laid on the top of the stake to obtain distance. If this is impossible, a plumb bob is used. The cord is bent over the proper graduation as before, the tension applied, and the

Fig. 3-15. Holding the plumb-bob cord at a tape graduation.

swing damped out by moving the tape up and down so that the point of the bob taps the stake. The head tapeman should keep the bob over the pencil line. The exact point is marked with the point of the bob and checked if necessary. If only one line point has been set, it is sometimes necessary to check the distance mark for line. When the tape can be laid on the stake, the edge of the tape can be used to transfer the line to the correct distance mark. A tack is driven at the final mark. Frequently the tack is again checked for line and distance (see Figs. 3-16 through 3-20).

The station number is marked on the stake with keel or on a guard stake set at a slant near it. The number of the stations should be checked by call. The head tapeman calls "station," and the rear tapeman calls the number of the station where he or she is standing.

The checks enumerated may seem rather excessive. It is the duty of the head tapeman to decide when they are necessary. He or she knows by experience whether conditions were proper for an accurate result. Too few or too many checks will often waste time. The speed and accuracy of the work are obviously entirely in the hands of the head tapeman. If the field party is shorthanded, the chief should take this position.

When the head tapeman is finished, if a full tape length has been measured, the rear tapeman drops the end of the tape and walks forward to the stake just set. In the meantime the head tapeman takes his or her equipment forward and drags the tape. When the zero end of the tape reaches the stake, the rear tapeman calls its station number, the head tapeman stops, and the process of setting a stake is repeated. The rear tapeman now handles the tape in the same manner as the head tapeman

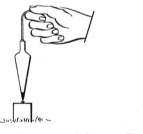 

Hold as close to bob as possible and keep bob point as close as possible to stake

To mark stake, settle bob on stake at proper point; then controlling bob as shown, seize bob and make hole with point

Fig. 3-16. Handling a plumb bob.

Fig. 3-17. Short hold on a plumb bob.

except that, instead of applying tension, he or she resists it.

Marks on Other Surfaces.   When the marks are made on masonry, the process is simpler. Pencil lines or scratches with the plumb-bob point are used for marks. A cross is chiseled at the mark if it must be permanent. Usually the mark is circled with keel to make it easy to find.

When the stake strikes an obstruction before it is firm, the earth is cleared away and the mark is made on the obstruction.

When the surface is hard and irregular, such as a macadam road, a heavy nail can be driven as a mark. Often a small piece of cloth or a roofing washer is placed on the nail to make it easy to find.

Fig. 3-18. Plumbing high in the wrong manner causes errors.

Fig. 3-19. Measuring for a tack with a short hold.

Fig. 3-20. Measuring for a tack with a long hold.

Signals Used in Giving Line

1. **Take line** means to point on the target designated. The rod or plumb bob is held horizontally and then placed over the point.
2. **Give line** or **line** means to give directions for bringing a target on line. To indicate this, the range pole or plumb bob is merely held approximately on line.
3. **Directions for line.** The transitman gives line by motions with the hand on that side of the body which is toward the direction of motion. The palm is held toward the person receiving the signal. Slow motions indicate large distances; quick motions indicate short distances. A handkerchief is often held in the hand to increase visibility. Sometimes the distance desired is signaled in hundredths of feet, the signals shown in Fig. 6-6 being used.

4. **Raise it** means that the plumb-bob string or other target cannot be seen. A handkerchief is moved up and down over the transit, or the transitman extends his or her arms downward at his or her sides and swings them outward and upward to meet over his or her head with the palms together and fingers pointing up. The tapeman should discover what is wrong and correct it. Usually the ground interferes, and more plumb-bob string should be exposed. Sometimes a better background is necessary, and the tapeman should place one leg behind the string or stand behind it. Sometimes the range pole should be substituted.

5. **Good** or **all right** means that the transitman is satisfied with alignment or other procedure. Both hands with the palms forward are moved up and down at the sides of the body. When visibility is poor, a handkerchief is waved over the head or in a large circle.

6. **Pick up the instrument.** To avoid lost time, the transitman should **never** pick up the instrument unless directed to do so by the chief. The signal is a quick upward motion with both hands palms up.

Taping on Smooth Surfaces.   Taping with the tape lying fully supported on smooth surfaces like paved roads, streets, or sidewalks is more accurate than taping with the tape supported at the ends only and using plumb bobs. When the tape is supported at the ends only, no matter how great a tension is used, the tape always sags to a certain extent. A slight change in tension will change the sag and thus change the distance between the zero and the mark used. In addition, the effect of wind and the swing of the plumb bobs are difficult to control, and further errors are introduced. When the tape is in direct contact with the supporting surface, these difficulties are avoided. Usually a tension of 10 lb is used, and the head tapeman must make sure that the tape is not affected by friction.

Since smooth surfaces are seldom level, elevations must be determined by leveling at each end of the distance measured, at all changes in slope, and at every 50 to 100 ft where the slope changes gradually. The plus where each elevation is determined is recorded. With these data, the slope corrections can be computed.

On long uniform slopes, the slope can be measured by measuring its vertical angle with a transit.

When there is a gap or sag in the pavement so that the tape is unsupported for not more than 20 ft, the effect on measurement is negligible.

CORRECTIONS FOR ACCURATE TAPING

In plane surveying, there are three tape corrections if the tape has been properly standardized:

Tape correction C_t for the actual length of the tape
Temperature correction C_f for the expansion of the tape
Slope correction C_h for reduction to horizontal

If the tension or the support of the tape is not the same when the tape is used as it was when standardized, there is a sag correction C_s and a tension correction C_p. See Appendix B.

3-9. Accuracy of a Tape.   Tape manufacturers make 100-ft tapes that are very nearly correct in length at 68°F when supported throughout and under a tension of 10 lb. When a tape is supported at two or more single points so that it sags between these points of support, the "length of the tape" is the sum of the straight-line distances between the supports. A 100-ft tape supported at the zero and 100-ft points under a tension of 20 lb is usually very nearly the same length as when it is supported throughout under a tension of 10 lb, as will be discovered when it is standardized.

In use, tapes tend to change length. They wear and thus become thinner and lighter so that they stretch more and sag less and thus become longer for both reasons. When a tape is kinked or repaired, its length changes. Molecular changes in a tape made of new steel sometimes cause it to change slightly in length. These changes are quite small and of little importance in many types of surveys. However, whenever accurate results are necessary, tapes must be compared with a **standard tape** at frequent intervals. The actual length of the tape is thus determined and the tape is said to be **standardized.**

A Standard Tape.   Every surveyor or survey organization required to do accurate work should have a standard tape available for checking working tapes from time to time. The standard tape must be purchased from a manufacturer by special order to conform with the following specifications:

1. The ribbon shall be of steel from old, seasoned stock about $\frac{1}{4}$ in. wide and about 0.014 in. thick.
2. Marks shall be placed at the 0 and 100-ft points only. Both marks must be on the body of the ribbon and not at either end. They shall extend from the center line of the ribbon to the edge farthest from the observer when facing the tape with the 0 end in the left hand.
3. The marks shall be about 0.004 in. wide.
4. The tape shall be standarized by the National Bureau of Standards, fully supported at a tension of 10 lb. Its length when supported at the zero and 100-ft points, at a tension of 20 lb, and at 68°F can be then computed as shown in Appendix B.

Such a tape is usually less expensive than a working tape. The Bureau of Standards will mark the tape with an identifying number and give a

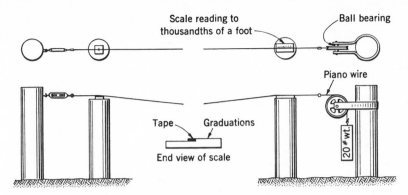

Fig. 3-21. Base for comparing tapes supported at the zero and 100-ft marks.

certificate stating the straight-line lengths between the end marks to the nearest 0.001 ft under the conditions described in paragraph 4 above.

To Compare a Working Tape with a Standard Tape. Every working tape used for accurate measurement must carry an identifying mark. The best is a letter stamped on the flat part of the end loop.

The essential elements in making the comparison are the following:

1. It will be found that it is next to impossible to establish two permanent marks that remain exactly 100 ft apart. If permanent marks are used, one mark must be a single line and the other a scale graduated in units of 0.002 ft. See Figs. 3-21 and 3-22.
2. It will also be found that, indoors, drafts keep the temperature

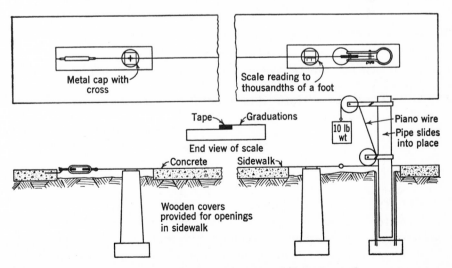

Fig. 3-22. Base for comparing tapes fully supported.

changing. The ideal location is in the shade on the north side of an unheated building.

3. Before comparison, the working tape must be cleaned and the two tapes placed together to attain the same temperature. The standard tape is first placed on the testing base, and the zero end is adjusted, using the turnbuckle, until it agrees with the single mark. The scale is read at the 100-ft mark, and, with the known length of the tape, the scale reading for 100 ft is computed.

4. The working tape must be supported exactly as it is to be used in the field. If a tape thermometer (Sec. 3-1) is to be used, it should be attached to the tape in its normal position. It is convenient to find the length of the tape both supported throughout and supported at the ends only.

   Since the standard tape and the working tape are both steel, no temperature corrections are necessary if the temperature remains the same throughout the test.

5. Usually the length of the working tape between the 0 and 100-ft marks is recorded under a tension of 10 lb when the tape is fully supported and 20 lb when supported at the ends only.

   Some surveyors prefer to find the tension which gives the tape the correct length, to eliminate office computations. Often, however, for some tapes, the required tension is too small to straighten out the tape or to eliminate enough of the sag. Either will give inaccurate results. Sometimes the tension required is too large to be practical. Also, if this system is used, the head tapeman must use different tensions with different tapes, so that it is never certain whether the correct tension has always been used.

Figures 3-21 and 3-22 show ideal arrangements for testing tapes. Much simpler methods are satisfactory.

Test for Spring-balance Handle. The spring-balance handle should be tested when horizontal. Attach it to a weight by a wire over a pulley.

3-10. Tape Correction. For a 100-ft tape, the tape correction per 100 ft, C_t, is computed as follows:

$$C_t = L_s - 100$$

where L_s is the actual length of the tape by standardization. An example for a tape identified as tape A is shown in Table 3-1.

3-11. Correction for Temperature. Air temperature readings will give the temperature of the tape on hazy or cloudy days. When the sun is shining, a tape thermometer is necessary (Fig. 3-3). It is designed so that the bulb comes in contact with the tape. The device is clipped to the tape and firmly bound to it with white adhesive tape near the forward

Table 3-1

Tape	A	A
Support	Throughout	0 + 100'
Tension	10 lb	20 lb
L_s	100.003	99.998
	−100.000	−100.000
Cor. per 100'	+0.003'	−0.002'

end, where it can be easily read by the head tapeman, where it creates very little extra sag, and so that it is off the ground when the tape is dragged forward. The average temperature for the measurement is determined by several readings, sometimes for every time the tape is used.

The correction for temperature can be computed by one of three methods, as given below.

Example. Apparent length = 1407.61, temperature = −10°F
By the Thermal Coefficient of Expansion of Steel per Degree F:

$$C_f = 0.00000645l(F − 68°)$$

where C_f = correction, ft
$\quad$ l = apparent length, ft, as observed
$\quad$ F = temperature, °F

$$C_f = (0.00000645)(1408)(−10 − 68) = −0.708'$$
$$\text{Corrected length} = 1407.61 − 0.71 = 1406.90'$$

By the Approximate Formula:

$$C_f = 0.0001l \frac{F − 68}{15.5}$$
$$= 0.1408 \frac{−10 − 68}{15.5} = −0.709'$$

$$\text{Corrected length} = 1407.61 − 0.71 = 1406.90'$$
By Table 3-2:
$$C_f = 0.01lT$$
where T = value found in table.

$$C_f = (14.08)(−0.050) = −0.704'$$
$$\text{Corrected length} = 1407.61 − 0.70 = 1406.91'$$

See also Table 3-2 below or Table VI at back of book.

Table 3-2. Interval Table for Temperature Corrections* for 100-ft Steel Tape to Find Actual Distances Measured

Deg F	Corrections per 100 ft	Deg F	Corrections per 100 ft	Deg F	Corrections per 100 ft
-13		$+41$		$+95$	
	-0.050		-0.015		$+0.020$
-6		$+49$		$+103$	
	-0.045		-0.010		$+0.025$
$+2$		$+56$		$+111$	
	-0.040		-0.005		$+0.030$
$+10$		$+64$		$+118$	
	-0.035		-0.000		$+0.035$
$+18$		$+72$		$+126$	
	-0.030		$+0.005$		$+0.040$
$+25$		$+80$		$+134$	
	-0.025		$+0.010$		$+0.045$
$+33$		$+87$		$+142$	
	-0.020		$+0.015$		$+0.050$
$+41$		$+95$		$+149$	

* Based on a coefficient of expansion of 0.00000645 per °F. Multiply the value in Table 3-2 by the number of hundreds of feet.

Example: 1232.48 ft at $+15$°F

$$(12.32)(-0.035) = -0.43 \qquad 1232.48 - 0.43 = 1232.05 \text{ ft}$$

3-12. Correction for Slope

$$C_h = -\frac{h^2}{2l}$$

where C_h = correction, ft

h = difference in elevation of the two ends of the measurement, ft

l = length on slope, ft

Example 1.   Slope length = 1407.61 ft, difference in elevation = ± 12.3 ft

$$C_h = -\frac{(12.3)^2}{2(1408)} = -0.054$$

If the angle of slope is measured,

$$C_h = -l \text{ vers } \alpha \qquad \text{or} \qquad -l(1 - \cos \alpha)$$

where C_h = correction, ft

l = slope distance, ft

α = angle of slope

Example 2.   Length on slope = 1407.61, slope = $\pm 0°30'$

$$C_h = -(1408)(0.0000381) = -0.054$$
$$\text{Corrected length} = 1407.61 - 0.05 = 1407.56$$

See also Tables VII and VIII at the back of the book.

3-13. Combining All Corrections.   All these corrections are so small that an accurate result can be obtained by adding them algebraically.

Example.   Assume that the tape was found to be 100.004 ft long when compared with the standard tape. The apparent measured length was 1407.61 ft, the average temperature was $-10°F$, and the difference in elevation of the two ends was $+12.3$ ft The true horizontal length would be computed as follows:

$$C_t, \text{ tape correction} = (14.08)(+0.004) = +0.056$$
$$C_f, \text{ temp. correction} = (14.08)(-0.050) = -0.704$$
$$C_h, \text{ slope correction} = -(12.3)^2/2(1408) = \underline{-0.054}$$
$$\text{Algebraic sum} = -0.702$$
$$\text{Corrected length} = 1407.61 - 0.70 = 1406.91'$$

3-14. The Sign of a Correction to Taping.   The sign of a correction to taping is confusing to most people (including the author). It is best to check the rule given below and then to memorize it. The rule and the discussion here are based on the tape's being too short. Obviously, if the tape is too long, the correction has the opposite sign.

Rule. If the tape is too short, subtract.

When the tape is too short, too many tape lengths will fit into the distance measured so that the recorded distance will be too great.

When a distance is measured, the value **indicated by the tape** is recorded. This is the value that must be corrected to find the true length.

Assume that two monuments were known to be exactly 100 ft apart. Suppose this distance was measured with a tape that was **too short.** For example, assume its length to be 99.996 ft. Note Fig. 3-23. The zero of the tape would be held at A. The 100-ft mark would come at M, where a ground mark would be made and **called 100 ft.** A farther distance to B would be measured and found to be 0.004 ft. The total distance would be recorded as 100.004 ft. The correction of 0.004 ft would have to be **subtracted;** thus

$$100.004 - 0.004 = 100'$$

This proves the rule. It must be remembered that it is the **recorded distance** that is corrected.

It is obvious that, if the tape is **too long,** the correction is **added** to the recorded distance.

A further complication occurs when a **required distance** is to be **laid out.** Assume that exactly 100 ft is to be laid out with the tape that

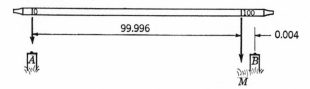

Fig. 3-23. The sign of a taping correction.

has a length of 99.996 ft. Obviously 0.004 must be added to the tape length. See Fig. 3-23. The rule operates as follows. The ground mark *M* is at 100 ft according to the tape. The tape correction, 0.004 ft, is subtracted and thus only 99.996 ft is the true value. Therefore the distance must be extended by 0.004 ft.

3-15. Conditions That Create a Short Tape.   The tape is snort, so that the correction is minus, under three conditions:

1. When, compared with the standard tape, its length is less than 100 ft.
2. When the temperature is less than 68°F.
3. When measurement is made on a slope. This is true for both upward and downward slopes. When a tape exactly 100 ft long is sloping in use, the horizontal projection is less than 100 ft. The tape indication, therefore, is 100 ft, but the true distance is less than 100 ft, which is exactly the same as using a short tape.

3-16. Tensions Used for Less-than-tape-length Measurements. When a tape is standardized and used supported at 0 and 100 ft, at some tension *t*, it is obvious that when shorter lengths than 100 ft are measured, the tension should be reduced. The rule is: Support the tape at zero and at the distance measured and apply a tension equal to the standard tension divided by 100 and multiplied by the distance measured. This rule gives very nearly perfect results. The maximum error of this rule for an average tape is about ±0.005 ft.

Example.   Given: the tension $t = 20$ lb when the tape was standardized. To measure 80 ft: support the tape at 0 and 80 and use 80 per cent of 20 lb = 16 lb tension.

When the tape is used fully supported, no change in tension is required for a less-than-tape-length measurement.

3-17. Other Corrections.   If the tape is used with a type of support and/or a tension for which it has not been standardized, a new tape correction must be determined. This can be done by standardizing the tape under the new condition or by computation (see Appendix B).

3-18. Electronic Distance Measurement. Electronic distance measurement devices are coming more and more into use. Several types are available with slightly differing advantages. Figure 3-24 shows one type. All are very accurate, and all measure slope distances between two pieces of equipment. One type has been used to measure distances up to 40 miles. They can be used for triangulation base lines, but their chief use is for traverses composed of many courses of over 400 or 500 ft, or when taping is impossible or too slow as over water, or when stadia is not sufficiently accurate. See Chaps. 5 and 8.

The general principle of all of these devices is much the same.   They

Fig. 3-24. A laser Geodimeter attached to a Wild T-2 to form the Wild DI-10 Distomat System. The Geodimeter uses light and a reflector. One type of reflector is shown at the right. The reflector consists of three corner prisms that reflect light to the source of the light even when aimed 20° out of line. (*Thorpe-Smith, Inc., Falls Church, Va.*)

project a light beam, a low-power laser beam, or a high-frequency radio beam on which is imposed (modulated by), or originates as, very short waves in the order of 10 to 20 ft in length in much the same way music or voice waves modulate radio frequencies. The length of these short waves is regulated to an extremely high precision by a crystal-controlled generater. These are picked up at the far end of the distance and retransmitted or reflected back to the original station. There the phase[1] of the incoming waves is determined by comparison with the known outgoing phase. The difference in phase gives the fraction of the known wavelength by which the distance exceeds some number of complete wavelengths. See Fig. 3-25. Accordingly, the length of travel of the waves is known to

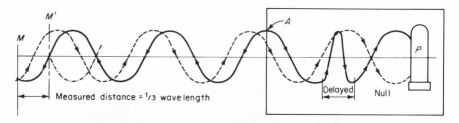

Fig. 3-25. Schematic illustration of electronic distance measurement. At *A*, within the instrument, the measurement wave is introduced. It travels in both directions, into the instrument toward *P*, a sensing device, and out toward the other end of the distance. At *M*, the equivalent location of a mirror or a retransmitting device, it is returned to the instrument and toward *P*. The original wave from *A* is delayed by an adjustable resolver on its way to *P*. The resolver delay is adjusted until the two waves cancel each other as shown. The amount of adjustment shown indicates that if *M* were at *M'*, no delay would be required. The actual distance here would be one-third of a wavelength longer. The resolver control is graduated to read in distance, based on the known wavelength.

consist of this short measured distance plus an unknown number of wavelengths. The number is determined by measuring the distance approximately by usually a series of two wavelengths, for example, 300 and 3,000 ft in length, respectively.

Figure 3-26 shows a special Geodimeter in use.

Since the length of the waves depends on the frequency and speed of travel through the air, it is slightly affected by the index of refraction

[1] By phase is meant any particular part of the waves.

Fig. 3-26. The Distomatic System in use by members of Thorpe-Smith, Inc. (*Thorpe-Smith, Inc., Falls Church, Va.*)

of the atmosphere. Very slight corrections must therefore be computed based on the barometric pressure and the temperature of the path of the beam. Since the measurement is sloping, the slope must be determined and the measurement reduced to the horizontal. Low-power laser beams have other uses in surveying (see Chap. 9).

3-19. Laser Beams Are often Used for Alignment. Laser beams are especially used over water to direct dredging and over land to direct tunneling and pipe laying. Both Dietzgen and Berger manufacture such an instrument The Dietzgen laser device is shown in Figs. 3-27 and 3-28. The Berger instrument is shown in Fig. 3-29. A modern Distomat system is shown in Fig. 3-30.

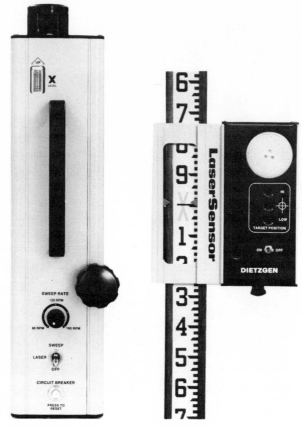

Fig. 3-27. The Dietzgen system. A laser is part of the device on the left which can be leveled horizontally and vertically. A rotating prism turns the beam so that it generates a plane as shown in Fig. 3-28.

Problems

3-1 to 3-6. Find the corrected lengths from the following recorded values. Use the approximate formula for temperature correction and check by Table 3-2. Slide-rule accuracy is sufficient.

Problem	Dist. recorded, ft	Tape length, ft	Av. temp., °F	Diff. in elev. or slope angle between points stated
3-1	1209.17	100.032	98	0–700 = 32.0 ft; 700–end = 10.0 ft
3-2	982.75	99.996	13	0–400 = 21.0 ft; 400–end = 8.0 ft
3-3	2064.61	99.981	40	0–1200 = 26.8 ft; 1200–end = 4.0 ft
3-4	1062.14	100.016	82	0–200 = 1.6 ft; 200–end = 5.6 ft
3-5	4041.86	99.987	32	0–700 = 3°08′; 700–end = 2°07′
3-6	3259.58	100.008	87	0–1000 = 1°09′; 1000–end = 1°03′

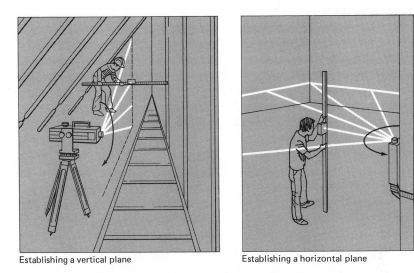

Establishing a vertical plane   Establishing a horizontal plane

Fig. 3-28. The Dietzgen laser device in use. (The Dietzgen Co.)

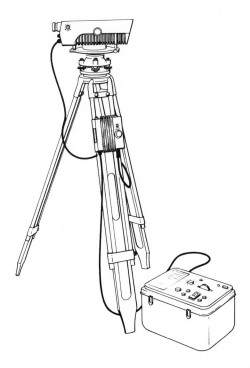

Fig. 3-29. The laser source attached to a level. Berger system. (Berger Instrument Div.)

Fig. 3-30. A more modern Distomat system in use by an observer of the Thorpe-Smith organization. (Thorpe-Smith Inc., Falls Church, Va.)

Suggested Field Exercises

3-1. To Standardize Tapes and Tension Handles. Assign a marked tape and tension handle to each field party of two. Find the length of each tape supported at 0 and 100 ft, with 20 lb tension. Test the tension handles.

3-2. Taping Practice. Set out a pair of stakes and tacks about 450 ft apart on ground that changes slope gradually. Assign a pair of stakes to each field party of two. Place range poles for alignment.

a. Measure distance back and forth several times using plumb bobs and spring-balance handle so that the feel of 20 lb is acquired.

b. Measure without spring-balance handles.

Corrections. Correct each measurement for tape correction and temperature.

List all corrected distances and compare them.

4

The Transit

CONSTRUCTION OF THE TRANSIT

4-1. The Engineer's Transit. The transit, called a **theodolite** in certain cases, is the key instrument in nearly every type of surveying operation. It follows that no engineering project of any importance can be designed or constructed without it. It is essential, therefore, that every surveyor should completely understand its construction, the principles upon which it is based, and the operations that it will perform. In addition, the surveyor must be able to use the instrument skillfully and rapidly so that he or she can perform every operation of which the transit is capable, to the accuracy required, with a minimum expenditure of time.

Some types of transits are designed for special purposes. The engineer's transit, however, is designed for nearly every operation required and therefore is the type of **transit** used most extensively in the United States (see Fig. 2-5). When properly operated, it will measure horizontal angles to any required accuracy, vertical angles to about ± 10 seconds,

Fig. 4-1. The three parts of the transit: alidade, circle, and leveling head. (*Keuffel & Esser Co.*)

and elevations to third-order accuracy. When it is equipped with **stadia hairs,** described in Chap. 8, horizontal and vertical distances can be measured to the accuracy required for mapping; and, when it is equipped with a compass, magnetic bearings can be determined to about ±5 minutes. To attain these accuracies, angles and magnetic bearings must be measured by certain procedures, described later in this chapter. The

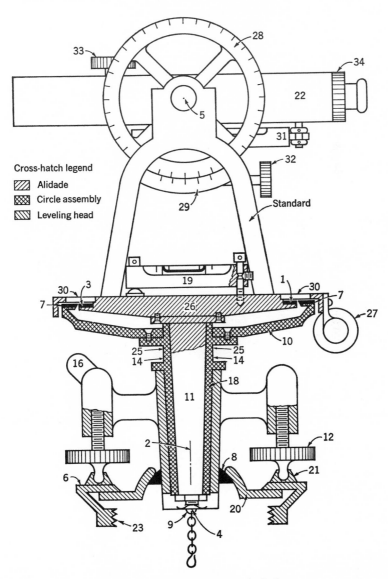

Cross-hatch legend

- ▨ Alidade
- ▩ Circle assembly
- ◩ Leveling head

Fig. 4-2. Principles of transit design.

Key for Figs. 4-2 and 4-3

1. A vernier
2. azimuth axis
3. x vernier
4. center of half ball
5. elevation or horizontal axis
6. footplate
7. graduations of horizontal circle
8. half ball
9. half ring
10. horizontal circle (lower plate)
11. inner center, or alidade spindle
12. leveling screw
13. lower clamp screw
14. lower clamp drum contact
15. lower tangent screw for slow motion
16. nub for lower clamp
17. nub for upper clamp
18. outer center
19. plate level
20. shifting plate
21. shoe
22. telescope
23. threads for tripod
24. upper clamp screw
25. upper clamp drum contact
26. upper plate
27. upper tangent screw for slow motion
28. vertical circle
29. vertical-circle vernier
30. window, glass
31. telescope level
32. vertical tangent screw for slow motion
33. focusing screw for focusing on object sighted
34. eyepiece focusing ring

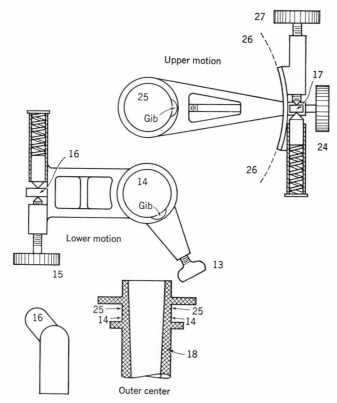

Fig. 4-3. Details of typical clamps and tangent (slow-motion) screws. The tops of the clamps are shown. They are turned down and placed around the outer center where shown by the numbers 14 and 25.

only transit covered in this book is the engineer's transit, as, once a surveyor has mastered this instrument, she or he can use any type of transit without difficulty. The theodolite is covered in Chap. 10.

4-2. The Three Fundamental Parts.   A transit consists of three fundamental parts: the **alidade** at the top, the **horizontal circle** in the middle, and the **leveling head** at the base (see Figs. 4-1 through 4-3).

The three parts are operated by two clamps, each equipped with a slow motion. The upper clamp clamps the horizontal circle to the alidade, and the lower clamp clamps the horizontal circle to the leveling head. When a clamp has been tightened, the appropriate slow-motion (tangent) screw can be used to make a fine setting.

The Alidade.   The alidade is mounted on a tapered spindle called the **alidade spindle,** or **inner center** (11). The essential parts of the alidade are the **telescope** (22), which is actually a telescopic sight that rotates in a vertical plane on the horizontal or **elevation axis** (5), and the **A and B verniers** (1 and 3), which act as indexes for reading the horizontal circle.

A **vertical circle** (28) is mounted on the telescope axis, which turns with the telescope. It is read with a **vertical-circle vernier** (29) which is mounted on one standard. It is adjusted to read the **vertical angle** between the line of sight and a plane perpendicular to the vertical or **azimuth axis** (2).

Two **plate levels** (19) are mounted horizontally, at right angles, on or near the **upper plate** (26). They are used to place the azimuth axis in the direction of gravity. A **telescope level** (31), which is a sensitive spirit level, is attached to the underside of the telescope. Usually a compass is mounted on the upper plate.

The Horizontal Circle.   The horizontal circle (lower plate, 10) is mounted on a **hollow** tapered spindle or **outer center** (18), the inner surface of which acts as a bearing for the alidade spindle; the outer surface turns in a bearing in the leveling head. This arrangement is called the **double center.** The horizontal circle is graduated in degrees and usually halves or thirds of a degree and numbered throughout, every 10 degrees, usually both clockwise and counterclockwise, starting from a common zero (see Fig. 4-10). It is read to varying degrees of precision from 1 minute to 10 seconds by two verniers mounted on the alidade 180° apart.

The Leveling Head.   The leveling head contains the tapered bearing for the outer center. Four **leveling screws** (12) are threaded into the arms of the leveling head and press **shoes** (21) down against the **footplate** (6). This action tends to raise the leveling head and thus pulls a **half ball** (8), attached to the end leveling-head bearing, upward into a socket in the **shifting plate** (20), which in turn is pulled upward against the underside of the footplate. At the bottom of the footplate are the threads (23) by which the instrument is screwed to the tripod. When the leveling screws are loosened, the shifting plate drops and the whole upper assembly can be shifted anywhere within a circle of about $\frac{3}{8}$ in. in diameter, so that the instrument can be placed exactly in the desired horizontal position.

A small chain, with a hook at the lower end to hold the plumb-bob cord, hangs from a small, half ring (9) attached to a cap which is screwed to the lower end of the leveling-head bearing. In a well-designed instrument, the ring is placed at the center of curvature of the half ball (4). In leveling the instrument, the whole assembly above the footplate rotates slightly around the center of curvature of the half ball. If, as is often the case, the half ring is too low, the plumb bob is moved horizontally when the instrument is leveled and thus moved off the point where it had been originally placed (see Fig. 4-27).

4-3. The Azimuth Axis.   The alidade spindle, the outer center, and the leveling-head bearing combine to form the vertical or azimuth axis. Thus the alidade and the circle can turn in azimuth independently of each other.

4-4. The Elevation Axis.   Two journals, one at each end of the telescope axle (the horizontal axle which supports the telescope) fit in bearings at the tops of the standards and thus form the horizontal or elevation axis. When the telescope is aimed up and down, it turns on this axis.

4-5. The Motions.   Three **motions** control the movements of the transit. Each motion consists of a **clamp** and a **tangent screw** (see Figs. 4-2 through 4-4). When the clamp is tightened, a **gib** is forced against a drum on the circle assembly or on the telescope axle. The tangent screw then becomes operative and provides a slow motion between the two parts clamped together. The **lower motion** joins the horizontal circle and the leveling head. The **upper motion** joins the

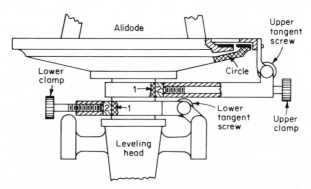

Fig. 4-4. Schematic view of the clamps and tangent screws. 1—drum on circle assembly; 2—clamp gib.

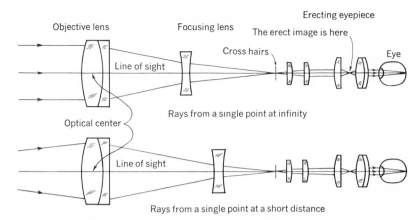

Fig. 4-5. Modern telescopic sight with an internal focusing lens. All three rays shown come from the single point at which the telescopic sight is pointed.

horizontal circle and the alidade. The **vertical motion** joins the telescope axle with the standard and thus controls the vertical angle of the telescope.

Each tangent screw acts on a nub (16 and 17) between it and an opposing spring. When a tangent screw is turned, the relative positions of the two parts clamped together are changed slightly. Their operation is best understood by a careful study of Figs. 4-2 through 4-4. Only the tangent screw of the vertical motion is shown.

4-6. The Telescopic Sight. The modern telescopic sight (see Fig. 4-5) consists of the following: (1) a **reticle** (or reticule), which provides the **cross hairs** near the rear of the telescope tube; (2) a microscope, or **eyepiece,** which magnifies the cross hairs and must be focused on them according to the eyesight of the observer; (3) an **objective lens** at the forward end of the telescope, which forms an image within the telescope; and (4) a **focusing lens,** which can be moved back and forth to focus the image on the cross hairs.

Since the image formed by the objective lens is inverted, the eyepieces of most transits are designed to erect the image. Telescopes which erect the image are called **erecting telescopes;** the others are called **inverting telescopes.**

When the image is focused on the cross hairs, the cross hairs become part of the image so that, when the observer looks through the eyepiece, he sees the object magnified about 24 times with the cross hairs apparently engraved on it.

To Focus a Telescopic Sight. Three steps are required to focus a telescopic sight for greatest accuracy. They are described in the next paragraphs. Figure 4-6 illustrates the process.

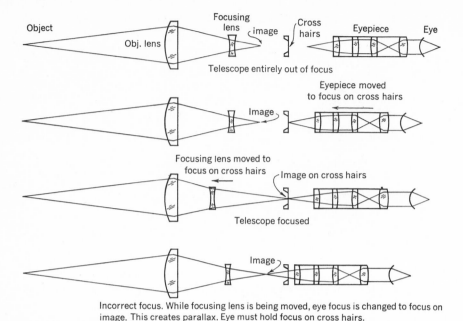

Incorrect focus. While focusing lens is being moved, eye focus is changed to focus on image. This creates parallax. Eye must hold focus on cross hairs.

Fig. 4-6. Principle of focusing a telescopic sight.

1. Aim the telescope at a bright, unmarked object, such as the sky, and regulate the eyepiece until the cross hairs are in sharp focus. Since the eye can change focus itself, there is always a short range in the movement of the eyepiece within which this condition can be satisfied.

2. Aim the telescope at the object to be viewed and, while keeping the eye focused on the cross hairs, regulate the focusing lens until the object is clear. This should occur only when the image is on the plane of the cross hairs, as this is the only plane where the eyepiece focus is sharp. If the observer allows himself to look at the image, instead of at the cross hairs, while he is regulating the objective focus, his eye focus may change slightly so that the image is seen clearly a short distance in front of or behind the cross hairs. The cross hairs will then not be in perfect focus, but the difference may not be noticeable. When the image and the cross hairs are **simultaneously** in apparently good focus, the plane of the image and the plane of the cross hairs must be very nearly coincident.

When the image is not exactly on the plane of the cross hairs, the cross hairs will move across the image when the eye is moved left and right or up and down, just as is the case when two objects at different distances are observed with the naked eye. Under these conditions, **parallax** exists, and the direction of the sight is not fixed.

3. Eliminate parallax. To accomplish this, move the eye up and down or left and right. If the cross hairs appear to move with respect to the object sighted, change the focus of the objective until the apparent motion is reversed. Continue focusing back and forth, reducing the apparent motion each time until it is eliminated. It may then be necessary to adjust the eyepiece slightly to make the image and the cross hairs appear clear-cut.

Theoretically, the parallax should be eliminated by this method each time the objective focus is changed. However, when the eyepiece has been set for a particular observer after the parallax has been once eliminated, it is common practice to keep the eyepiece in this position throughout the work and to rely on focusing the objective so that both the cross hairs and the object are in sharp focus simultaneously, to eliminate parallax.

The Line of Sight.   A straight line from any point on the image through the optical center of the objective lens will strike a corresponding point on the object. A straight line from the cross hairs through the optical center of the lens will strike the point on the object where the observer sees the cross hairs apparently located. Thus the **line of sight** of a telescopic sight is defined by the cross hairs and the optical center of the objective. As stated above, when a telescopic sight is properly focused, the observer can move his eye slightly without changing the position of the cross hairs on the object. This differs in principle from a rifle sight, for the eye must be accurately aligned with the latter in order to determine where it is pointing. The telescopic sight on a transit also magnifies the object about 24 diameters. The diameter of the field of view is therefore very small, about 1° or 1.75 ft at 100 ft.

4-7. The Telescope Level.   A sensitive spirit level is attached to the underside of the telescope and is adjusted so that, when the bubble is centered, the line of sight is level. This makes it possible to use the transit as a level. See Chap. 6.

A **spirit level** consists of a glass vial partly filled with a very-low-viscosity spirit such as alcohol or ether. The inside is ground to a barrel-shaped surface that is symmetrical with respect to a longitudinal axis, as indicated in exaggerated form in Fig. 4-7. The vial is mounted in a

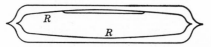

Fig. 4-7. A level vial showing the curvature exaggerated.

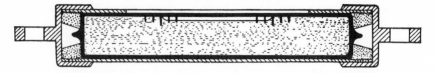

Fig. 4-8. Cross section of a level tube showing the mounting of the vial. Dotted material is plaster of paris. (*C. L. Berger & Sons, Inc.*)

metal tube, as shown in Fig. 4-8. Several graduations at each end of the bubble are placed near, or are etched on, the vial so that, as the spirit expands or contracts with temperature variations, the ends of the bubble can be placed at corresponding graduations; thus, when the bubble is centered, the direction of the vial with respect to gravity is the same at any temperature. When the bubble is centered, it is said to "read zero."

4-8. The Graduation of the Horizontal Circle. (See Fig. 4-11.) The graduated circle is divided automatically on a large wheel. Modern dividing engines usually space the graduations very uniformly; but the circle can never be exactly centered on the wheel, and therefore the graduations on one part of the circle are usually slightly nearer together than the graduations on the opposite side of the circle. When an angle is read by averaging the readings of the two verniers 180° apart, the effect of this **eccentricity** of graduation is eliminated because, if the *A* vernier travels over graduations that are too near together, the *B* vernier will travel over graduations proportionally too far apart. The very fact of reading two verniers increases the accuracy by using an average instead of a single reading. For the most accurate results, therefore, both verniers should be used.

4-9. Verniers. Verniers, in general, are devices for determining readings smaller than the smallest division on the scale with which they operate. They consist of an auxiliary scale that is moved along the main scale. When the graduated circle is marked off in half degrees, i.e., divisions 30 minutes in length, it is usual to design the verniers so that the direction of the alidade can be read to 1 minute. In this case each vernier consists of a series of 30 divisions, each division being $\frac{1}{30}$ shorter than a division on the graduated circle, so that the whole vernier scale of 30 divisions covers exactly 29 divisions of the graduated circle.

Assume that the zero graduation of the vernier coincides with the 57°30′ graduation of the circle; the reading will be 57°30′ (see Fig. 4-9). When the alidade is turned 1 minute of arc ($\frac{1}{30}$ of a division on the circle), the next graduation on the vernier will coincide with a graduation on the circle. The reading will then be 57°30′ plus 1 minute (as shown by the vernier), i.e., 57°31′. Likewise, when the alidade is turned 7 minutes of arc, the seventh-minute graduation on the vernier will coincide with a graduation of the circle, and the angle will be read 57°37′, etc. The vernier described above is called a 1-minute vernier, and a transit with such a vernier is called a 1-minute transit.

Ordinarily **double verniers** are used. These have complete sets of divisions running both ways from a common zero line. With such verniers, directions can be read clockwise or counterclockwise whenever desired. Since the verniers are placed to be read from the part of the circle nearest the observer, a clockwise angle is read from right to left. In that case the set of divisions on the vernier to the left of the central zero mark is used (see Fig. 4-10).

The patterns of lines on the graduated circles and the verniers have become standardized. The circle described above will have three lengths of lines. The longest lines mark the 5° graduations, the lines of the next length mark the 1° graduations, and the shortest lines are used for the ½-deg positions. The 10° positions are numbered. Each 10° position (except zero) has two numbers, one for the clockwise direction and one for the counterclockwise direction. Other systems of enumeration are used (see Fig. 4-11).

Two lengths of lines are used for the 1-minute verniers described above, the longer length to mark the 5-minute graduations and the shorter to mark the 1-minute positions. The 10-minute positions are appropriately numbered.

The arrangement of vernier and scale described above is probably the most generally used. Another common form is the 20-second vernier (see Fig. 4-12). The circle is divided into thirds of a degree (20 minutes), and the vernier is divided into 20 one-minute divisions, each of which is in turn divided into thirds (20 seconds). Since it is necessary to show each 20-second movement of the alidade, and the smallest division of the circle is ⅓ deg, a movement of $\frac{1}{60}$ of the circle divisions must be shown (20 seconds is $\frac{1}{60}$ of ⅓ deg). Hence each of the smallest divisions on the

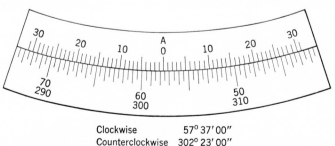

Clockwise   57° 37′ 00″
Counterclockwise   302° 23′ 00″

Fig. 4-9. Double 1-minute vernier.

GRADUATED 30 MINUTES READING TO ONE MINUTE
DOUBLE DIRECT VERNIER

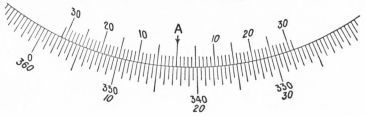

GRADUATED 20 MINUTES READING TO 30 SECONDS
DOUBLE DIRECT VERNIER

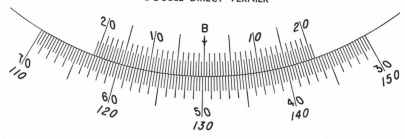

GRADUATED TO 15 MINUTES READING TO 20 SECONDS
DOUBLE DIRECT VERNIER

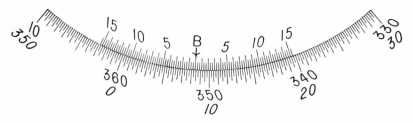

Fig. 4-10. Typical double verniers and scales (*Keuffel & Esser Co.*)

vernier (the 20-second divisions) is $\frac{1}{60}$ shorter than each of the smallest divisions on the circle (the $\frac{1}{3}$-deg divisions). Note also Figs. 4-13 and 4-14.

Vernier Readings. As is the case of all scales, vernier readings should be estimated to a higher degree of precision than the reading of the scale. Consider a **1-minute** vernier that at first reads 60°20′ exactly (see Fig. 4-15a). As the alidade is turned, the angle will increase gradually

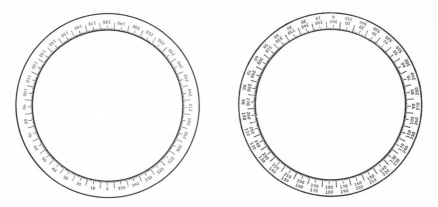

Fig. 4-11. Usual methods of marking the horizontal, graduated circle. The numbers are often slanted to show the direction of measurement.

from 60°20′ to 60°21′. When it has moved halfway, the vernier line representing 20 minutes will have moved beyond the line on the graduated scale corresponding to it, but the line representing 21 minutes will not have reached the graduated line with which it will correspond. **Both** lines will be **between** adjacent lines on the graduated circle. If the pair of vernier lines appear to be equally spaced between the two lines on the circle, the reading is 60°20′30″, as in Fig. 4-15c. If the 20-minute mark

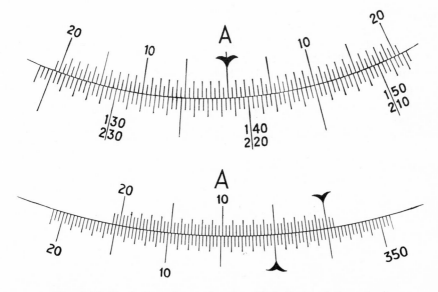

Fig. 4-12. Typical verniers and scales. (*C. L. Berger & Sons, Inc.*)

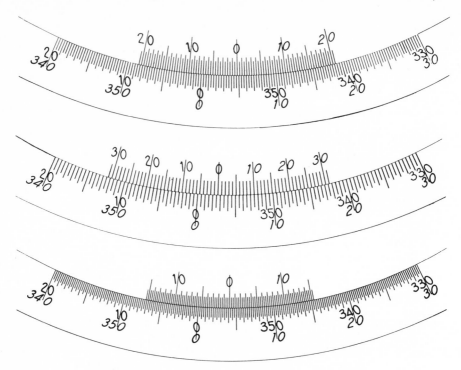

Fig. 4-13. Verniers and scales. (*W. and L. E. Gurley.*)

has only just moved off its line but the 21-minute line has quite a distance to go, the reading is 60°20′15″, as in Fig. 4-15*b*. When the 20-minute line has moved well beyond its line and the 21-minute line has almost reached its line, the reading is 60°20′45″ (see Fig. 4-15*d*). Closer estimates are not used. Verniers reading to 20 seconds are read to 10 seconds and no closer. In general, more precise verniers are read to one-half the least reading.

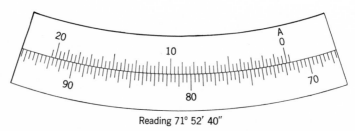

Reading 71° 52′ 40″

Fig. 4-14. Single 20-second vernier.

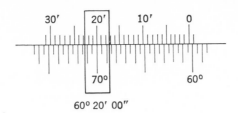

60° 20′ 00″

The rectangle above is enlarged below

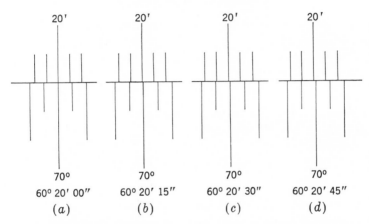

60° 20′ 00″	60° 20′ 15″	60° 20′ 30″	60° 20′ 45″
(a)	(b)	(c)	(d)

Fig. 4-15. Reading a 1-minute vernier to 15 seconds.

4-10. Geometry of the Transit. The following geometric requirements are necessary in a transit. As no device can be made or adjusted perfectly, when accurate results are required a field procedure must be used (as described later) which effectively eliminates any residual errors in these requirements. See also Appendix C.

1. The inner and outer spindle and the bearings of the horizontal axis must be so perfectly fitted that the instrument turns about geometric lines, not cylinders or cones.
2. The inside and outside tapered surfaces of the outer center must be concentric. Unless this condition is met, the alidade will not remain level when the circle is turned.
3. The vertical axis, horizontal axis, and line of sight must meet in a point called the instrument center.
4. The horizontal axis must be perpendicular to the vertical axis.
5. The line of sight must be perpendicular to the horizontal axis. The lens used for focusing must therefore be guided so that, when it is moved, it does not change the direction of the line of sight.

6. The plate bubbles must center when the vertical axis is vertical.
7. The telescope bubble must center when the line of sight is horizontal.
8. The horizontal graduated circle must be concentric with and perpendicular to the vertical axis.
9. The graduations on the horizontal graduated circle must be concentric with the vertical axis.
10. The vertical circle must be concentric with and perpendicular to the horizontal axis.
11. The graduations on the vertical circle must be concentric with the horizontal axis.
12. The vertical circle must read zero when the line of sight is perpendicular to the vertical axis.

When high accuracy is required, the instrument is operated so that if any of these geometric requirements except loose bearings are not met exactly, the errors they introduce are neutralized and their effects are thus eliminated.

4-11. The Compass.   On most transits a compass box is mounted on the upper plate between the standards (see Figs. 4-1 and 4-16). The compass needle has at the center a conical jewel bearing which rests on a hardened steel pivot. When not in use, the needle is lifted off the pivot by the lever and screw indicated. When the needle is in place, a slight jar may cause the jewel to damage the fine point of the pivot so that the needle becomes sluggish. Loss of magnetism seldom occurs, so that an insensitive needle usually indicates a dull pivot. On the needle is a small coil of brass wire that may be moved along the needle to balance the effect of the dip of the earth's magnetic field. The needle, of course, aligns itself with the horizontal component of the earth's magnetic field, which is usually called the **magnetic meridian.** The circle which surrounds the needle is graduated in degrees and half degrees and numbered to show the bearing of the line of sight of the telescope (indicated by the arrow pointing to the mark). Since the circle turns with the alidade and not with the compass needle, the east and west indications must be as shown in Fig. 4-16.

The Earth's Magnetic Field.   The magnetic field of the earth can be approximately described as the field that would result if a huge bar magnet were embedded within the earth, with one end far below the surface in the Hudson Bay region and the other end in a corresponding position in the Southern Hemisphere. The lines of force follow somewhat irregular lines running from the south magnetic pole to the north magnetic pole. They are approximately parallel with the earth's surface at the equator and dip downward toward the poles. The field is slowly changing in its general direction, and its direction is slightly affected by the position of the sun and by changes in radiation from the sun. A compass needle will therefore point exactly north only by chance. At any given time, at any point on the earth's surface, the true geographic bearing of the needle is called the **magnetic declination.** Accordingly, when the needle points N10°W the declination is said to be 10° west.

Changes in Declination. Changes in the earth's magnetic field cause four types of **variations** in declination: secular, diurnal, annual, and irregular.

The **secular variation** is a long swing not well understood. The

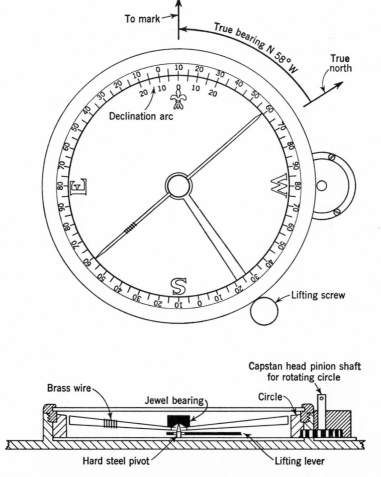

Fig. 4-16. A compass box. The circle is set for declination 10° west. Compass reads N58°W, although the line of sight is only 48° from magnetic north.

direction of the swing and its rate are different at different parts of the earth. Within the United States the maximum rate is about 5 minutes of arc per year. The U.S. Coast and Geodetic Survey publishes every 5 years a chart, called an **isogonic chart,** which shows lines of equal

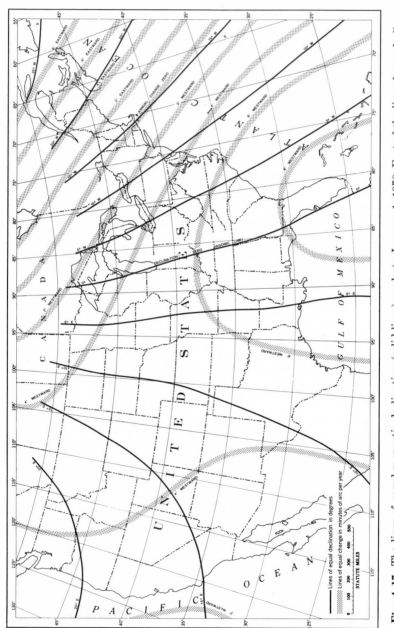

Fig. 4-17. The lines of equal magnetic declination (solid lines) apply to January 1, 1970. East of the line of zero declination (the agonic lines), the north end of the compass needle points west of north; west of that line it points east of north. The stippled lines give the annual rate and direction of the movement of the north end of the needle. (*U.S. Department of Commerce, Environmental Sciences Services Administration, Coast and Geodetic Survey.*)

declination and lines of equal annual secular variations in declination (see Fig. 4-17). These charts thus provide a means of estimating the declination at any time at any point in the United States.

The **diurnal variation** is a swing of about 4 to 10 minutes of arc, depending on the locality. At night the needle is quiescent in its mean position. It swings east 2 to 5 minutes in the morning and west 2 to 5 minutes in the afternoon.

The **annual variation** is a swing of about 1 minute back and forth during the year.

Irregular variations occur during magnetic storms.

Local Attraction. Any deposit of magnetic material will disturb the direction of the magnetic meridian. This deposit may be a large area of magnetic ore or a small iron object near the compass. The effect of such a disturbance is called **local attraction.**

To Use the Compass. To determine the bearing of a line with respect to the magnetic meridian, free the compass needle and aim the line of sight along the line. When the needle comes to rest, the position of the north end of the needle gives the bearing. The north end of the needle can be recognized by the position of the brass coil of wire. In the Northern Hemisphere it must be on the south end of the needle and in the Southern Hemisphere on the north end of the needle. The bearing can be estimated to the nearest 15 minutes. The needle should be raised from its pivot as soon as the work is completed.

The Declination Arc. On most transits, the compass circle can be rotated so that, when the declination is set off, the needle will read true instead of magnetic bearings. When the declination is 7° west, for example, the zero of the circle is moved 7° to the left so that, when the telescope is pointed true north, the needle will read zero. To set off the declination, the compass circle is rotated with the capstan-headed pinion shaft shown in Fig. 4-16. A scale is provided to indicate the angle. The scale is known as the **declination arc.**

OPERATION OF THE TRANSIT

4-12. Basic Method of Measuring an Angle. The simplest procedure for measuring a horizontal angle is given in the following steps:

1. Set up and level the transit over the point where the angle is to be measured.
2. Set the *A* vernier at zero with the upper motion.
3. Aim at the point that marks the left-hand side of the angle, using the lower motion.

4. Free the upper clamp and aim at the point that marks the right-hand side of the angle, using the upper motion.
5. Read the clockwise angle with the *A* vernier.

A good transitman should know a great deal more about the use of the transit than the bare information covered in these five steps. He or she should thoroughly understand the concepts and be capable of carrying out all the operations covered in the remainder of this chapter.

The secret of **successful use of the transit** is the formation of a set of standard habits based upon the ever-present necessity for speed and upon a clear knowledge of exactly what the transit does. Poor results are due mainly to ignorance, although partly to clumsiness.

4-13. Preliminary Procedure. *To Place the Transit on the Tripod.* Adjust the friction of the tripod legs at the tripod head. The legs should fall slowly of their own weight from a horizontal position. If a wide-framed tripod with metal hinges is used (Fig. 4-18*B*), the friction should be adjusted so that it is just possible to notice the friction when the legs are moved by hand. Set up the tripod with the legs well spread and pressed firmly into the ground. If the surface is hard, each tripod shoe should be placed in an indentation in the surface. Remove the instrument from the case, and, lifting it by the base, immediately screw it firmly on the tripod. Remove the dust cap from the objective lens and replace it with the sunshade. The sunshade should always be used. It improves the mechanical balance of the telescope and prevents any glare caused by the sun striking the objective. It improves visibility, even when there is very little light, as it does not interfere with any useful light rays but it helps to eliminate unfocused light which tends to dim the image.

To Carry the Transit.   Hold the transit on the shoulder in a horizontal

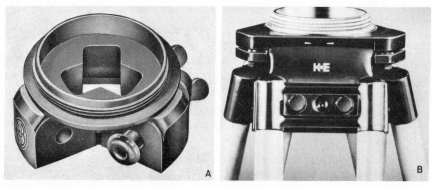

Fig. 4-18. (*A*) The usual type of tripod head. (*C. L. Berger & Sons, Inc.*) (*B*) A tripod head for a wide-framed tripod. (*Keuffel & Esser Co.*)

Fig. 4-19. Carrying an instrument under obstructions.

position, instrument to the rear, and balanced to carry the weight of the arm. When overhead obstructions exist or when going through a doorway, carry the transit under the arm, balanced in a horizontal position with the instrument forward (Fig. 4-19).

To Set Up the Transit over a Point.   Proceed according to the following steps:

1. Stand about 2 ft downhill, facing the point over which the transit is to be set up. Seize two legs, and place the third leg on the ground about 2 ft uphill from the point. Pull the other two legs outward and backward, and place them on the ground so that the footplate is nearly level. If the tripod head is not level after placing the legs, stand facing the tripod where the slope of the head from left to right is greatest, and swing the nearest leg left or right until the fault is corrected. Avoid placing a tripod leg on line.

   These steps are very quickly accomplished, but they are **the key steps in making a rapid setup** (Figs. 4-20 through 4-22).
2. Attach the plumb bob. Center the instrument on the footplate. Move the transit bodily without changing the relative position of

Fig. 4-20. Stand 2 ft downhill.

Fig. 4-21. Seize two legs and place the third leg 2 ft uphill.

the legs, so that the bob hangs within 2 or 3 in. of the point (Fig. 4-23).

3. Push each leg firmly into the ground. To accomplish this, walk around the transit to each leg, and grasp the leg about 18 in. from the foot. Raise or lower the plumb bob until it hangs about 1 in. above the point. It will probably be about 2 or 3 in. to one side of the point.

Choose the leg that is most nearly on the opposite side of the point from the plumb bob. By pushing this leg farther into the ground or moving it outward and then pushing it into the ground, move the plumb bob until it is exactly opposite a second leg. Move the second leg until the bob comes within ¼ in. of the point (Figs. 4-24 through 4-26).

In setting up on a pavement or on masonry the legs can be moved in either direction, thus simplifying the procedure. The points of the tripod shoes should be placed in cracks or other indentations to prevent slipping. On smooth hard surfaces small notches must be cut for the points with a cold chisel.

If the setup is an important one and a conventional tripod is

Fig. 4-22. Pull two legs outward and backward, placing them on the ground so that the **footplate is nearly level.**

Fig. 4-23. Move the transit bodily until the bob hangs within 2 or 3 in. of the point, and push the legs firmly into the ground.

Fig. 4-24. After pushing each leg firmly into the ground, move the leg most nearly opposite the plumb bob outward until the bob is opposite a second leg.

Fig. 4-25. Move the second leg outward until the bob moves within ¼ in. of the point.

being used, relieve any residual friction in the tripod-leg hinges by
loosening and retightening the tripod-hinge thumbscrews. If friction
remains, it may suddenly relieve itself. This will cause the transit
to jump slightly off line and slightly out of level.

4. Loosen two **adjacent** leveling screws. This loosens all the leveling
   screws. Level the instrument roughly without setting the screws
   tighter. Preliminary rough leveling is necessary when the plumb line

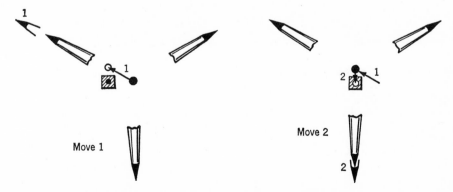

Fig. 4-26. Method of moving the legs.

hangs from a point below the center of the half ball (see Fig. 4-27). To level an instrument, turn it until the plate levels are in line with pairs of **opposite** leveling screws. Then turn the leveling screws according to the old rule: "Thumbs in, thumbs out, the bubble follows the left thumb" (see Fig. 4-28). Slide the head until the plumb bob is over the point, i.e., until the center of the small ellipse in which it swings is over the point (Fig. 4-29). If there is wind, protect the bob itself from the wind; the string is not much affected. Wait for lulls in the wind for final positioning.

5. Level the instrument. The leveling screws should have been left rather loose when the instrument was shifted over the exact point. While leveling accurately, the screws should be tightened gradually as the leveling progresses. The tightness can be regulated by the relative motions of the pair of screws being used. Leave the screws firm but not bound. If they are too tight, they deform the leveling-head bearing and spoil the accuracy of the instrument. If they bind in leveling, loosen an **adjacent** screw. It must be remembered that, when one pair of leveling screws is being used, the other pair is often being forced tightly against the footplate or else being forced to slide over the surface of the footplate.

When the bubbles have been centered, turn the instrument 180°

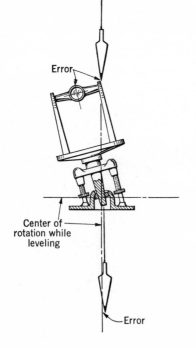

Error

Center of rotation while leveling

Error

Fig. 4-27. Position of transit when not level. Unless some transits are level, they cannot be set up precisely over or under a point.

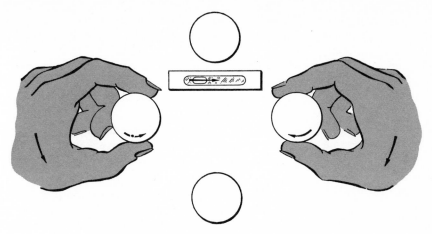

Fig. 4-28. Operating the leveling screws.

in azimuth. If the bubbles do not again center, the levels are not in adjustment. Bring the bubbles halfway toward the center with the leveling screws. The vertical axis will now be vertical (the desired condition), and the bubbles will remain in this new position in whatever direction the instrument is pointed.

Fig. 4-29. After rough leveling, slide the head until the bob is over the tack.

If this procedure is followed, setting up the transit takes 1 or 2 min. The usual mistake is to disregard the importance of steps 1, 2, and 3.

Note: Accurate work is impossible when the setup is on ice, frozen ground, asphalt, or tar. The tripod legs slowly sink and thus throw the instrument out of level and position.

4-14. To Measure a Horizontal Angle.   *Precautions to Be Observed While Operating the Transit.*   Once the transit has been set up, do not touch it or allow anything to touch it except when and where it is necessary for operating it. Never straddle the legs, but always stand between them. Be particularly careful not to kick or touch the tripod while walking around it.

To Focus the Eyepiece.   The eyepiece must be focused according to the eyesight of the observer. Errors due to parallax will occur if this is neglected. Focus the eyepiece, eliminate the parallax, and check the eyepiece focus as described in Sec. 4-6.

Use of the Clamps.   To set an angle on the circle or to point a target, first choose the proper motion to use, loosen the clamp, rotate the instrument approximately into position, clamp it, and make a fine adjustment with the tangent screw. Some authorities advise completing the aim with a final clockwise motion of the tangent screw, thus turning the instrument with the screw rather than allowing the opposing spring to turn the instrument.

Once the exact setting or pointing has been accomplished with a tangent screw, **never** tighten the corresponding clamp. Tightening a clamp **always** slightly changes the setting or the aim.

To Set the Vernier at Zero. Step 1.   Loosen **both** clamps. Turn the instrument so that there is good light on the *A* vernier. By pressing the finger up against the bottom of the horizontal circle, turn the circle until the zero is nearly in position at the zero of the *A* vernier, clamp the upper clamp, and, using a reading glass, bring the zero into exact position with the tangent screw. Make the two lines adjacent to the zero on the vernier have equal offsets from their opposites on the scale. See Fig. 4-30.

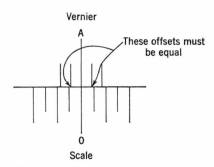

Fig. 4-30. Setting the vernier at zero.

Fig. 4-31. Using the reading glass.

To Use the Reading Glass. Steady the hand by touching the upper plate with the little finger and hold the glass about 2 in. from the vernier **in line with the graduations being read.** The eye must be placed on this line, and the plane of the reading glass must be perpendicular to the line from the eye to the graduation. See Fig. 4-31.

Read the *B* vernier, unless good accuracy is not required. When the *B* vernier is used, the effect of eccentricity of the circle is eliminated and, since the two readings are averaged, a more accurate reading results. As it takes very little extra time, it is an inexpensive way to increase accuracy.

Step 2.   Point at the initial point, using the lower motion. The initial point should be the left-hand side of the angle to be measured unless it is essential that the other side be used.

To Point the Instrument.   By looking over the telescope, aim approximately at the target. When the target is brought into the field of view, tighten the proper clamp, move the telescope up or down by hand until the **horizontal** cross hair is near the target, and then bring the vertical cross hair on the target with the proper tangent screw. In general, avoid using the vertical motion while measuring horizontal angles.

Step 3.   Aim at the second point, using the upper motion. When the upper clamp is free, the alidade may be turned in **either direction** to bring the aim to the second point.

Step 4.   Loosen the lower clamp. This ends the cycle.

To Read the Angle. Step 5.   Read the clockwise angle on both verniers. The A vernier is sufficient if good accuracy is not essential. If only the A vernier is used throughout, the reading of the A vernier is taken as the value of the angle.

If the B vernier is also used, the value of the angle is the average of the A and B readings minus the **initial** reading, which is the average of the A and B readings when the A vernier is set at zero.

To Read the Vernier.   Estimate the reading of the vernier to minutes with the naked eye, and then make the complete reading with the aid of the reading glass. Hold the glass as described for setting the vernier at zero. When both the A and B verniers are used, only the minutes and seconds are read on the B vernier. The degrees, minutes, and seconds on the A vernier are recorded, and only the seconds on the B vernier.

Summary.   To measure a horizontal angle:

1. Free both motions and set the A vernier at zero with the upper motion (read the B vernier if necessary).
2. Point at the initial (left-hand) point, using the lower motion.
3. Point at the second point, using the upper motion.
4. Free the lower motion.
5. Read the A vernier (and the B vernier if necessary) **clockwise.**

Field Records.   Usually only the A vernier is used. In this case almost any form of record of the angle is satisfactory. When the B vernier is also read, the standard form of record is that shown in Fig. 4-32. The explanation of the record for the first two angles in the figure is shown in Table 4-1. Check carefully.

Note: The A-vernier readings are recorded in full. Only the seconds in the B reading are recorded. When a B reading is 1 minute greater than the minutes in the A record, 60 seconds is added to the seconds in the B record. When the B reading is 1 minute less than the A record, a line or **bar** is placed over the seconds in the B record.

Example

A record	B reading	B record	Averages
85°10′45″	265°11′15″	75″	=85°11′00″
15 20 15	195 19 30	$\overline{30}$	=15°19′00″

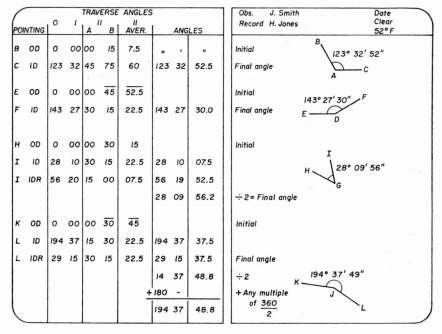

Fig. 4-32. Standard form of field notes.

This is a very simple system, but sometimes it is difficult to become accustomed to it. When making the small computations required, it is best to write out the angles in their usual form on scrap paper until the system becomes familiar.

Table 4-1

Angle A-BC—Telescope Direct (D.) Throughout

Pointing	A vernier	B vernier	Average
B (initial)(D.)	0°00′00″	180°00′15″	0°00′07.5″
C (D.)	123 32 45	303 33 15	123 33 00
To compute angle, subtract initial average			123 32 52.5

Angle D-EF—Telescope Direct (D.) Throughout

Pointing	A vernier	B vernier	Average
E (initial)(D.)	0°00′00″	179°59′45″	359°59′52.5″
F (D.)	143 27 30	323 27 15	143 27 22.5
To compute angle, subtract initial average			143 27 30.0

4-15. Measuring Angles by Repetition. The accuracy of the measurement of an angle can be greatly increased by the method of repetition. Every surveyor should thoroughly understand this procedure. The method automatically eliminates, or substantially reduces, all residual errors in the geometric requirements of a transit except loose bearings. In addition, it reduces the errors due to the fact that the verniers cannot be read very precisely.

Procedure for Doubling the Angle. Note that after an angle has been turned once and the lower motion has been loosened (step 4 in Sec. 4-14, Summary), the value of the angle is held at the A and B verniers. No matter how the alidade is turned, these readings remain. Even when the telescope is turned upside down (i.e., reversed), there is no change.

It follows that, if the telescope is reversed and then aimed at the left-hand point, the vernier will still give the value of the angle. Accordingly, after step 5 has been completed, reverse the telescope and aim at A with the lower motion.

Then, if the telescope is aimed at the second point with the upper motion, there will be added to the original reading a second value of the angle. This is shown schematically in Fig. 4-33. Reversing the telescope eliminates many of the residual errors in the geometric requirements of the transit.

Obviously, the angle read must be divided by 2 to obtain the desired angle. This cuts the reading errors in half.

The verniers are read after the first turn to give an **approximate value of the angle.** This is used as a rough check on the final angle computed.

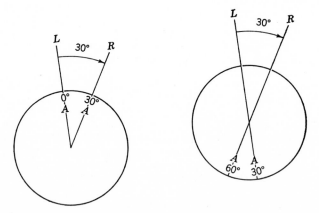

Fig. 4-33. Repeating an angle 1 D.R. The A vernier starts at zero and is moved to 30°. While the upper clamp is still tight, the telescope is reversed and pointed at L so that the A vernier still remains at 30° as in the second figure. When the angle is turned again, the A vernier moves to 60°.

The last two angles in Fig. 4-32 show the records of an angle turned twice. The explanation of the record is shown in Table 4-2. Check carefully.

Table 4-2

Angle G-HI—Telescope Direct (D.) and Reversed (R.)

Pointing	A vernier	B vernier	Average
H (initial)(D.)	0°00′00″	180°00′30″	0°00′15″
I (one)(D.)	28 10 30	208 10 15	28 10 22.5
To compute angle, subtract initial average			28 10 07.5
I (one)(D.R.)	56°20′15″	236°20′00″	56°20′07.5″
To compute angle, subtract initial average			56 19 52.5
Divide by number of repetitions. Final			28 09 56.2
Compare with one (D.) 28°10′07.5″			

Angle J-KL—Direct (D.) and Reversed (R.)

Pointing	A vernier	B vernier	Average
K (initial)(D.)	0°00′00″	179°59′30″	359°59′45″
L (one)(D.)	194 37 15	14 37 30	194 37 22.5
To compute angle, subtract initial average			194 37 37.5
L (one)(D.R.)	29°15′30″	209°15′15″	29°15′22.5″
To compute angle, subtract initial average			29 15 37.5
Divide by number of repetitions.			14 37 48.8
Compare with one (D.) 194°37′22.5″			

The total angle contains a whole circle (360°) in addition to the angle read, or, for half the final angle, 180°. This must be added. $\qquad$ +180

$\qquad\qquad\qquad\qquad\qquad\qquad$ Final   194°37′48.8″

More Repetitions. If, when the lower clamp has been released, the procedure is again repeated as many times as desired, and the reading divided by the number of turns, the result will be still more accurate. Usually the angle is turned once, twice, six times, or twelve times. The telescope is reversed after half the turns are completed. There is little advantage in more than 12 repetitions. For more accuracy, make two or more sets of 12. When two sets are used, start the second set at 180° instead of zero. For three sets, start at 0, 120°, 240°, etc.

The names and symbols for these operations are shown in Table 4-3.

Table 4-3

No. of turns	Operation	Symbol
1	Once direct	1 D.
2	Once direct and once reversed	1 D.R.
6	Three direct and three reversed	3 D.R.
12	Six direct and six reversed	6 D.R.

Steps in Procedure

1. Free both motions.
2. Set the *A* vernier at zero with the upper motion. Read the *B* vernier.
3. Point at the left-hand point, using the lower motion.
4. Point at the second point, using the upper motion.
5. Free the lower motion and **call out loud, "one."** If the number is called out loud, the number of turns will be remembered. The call should be made when the lower clamp is loosened.
6. Read the *A* and *B* verniers.
7. Reverse the telescope if the number called is half the number of turns to be used. In either case point the left-hand point with the lower motion.
8. Repeat as desired, but do not read the verniers again until the total number of turns has been completed.

Figure 4-34 shows the field record for two angles, each turned three times direct and three times reversed (3 D.R.). The reader should write out the actual readings which gave these figures.

Note that in measuring an angle 3 D.R., the circle may be overrun several times. Accordingly, a certain number of units of 360° should be added to the reading of the angle; or, which is more convenient, a certain number of units of 60° (which is $\frac{1}{6} \times 360°$) must be added to one-sixth of the angle read. The number of units of 60° is chosen which is necessary to make the final angle nearly equal to the value of 1 D. In Fig. 4-34, for angle *A-BC*:

$$\text{Angle read} = \quad 21°17'45.0''$$
$$\text{Dividing by the number of turns (6)} = \quad 3\ 32\ 57.5$$
$$\text{Adding } 2 \times 60° = \underline{120}$$
$$\text{Final angle closely equal to 1 D.} = \overline{123°32'58''}$$

Dividing by 6 may be accomplished as follows: Divide the degrees by 6, and use the remainder as the first digit of the minutes in the quotient. Divide the minutes by 6, using the quotient as the second digit of the

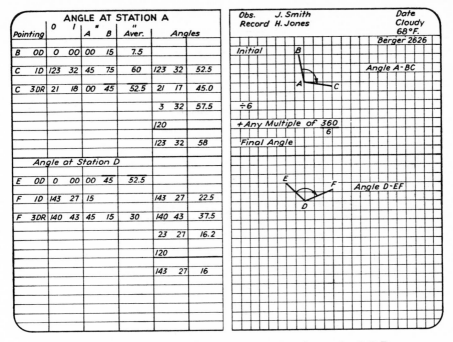

Pointing	0	′	″ A	″ B	Aver.	Angles		
ANGLE AT STATION A								
B OD	0	00	00	15	7.5			
C ID	123	32	45	75	60	123	32	52.5
C 3DR	21	18	00	45	52.5	21	17	45.0
						3	32	57.5
						/20		
						123	32	58
Angle at Station D								
E OD	0	00	00	45	52.5			
F ID	143	27	15			143	27	22.5
F 3DR	140	43	45	15	30	140	43	37.5
						23	27	16.2
						/20		
						143	27	16

Obs. J. Smith   Date
Record H. Jones   Cloudy
                        68°F.
                        Berger 2626

Initial — B — Angle A-BC — A — C

÷6

+Any Multiple of 360 — 6

Final Angle

E — F — Angle D-EF — D

Fig. 4-34. Examples of field notes from measuring angles 3 D.R.

minutes and the remainder as the first digit of the seconds. Divide the seconds by 6, using the result to fill out the seconds. For example, to divide 291°29′15″ by 6,

$$291°29′15″$$

Step 1	48 3	(291 ÷ 6 = 48 and 3 over)
Step 2	48 34 5	(29 ÷ 6 = 4 and 5 over)
Step 3	48 34 52.5	(15 ÷ 6 = 2.5)

The Check against Blunders. The final angle should agree with the 1 D. value within about 15 seconds, no matter how many times the angle is turned.

Experience indicates that the results of 3 D.R. can be relied upon to be within 3 to 6 seconds of the true value, depending on the skill of the observer.

4-16. Closing the Horizon. Usually in triangulation and occasionally in traverse, more than one angle is measured at a single station. A quick and useful check can then be obtained by measuring the unused angle that completes the circle, or **closes the horizon.** When this angle is measured, the angles can be adjusted so that their sum equals 360°. The same increment should be applied to each angle (including the unused

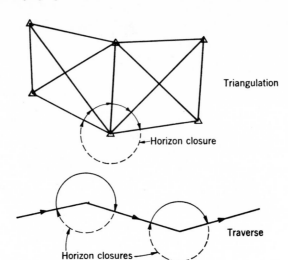

Triangulation

Horizon closure

Traverse

Horizon closures

Fig. 4-35. Horizon closures. Only the angles shown by full lines are required. The angles shown by dotted lines close the horizon so that station adjustments.can be made.

Pointing		O ' A	" B	" Aver.	Angles								
			STATION A										
B	OD	0 0	00	15	7.5								
C	ID	74 13	45			74	13	37.5					
C	3DR	85 22	15	30	22.5	85	22	15					
						14	13	42.5					
						74	13	42.5					
C	OD	0 0	00	30	45								
D	ID	158 48	00			158	48	15					
D	3DR	232 49	30	15	22.5	232	49	37.5					
						38	48	16.2					
						158	48	16.2					
D	OD	0 0	00	45	52.5								
B	ID	126 57	45			126	57	52.5					
B	3DR	41 47	00	45	52.5	41	47	00					
						6	57	50					
						126	57	50					

Obs.   J. Smith   Date
Record H.Jones   Clear
                                     60°F.

Sta. Adj.

Adj. Angles

A - BC =				
74	13	'42.5	+3.8	74° 13' 46.3"
A - CD =				
158	48	16.2	+3.8	158° 48' 20.0"
A - DB =				
126	57	'50	+3.7	126° 57' 53.7"
358	118	108.7	+11.3	358° 118' 120.0"

Fig. 4-36. Field notes for a horizon closure and a station adjustment.

angle), as the chance for error is the same for each angle, despite its size. Such an adjustment is called a **station adjustment** (see Fig. 4-35).

Often this procedure is applied to traverse angles where only one angle is used.

Figure 4-36 shows the **field notes that result from a horizon closure.**

4-17. Caution. When the two targets which mark the two sides of an angle differ by 20° or more in vertical angle, a special procedure is required (see Secs. 4-18 and 4-19). The following pairs of targets differ by 20° in vertical angle:

$$A + 5° \qquad A + 5° \qquad A - 5°$$
$$B - 15° \qquad B + 25° \qquad B - 25°$$

SPECIAL OPERATIONS

4-18. To Level the Transit Precisely. When the vertical axis of the transit does not accurately coincide with the direction of gravity, errors are introduced in the horizontal and vertical angles measured. The plate levels indicate the direction of the vertical axis with sufficient accuracy for the requirements of most observations. However, the plate levels should not be relied on when (1) the horizontal angle is to be measured between two points separated by an angle of elevation of 20° or more, (2) a vertical angle is to be measured more accurately than to the nearest 20 seconds, or (3) the transit is used to establish a vertical plane for controlling a vertical column, steel erection, etc.

For these requirements the transit must have a firm support. Stakes must be driven to support the legs if the ground is springy. The transit is leveled in the usual way, and the following procedure is carried out:

1. Tighten the upper clamp and free the lower clamp. Do not use these clamps again until leveling is complete.
2. By looking over the telescope, aim approximately at some object at about the same elevation as the telescope and in line with two opposite leveling screws.
   a. Center the telescope bubble, with the vertical motion. Note Fig. 4-37.
   b. Turn 180° in azimuth. By sighting backward over the telescope, aim at the object selected.
   c. Bring the bubble halfway toward the center with the vertical motion.
   d. Turn back to 0° azimuth by sighting over the telescope at the object.

   e. Center the bubble with the leveling screws. Repeat until the bubble centers in both positions.

   f. Turn approximately 90° and center the bubble with the leveling screws.

   g. Check the position of the bubble at approximately 270, 0, 180, and 90°.

When the bubble centers in all these positions, the outer bearing, i.e., the bearing between the circle and the leveling head, is in the direction of gravity.

4-19. To Measure a Horizontal Angle between Points That Differ by Over 20° in Vertical Angle. Level precisely, as described above. Measure the angle twice. Start the first measurement with the *A* vernier at zero, and start the second measurement with the *A* vernier at 180°. Each measurement should be based on the required number of repetitions. Use the average of the results.

If the inner bearing is not aligned with the outer bearing, the second measurement reverses the effect of this error and thus cancels it.

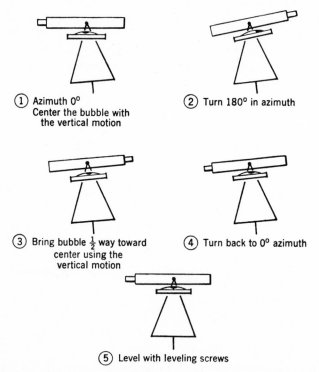

Fig. 4-37. Leveling an instrument by means of the telescope level.

To Measure a Vertical Angle.   Aim the cross hairs at the point, using the vertical motion, and read the vertical-circle vernier. If the instrument is not known to be in adjustment, the observation should be made direct and reversed and the average used. This eliminates any error in the position of the vernier that might introduce an index correction.

If the instrument has only a vertical arc, instead of a vertical circle, so that the vertical angle cannot be read when the telescope is reversed, level the instrument precisely, as described above. Note the reading of the vertical arc at the end of this process and before the telescope is changed in elevation. If the arc does not read zero, take this reading and note its sign. This is the index error. This value with its sign changed is the index correction. Apply this correction to all vertical angles observed with that instrument.

To Measure a Precise Vertical Angle.   The instrument should be leveled precisely, by using the telescope level. A method must be chosen that will obtain an average of several readings of the vernier. Reading the vernier after each of several pointings will not accomplish the result, for the reading will probably be the same each time. The desired result is best accomplished by utilizing the **stadia cross hairs.** Most instruments have supplementary horizontal cross hairs, one above and one below the center hair. They are used for stadia measurements described in another chapter. First clamp the upper motion, and use only the lower motion for direction. Read the vertical angles obtained by aiming the three cross hairs successively at the point, both direct and reversed. Use the average of the six values obtained.

If vertical angles and horizontal angles are to be measured simultaneously, use two sets of 3 D.R. and, while aiming, use first one and then another stadia hair in rotation so that six vertical angles are obtained at each point observed.

4-20. To Use Compass Bearings.   Many transits are equipped with a magnetic compass, and many other types of instruments depend on the compass alone to determine direction. The compass is used to determine directions in transit-stadia surveys, plane-table surveys, preliminary route surveys, and in many forms of rapid mapping. It is used to check transit angles, and, by applying the correction for declination, it is used to determine the approximate direction of true north.

Certain operations can be performed with a compass which are impossible with most angle-measuring instruments. Many of these are especially useful for surveys for maps.

1. In general, no backsights need be taken, since directions are determined with respect to the magnetic meridian at each instrument station.

2. Small obstacles can be passed by merely placing the instrument beyond the obstacle as nearly on line as can be estimated.
3. Although magnetic bearings are not very accurate, they are usually as accurate as is necessary for many types of mapping.
4. Accidental errors in determining direction do not accumulate along a compass traverse, since the directions are measured from the magnetic meridian at each instrument station.
5. When a blunder is made in determining the direction of a traverse course, it affects the direction of none of the following courses.

When there happens to be local attraction at a station, all the directions measured at that station will be in error by the amount of the local attraction. The angle between any two bearings, however, will be correct despite any local attraction.

When local attraction is suspected at any station, a backsight is taken to the previous station. If the back bearing differs by 180° from the forward bearing taken at the previous station, both stations can be assumed to have no local attraction. If the difference is not 180° at some station, all bearings observed from that station can be corrected accordingly.

The following example shows how this principle is applied to the bearings of a compass traverse:

Example.   Figure 4-38 shows the observed compass bearings in a compass traverse. At each station the full line indicates the direction of the magnetic meridian without local attraction, and the dotted arrow shows the actual direction of the compass needle.

1. Note that on *D-E* the back and forward bearings check and therefore no local attraction exists at *D* or *E*. Thus bearing *E-F* is correct.
2. Bearing *F-E* is too large by 30 minutes so that the needle must be as shown.
3. Since the needle is as shown, bearing *F-A* is too large by 30 minutes and hence must be S18°15′W.
4. Since bearing *A-F* checks with corrected *F-A*, there is no local attraction at *A*, and bearing *A-B* is correct.
5. This procedure is carried around the traverse, which results in the following corrected bearings:

E-F	N78°30′E	*B-C*	N22°15′E
F-A	S18 15 W	*C-D*	N84 15 W
A-B	N64 30 E	*D-E*	S12 00 W

If there are no errors in the observed compass bearings, the arithmetic will check at the completion of the computation.

Sometimes it is necessary to establish the direction of a compass bearing determined some years ago when the declination was different. In

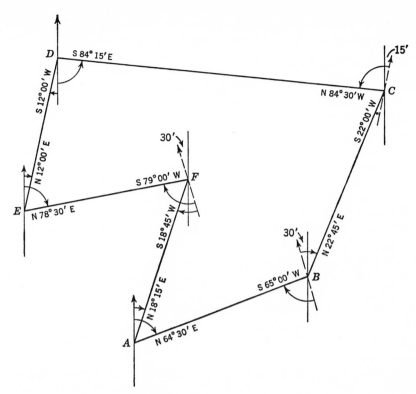

Fig. 4-38. The effect of local attraction in a traverse.

this case, the difference between the present declination and the former declination is estimated from the isogonic charts which apply (see Sec. 4-11).

To Determine an Accurate Compass Bearing. The procedure for determining an accurate compass bearing is the following:

First make sure that the north point (the zero) on the compass circle is exactly at the zero of the declination arc.

Set up where there is no local attraction. Aim at any well-defined point. Read the angle at each end of the needle. Aim at the point with the instrument reversed, and again read the angle at each end of the needle. Average the four angles and add the proper letters for the direction of the point used.

To read the needle, place the eye exactly in line with the needle and estimate each reading to the nearest 5 minutes.

The process can be repeated as often as desired to improve the accuracy. The value obtained can be corrected for the time of day and time of year by values that can be obtained from the Director, U.S. Coast and Geodetic Survey.

4-21. Care of the Instrument.   Complete directions for instrument care are beyond the scope of this text. The following rules, plus a recognition of the delicacy of the instrument, will usually prevent damage. The most important rule is to prevent falls. A fall will always result in the need for extensive repairs or will destroy the instrument entirely. The rules apply both to the transit and to the level.

1. **Handle the instrument by the base** when not on the tripod. This prevents deflecting the more delicate parts.
2. **Never stand the tripod on a smooth surface.** The legs may slip outward.
3. **Always stand the tripod up carefully.** The legs must be wide and firm even when the setup is not to be used for observations. The wind or a slight touch may knock it over.
4. **Never leave the instrument unattended** unless special precautions are made for its protection.
5. **Never subject the instrument to vibration,** which ruins the adjustments. Most instrument cases have large rubber feet, which absorb vibration if the rest of the case is free from contacts.
6. **Never force the instrument.** If the telescope or alidade does not turn easily, do not continue to use the instrument. Such use might damage a bearing.
7. **Keep the instrument in its case.** This usually guarantees protection.
8. **Place it in the case so that the only contact is with the base.** Keep all three transit clamps tight. This reduces chances for vibration. Some cases have felt-covered contact points, which are safe.
9. **Keep the instrument free from dust and rapid temperature changes.** Dust ruins the finish and the bearings. Temperature ranges introduce moisture into the telescope tube. The moisture will fog the telescope, and the telescope must be dismantled to remove it.
10. **If the instrument is wet, let it dry.** Do not dry it, for this ruins the finish and smears the glass and graduations.

Problems

4-1 to 4-6. Construct a vernier and scale for the following requirements. Make the lines the proper lengths and include the proper numbering. Instead of an arc, use a straight line between the vernier and the scale, and space the graduations by eye.

Problem	Smallest circle graduation	Least reading of vernier	Clockwise reading
4-1	30 minutes	1 minute	51°44′00″
4-2	30 minutes	1 minute	21 43 00
4-3	20 minutes	20 seconds	125 32 20
4-4	20 minutes	30 seconds	150 07 40
4-5	20 minutes	30 seconds	68 48 30
4-6	20 minutes	30 seconds	233 52 30

4-7 to 4-18. For each of the geometric conditions of the transit listed below, state the following: (*a*) What angles will be affected most? (*b*) What angles will be affected slightly? (*c*) In each case, will the angular error be greater when the points observed are nearer the transit? (*d*) What field procedure will eliminate the error introduced?

4-7. Line of sight above the horizontal axis

4-8. Line of sight to the left of the vertical axis

4-9. The horizontal axis behind the vertical axis

4-10. The horizontal axis not perpendicular to the vertical axis

4-11. The line of sight not perpendicular to the horizontal axis

4-12. The objective slide not perpendicular to the horizontal axis

4-13. The plate bubbles not reading zero when the vertical axis is vertical

4-14. The telescope bubble not reading zero when the line of sight is horizontal

4-15. The graduations on the horizontal graduated circle not concentric with the vertical axis

4-16. The graduations on the vertical circle not concentric with the horizontal axis

4-17. The vertical circle reading a +5 minutes when the line of sight is perpendicular to the vertical axis

4-18. The inner bearing not aligned with the outer bearing

4-19. Define a magnetic meridian.

4-20. At any position on the earth, what variations occur in the direction of the magnetic meridian?

4-21 to 4-26. Complete the field notes and make the station adjustments for the following sets of data from horizon closures. The method of 3 D.R. was used.

4-21

°	′	″ A	″ B
0	0	00	$\overline{45}$
92	13	30	30
193	21	15	30
0	0	00	15
156	47	30	60
220	46	15	30
0	0	00	$\overline{30}$
110	58	45	15
305	52	00	$\overline{30}$

4-22

°	′	″ A	″ B
0	0	00	30
117	58	15	$\overline{45}$
347	47	00	$\overline{45}$
0	0	00	15
174	19	30	45
325	57	15	$\overline{30}$
0	0	00	15
67	43	00	$\overline{45}$
46	16	15	$\overline{45}$

4-23

°	′	″ A	″ B
0	0	00	30
177	45	15	45
346	31	15	45
0	0	00	$\overline{30}$
81	57	00	00
131	43	30	15
0	0	00	$\overline{45}$
72	29	15	$\overline{45}$
74	54	15	30
0	0	00	15
27	48	45	15
166	50	15	0

4-24

°	′	″ A	″ B
0	0	00	$\overline{45}$
83	54	15	45
143	27	30	30
0	0	00	30
23	02	30	30
138	14	15	45
0	0	00	15
18	09	30	60
108	59	00	30
0	00	00	$\overline{30}$
234	52	45	75
329	19	15	$\overline{45}$

4-25

°	′	″ A	″ B
0	00	00	15
158	22	30	15
230	13	30	45
0	00	00	$\overline{45}$
142	17	00	30
133	44	00	$\overline{45}$
0	00	00	30
59	21	00	$\overline{45}$
356	04	00	15

4-26

°	′	″ A	″ B
0	00	00	20
82	10	20	30
133	02	00	$\overline{40}$
0	00	00	$\overline{40}$
67	29	00	$\overline{50}$
44	54	40	30
0	00	00	$\overline{50}$
210	20	40	50
182	05	00	$\overline{50}$

4-27. A six-sided closed traverse was surveyed over the stations *ABCDEF*. At each station the back bearing of the previous course and the forward bearing of the next course were observed, with the results given below. Compute the forward bearings for all six courses, corrected for local attraction.

Compass sta.	Point sighted	Bearing
A	F	N10°15′W
A	B	S72 00 E
B	A	N73 00 W
B	C	N64 30 E
C	B	S62 45 W
C	D	N 3 00 W
D	C	S 1 30 E
D	E	S81 15 W
E	D	N82 00 E
E	F	N77 45 W
F	E	S77 15 E
F	A	S10 15 E

Suggestion: Draw a sketch of the traverse, and at each station indicate the direction of the magnetic meridian and the relative direction of the compass needle as affected by local attraction.

4-28. Similar to Prob. 4-27 but a five-sided traverse.

Compass sta.	Point sighted	Bearing
A	E	S88°30′E
A	B	S22 15 E
B	A	N22 45 W
B	C	S40 15 E
C	B	N40 15 W
C	D	N51 45 E
D	C	S50 15 W
D	E	N31 45 W
E	D	S32 15 E
E	A	S89 00 W

Suggested Field Exercises

4-1. To Set Up the Transit. Each student should have a transit. The instructor should demonstrate each step and see that every student has followed it before proceeding to the next step. After one or two demonstrations, the students should continue making setups of their own.

Sloping ground should be chosen for the exercise. Each student should have a nail to mark the point over which the setup is to be made, and the nail should be moved a few inches between each setup.

4-2. To Measure Horizontal Angles. Three easily recognized targets should be chosen or established. They should be white, about three times the apparent width of the cross hair (a total of 10 seconds), and placed against a dark background. The three angles that close the horizon should be measured first by a set of 1 D., then 1 D.R., and then 3 D.R. The station adjustment should be computed for each set.

One transit should be assigned to each student or pair of students. Points should be marked for setup positions nearly in a straight line about 10 ft apart, arranged so that there is a clear sight from each to all three targets.

The instructor should check each zero setting and angle reading at first; and, if working in pairs, both students should check all settings and readings.

Compass bearings should be taken during the work. Accurate compass bearings, described in Sec. 4-20, can be taken if desired. The compass angles should be computed and checked against the transit values.

4-3. To Measure Vertical Angles and Horizontal Angles between Points of Large Difference in Height. Have the instruments set up where one or, if possible, two high points can be observed at vertical angles greater than 30°. Place a target at about the level of the instruments.

The students should measure the vertical angles to each of the three points and the horizontal angles between the ground point and each high point as follows:

1. By simple methods, twice; the instruments should be thrown out of level and releveled between the two operations.
2. By accurate methods, as described in Secs. 4-18 and 4-19.

5

Traverses and Elementary Triangulation

TRAVERSES

5-1. Definitions. A traverse consists of a continuous series of lines called **courses** running between a series of points called **traverse stations.** The lengths of the lines and the angles between them are measured. Traverses can be either open or closed. Open traverses end without closure. They cannot be properly checked and therefore are not recommended. Closed traverses are of two kinds, loop traverses and connecting traverses. Loop traverses close on themselves (Fig. 5-3), and connecting traverses begin at a known direction and position and end at another known position and direction (Fig. 5-7). Thus both the angles and the measured lengths in a closed traverse may be checked.

5-2. The Use of Traverses. Traverses, like triangulation surveys, are used to find the accurate positions of a relatively few marked points called stations. From this system of stations, many less precise measurements can be made to features to be mapped or located, without accumulating accidental errors. Traverses thus usually serve as control surveys. When plans for construction are drawn, the stations can again be used as beginning points from which to lay out work. Traverses are cheaper and more effective than triangulation in small areas or where many obstacles interfere with sight lines.

Thus, when new construction of any kind is desired, a system of traverse stations in the area involved must be established and surveyed at the outset. Efforts to avoid this operation because of ignorance or inertia are usually costly, retard the work, and often cause serious revision of plans.

5-3. Field Work.   The positions of the traverse stations are chosen so that they are as near as possible to the objects to be located without

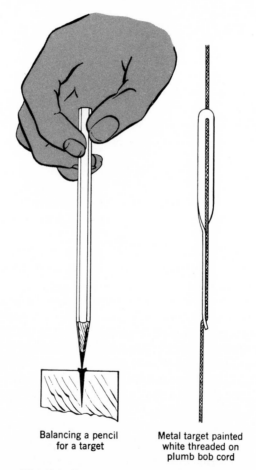

Balancing a pencil
for a target

Metal target painted
white threaded on
plumb bob cord

Fig. 5-1. Targets for transit observation.

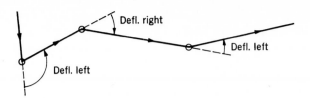

Fig. 5-2. Deflection angles.

unduly increasing the work of measuring the traverse. They are usually marked by stakes with tacks or by stone or concrete monuments set nearly flush with the ground, with a precise point marked on the top by a chiseled cross, drill hole, or special bronze tablet.

The angle and length measurements are made as described elsewhere. Signals must be placed at the stations so that the rear tapeman can align the taping and for angle measurement. Many types of signals have been devised. Usually a range pole stuck in the ground is used for taping, and a range pole held carefully balanced on the point is used for measuring angles. When the courses are short, a plumb bob is held over the point, or a pencil is balanced on it for angle measurements (see Fig. 5-1). Considerable time can be saved by rigging a device to support one of these signals in place so that it is not necessary to hold it.

Forward Direction. For purposes of consistency, it is necessary to assume which is forward and which is backward for any traverse. The order in which it is measured is usually called the forward direction. Loop traverses should, if possible, be measured counterclockwise around the loop.

Direction of Angle Measurement. The angles of a traverse should be measured clockwise from the backward direction to the forward direction. This is the most rapid field method and the least likely to introduce blunders. Some engineers recommend measuring deflection angles. They are often used in highway surveys and in other connecting or open traverses. A deflection angle is the angle between the forward prolongation of the back course and the forward course (see Fig. 5-2). It can also be defined as the change of direction of the traverse at a station. Unless the directions of the deflection angles left or right are properly recorded, a blunder will result.

To Measure a Deflection Angle. Measure the angle as previously described and subtract 180° from the result. If the difference is plus, it is a right deflection angle; if minus, left. Section 5-4 describes how deflection angles are used in azimuth computation. To compute bearings, use sketches as shown in Fig. 5-4.

Some surveyors recommend the following:

1. Set the vernier at zero.
2. Backsight with the telescope reversed, using the lower motion.
3. Foresight with the telescope direct, using the upper motion.
4. Record the clockwise or counterclockwise angle, whichever is less than 180°. If clockwise, it is a right deflection; if counterclockwise, left.

    This procedure introduces a systematic error of the instrument. To cancel this error, repeat the angle as follows:

5. Leave the vernier as it is after step 4 and backsight with the telescope direct, using the lower motion.
6. Foresight with the telescope reversed, using the upper motion.

5-4. The Computation of a Traverse.   Usually it is necessary to reduce the traverse field data to the form of the rectangular coordinates of the stations. This is accomplished by the type of computation described in the following paragraphs.

As a guide to computation, a sketch of the traverse should be drawn approximately to scale, showing the names of each of the traverse stations (see Fig. 5-3). If it is plotted carefully, it is a check against blunders in the survey.

Computation of Direction.   One of the first operations in computing a traverse consists in computing the directions of successive courses by applying a traverse angle to the direction of one course to obtain the direction of the following course. When bearings are used to express direction, the best method of accomplishing this is to draw a sketch for

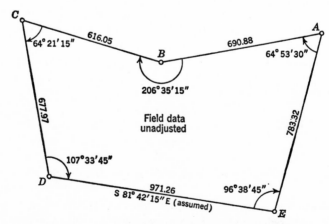

Fig. 5-3. Example of a loop traverse showing unadjusted field data.

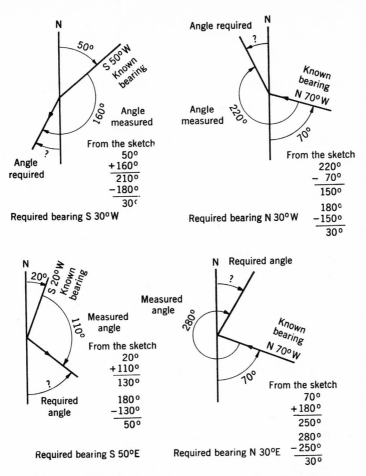

Fig. 5-4. Typical sketches for the computation of bearings.

each station, showing the meridian and the two courses involved (see Fig. 5-4). The required arithmetic will then be evident. Sketches may also be relied on when azimuths are used, but if the angles are measured as recommended, considerable time may be saved by using a rule for computing direction.

Azimuth Rule (see Fig. 5-5).

$$B = A + C \pm 180°$$

where B = azimuth of next course
A = azimuth of previous course
C = traverse angle

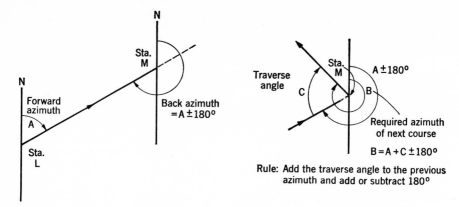

Fig. 5-5. The azimuth rule.

It is evident that $A \pm 180°$ is the back azimuth of the previous course. C can then be added, for it is measured in such a way that it expresses the increase in azimuth from the back azimuth of the previous course to the forward azimuth of the next course. It may be necessary sometimes to add or subtract 360° to avoid minus angles or angles that are over 360°.

Examples of Traverse Computation. The computations of a loop traverse and of a connecting traverse are here given in detail.

A Loop Traverse

Loop Traverse. Figure 5-3 illustrates a loop traverse. The steps required for its computation are listed in order.

1. Compute the angular error. The sum of the angles should be

$$(n - 2)180° = 540°$$

A	64°53′30″
B	206 35 15
C	64 21 15
D	107 33 45
E	96 38 45

Sum = 540°02′30″

Error per angle: 2′30″ ÷ 5 = 30″

Assume that an angular error of 1 minute per angle is allowed. The angular measurement is thus acceptable.

Table 5-1. Bearing Computation

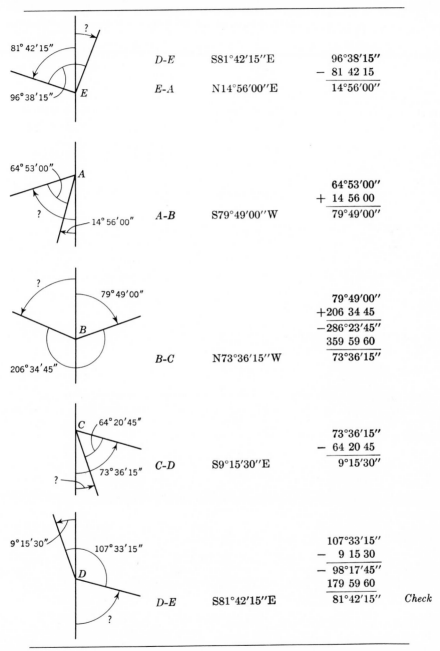

D-E	S81°42′15″E	96°38′15″
		− 81 42 15
E-A	N14°56′00″E	14°56′00″

		64°53′00″
		+ 14 56 00
A-B	S79°49′00″W	79°49′00″

		79°49′00″
		+206 34 45
		−286°23′45″
		359 59 60
B-C	N73°36′15″W	73°36′15″

		73°36′15″
		− 64 20 45
C-D	S9°15′30″E	9°15′30″

		107°33′15″
		− 9 15 30
		− 98°17′45″
		179 59 60
D-E	S81°42′15″E	81°42′15″ *Check*

Table 5-2. Form for Computation of Coordinates. Adjustment by Compass Rule

Sta.	Corrected bearings / Lengths	logs	Unadjusted Lat.	Dep.	Corrections Lat.	Dep.	Adjusted Lat.	Dep.
							Coordinates	
A	S79°49′00″W	2.086880					857.00	1371.35
		9.247478						
	690.88	2.839402	−122.15	−679.99	+0.13	−0.19	−122.02	−680.18
		9.993104						
B		2.832506					734.98	691.17
B	N73°36′15″W	2.240283						
		9.450667						
	616.05	2.789616	+173.89	−591.00	+0.11	−0.17	+174.00	−591.17
		9.981970						
C		2.771586					908.98	100.00
C	S9°15′30″E	2.825516						
		9.994305						
	677.97	2.831211	−669.14	+109.08	+0.12	−0.19	−669.02	+108.89
		9.206519						
D		2.037730					239.96	208.89
D	S81°42′15″E	2.146554						
		9.159219						
	971.26	2.987335	−140.14	+961.10	+0.18	−0.27	−139.96	+960.83
		9.995432						
E		2.982767					100.00	1169.72
E	N14°56′00″E	2.879018						
		9.985079						
	783.32	2.893939	+756.86	+201.86	+0.14	−0.23	+757.00	+201.63
		9.411106						
A		2.305045					857.00	1371.35
		Plus sums	+930.75	+1272.04				
		Minus sums	−931.43	−1270.99				
Sums	3739.48		− 0.68	+ 1.05	+0.68	−1.05		

Error in dept. + 1.05

Total error = 1.25   Error in lat. − 0.68

2. **Adjust the angles.** Give the same correction to each angle, as the chance for error is the same.

A	64°53′30″	− 30″ =	64°53′00″
B	206 35 15	− 30 =	206 34 45
C	64 21 15	− 30 =	64 20 45
D	107 33 45	− 30 =	107 33 15
E	96 38 45	− 30 =	96 38 15
		Sum =	540°00′00″ *Check*

3. **Compute bearings.** Starting with an assumed or known bearing (in this case $D-E =$ S81°42′15″E), compute the bearings by applying the corrected angles successively. (See Table 5-1.)

4. **Compute the latitudes and departures.** (See Table 5-2.)

$$\text{Latitude} = \Delta y = \text{length multiplied by cosine of bearing}$$
$$N = \text{plus}$$
$$S = \text{minus}$$
$$\text{Departure} = \Delta x = \text{length multiplied by sine of bearing}$$
$$E = \text{plus}$$
$$W = \text{minus}$$

If a computing machine is available, natural functions should be used. The form shown here is for logarithmic computation. The log of the length is placed at the middle of the column for each course, the log cosine bearing is placed directly above it, and the log sine bearing is placed below it. The two upper logs are added to obtain the log of the latitude, and the two lower logs are added to obtain the log of the departure. The form for machine computation is slightly different. It is given later for a connecting traverse.

5. **Compute the error.** Since the traverse begins and ends at the same point, the sum of the latitudes and the sum of the departures should both be zero. By adding the columns, the errors can be found. The error in latitude is −0.68; the error in departure is +1.05. The **total error** is evidently the square root of the sum of the squares of these values.

$$\text{Total error} = \sqrt{(-0.68)^2 + (+1.05)^2} = 1.25$$

6. **Compute the measure of accuracy or simply the accuracy.** This is the ratio of the total error to the total length of the survey. The sum of lengths of the courses is 3739.48; hence

$$\text{Accuracy} = 1.25 : 3739.48 = 1 : 2992$$

The accuracy of the usual transit traverse is about 1:3000. If the ratio is larger, a blunder probably exists, and the survey is rejected. In this case, the survey is accepted.

7. **Adjust the latitudes and departures.** This consists of a reasonable process that makes their respective sums equal zero. There are two methods.

   a. **The compass rule.** Apply corrections in proportion to the lengths of the courses (see Table 5-3):

$$\text{Cor.} = \frac{C}{L} S$$

where Cor. = correction to a latitude (or departure)
C = total error in sum of latitudes (or departures) with sign changed
L = total length of survey
S = length of particular course

The compass rule is considered more mathematically correct than the transit rule. However, it changes the latitudes and departures in such a way that both the bearings and lengths of the courses are changed.

   b. **The transit rule.** Apply corrections to the latitudes in proportion to the lengths of the latitudes and to the departures in proportion to the lengths of the departures (see Table 5-4):

$$\text{Cor.} = \frac{C}{L} S$$

where Cor. = correction to a latitude (or departure)
C = total error in sum of latitudes (or departures) with sign changed
L = sum of latitudes (or departures) without regard to sign
S = length of particular latitude (or departure)

The transit rule changes the latitudes and departures in such a way that the lengths of the course are changed slightly but the bearings remain almost the same.

Both rules. The sum of the corrections for latitudes must be equal to the error in latitude with its sign changed. The sum of the corrections for departures must likewise be equal to the error in departure with its sign changed. Because of rounding off the computed corrections, it is sometimes necessary to change slightly one or two corrections to create this relationship. Usually the changes are applied to the largest values. The final results are entered in the form (see Tables 5-2 and 5-5).

Table 5-3. Computation of Corrections by Compass Rule

Course	Cor. to latitudes	Cor. to departures
AB	$\dfrac{0.68}{3739} \times 691 = 0.13$	$\dfrac{-1.05}{3738} \times 691 = -0.19$
BC	$\dfrac{0.68}{3739} \times 616 = 0.11$	$\dfrac{-1.05}{3738} \times 616 = -0.17$
CD	$\dfrac{0.68}{3739} \times 678 = 0.12$	$\dfrac{-1.05}{3738} \times 678 = -0.19$
DE	$\dfrac{0.68}{3739} \times 971 = 0.18$	$\dfrac{-1.05}{3738} \times 971 = -0.27$
EA	$\dfrac{0.68}{3739} \times 783 = 0.14$	$\dfrac{-1.05}{3738} \times 783 = \overset{-0.23}{-\cancel{0.22}}$
Sums	0.68	-1.05

Table 5-4. Computation of Corrections by Transit Rule

Course	Cor. to latitudes	Cor. to departures
AB	$\dfrac{0.68}{1862} \times 122 = 0.04$	$\dfrac{-1.05}{2543} \times 680 = -0.28$
BC	$\dfrac{0.68}{1862} \times 174 = 0.06$	$\dfrac{-1.05}{2543} \times 591 = -0.24$
CD	$\dfrac{0.68}{1862} \times 669 = 0.24$	$\dfrac{-1.05}{2543} \times 109 = -0.04$
DE	$\dfrac{0.68}{1862} \times 140 = 0.05$	$\dfrac{-1.05}{2543} \times 961 = \overset{-0.41}{-\cancel{0.40}}$
EA	$\dfrac{0.68}{1862} \times 757 = \overset{0.29}{\cancel{0.28}}$	$\dfrac{-1.05}{2543} \times 202 = -0.08$
Sums	0.68	-1.05

8. **Compute the adjusted latitudes and departures.** Add the corrections algebraically to the unadjusted latitudes and departures.
9. **Compute the coordinates.** Choose coordinates such that all coordinates will be plus. In the example, point E, the most southerly point, is given a y, or north, coordinate of 100; and the point C, the

Table 5-5. Adjustment of the Same Survey (as given in Fig. 5-3) by Transit Rule

Station	Unadjusted		Corrections		Adjusted	
					Lat.	Dep.
	Lat.	Dep.	Lat.	Dep.	Coordinates	
A					857.15	1371.51
	−122.15	−679.99	+0.04	−0.28	−122.11	−680.27
B					735.04	691.24
B						
	+173.89	−591.00	+0.06	−0.24	+173.95	−591.24
C					908.99	100.00
C						
	−669.14	+109.08	+0.24	−0.04	−668.90	+109.04
D					240.09	209.04
D						
	−140.14	+961.10	+0.05	−0.41	−140.09	+960.69
E					100.00	1169.73
E						
	+756.86	+201.86	+0.29	−0.08	+757.15	+201.78
A					857.15	1371.51
Absolute sums	1862.18	2543.03				
Plus sums	+930.75	+1272.04	+0.68	−1.05		
Minus sums	−931.43	−1270.99				
Sums	− 0.68	+ 1.05				

most westerly point, is given an x, or east, coordinate of 100. The coordinates are computed by successive algebraic addition of the adjusted latitudes and departures. An arithmetic check is obtained when the computation is carried around to the starting point, which should have the same coordinates as before. In Tables 5-2 and 5-5, note the coordinate values opposite the letters in the computation form. See also Fig. 5-6 (based on compass rule).

A Connecting Traverse

Connecting Traverse Computation. Figure 5-7 illustrates the required sketch of a connecting traverse. It begins at the known positions of

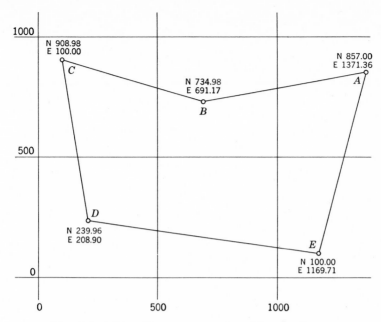

Fig. 5-6. Adjusted loop traverse plotted by coordinates.

triangulation stations Dog and Cat. It closes on Cow and the direction Cow-Ox. The data on the figure are triangulation data and the field data. The positions, i.e., the coordinates, of the triangulation stations must be held fixed and the traverse adjusted to them.

1. **Compute the angular error.** Compute the directions of the fixed lines upon which the traverse begins and closes.

$$\text{tan direction Cat-Dog} = \frac{807.53 - 1000.00}{1640.11 - 1200.00} = -0.437322$$

    Cat-Dog, bearing S23°37′15″E azimuth 156°22′45″

$$\text{tan direction Cow-Ox} = \frac{1407.36 - 1651.71}{524.85 - 66.26} = -0.53283$$

    Cow-Ox, bearing S28°03′00″E azimuth 151°57′00″

Starting with a known direction, compute the directions of the courses by applying the field angles successively (see Table 5-6). Either bearings or azimuths may be used. Azimuths should be avoided if tables of functions of angles up to 360° are not available.

Note: Northwest and southeast bearings are minus angles.

By either method, the error, in the direction the angles were measured, i.e., clockwise, is +1′30″ or +18″ per angle. If it is

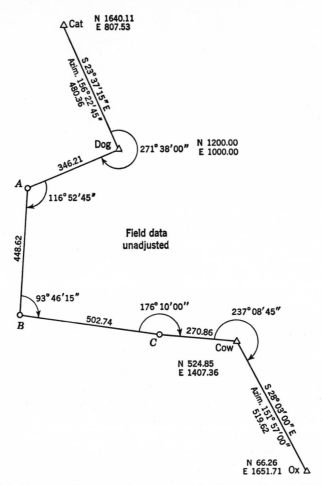

Fig. 5-7. Example of a connecting traverse showing unadjusted field data.

assumed that an error of 30 seconds per angle is allowed, the angular measurement is acceptable.

2. **Adjust angles.** Give the same correction to each angle, as the chance for error is the same.

Dog	271°38'00"	− 18"	=	271°37'42"
A	116 52 45	− 18	=	116 52 27
B	93 46 15	− 18	=	93 45 57
C	176 10 00	− 18	=	176 09 42
Cow	237 08 45	− 18	=	237 08 27

3. **Compute directions.** (See Table 5-7.)

Table 5-6. Computation of Angular Error

By bearings with angles		By azimuths with angles	By azimuths with defl. angles
S 23°37′15″E	Cat-Dog	156°22′45″	156°22′45″
+271 38 00		+271 38 00	+ 91 38 00
248 00 45		428 00 45	
−180		−180	
S 68 00 45 W	Dog-*A*	248 00 45	248 00 45
+116 52 45		+116 52 45	− 63 07 15
184 53 30		364 53 30	
−180		−180	
S 4 53 30 W	*A*-*B*	184 53 30	184 53 30
+ 93 46 15		+ 93 46 15	− 86 13 45
98 39 45		278 39 45	
−179 59 60		−180	
S 81 20 15 E	*B*-*C*	98 39 45	98 39 45
+176 10 00		+176 10 00	− 3 50 00
94 49 45		274 49 45	
−179 59 60		−180	
S 85 10 15 E	*C*-Cow	94 49 45	94 49 45
+237 08 45		+237 08 45	+ 57 08 45
151 58 30		331 58 30	
−179 59 60		−180	
S 28 01 30 E	Cow-Ox	151 58 30	151 58 30
−S28 03 00 E	Cow-Ox fixed	151 57 00	−151 57 00
+ 1′30″	Error	+ 1′30″	+ 1′30″

4. **Compute the latitudes and departures.** The form for computation, using natural functions, is shown in Table 5-8. It should be used only when computing machines are available.
5. **Compute the error.**

$$\text{Total error} = \sqrt{(0.21)^2 + (0.38)^2} = 0.43$$

6. **Compute the measure of accuracy.**

$$\text{Accuracy} = 0.43:1568.43 = 1:3648$$

7. **Compute the corrections to latitudes and departures.** (See Table 5-9.)
8. **Compute the adjusted latitudes and departures.** See form in Table 5-8.

Table 5-7. Computation of Directions

By bearings		By azimuths
S 23°37'15"E Cat-Dog		156°22'45"
+271 37 42		+271 37 42
248 00 27		428 00 27
−180		−180
S 68 00 27 W Dog-*A*		248 00 27
+116 52 27		+116 52 27
184 52 54		364 52 54
−180		−180
S 4 52 54 W *A*-*B*		184 52 54
+ 93 45 57		+ 93 45 57
98 38 51		278 38 51
−179 59 60		−180
S 81 21 09 E *B*-*C*		98 38 51
+176 09 42		+176 09 42
94 48 33		274 48 33
−179 59 60		−180
S 85 11 27 E *C*-Cow		94 48 33
+237 08 27		+237 08 27
151 57 00		331 57 00
−179 60 00		−180
S 28 03 00 E Cox-Ox		151 57 00
S 28 03 00 E Cow-Ox fixed		151 57 00
0 Check		0

Table 5-8. Form for Machine Computation of Coordinates

Sta.	Corrected bearings Lengths	cos sin	Unadjusted Lat.	Dep.	Cor. Lat.	Dep.	Adjusted Lat. Coordinates	Dep.
Dog	S68°00'27"W	0.37449					1200.00	1000.00
A	346.21	0.92723	−129.65	−321.02	−0.04	−0.08	− 129.69	− 321.10
A	S4°52'54"W	0.99637					1070.31	678.90
B	448.62	0.08510	−446.99	− 38.18	−0.06	−0.11	− 447.05	− 38.29
B	S81°21'09"E	0.15036					623.26	640.61
C	502.74	0.98864	− 75.59	+497.03	−0.07	−0.12	− 75.66	+ 496.91
C	S85°11'27"E	0.08384					547.60	1137.52
Cow	270.86	0.99648	− 22.71	+269.91	−0.04	−0.07	− 22.75	+ 269.84
Sums	1568.43		−674.94	+407.74	Cow		524.85	1407.36
	Coord. diff.		−675.15	+407.36				
	Error		+ 0.21	+ 0.38				

Table 5-9. Computation of Corrections by Compass Rule

Course	Cor. to latitudes	Cor. to departures
Dog-A	$\dfrac{-0.21}{1568} \times 346 = \overset{-0.04}{\cancel{-0.05}}$	$\dfrac{-0.38}{1568} \times 346 = -0.08$
A-B	$\dfrac{-0.21}{1568} \times 449 = -0.06$	$\dfrac{-0.38}{1568} \times 449 = -0.11$
B-C	$\dfrac{-0.21}{1568} \times 503 = -0.07$	$\dfrac{-0.38}{1568} \times 503 = -0.12$
C-Cow	$\dfrac{-0.21}{1568} \times 271 = -0.04$	$\dfrac{-0.38}{1568} \times 271 = -0.07$
Sums	-0.21	-0.38

9. **Compute the coordinates.** Beginning with the fixed coordinates at the start of the traverse, compute the coordinates of each station by successive algebraic addition. An arithmetic check is obtained when the computed coordinates of Cow agree with its fixed coordinates. See also Fig. 5-8.

Final Operations

Plotting the Traverse. Traverses may be plotted by protractor and scale. If the coordinates are available, much greater accuracy is obtained

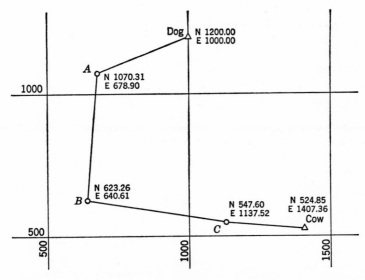

Fig. 5-8. Adjusted connecting traverse plotted by coordinates.

when the stations are plotted by coordinates. When the stations are plotted, light connecting lines should be drawn to represent the traverse lines. The plotted traverse should then be checked by scaling the lengths of the courses and by measuring the traverse angles with a protractor. The results are compared with the original field notes.

The coordinates of each station should be printed near the station on the map for ease in using them.

Topographic Features. Objects and other topographic features that are required for the map are located by field measurements from the traverse lines. These measurements are made from traverse stations or points set at known distances along the traverse lines. They consist of the measurement of any convenient combinations of angles and distances. Any two of these measurements will locate a point. The subject is covered in Chap. 7.

To Compute the New Bearings and New Lengths of the Courses. It is sometimes necessary to compute the new bearings and lengths of the courses that result from the adjustment of the traverse. The following formulas are used:

$$\tan B = \frac{d}{l}$$

$$L = \frac{l}{\cos B} \qquad \text{or} \qquad L = \frac{d}{\sin B}$$

where B = new bearing
d = corrected departure
l = corrected latitude
L = new length

Use the formula containing l when l is larger. If d is larger, use the formula containing d. If either l or d is unavailable, compute them from the differences in the coordinates.

Example. To compute the adjusted bearing and length of the course *DE*. From the adjustment by the compass rule in Table 5-2:

$$\tan B = \frac{d}{l}$$

$$\log d,\ 960.83 = 2.982647$$

$$- \log l,\ -139.96 = \underline{2.146004} \quad (-)$$
$$\log \tan B = \overline{0.836643} \quad (-)$$

$$B = \text{S}81°42'44''\text{E}$$

$$L = \frac{d}{\sin B}$$

$$\log d,\ 960.83 = 2.982647$$

$$\log \sin B,\ 81°42'44'' = \underline{9.995441}$$
$$\log L = \overline{2.987206}$$

$$L = 970.97$$

From the adjustment by the transit rule in Table 5-5:

$$\tan B = \frac{d}{l}$$

$$\log d, 960.69 = 2.982583$$
$$-\log l, 140.09 = 2.146407$$
$$\log \tan B = \overline{0.836176}$$

$$B = S81°42'12''E$$

$$L = \frac{d}{\sin B}$$

$$\log d, 960.69 = 2.982583$$
$$-\log \sin B, 81°42'12'' = 9.995431$$
$$\log L = \overline{2.987152}$$

$$L = 970.85$$

Computation of Area Enclosed in a Traverse. The simplest and most accurate method of computing the area enclosed in a traverse is shown in Table 5-10. The procedure is as follows:

1. If the latitudes and departures of the traverse stations are not known, compute them from the coordinates.
2. Arrange the courses in order, beginning with the course that starts at the most westerly traverse station (see column headed "Course"); enter the latitudes and departures as shown. South latitudes and west departures are minus.
3. Compute the **double meridian distances** (the DMD's) according to the following rules, as shown:

Table 5-10. Computation of the Area of the Traverse
(see Fig. 5-6 and Table 5-2)

Course	Latitude	Departure	DMD	Double areas
CD	−669.02	+108.89	+ 108.89 + 108.89 + 960.83	−72,850
DE	−139.96	+960.83	+1178.61 + 960.83 + 201.63	−164,958
EA	+757.00	+201.63	+2341.07 + 201.63 − 680.18	+1,772,190
AB	−122.02	−680.18	+1862.52 − 680.18 − 591.17	−227,265
BC	+174.00	−591.17	+ 591.17	+102,864

Double area, sq ft = +1,409,986
Area, sq ft = 704,993
÷ 43,560 sq ft = 16.18 acres

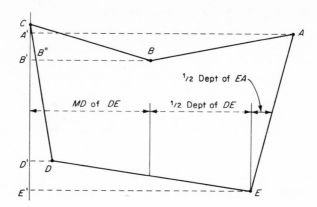

a. The DMD of the first course is equal to the departure of that course.

b. The DMD of any other course is equal to the DMD of the previous course plus the departure of the previous course plus the departure of the course itself.

c. The DMD of the last course should equal the departure of that course with its sign changed.

4. Compute the **double areas.** A double area of a course is the product of the latitude and the DMD of that course.

5. The sum of the double areas divided by 2 is the area.

6. If the area is in square feet compute the acreage.

$$\text{Acres} = \frac{\text{sq ft}}{43{,}560}$$

The Chain.   Many old surveys are measured in **chains:**

$$1 \text{ chain} = 4 \text{ rods} = 66 \text{ ft}$$
$$1 \text{ chain} = 100 \text{ links}$$
$$10 \text{ square chains} = 1 \text{ acre}$$

If the area is in square chains, divide by 10 to find the acreage.

Explanation.   The figure for Table 5-10 shows a plot of the traverse. Through *C* (the most westerly point), draw a north-south reference line. Drop a perpendicular to this line from each traverse point, *AA'*, *BB'*, and so forth, forming a triangle or a trapezoid for each course. The sum of the areas from courses having plus latitudes minus the sum of those having minus latitudes is the area of the traverse.

This is obvious for the figure *B'E'EABB'*. It is also true for the figure

CB'B. (From *C'B'B* is subtracted *C'B'B''*.) Together these figures make up the traverse area.

Any reference line can be drawn, preferably north-south or east-west, and a traverse of any shape can be similarly analyzed.

A meridian distance (MD) of a course is defined as the distance from the reference line to the midpoint of the course. It is equal to half the sum of the departures at the ends of the course. Also,

MD of first course = ½ departure of first course
MD of following courses = MD of previous course plus ½ departure of previous course plus ½ departure of course itself, with due attention to signs

The area formed by any course as described above is the product of the MD of that course and the latitude of that course with due attention to signs.

To avoid dividing by 2 so frequently, the equations above are multiplied by 2, thus:

DMD of first course = departure of first course
DMD of following courses = DMD of previous course plus departure of previous course plus departure of course itself

The sum of the double areas thus computed is twice the area of the traverse.

The Meter.　The metric system is gradually coming into use for certain surveys.

$$1 \text{ km} = 1000 \text{ m} = 0.62137119 + \text{ miles}$$
$$1 \text{ m} = 100 \text{ centimeters} = 1000 \text{ mm} = 3.2808399 - \text{ ft}$$
$$1 \text{ mile} = 5280 \text{ ft} = 1.609344 \text{ km exactly}$$
$$1 \text{ ft} = 0.3048 \text{ m exactly}$$

Electronic Computers.　Most electronic computers can be programmed to compute traverses. When one of these is available, a third method in addition to the compass and transit methods of traverse adjustment can be used, which is known as the Crandall[1] method. This leaves the bearings unchanged and utilizes the method of least squares to adjust the lengths of the courses by the minimum amounts necessary to provide a closure. It is questionable whether or not a method should be used which does not include corrections to the bearings.

[1] Prof. C. L. Crandall, *Trans. A.S.C.E.*, vol. XVI, p. 453.

ELEMENTARY TRIANGULATION

5-5. The Use of Elementary Triangulation. While advanced forms of triangulation are the most important means of establishing control over large areas, they are seldom utilized in the major portion of surveying operations. A simple system of a few well-placed triangles will often, however, greatly increase the over-all accuracy of the traverse net with a minimum expenditure of time and labor. See Fig. 5-9 in which triangulation is used to control three traverse nets and to form the connections between them.

There are many types of triangulation schemes and methods of computing the coordinates they establish. Figure 5-10 shows a system of single triangles, and Fig. 5-11 shows other types. The system of single

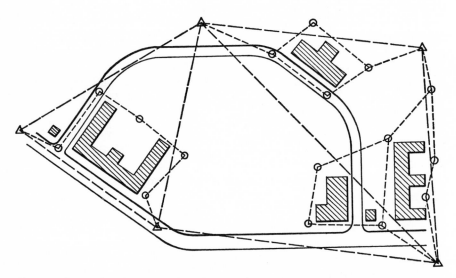

Fig. 5-9. Survey for plant extension, showing the scheme of triangulation and traverse stations.

triangles only will be described, for the other types require methods of adjustment beyond the scope of this text.

5-6. A System of Single Triangles. The triangulation stations should be placed around the exterior of the area to be surveyed. They are arranged so that the triangles formed are as nearly equilateral as possible to give the maximum **strength** and hence the most accurate results. A minimum of two sides of the system is measured to serve as a base and a check base, as indicated in Fig. 5-10 by AB and GH. All the angles are measured and station adjustments completed.

The resulting angles are then adjusted so that the sum of the angles in each triangle equals 180°. Equal increments are applied to the three angles of each triangle to obtain this adjustment.

The lengths of the sides are computed by the sine formula, beginning with the length of a measured base AB. This will result in a computed as well as a measured length for the check base GH. If these two values

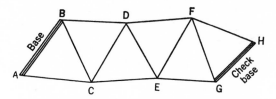

Fig. 5-10. System of single triangles.

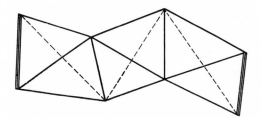

System of overlapping triangles in quadrangles.
If the dotted lines are omitted this becomes a
system of single triangles

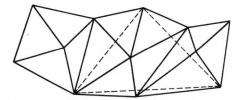

System of central point figures
with an overlapping triangle

Fig. 5-11. Other systems of triangles.

check within the accuracy desired, the results can be allowed to stand. If desired, all the computed lengths of the sides and of the measured base can be increased or decreased by the same ratio (or by adding or sub-

tracting a small logarithmic increment) so that the final value of the
check base will be about the average of the original computed value
and the measured length.

The bearing of one of the sides is assumed or determined, and the other
bearings are computed by using the adjusted angles. The station coordinates are computed by the methods used for traverses. As the figure is
geometrically consistent after adjustment, any route through the triangles
will give the same results.

When the coordinates of the triangulation stations have been computed, they are thereafter held fixed. All traverses tied to them are
adjusted, to close on them, as described under Connecting Traverse
Computation in Sec. 5-4.

Example.   The field data and the computation required for the triangulation
system shown in Fig. 5-9 are given in Fig. 5-12 and Tables 5-11 to 5-14. Figure 5-12
shows the sketch and the field data after station adjustment. Table 5-11 gives the
form for machine computation; Table 5-12 gives the form for logarithmic computation.

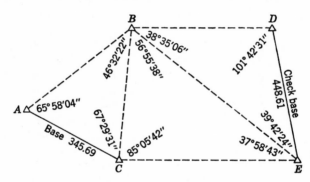

Fig. 5-12. Triangulation system shown in Fig. 5-9. Sketched for computation. The
values given are the field measurements. The angles are the values after station
adjustment.

The angles of each triangle are entered in a special order. The angle opposite the
known side is entered first. Second is the angle opposite the side that is not part of
the following triangle, and third is the angle opposite the side that is used as the
known side of the following triangle.

In the example, the computed value of the check base was found to be 0.04 ft
smaller than its measured length. It was then assumed that the original base and
all the sides computed from it were too small by half the ratio of 0.04 divided by the
measured length of the check base. Accordingly each side was corrected by adding
the product resulting from its length multiplied by the ratio 0.02/448.568.

The coordinates are computed by using a traverse that extends around the perimeter, as shown in Tables 5-13 and 5-14. The results would be the same by any route,
but this route utilizes sides not previously used and thus provides a check. The
traverse should check exactly. In the example, slight errors caused by rounding off
appear. These are eliminated by changing one latitude and two departures, as shown.

Table 5-11. Triangulation Form for Machine Computation

Angles				Sines	Formulas	Sides		Cor.	Final sides
						AC	345.69	+0.015	345.70
B-CA	46°32′22″	+1″	23″	0.72585	*AC*/sin *B-CA*		476.255		
C-AB	67°29′31″	+1″	32″	0.92383	*x* sin *C-AB*	*AB*	439.979	+0.020	440.00
A-BC	65°58′04″	+1″	05″	0.91332	*x* sin *A-BC*	*BC*	434.973	+0.019	434.99
Sums	179°59′57″	+3″	60″						
E-CB	37°58′43″	−1″	42″	0.61536	*BC*/sin *E-CB*		706.859		
B-EC	56°55′38″	−1″	37″	0.83798	*x* sin *B-EC*	*CE*	592.334	+0.026	592.36
C-BE	85°05′42″	−1″	41″	0.99634	*x* sin *C-BE*	*BE*	704.272	+0.031	704.30
Sums	180°00′03″	−3″	00″						
D-EB	101°42′31″	−1″	30″	0.97919	*BE*/sin *D-EB*		719.239		
E-BD	39°42′24″	0	24″	0.63886	*x* sin *E-BD*	*BD*	459.493	+0.020	459.51
B-DE	38°35′06″	0	06″	0.62367	*x* sin *B-DE*	*DE*	448.568	+0.020	448.59
Sums	180°00′01″	−1	00						

$$DE \quad \text{Computed} \quad 448.57 \qquad \text{Correction} + \frac{0.02}{448.568} = +0.0000446 \text{ per ft}$$

$$DE \quad \text{Measured} \quad 448.61$$
$$\text{Error} \quad \underline{-0.04}$$

5-7. Lengths and Bearings from Coordinates.   It is often necessary to compute the length and bearing of a straight line that connects two points whose coordinates are known. Assume that the two points are A and B with the coordinates x_A and y_A, and x_B and y_B, respectively. The formulas are the following:

$$\Delta x = x_B - x_A \qquad \Delta y = y_B - y_A$$
$$\tan \text{bearing} = \frac{\Delta x}{\Delta y} \qquad \cot \text{bearing} = \frac{\Delta y}{\Delta x}$$

(Choose the formula which makes the numerator smaller than the denominator.)

$$\text{Length} = \frac{\Delta x}{\sin \text{bearing}} \qquad \text{Length} = \frac{\Delta y}{\cos \text{bearing}}$$

(Choose the formula that has the larger numerator.)

$$\text{Check: } (\text{length})^2 = (\Delta x)^2 + (\Delta y)^2$$

Example.   To find the bearing and the length of the line from C to D in Fig. 5-12: from Table 5-14,

$$
\begin{array}{lcc}
 & x & y \\
D & 1833.38 & 1170.57 \\
C & -1283.18 & -\ 801.71 \\
\hline
 & \Delta x\ 550.20 & \Delta y\ 368.86
\end{array}
$$

$$\text{cot bearing} = \frac{368.86}{550.20} = 0.67041$$

$$\text{Bearing} = 56°09'42''$$

$$\text{Length} = \frac{550.20}{0.83061} = 662.40$$

$$\text{Check: } 438{,}774 = 302{,}720 + 136{,}058$$

Error only in sixth significant figure.

Table 5-12. Triangulation Form for Logarithmic Computation

Angles and sides				Logs; log sines	Cor. logs	Final logs	Final sides
AC	345.69			2.538686	+22	2.538708	345.71
B-CA	46°32′22″	+1″	23″	−9.860848			
C-AB	67 29 31	+1	32	2.677838 +9.965591			
AB				2.643429	+22	2.643451	440.00
A-BC	65 58 04	+1	05	+9.960623			
BC				2.638461	+22	2.638483	434.99
	179°58′57″	+3″	60″				
E-CB	37°58′43″	−1″	42″	−9.789131			
B-EC	56 55 38	−1	37	2.849330 +9.923232			
CE				2.772562	+22	2.772584	592.36
C-BE	85 05 42	−1	41	+9.998406			
BE				2.847736	+22	2.847758	704.30
	180°00′03″	−3″	00″				
D-EB	101°42′31″	−1″	30″	−9.990869			
E-BD	39 42 24	0	24	2.856867 +9.805404			
BD				2.662271	+22	2.662293	459.51
B-DE	38 35 06	0	06	+9.794958			
DE				2.651825	+22	2.651847	448.59
	180°00′01″	−1″	00″				

		Logs
DE computed		2.651825
DE measured, 448.61		2.651869
Error in logs		−44
Correction log		+22

Table 5-13. Triangulation. Bearing Computation

Course	Bearing	Angles
AC	−S 55°00′00″E	67°29′32″
	152°35′13″	85°05′41″
	−97°35′13″	152°35′13″
	179°59′60″	
CE	S 82°24′47″E	37°58′42″
	−77 41 06	39°42′24″
ED	−N 4°43′41″W	77°41′06″
	101°42′30″	
	−96°58′49″	
	179°59′60″	
DB	−N 83°01′11″W	38°35′06″
	142°03′06″	56°55′37″
BA	S 59°01′55″W	46°32′23″
	65 58 05	142°03′06″
	125°00′00″	
	180 00 00″	
AC	−S 55°00′00″E	*Check*

Table 5-14. Triangulation. Computation of Coordinates

Sta.	Bearings / Lengths	Cosine / Sine	Coordinates Lat.	Coordinates Dep.
A	S° 5500′00″E	0.57358	1000.00	1000.00
C	345.70	0.81915	−198.29	+283.18
C	S 82°24′47″E	0.13203	801.71	1283.18
			0	8
E	592.36	0.99124	− 78.2⊀	+587.1⊀
E	N 4°43′41″W	0.99659	723.51	1870.36
D	448.59	0.08243	+447.06	− 36.98
D	N 83°01′11″W	0.12153	1170.57	1833.38
				0
B	459.51	0.99259	+ 55.84	−456.1⊀
B	S 59°01′55″W	0.51456	1226.41	1377.28
A	440.00	0.85746	−226.41	−377.28
A			1000.00	1000.00

Final Coordinates

Sta.	North	East
A	1000.00	1000.00
B	1226.41	1377.28
C	801.71	1283.18
D	1170.57	1833.38
E	723.51	1870.36

Problems

5-1 to 5-4. In each of these problems are given the field measurements of the interior angles of a loop traverse of 12 sides. First adjust these angles. The bearing of one side is given. With this bearing and the adjusted interior angles, draw a sketch of the traverse. **The alphabetical order of the stations gives the forward, counterclockwise direction around the loop.** Looking forward around the loop, the interior angles are on the left.

The lengths of the courses have no effect on the results, so they can be made any convenient lengths.

Interior Angles

Sta.	5-1	5-2	5-3	5-4
A	210°30′	303°30′	213°05′	54°08′
B	61 31	89 33	49 55	216 54
C	299 27	56 27	270 48	56 55
D	45 06	144 17	130 17	127 28
E	194 55	279 07	60 42	263 17
F	88 11	152 13	297 53	55 02
G	153 00	58 03	112 18	150 07
H	329 35	226 07	157 37	117 35
I	41 40	44 16	61 14	308 06
J	107 15	304 22	303 52	60 07
K	208 55	84 38	90 12	88 57
L	60 07	57 51	52 31	301 12
Bearings	DE: S21°30′E	FG: N77°49′E	KL: N61°09′W	BC: S22°18′W

5-5 to 5-10. The field data and the fixed data are given below for each of six traverses. The forward direction is given by the alphabetical order of the station names. Each angle is measured clockwise from the back direction to the forward direction so that they are on the left of the traverse looking forward.

In each problem, draw a sketch, compute the accuracy, and compute the final coordinates. Adjust by the compass rule.

5-5. Loop traverse.

Sta.	Traverse angle	Length, ft
A	91°18′	AB 554.09
B	94 28	BC 425.31
C	109 52	CD 426.05
D	102 26	DE 345.28
E	142 06	EA 322.21

Bearing BC: S3°11′E
Coord. B: N 1000.00, E 1000.00

5-6. Same as Prob. 5-5 but bearing BC: N9°17′W.

5-7. Loop traverse.

Sta.	Traverse angle	Length, ft
A	96°05′	AB 560.27
B	95 20	BC 484.18
C	65 15	CD 375.42
D	216 22	DE 311.44
E	67 08	EA 449.83

Bearing EA: S10°14′E
Coord. E: N 1000.00, E 1000.00

5-8. Same as Prob. 5-7 but bearing EA: N18°53′E.

5-9. Connecting traverse (work to nearest minute).

Sta.	Angle	N coord.	E coord.	Course	Length, ft
Ash		1336.35	1050.47		
Fir	86°33'	1000.00	1000.00	Fir-*G*	347.15
G	223 55			*G-H*	449.82
H	114 48			*H*-Oak	144.76
Oak	141 36	670.23	1780.27		
Pine		945.97	1975.74		

5-10. Same as Prob. 5-9, except coordinate values:

Sta.	N coord.	E coord.
Ash	1266.05	1211.88
Fir	1000.00	1000.00
Oak	324.28	1510.85
Pine	465.34	1818.00

5-11, 5-12. Compute the final lengths of the sides of the triangulation nets shown. For Prob. 5-11 use Fig. 5-13; for Prob. 5-12 use Fig.5-14.

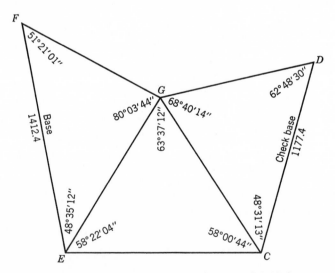

Fig. 5-13. Data for Prob. 5-11. Unadjusted field data.

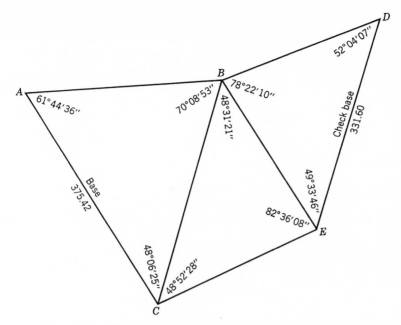

Fig. 5-14. Data for Prob. 5-12. Unadjusted field data.

Suggested Field Exercise

Small traverses or small triangulation nets can be laid out as separate problems. It is far more satisfactory, however, to select an area for a complete mapping project. A triangulation net, made up of single triangles, should be established to which traverses can be connected much as shown in Fig. 5-10. The triangulation coordinates should be held fixed, and the traverse connections adjusted to fit them.

Certain important topographic features like property lines or buildings where precise measurement may be necessary can be connected to the control by transit and tape.

Later elevations by hand levels and precise levels and locations by plus and offset or connections by stadia can be measured.

As there is some advantage in measuring a loop traverse by transit and tape or by stadia, which is independent of other control, these might be carried out as preliminary exercises.

6

Leveling

PRINCIPLES OF LEVELING

6-1. Importance of Leveling. The importance of leveling cannot be overestimated. Gravity plays such an important part in every human endeavor that, with a very few minor exceptions, it must always be considered in every form of design and construction.

6-2. The Level Instrument. A surveying level consists of a high-powered telescopic sight with a sensitive spirit level attached to it and adjusted so that, when the bubble is centered, the line of sight is level. The whole is mounted on a vertical spindle which fits in a bearing in the leveling head. Both three- and four-screw leveling heads are used.

Several self-leveling or so-called automatic instruments are available. When one of these is set up so that the bubble of a circular level is centered, a pendulum-prism system automatically levels the line of sight.

The six types of level used in surveying are listed below in the order of the accuracies usually expected of them, beginning with the least accurate.

1. The hand level (Sec. 7-3)
2. The transit telescope level (Secs. 4-7 and 6-11)
3. The wye level (Sec. 6-11)
4. The dumpy level (see below)
5. The tilting level (Sec. 6-11)
6. The self-leveling level (Sec. 6-11)

Slight changes in design of the last two will change their relative expected accuracies.

Except for minor changes in the operation of these instruments, leveling is carried out in much the same way with all of them. The method described here uses the dumpy level for an example, as this type of instrument is employed more than any other. The methods of operating the other types of instruments are described in later paragraphs.

The Dumpy Level. Figure 6-1 shows a typical dumpy level, and Fig. 6-2 shows the principle of its operation. There are four leveling screws, and the direction in azimuth can be controlled by the clamp and tangent screw, which are seldom used except in a wind.

6-3. The Rod. There are many types of level rods. The most useful for the type of work covered in this text is described here in detail. The level rod is a wooden rod graduated upward from zero at the bottom and provided with a movable metal target that can be clamped where desired. The rod should be graduated in hundredths of a foot. It is

Fig. 6-1. A dumpy level.

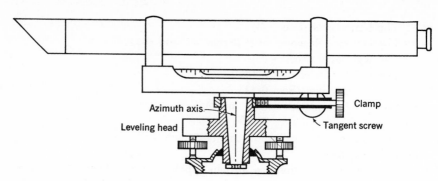

Fig. 6-2. The principle of the dumpy level. The level vial is adjusted so that the bubble centers when the azimuth axis is in the direction of gravity, and the line of sight is adjusted so that it is level when the bubble is centered.

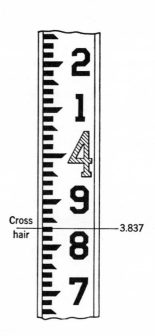

Fig. 6-3. A level rod and method of reading a rod with the customary graduation pattern. (Keuffel & Esser Co.)

made in two parts. The rear section can be slid upward, and when **fully extended** the front face of the rod reads continuously from 0 to 12 or 13 ft (Fig. 6-3). The graduations should be heavy and clean, so that they may be read through the telescope by the instrumentman. In other words, the rod should be **self-reading**. On most self-reading rods the graduations are 0.01 ft wide and spaced 0.01 ft apart. The dividing line between the black graduations and the white face of the rod is the exact hundredth. A standard design for these graduations is used (see Figs. 6-3, 6-4, and 6-5).

The top of the front face of the rod (from 6.75 ft upward to about 7.20 ft) is attached to the back section (see Fig. 6-5). The back face of the back section of the rod is graduated downward from about 7 to 12 or 13 ft. As the back section is slid upward, it runs under an index mark and vernier. The reading at the index indicates the height of a certain mark, usually the 7-ft mark on the front face. Thus, if the target is set at the proper mark and the back section of the rod is partly raised, the height of the target above the ground is indicated by the index on the rear face. A clamp is provided to hold the back section in place.

A stop is provided to prevent the rod from coming apart when it is extended too far. The stop is often placed so that it stops the rod when the readings are continuous from bottom to top. Sometimes it is not so placed, and sometimes it is knocked out of position by long use. The rodman should make sure that the index at the back of the rod reads exactly 12 or 13 ft (whichever applies) when he or she sets the rod in its extended position. The stop should be used only when he or she is certain that it stops the rod in the proper position.

6-4. Benchmarks. Benchmarks are marked points of known elevation above any datum. They should be objects that are easily recognized, easily found, not likely to move, marked by an identifying mumber, round on top, and set low with respect to the surrounding ground. The best are cross marks or bronze tablets set in old masonry with good foundations, as these are more likely to retain their elevation permanently. But almost any object can serve as a benchmark. Monuments set flush with the ground, certain parts of fire hydrants, nails in trees, and even stakes are used.

Benchmark Leveling. Benchmark leveling, sometimes called **differential leveling,** is the process of determining the elevation of a series of benchmarks. It is always a control survey and therefore usually performed with considerable precision.

System of Benchmarks. A system of benchmarks is always in demand from the moment any work is contemplated and throughout the entire life of the project. Benchmarks should be established, if possible, before leveling is required for the original map. Thereafter they should be maintained for mapping, construction, future changes, and for maintenance. When they are available, all leveling work can be kept on the same datum by beginning the work at the nearest benchmark. Thus elevations can be established according to the exactly same datum used in the plans. Old and new plans will agree, leveling work can be checked whenever a benchmark is passed, and the very number of them ensures the permanence of the datum they establish. At least three benchmarks should always be established for any project so that if one is disturbed the pair that check will be known to be correct.

LEVELING PROCEDURES

6-5. To Set Up the Level.   As with the transit, spread the legs and place them so that the footplate is level. Walk around the instrument, pushing each leg firmly into the ground. Loosen two adjacent screws. Turn the telescope over a pair of opposite leveling screws. Level until the bubble crosses the center of the tube. Turn the telescope over the other pair of leveling screws. Bring the bubble to within two divisions of the center. Repeat over the first pair. The instrument is now only approximately leveled; but, as will appear later, this is sufficient for the setup. See also Sec. 4-13.

6-6. To Operate the Level.   As with the transit, the eyepiece must be focused on the cross hairs (see Sec. 4-6). After the instrument is set up, do not touch it or allow anything to touch it except when and where necessary for operating it. Never straddle the legs, but always stand between them; and be particularly careful not to kick or touch the tripod while walking around the instrument.

6-7. To Handle the Rod.   Clamp the target at the 6.5- or the 7-ft mark, whichever applies. **At all times** keep the rod standing on the benchmark or turning point, except when actually moving or computing. Keep it **balanced** and with the front face turned toward the instrument. When the instrument is set up or about to be set up in a position that requires the extended rod (**high rod**), raise the rod **all the way,** i.e., until the index on the back of the rod reads exactly 12 or 13 ft (whichever applies), and clamp it in position. If it is raised part way, the graduations are not continuous and a blunder will result. Keep the eyes on the levelman at all times, unless performing some necessary function that interferes. Always lower the rod to carry it.

6-8. To Take a Rod Reading.   As with the transit, it is necessary to sight over the top of the telescope to direct it toward the rod.

1. **Focus on the rod** (and bring the vertical cross hair near the rod).
2. **Level precisely** with the pair of opposite leveling screws that most nearly points toward the rod.

3. **Read the rod** (without moving the feet).
4. **Check the bubble.**
5. **Record the reading.**
6. **Give the reading to the rodman** by voice or signal, naming all the digits read. For example, in reading to thousandths of a foot, 5.1 is read, "five point one oh oh."
7. The rodman, **while still balancing the rod,** will point to the exact reading with a pencil. If she or he cannot learn to do this (it is an art) or if the reading is out of reach, the target will be set at the exact reading. The target is provided with a vernier, which may be used to aid in setting the target (see Fig. 6-4). If the reading is above the rod, the rodman sets the target at the 6.5 or 7.0 mark (whichever

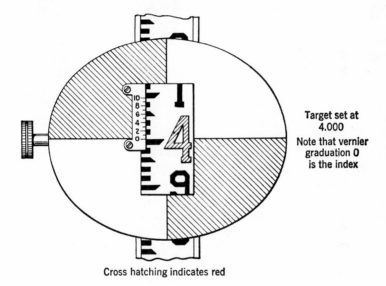

Target set at
4.000
Note that **vernier**
graduation **0**
is the **index**

Cross hatching indicates **red**

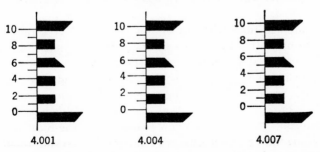

Fig. 6-4. The target and the method of reading the target vernier.

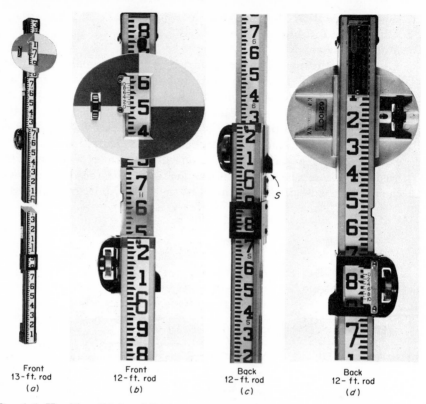

Front
13-ft. rod
(*a*)

Front
12-ft. rod
(*b*)

Back
12-ft. rod
(*c*)

Back
12-ft. rod
(*d*)

Fig. 6-5. Handling "high rod." (*a*) Set target at 7 ft on 13-ft rod, or (*b*) set target at 6.5 ft on 12-ft rod. (*c*) Slide up the rear section of the rod until stopped by the adjustable stop *S*. The reading will now be continuous from 0 to 13 (or 12) ft. Clamp. (*d*) Set the back of the rod at the reading given by the levelman, in this case, 6.776 ft. Note that the numbers read downward. (*Keuffel & Esser Co.*)

applies) and extends the rod to the proper mark on the back of the rod. A vernier, similar to the target vernier, is available for this setting if desired.

8. The levelman will note whether or not the pencil or target comes on the cross hair. If satisfied, he or she calls or signals "all right" and both levelman and rodman record the reading.

　　If not satisfied, he or she reads the rod again and gives another reading or the same reading, as the case may be. The rodman resets the pencil or target. If high rod was used, the levelman will call or signal "high rod" instead of the new reading, and the rodman resets the rod by extending it **all the way** before the second reading is taken.

The levelman will often notice that there is a slight discrepancy between the first reading and the pencil or target position in reading to thousandths. The first reading is the correct one if the difference is 0.003 ft or less. If more, the reading is repeated, for a blunder has been made.

When there is a wind, balancing the rod does not ensure verticality. The levelman will have the rodman plumb the rod according to the vertical cross hair and wave the rod backward and forward slightly. The lowest point touched by the cross hair is read.

Communication should be by voice, if possible. Often, however, leveling is done near construction where the noise makes signals necessary.

Signals should be chosen that are easy to see and to remember. Often the best signal is an imitation in pantomime of the action desired. The signals given below are suggested.

1. **All right.** Hands outstretched sideways, palms forward and moved up and down together.
2. **Plumb the rod.** Hand over head, elbow straight, palm forward and inclined in the proper direction.
3. **Wave the rod.** Both hands over head, palms forward, swung back and forth together.
4. **High rod.** Both hands extended outward to the sides, palms up, and the arms moved up to vertical together.
5. **Raise for red.** When the footmark is invisible, the levelman reads and memorizes the tenths, hundredths, and thousandths and then calls "raise for red" or extends one hand forward, palm up, and raises it a little. The rodman lifts the rod slowly and exactly vertically. The footmark is read when it appears.
6. **Take, or this is, a turning point.** One hand moved in a horizontal circle over the head.
7. **Kill the target.** Hand in front of the body, palm down, and moved up and down quickly. Sometimes the target covers the part of the rod that must be read. This signal is then given.
8. **Kill the brass.** Same signal as "high rod." Sometimes the brass strip that is attached to the rear half of the rod at the bottom and fits around the front of the rod conceals the reading. By partly extending the rod the brass is moved upward out of the way. The rodman can always judge by the relative positions of the instrument and the rod whether "high rod" or "kill the brass" is meant.
9. **Turn the rod around.** A small horizontal circle made with the forefinger. It is given when the back or side of the rod is turned toward the instrument.

10. **Turn the rod right end up.** An imitation of the motion of turning the rod right end up with two hands.

See Fig. 6-6 for signals for rod readings.

6-9. Benchmark-leveling Procedure. Figures 6-7 through 6-9 illustrate a benchmark-leveling procedure. They should be carefully followed while this description is being read.

Both the levelman and the rodman should have a field book. The one kept by the levelman is the level book; the other is the **peg** book.

Fig. 6-6. Signals for numbers. Either side of the body can be used for any number.

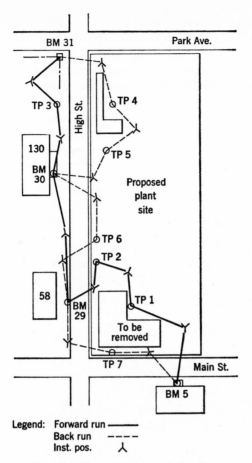

Fig. 6-7. Plan of benchmark leveling.

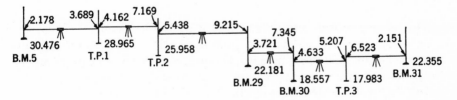

Fig. 6-8. The principle of benchmark leveling.

The work begins at a benchmark of known elevation, in this case B.M. 5. Both levelman and rodman record B.M. 5 in the station column; the known elevation, 30.476, in the elevation column; and a description of the benchmark on the right-hand page; all three on the same line.

The rodman holds the rod on the B.M. The levelman sets up where the rod can be clearly observed and not more than 150 or 200 ft away. He or she should never set up on frozen ground or on an asphalt pavement, as the instrument will slowly sink into such a surface. This changes the height of instrument (H.I.). This movement is usually quite obvious, as the instrument will keep going out of level as it sinks. It is next to impossible to become aware of it with a self-leveling level.

B.M. LEVELING – HIGH ST., MAIN TO PARK						π Smith Date
Sta	+	HI	–	Rod	Elev.	Rod Jones Fair, No Wind 76°F.
BM 5	2.178	32.654			30.476	Level Berger 12978
						Precise B.M. Disk Set in Top Step of Entrance #125 Main St.
TP 1	4.162	33.127	3.689		28.965	
TP 2	5.438	31.396	7.169		25.958	
BM 29	3.721	25.902	9.215		22.181	"R" in Corey F.H. Opp. #58 High St.
BM 30	4.633	23.190	7.345		18.557	X in Stone Top Step #130 High St.
TP 3	6.523	24.506	5.207		17.983	
BM 31	4.528	26.883	2.151		22.355	▢ in Conc. Base Iron Fence S.W Cor. High St. and Park Ave.
TP 4	5.812	26.517	6.178		20.705	
TP 5	6.218	29.011	3.724		22.793	
BM 30	7.083	25.646	10.448		18.563	
TP 6	5.578	27.053	4.171		21.475	
BM 29	9.511	31.708	4.856		22.197	Arith. Ck 30.476
TP 7	8.235	33.622	6.321		25.387	+ 73.620
						104.096
BM 5			3.139		30.483	– 73.613
						30.483
	73.620		73.613			Error +.007

Fig. 6-9. Form of notes used with benchmark leveling.

The reading of the rod, 2.178, is taken by the levelman, checked, and recorded by both surveyors in the same line as the B.M. 5 and in the plus column. The rodman then paces the distance to the level. An equal distance in the desired direction is paced and a temporary point called a **turning point** (T.P.) is selected to carry the line of levels forward. The turning point must have the following characteristics:

1. The rod, when held on it, will be visible from the level.
2. It must be firm and round on top. If no satisfactory object can be found, a stake is driven to serve as a T.P.

The point chosen is covered with keel (lumber crayon) and marked with the number of the T.P. immediately after it is selected.

As with a transit, the field procedure must be designed to eliminate effectively any possible residual errors in the instrument. If the line of sight is not **exactly** level when the bubble is centered, but slopes either up or down, it will slope by the same amount whenever the bubble is centered. When the horizontal lengths of the two sights, taken from an instrument position, are the same, the line of sight will strike the rod

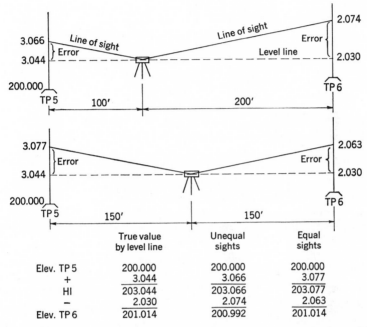

	True value by level line	Unequal sights	Equal sights
Elev. TP 5	200.000	200.000	200.000
+	3.044	3.066	3.077
HI	203.044	203.066	203.077
−	2.030	2.074	2.063
Elev. TP 6	201.014	200.992	201.014

Fig. 6-10. When the plus sight and the minus sight are the same horizontal length, the error of adjustment of the level is canceled, as the error on both rod positions is the same.

held on either one by exactly the same error in height. Since one of the sights is a plus sight and the other is a minus sight, the two errors will cancel (see Fig. 6-10).

The most difficult operation for the inexperienced levelman is to choose the proper location for the instrument when working **downhill and uphill.** In working downhill, there is a tendency to set up the level too far downhill so that it is below the foot of the rod. In working uphill, the level is often set up too far uphill where, while the plus sight may be observed, the length of the sight is so great that the length of the following minus sight cannot be made equal to it (see Fig. 6-11).

An experienced rodman can estimate the distance without pacing.

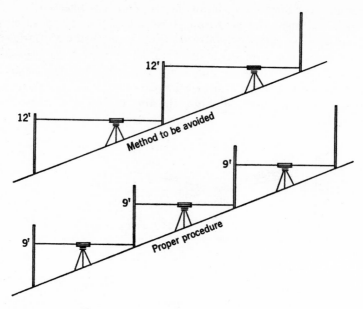

Fig. 6-11. Leveling on a uniform slope.

While the rodman is so engaged, the levelman computes the H.I. by adding the plus rod reading, 2.178, to the elevation of the B.M., 30.476, and records the result, 32.654, in the H.I. column **beside the plus reading that gave it.**

The rodman holds the rod on T.P. 1, and the reading 3.689 is read and checked. It is recorded by both surveyors in the minus column on the next line down, and the note T.P. 1 is recorded on that line in the station column. The levelman picks up the level and moves forward. Meanwhile the rodman computes the H.I. and subtracts 3.689 from the H.I. to find the elevation of T.P. 1, 28.965, which is recorded in the line with T.P. 1 but in the elevation column. The rodman must hold the rod on T.P. 1 as soon as possible after computation so that the levelman can choose a location for the instrument from which the rod can be observed.

The levelman sets up, and the plus reading 4.162 is taken, checked, and recorded by both surveyors in the plus column and in the line of T.P. 1. While the rodman is pacing the new distance to the instrument, the levelman computes and records the elevation of T.P. 1 and the new H.I., 33.127.

When the rodman reaches the levelman after pacing the distance, both check their values for the elevation of T.P. 1.

The process continues thus.

Fig. 6-12. A tilting level.

Note the following:

1. All **rod readings** are recorded by both surveyors as soon as they are checked. All rod readings are recorded in the line that refers to the **point sighted.**
2. All arithmetic is carried out **as far as is possible** with the readings available. The computations are made by each person while the other is moving.
3. All work must stop until both agree on the elevation of each T.P.
4. Note that the rod column is not used in benchmark leveling. It may be omitted.
5. The arithmetic can be checked by adding the plus and minus columns separately and applying the sums algebraically to the original benchmark. The elevation of the final benchmark should be obtained by this procedure. See Fig. 6-9.
6. Note finally that **every B.M. must be used as a T.P.**

6-10. Importance of Rules.   While some of the procedures described may seem arbitrary, if a single one is omitted or the operation is out of order, experience has shown that errors will occur in the work. It should be noted that the three fundamental rules for avoiding errors are carried out (see Sec. 1-5).

6-11. Operation of Other Types of Level.   *The Wye Level.*   In a wye level, the telescope is mounted in wyes so that it can be rotated around its longitudinal axis. This makes it possible for one person to adjust the instrument without assistance. It is more expensive than a dumpy level and is considered not to be so good an instrument. It is operated exactly like a dumpy level.

The Transit as a Level.   The transit is often used for minor leveling operations. The procedure is as follows:

Set up and level the instrument accurately with the plate levels. Aim and focus on the rod. Center the telescope bubble with the vertical motion, read the rod, and check the bubble.

A Tilting Level.   In a tilting level, shown in Figs 6-12 and 6-13, the telescope, with its attached spirit level, is mounted on a horizontal (tilting) axis so that it can be tilted up and down through a few degrees with a micrometer screw acting against a spring. The tilting axis should intersect the line of the horizontal axis prolonged. Attached to the leveling head is a circular level.

To operate this level, set up the instrument and center the circular bubble with the leveling screws. This places the vertical axis nearly in the direction of gravity so that the line of sight swings in nearly a horizontal plane. Focus on the rod and center the telescope level bubble

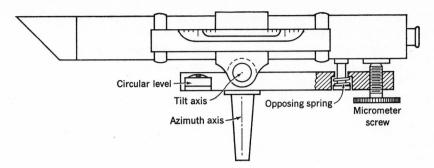

Fig. 6-13. The principle of the tilting level. The level vial is adjusted so that the bubble centers when the line of sight is level. The azimuth axis is placed nearly in the direction of gravity with the leveling screws according to the circular level. The micrometer screw, acting against the opposing spring, tilts the telescope on the tilt axis. When the micrometer screw is set at the reversing point, the line of sight is perpendicular to the azimuth axis.

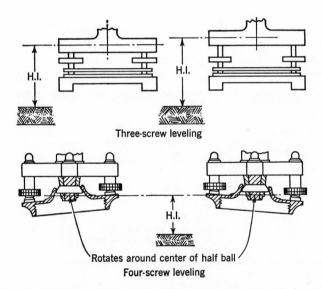

Fig. 6-14. With a three-screw leveling head, turning any screw will change the height of instrument. With a four-screw leveling head, turning the leveling screws has no effect on the height of instrument. (*Keuffel & Esser Co.*)

with the micrometer screw. Then proceed as before.

   Three-screw Levels. Some tilting levels and nearly all automatic levels are approximately leveled, according to the circular level, by three leveling screws instead of four. Any one of the screws can be used separately. The bubble moves toward any screw turned clockwise. Never turn a leveling screw of a three-screw leveling head once a sight has been taken, as this will change the H.I. (see Fig. 6-14).

The Two-Screw Instrument. Today many types of levels have only two leveling screws. This is definitely an improvement. If, during a particular setup, the instrument goes slightly out of level, the level of the instrument can be brought back by slightly releveling without materially changing the height of the instrument.

The two screws are placed as on the three-screw instruments. The third screw is replaced by a stationary support.

Like all other types of levels (except the automatic instruments) the spirit level is attached to the telescope. To level the instrument, turn the telescope over one leveling screw and the stationary support and center the bubble. Then turn the telescope over the other leveling screw and the stationary support, and center the bubble. Repeat until the bubble centers in both positions.

The Coincidence Bubble. Most tilting levels have coincidence bubbles. An optical system makes it possible to observe the telescope level bubble in a small window beside the main telescope, as shown in Fig. 6-15. When the micrometer screw is turned, the two ends of the image of the bubble move in opposite directions. The device is adjusted so that when the two ends are brought in coincidence, the line of sight is level. This arrangement makes it possible to center the bubble with great accuracy.

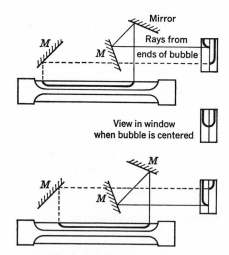

Fig. 6-15. One arrangement of mirrors that creates a coincidence bubble. Light is reflected up through the vial by a large reflecting surface, not shown.

The Automatic (Self-leveling) Level. See Fig. 6-16. With this instrument the circular level should be carefully centered. Thereafter the pendulum device takes over, so that no further leveling is required (Fig. 6-17).

Fig. 6-16. The Zeiss Ni 2 self-leveling level. (*Keuffel & Esser Co.*)

Sometimes the pendulum sticks. To make sure that it is free, after the circular level is approximately centered, turn one of the leveling screws quickly in one direction and back while looking through the telescope. If the line of sight heaves up and down once or twice, the pendulum is free.

6-12. Important Rules. The following rules should be kept continuously in mind during leveling:

1. Balance the horizontal lengths of the plus and minus sights from a single setup.
2. Have the bubble centered when the reading is taken.
3. Turn on all B.M.'s; i.e., use them as T.P.'s.
4. Keep the rod balanced on the point, facing the instrument, and watch the levelman **at all times.**
5. Mark all T.P.'s before they are used.

6-13. The Uses and Method of Leveling. The determination of elevations with a surveying instrument, better known as **running levels** or simply **leveling,** as distinguished from benchmark leveling, is usually the most important operation in the field. The procedure for running levels is always the same. It consists of benchmark leveling with one step added. At each height of instrument established by benchmark leveling, rod readings (rod shots) may be taken on as many points as desired. The elevation of each of these points is computed by subtracting the rod reading from the proper height of instrument. Running levels, therefore, usually consists of carrying a line of benchmark levels from

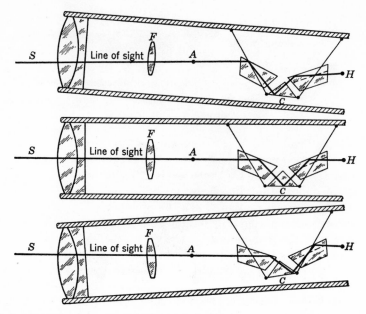

Fig. 6-17. Schematic diagram showing the operation of the Zeiss level. C is the compensator. Note that it swings backward or forward as the telescope is tilted and thus keeps a level line of sight S on the cross hairs H. The line of sight is refracted by the objective and by the focusing lens, too little to show on the diagram. The optics are designed so that the line of sight S prolonged intersects the optical axis at the *anallactic point A*. A moves a short distance along the optical axis as the focus is changed, and thus it very slightly changes the H.I. when the telescope is tilted. Also, when the telescope is tilted, unless A is exactly over the vertical axis, turning the telescope in azimuth will change the H.I. very slightly. These effects are negligible except in very precise, very short sights.

a benchmark to the vicinity of the work, taking several rod shots from each height of instrument there, and, finally, carrying the line to another benchmark, or back to the original benchmark, for a check. The plus and minus sights by which the desired heights of instrument are determined constitute a control survey and for this reason must be read more precisely and usually to more decimals than the rod shots. The rod shots should be taken only to the decimals of a foot necessary to the work in hand. They are taken by the same procedure as the plus and minus readings in B.M. leveling, although the check by the rodman is sometimes omitted.

6-14. Rod Shots.   Rod shots taken to points on rough ground are read to the nearest tenth of a foot. This gives results to ± 0.05 ft. When ± 0.1 ft is sufficiently accurate, and more than three or four rod shots are taken at one instrument position, the instrument is handled by a special procedure to increase the speed of the work. Briefly, once the plus sight is taken to obtain the H.I., the azimuth axis is set vertical according to the telescope level. Thereafter, the instrument is not leveled again for the rod shots, but the usual procedure is used for the minus reading on the next turning point. It is not so accurate to rely on the azimuth axis to keep the line of sight level as it is to center the bubble for each shot.

Special Procedures to Set the Axis Vertical.   The special procedures for the different instruments are given below.

1. **Dumpy level and wye level.** Center the bubble accurately with the telescope over each pair of opposite leveling screws.
2. **Transit.** Level with the plate levels. Aim the telescope in line with a pair of opposite leveling screws. Center the telescope level bubble with the vertical motion. Turn the alidade 180° in azimuth. If the bubble does not center, eliminate half the error with the leveling screws and the remainder with the vertical motion. Turn the alidade 90° and center the bubble with the leveling screws.
3. **Tilting level.** Find and mark the **reversing point** for the micrometer screw of the instrument. When the micrometer is set at this point, the instrument is handled like a dumpy level.

   To find the reversing point, level the instrument with the circular level. Turn the micrometer screw to approximately the center of its run. Record the reading of the micrometer drum or, if there is no graduated drum, mark the position of the micrometer screw with a pencil. Center the main bubble with the leveling screws. Turn the alidade 180° in azimuth. Center the bubble with the micrometer screw. Record this reading or mark the position with a pencil. The reversing point is the micrometer-screw position that is halfway between the two readings or marks. Compute and record this value or mark the reversing point.

   To check, set the micrometer at the reversing point, and level the instrument carefully with the leveling screws. The bubble should remain nearly centered when the alidade is turned in any direction.
4. **Self-leveling level.** No special method is required for this instrument.

APPLICATIONS OF LEVELING

6-15. Profile Leveling.   Profile leveling is the process of obtaining the elevations of a series of points along a continuous line. The line may be straight or curved or may turn at sharp angle points. The results are plotted in the form of a continuous vertical cross section called a **profile.** The vertical scale is almost always made greater than the horizontal scale, usually in the ratio of 10:1.

Profiles are required for the construction of roads, drives, sidewalks, curbs, gutters, fences, highways, tunnels, railroads, pipelines, sewers, drains, ditches, gas and water facilities, and the like.

Field Procedure for Profile Leveling. Figure 6-18 illustrates the plan and profile of an example of profile leveling. Figure 6-19 illustrates the corresponding field notes. The procedure is given below.

Marks are usually placed every 50 ft along the center line desired. Each 100-ft point is called a **station** and numbered from zero. Points between stations are numbered as a **plus,** i.e., the number of feet from the last station. The enumeration is written as shown in Fig. 6-18.

The level is set up near station 0 + 0. The plus reading 2.587 is taken on B.M. 5. This is added to the elevation 30.476 to obtain the H.I., 33.063. The rod is read on station 0 + 0, 4.2. This is called a **rod reading** or **rod shot** and is placed in the rod column. It is subtracted from the H.I. to obtain the elevation of station 0 + 0, 28.9. From this setup of the instrument, all rod shots are taken until the view is obstructed or a sight of over 150 ft is required. T.P. 1 is then established. A minus shot of 3.782 is taken on T.P. 1 and subtracted from the H.I. 33.063, giving the elevation 29.281 for T.P. 1. The instrument is moved, and the process is repeated between T.P. 1 and T.P. 2, etc. The work must end on a B.M. of known elevation so that a check may be obtained.

The elevation of each station is computed by subtracting the rod shot from the **proper** H.I. It is therefore essential that all the rods shots from one H.I. be recorded before the minus reading to the next T.P. Also, the

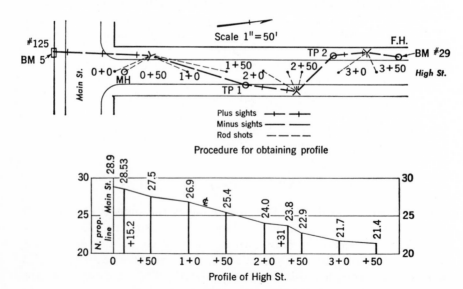

Fig. 6-18. Principles of profile leveling.

minus shot to the next T.P. should be taken after all the rod shots, so that, if the field check does not indicate a blunder, this is an immediate indication that the level was not disturbed at any H.I.

These two considerations dictate the order of procedure; i.e., all the rod shots shall be taken at any H.I. before the minus sight to the next T.P. is taken.

It is evident that profile leveling is identical with benchmark leveling except that at many H.I.'s a number of side or rod shots are taken. All the rules for B.M. leveling apply.

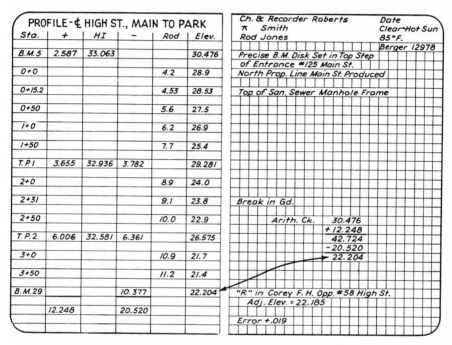

Fig. 6-19. Example of profile-leveling field notes.

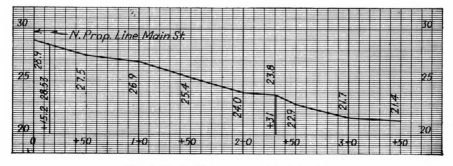

Fig. 6-20. Profile paper used for plotting.

Often no B.M. exists at the end of the work. It is then necessary to carry the levels back to the original B.M. by a series of turning points in order to obtain a field check. Often it is advisable to establish several B.M.'s on the way out. This can be accomplished by merely recording the description of turning points. These are useful for giving grades for construction. On the way back they should be used as turning points so that any blunders can be isolated.

Under no circumstances should leveling of any type be performed without starting on, or setting, at least one benchmark. If a benchmark of known elevation is not available, one should be set and given an arbitrary elevation. The benchmarks established on the original profile are later used as starting points for the leveling necessary to mark the proper elevations for construction.

Often the precision of profile leveling need not be so high as that of B.M. leveling. When the distance between benchmarks is short and the elevations are required only to the nearest tenth of a foot, the plus and minus shots should be taken to hundredths. In the illustration the plus and minus readings are taken to thousandths of a foot, as the elevation of a manhole is required to hundredths of a foot.

Rod shots taken on the ground, macadam roads, or surfaces that are not definite or smooth are usually taken to tenths. Sometimes they must be taken to hundredths, as on concrete roads or railroad rails. In that case the plus and minus shots are usually taken to thousandths.

The profile is plotted as shown in Figs. 6-18 and 6-20. The horizontal line at the bottom of the profile is given the highest elevation in round numbers that is still lower than the lowest point in the profile. For example, in the profile given, the bottom line could have been given the value of 10 ft or 15 ft and the profile plotted accordingly. The profile must be plotted exactly to scale, and the vertical scale should be two to twenty times (usually ten times) as large as the horizontal scale. Frequently specially prepared **profile paper** is used for plotting as shown in Fig. 6-20.

6-16. Leveling for a Plot Plan. When a building is planned, a considerable body of elevation data must be assembled. This includes the elevations of nearby sanitary sewers, subsurface and surface drains, water and gas mains, electric and telephone conduits, adjacent streets and rail connections, and the elevations of the entire ground surface of the property. This information placed on a plan of the property is called a **plot plan.** The plan must show the location of the items mentioned, the topography, and the boundaries of the property. The elevations are given by numbers lettered on the plan, and the position of the ground surface is usually shown by **contour lines,** i.e., lines connecting points of equal elevation.

Horizontal Measurements for a Plot Plan. When a land surveyor has marked the property lines, a rectangular grid is established by setting

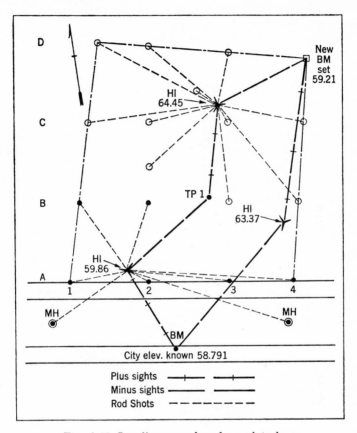

Fig. 6-21. Leveling procedure for a plot plan.

stakes or plaster laths, marked for identification, at intervals of 25, 50, or 100 ft at grid intersections and at the points where the grid lines intersect the boundaries (see Fig. 6-21). The objects desired are located by measurements from this system.

Leveling Procedure for a Plot Plan. The leveling is carried out exactly as for profile leveling (see Figs. 6-21 and 6-22), except that usually more rod shots can be observed from one instrument position. Rod readings are taken at each stake and wherever a **break** in the slope of the ground exists between stakes. The positions of these breaks are located by rectangular measurements from the stakes. Note Fig. 6-22. In this example two breaks are recorded. Breaks must not be omitted, for in drawing

PLOT PLAN PROP. A.G. SMITH					
Sta	+	HI	−	Rod	Elev.
BM	1.07	59.86			58.791
MH				11.62	48.24
MH				9.49	50.37
A1				5.4	54.5
A2				3.7	56.2
A3				1.9	58.0
A4				1.8	58.1
B1				3.4	56.5
B2				3.0	56.9
TP1	6.34	64.45	1.75		58.11
B3				5.1	59.3
B4				5.3	59.1
Brk				5.3	59.1
C1				5.4	59.0
C2				4.9	59.5
C3				4.0	60.4
C4				4.8	59.6
Brk				3.2	61.2
D1				4.7	59.7
D2				3.6	60.8
D3				4.6	59.8
BM-D4	4.16	63.37	5.24		59.21
BM				4.56	58.81
	11.57		11.55		

Ch & Recorder Roberts — Date Hazy 72°F.
π Smith
Rod Jones
"R" in Corey F.H. Opp. Property — Level — Berger 12978
West of Prop. Invert
East. " " "

Enumeration
D
C
B
A   1 2 3 4   Street

22' N from B2 — Arith Ck.
58.79
+11.57
70.36
−11.55

20' N from C, 30'E from 2 — 58.81

Top Conc. Prop. Mon. N.E. Cor. Prop.
Error +.02

Fig. 6-22. Form of field notes for a plot plan.

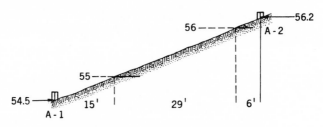

Fig. 6-23. Principles of interpolating contours.

the contours it is assumed that the slope is uniform between points where elevations have been determined.

Completing the Plot Plan. A scale drawing is made, showing the points where elevations are taken and the location of all objects required (see Fig. 6-24). The elevation of each point is lettered near it. Frequently the dot marking the point whose elevation is given is used as the decimal point in the elevation. It is convenient to write elevations along 45° lines.

Interpolation of Contour Lines. Since, from the method of making the survey, it is known that the ground slopes uniformly between adjacent elevation points, the position of the contour lines can be interpolated between them by making the distances proportional to the differences in elevation (see Fig. 6-23). These interpolations can be made by eye with sufficient accuracy for most plot plans. The contour lines are then drawn by connecting the interpolated positions by smooth curves

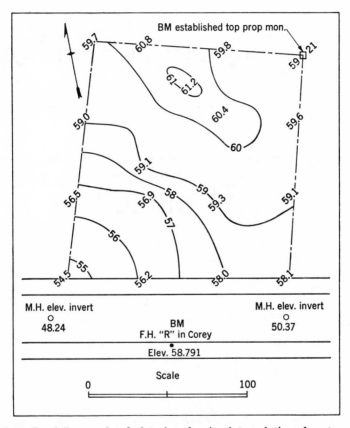

Fig. 6-24. Partially completed plot plan showing interpolation of contour lines.

(Fig. 6-24). In drawing the contour lines the following rules should be observed (see Fig. 6-25):

1. Contour lines never end, meet, or cross, except in the unusual case of a vertical or overhanging cliff.
2. Unless there are data to the contrary, contour lines must be uniformly spaced.
3. Contour lines must be drawn so that the ground higher than the

contour line is always on the same side of the contour line.

4. When contour lines indicate the sides of a depression in the ground with no drainage outlet, they are called **depression contours** and are marked as shown by the 96-ft contour that surrounds the number 3 in Fig. 8-10.

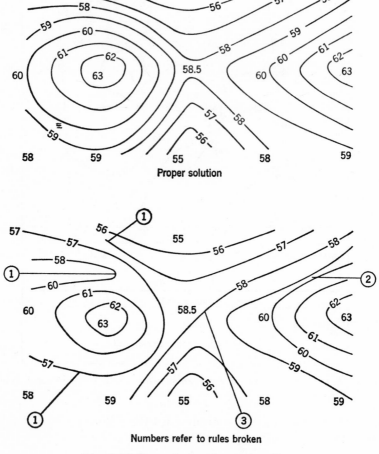

Fig. 6-25. Errors in drawing contour lines.

6-17. Other Applications of Leveling. The procedure for determining elevations should be clear from the foregoing applications. Determining scattered elevations or any other leveling problem can be handled in a similar manner.

Problems

6-1 to 6-6. These are plans of benchmark-leveling runs. Along each line representing a sight is given the rod reading that resulted from that sight. The numbering of the T.P.'s shows the direction of progress. Place the data in the form of field notes. Show the arithmetic check. Record the error.

6-1. Fig. 6-26.

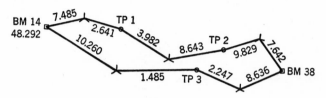

Fig. 6-26

6-2. Fig. 6-27.

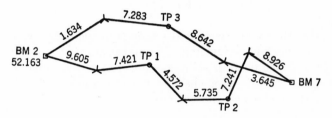

Fig. 6-27

6-3. Fig. 6-28.

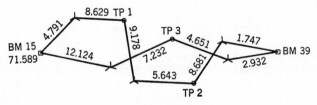

Fig. 6-28

6-4. Fig. 6-29.

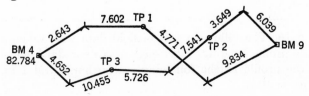

Fig. 6-29

6-5. Fig. 6-30.

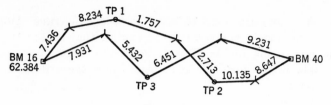

Fig. 6-30

6-6. Fig. 6-31.

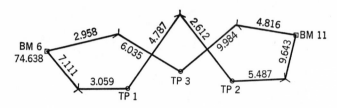

Fig. 6-31

6-7 to 6-12. Below are listed rod readings in the order in which they were taken in benchmark leveling. The elevation of the starting benchmark is given at the head of each column. The last reading is taken on the starting benchmark as a check. Give the complete form of notes. Show the arithmetic check and the errors.

6-7 74.36	6-8 67.428	6-9 59.27	6-10 39.372	6-11 86.43	6-12 91.646
6.48	8.562	11.36	7.546	8.47	1.732
5.72	4.077	5.32	3.642	9.61	9.276
1.06	9.714	1.87	6.851	7.38	4.231
2.38	2.394	10.24	5.420	6.52	8.227
8.67	4.758	2.65	8.221	5.36	5.524
9.22	11.645	6.23	6.013	8.27	4.619
0.27	2.625	4.68	10.687	9.22	9.063
8.13	6.755	5.27	8.746	10.31	6.247
6.42	8.481	8.41	1.632	6.36	8.362
1.75	9.262	7.59	11.103	7.41	0.559
5.23		10.36		9.22	
0.90		4.71		3.87	

6-13 to 6-16. The following sets of field data were taken in the order given during profile leveling. Place each set in standard field-book form, and draw the profile to the following scales: horizontal 1 in. = 100 ft, vertical 1 in. = 10 ft.

	Elev.	Pt. sighted	Rod	Pt. sighted	Rod	Pt. sighted	Rod
6-13.	B.M. 20	B.M. 20	3.516	T.P. 1	4.280	7 + 0	8.3
	50.312	0 + 0	2.0	4 + 0	3.9	8 + 0	9.9
		1 + 0	7.3	5 + 0	1.4	9 + 0	9.7
	B.M. 21	2 + 0	11.1	T.P. 2	1.201	B.M. 21	9.989
	43.047	3 + 0	10.4	T.P. 2	3.016		
		T.P. 1	6.872	6 + 0	4.2		
6-14.	B.M. 14	B.M. 14	4.674	T.P. 1	8.149	7 + 0	9.6
	35.792	0 + 0	7.1	4 + 0	4.0	8 + 0	6.6
		1 + 0	10.7	5 + 0	2.7	9 + 0	5.8
	B.M. 15	2 + 0	12.3	T.P. 2	9.614	B.M. 15	7.167
	34.680	3 + 0	7.8	T.P. 2	9.677		
		T.P. 1	6.842	6 + 0	6.8		
6-15.	B.M. 27	B.M. 27	9.39	T.P. 1	1.29	7 + 0	5.3
	64.81	0 + 0	10.8	4 + 0	0.2	8 + 0	3.1
		1 + 0	6.3	5 + 0	6.0	9 + 0	1.7
	B.M. 48	2 + 0	2.2	6 + 0	9.2	B.M. 48	3.12
	63.23	3 + 0	1.2	T.P. 2	6.53		
		T.P. 1	5.50	T.P. 2	2.91		
6-16.	B.M. 16	B.M. 16	5.628	T.P. 1	3.754	7 + 0	6.0
	65.238	0 + 0	12.8	4 + 0	12.8	8 + 0	5.4
		1 + 0	8.1	5 + 0	10.6	9 + 0	11.6
	B.M. 17	2 + 0	4.7	T.P. 2	8.121	B.M. 17	5.221
	63.534	3 + 0	8.4	T.P. 2	6.281		
		T.P. 1	4.037	6 + 0	6.5		

6-17 to 6-20. Draw a grid 6 in. wide by 7 in. long with 1-in. intersections, and place the given elevations at the intersections in the same arrangement as printed here.

6-17. Draw the 5-ft contours. No depression contours are necessary.

77.0	73.0	68.0	77.0	81.0	85.0	77.0
77.0	71.0	80.0	86.0	83.0	95.0	85.0
80.0	72.0	80.0	95.0	78.0	85.0	89.0
79.0	86.0	77.0	82.0	83.0	73.0	84.0
78.0	80.0	86.0	72.0	73.0	68.0	80.0
80.0	71.0	75.0	79.0	68.0	62.0	72.0
84.0	76.0	68.0	73.0	74.0	67.0	60.0
85.0	73.0	65.0	69.0	72.0	65.0	61.0

6-18. Draw the 5-ft contours. No depression contours are necessary.

22	17	28	40	47	52	57
27	24	22	33	41	46	51
35	32	28	27	34	42	49
45	41	37	31	37	43	49
50	52	48	40	36	42	50
45	46	46	42	44	49	52
34	30	38	45	50	55	60
23	34	45	50	55	60	65

6-19. Draw the 1-ft contours. No depression contours are necessary.

29.3	27.6	25.6	23.0	24.0	23.1	21.8
28.5	27.3	25.9	24.0	26.0	23.9	22.0
27.5	26.8	25.8	24.0	27.2	24.6	22.9
26.4	26.0	25.3	23.0	26.0	25.0	23.8
25.5	25.1	24.7	22.5	24.9	25.3	24.9
24.3	23.9	23.0	22.0	23.5	24.7	26.3
26.0	25.8	25.3	24.0	21.3	23.8	24.3
27.4	27.4	27.0	26.1	23.5	20.6	23.0

6-20. Draw the 1-ft contours. No depression contours are necessary.

50.0	50.5	50.5	49.6	48.4	50.0	52.0
49.2	49.6	49.6	49.3	48.7	49.3	50.2
49.7	49.0	48.9	48.8	49.5	48.8	49.3
51.2	50.8	50.1	50.3	50.5	49.0	48.2
50.2	50.2	50.1	49.8	49.4	48.4	49.0
48.0	48.3	48.4	48.6	48.4	50.0	51.0
50.7	50.5	49.7	48.3	50.0	51.0	52.0
52.7	51.8	50.4	48.1	50.3	51.4	52.6

Suggested Field Exercises

6-1. To Set Up and Take Sights. Guy up two level rods about 20 ft apart. Equip each student with a level or a transit and interchange when halfway through the exercise. Have each student set up, read both rods, and compute the difference in the two readings. Start about 20 ft from the rods, move back a yard or two, and repeat as often as possible. Make sure that each shot is taken exactly according to the proper procedure. By comparing the differences, each student can gauge his or her accuracy.

6-2. Benchmark Leveling. Arrange parties of two and interchange rodman and levelman halfway through the exercise. Starting with a benchmark of given or assumed elevation, have each party establish a series of benchmarks about 800 ft apart and return over the same benchmarks to the starting benchmark.

6-3. Run Profiles. Along some light-traffic road, or across open fields, mark out 100-ft stations with keel or lath. All the field parties can use the same layout.

Parties of two can be used, or a third person can be added to keep the peg book. Each party should have a 50-ft woven tape to measure the pluses of breaks in grade.

Each party should run out its own control benches or use the benchmark-leveling benches.

6-4. To Locate Contours. Establish a 50-ft grid system about 500 or 600 ft on a side, marked with laths, in an open field where there is a considerable range in elevation and irregular contours. Have the students carry in an elevation from a bench about 500 ft distant, set a bench at the grid, and check back on the original bench.

The same field-party arrangement and equipment can be used as for profile leveling.

One-foot contours should be drawn in the office.

7

Surveys for Maps

7-1. Types of Maps.   In general, there are two basic types of maps, area maps and strip maps. Area maps are essential for the development of areas like real estate projects, airfields, and plant layouts; strip maps, for constructing all forms of line transportation like highways, railroads, streets, and pipelines.

Control for an area map consists of loop traverses, and sometimes connecting traverses based on triangulation, together with a system of benchmarks connected by benchmark leveling.

Control for a strip map consists of a long single traverse and a line of benchmarks, both run along the approximate center line of the project.

Maps must show the positions and elevations of topographical features like buildings, roads, streams, and contour lines; so horizontal and vertical measurements must be made which connect or **tie** these features to the control systems.

Whenever the project is large enough, ties are made by photogrammetry. Large surveys can be made using aerial photographs for at least

170

half the cost of ground methods, and usually they result in at least twice as much information. However, aerial mapping requires systems of horizontal and vertical control on the ground, as well as ground ties to selected points that appear on the photographs. Both the control systems and the ground ties are made by the methods described in the following paragraphs, but the ties are usually longer, often extending half a mile from the control. It has been found that the necessary ground surveys are responsible for about half the cost of an aerial map.

7-2. Horizontal Ties. A complete horizontal tie must always consist of at least **two** measurements between the control and the point to be located. These measurements always consist of one of the following: two distances, an angle and a distance, or two angles. Figure 7-1 shows the various combinations. Sometimes one or more extra measurements are made for a check.

A Locus. Each measurement establishes a line on which the topographic point must be placed on the map. This line is a **locus** of the point. The place where the lines (loci) of two measurements cross is the location of the point. Note in Fig. 7-1 that these lines (or loci) are always either straight lines or circles. They are created as follows:

1. A distance measurement from a point on the control indicates that the topographic point is on a circle whose center is at the control point and whose radius is the distance measured.
2. A distance measurement from a line on the control indicates that the topographic point is on a straight line parallel to the control line and at the measured distance from it.
3. An angle measurement made at a point on the control indicates that the topographic point is on a straight line that extends from the point where the angle was measured and in the direction indicated by the value of the angle.
4. An angle measurement, made at the topographic feature between the directions of two control points indicates that the topographic feature is on a circle that passes through the two control points and the topographic point.

Strength of Horizontal Ties. A horizontal tie is strongest when the loci intersect at 90°. The more these angles depart from 90°, the weaker the tie. A weak tie is one in which the location on the map will be in error by considerably more than the error of measurement. Figure 7-2 shows some weak and strong ties and the error that will be caused by an error in the measurements.

Choosing Measurements for Horizontal Ties. Obviously, the measurements made for horizontal ties should be chosen so that the loci will

Method	Measurements	Loci
1 Polar coordinates Angle and distance	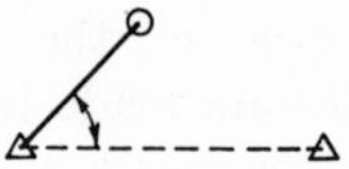	
2 Rectangular coordinates Plus and offset	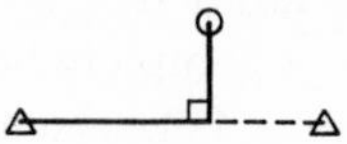	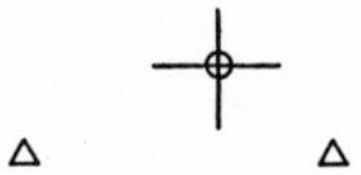
3 Focal coordinates Triangulation	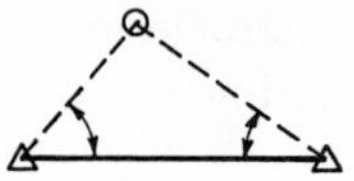	
4 Linear coordinates Two distances	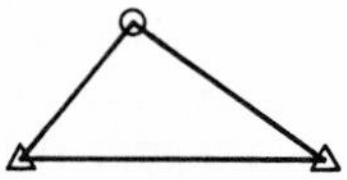	
5 Resection, 3 stations Three point method		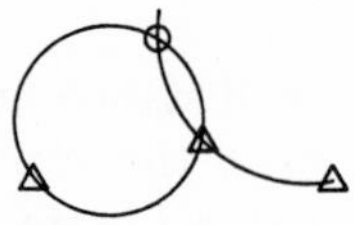
6 Resection, 2 stations Two point method	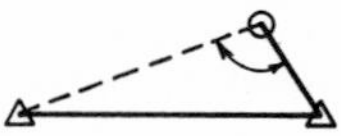	
7 Similar to No. 1 Angle and distance from a line	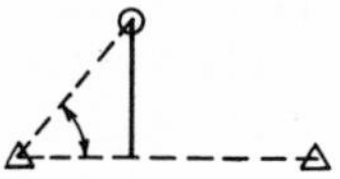	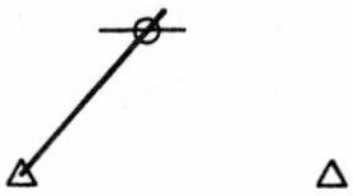
8 Similar to No. 2 and No. 4 2 distances from lines	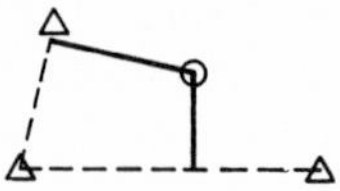	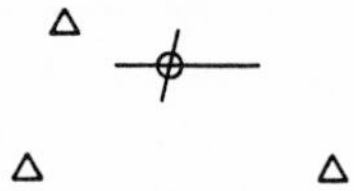
9 Similar to No. 6 Angle at point distance to a line	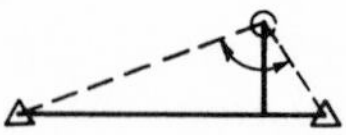	

Legend: Measured distance ———   Measured angle   Line of sight – – –

Fig. 7-1. Methods of making horizontal ties.

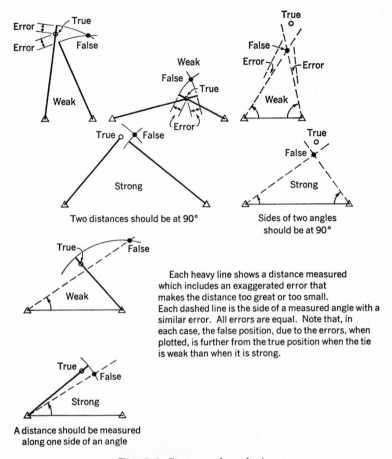

Fig. 7-2. Strong and weak ties.

intersect at an angle as close to 90° as possible. The following rules produce the strongest ties; the farther the departure from them, the weaker the tie. See Fig. 7-2.

1. Two distances should be at right angles.
2. A distance with an angle should be measured along one side of the angle.
3. Two angles measured from the control should add to 90°.

Ties Used Most Frequently.   The ties in Fig. 7-1 are shown in the order of their importance. Tie 1 is the most useful for an area map; Tie 2, for a strip map. These two types are used almost exclusively. The others are used under special circumstances.

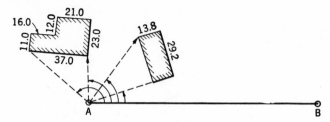

Fig. 7-3. Angle and distance measurements.

Angle and Distance.   Figure 7-3 shows the angle-and-distance method used to locate two buildings. Two corners of each building are located, and the building dimensions are measured along the sides of the buildings. If other buildings are to be built connecting the existing buildings, the angles must be measured very accurately and the distances must be measured with a steel tape. Otherwise a woven tape is used (Fig. 7-4). The field notes are shown in Fig. 7-5. By far the best method for angle and distance, however, is the **stadia method** described later.

Plus and Offset.   Figure 7-6 shows the plus-and-offset method. The traverse line is first marked off in stations which are **lined in** with the transit. The rear tapeman holds the zero of the tape first at $1 + 0$; the head tapeman places himself or herself on line with the line of marked stations and at the point on the line where he or she estimates a perpen-

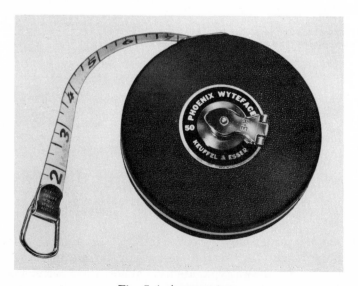

Fig. 7-4. A woven tape.

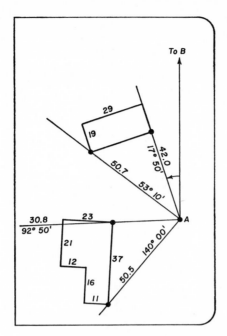

Fig. 7-5. Field notes for Fig. 7-3.

dicular from the traverse line would strike the building corner. Next, the plus (+70) is measured and then the 18.1 ft to the corner. The 18.1 ft is called an **offset,** in this case, a **left offset.** For the next corner the rear tapeman holds at 2 + 0, and the process is repeated. Finally the dimensions of the building are measured. The field notes are shown in Fig. 7-7.

Estimating the Perpendicular. The perpendicular can be estimated in a number of ways. Figure 7-8 shows a method of swinging the arms; Figs. 7-9 and 7-10 show the method of using the **double pentaprism.** When an accurate offset is required, a **swing offset** is used.

Swing Offset. Assume that it is required to measure a perpendicular from a point *P* to a line *AB* (see Fig. 7-11). Set up at *A*, and point on *B*.

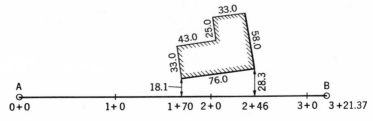

Fig. 7-6. Plus-and-offset measurements.

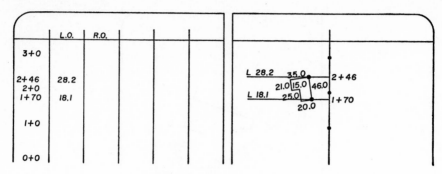

Fig. 7-7. Field notes for Fig. 7-6.

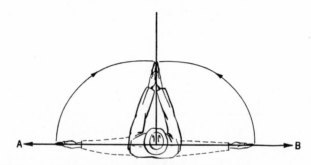

Fig. 7-8. Estimating a perpendicular.

Swing a tape or a leveling rod as shown, finding the shortest distance. When the tape is near the transit, the graduations are turned toward the transit and the least reading is noted by the transitman. Otherwise, a yellow pencil or target is held on the tape or rod and adjusted until it just touches the line of sight when the tape is swung back and forth. The final position of the pencil or target is used.

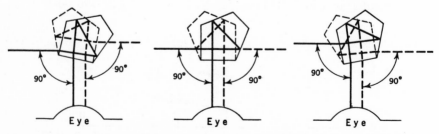

Fig. 7-9. The double pentaprism turns the line of sight 90° in both directions. Rotating the device does not change the right angles.

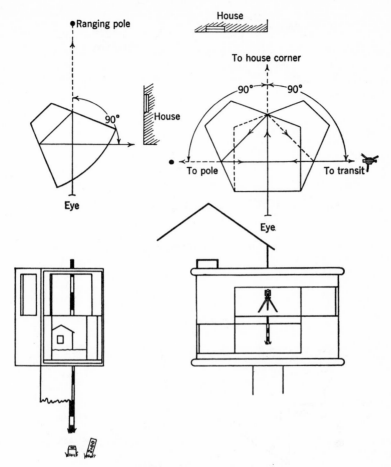

Fig. 7-10. The double pentaprism for estimating right angles. The observer moves forward and back until the objects and the ends of the line are aligned in the window of the pentaprism, and he or she moves left and right until the point to be located is in line with them.

To **establish line** at a certain perpendicular distance from *P*, the tape or rod is swung as before. The transitman points at the greatest distance reached by the proper mark on the tape or the rod.

Other Ties. The other ties in Fig. 7-1 are used under the circumstances given here.

Tie 3. When it is too difficult to tape to the point, as when it is across a river, or a road with heavy traffic.

Tie 4. For short distances between objects See Fig. 7-12.

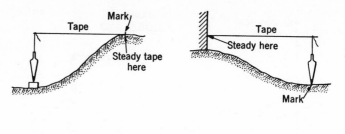

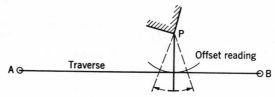

Fig. 7-11. A swing offset.

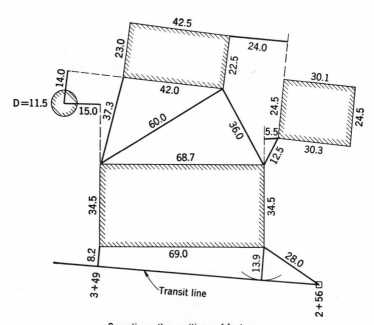

Sometimes the positions of features
are determined by measurements
from other features rather than
from the control line itself

Fig. 7-12. Locating one feature from another.

Tie 5. When the point can be reached but distances to control cannot be measured. Targets must be placed on the control. The point is located on the map by plotting the angles on tracing paper and fitting them to the control by trial and error. When they are in place, the point is pricked through the tracing paper.

Ties 6 through 8. When the distance from control to the point cannot be measured along the side of the measured angle because of a swamp, lake, or highway, etc.

Tie 9. When both distance measurements are obstructed.

Accuracy. The accuracy with which horizontal ties are measured depends on the purpose of the survey. When great accuracy is required, the lengths are measured with a steel tape and the numerical values are placed on the map. Any other distances required for the plans are computed rather than scaled from the map. However, most distances used for mapping are measured with a woven tape (see Fig. 7-4) or stadia.

7-3. Vertical (Elevation) Ties. Vertical ties for area maps are measured by leveling, as described in Chap. 6, but more often by stadia, described in Chap. 8. Leveling is used almost exclusively for strip maps, with occasional use of stadia.

Leveling for Strip Maps. When the project is too small to utilize photogrammetric methods, leveling is carried out by **cross sectioning.** This process is shown in Fig. 7-13. After the traverse and the line of benchmarks have been established along the approximate center line of the project, a short profile is measured at right angles to the traverse at each break and at each 50- or 100-ft point. These are carried far enough out on each side of the line to cover all possible earthwork (see Sec. 6-15). The right angles are estimated as described in Sec. 7-2. Elevations are taken at the center line, at the breaks, and at the ends of each profile. Offsets are measured to break, and the pluses and offsets to topographic features are determined. The leveling is carried by ordinary levels from benchmark to benchmark with side shots to the points where elevations are determined.

Frequently, however, only the center-line elevations are determined with the level. The positions of the contour lines are determined with **hand levels.**

The Hand Level. Figures 7-14 and 7-15 show the appearance and operation of a hand level. The instrument is adjusted so that, when the bubble is centered, the line of sight is horizontal. There is a horizontal line across the open end of the instrument tube which, together with the peephole, determines the line of sight. When the reflection of the bubble is centered on this line, the line of sight is horizontal. The instrument can be held in the hand and the bubble centered while the position of the line of sight on the rod is observed. Owing to the unsteady

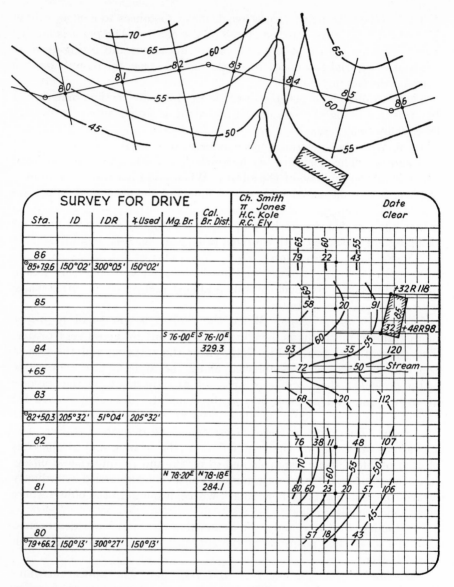

Fig. 7-13. Portion of route-survey field notes. The notes shown on the right-hand page are often kept in a large, separate field book.

support, magnification is not very helpful, and therefore a telescope is seldom incorporated. The levelman can seldom read the rod, and therefore a target is used. The rodman adjusts the rod at the direction of the level-

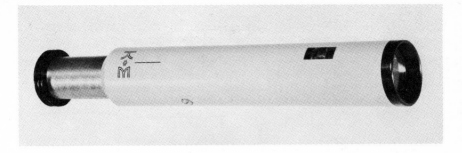

Fig. 7-14. Hand level.

man and then reads its position. The levelman must be careful to keep the instrument at the same height between plus and minus sights. The accuracy is about 3 ft $\sqrt{\text{miles}}$. The level is used in a special way for determining contours in a route survey.

Locating Contours with a Hand Level. Usually the position of each contour is found along a line perpendicular to the traverse at each station (see Fig. 7-16). An example is given of this procedure when used for locating 5-ft contours.

The lens and prism cover only half of the tube

View through the hand level showing the appearance
of the rod target when it has been placed at the same
elevation as the instrument

Fig. 7-15. Operation of the hand level.

The hand level is placed on a forked stick cut so that the line of sight is 5 ft from the ground. To work downhill the stick is first placed on the ground at the station. Assume that from the profile notes it is known that the station has the elevation 92.3 ft. The H.I. is 97.3. To find the

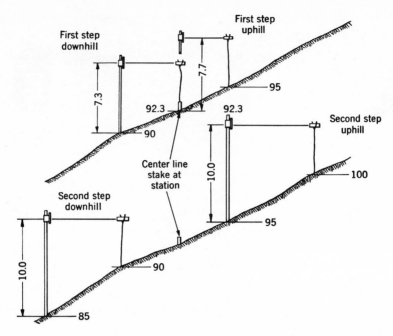

Fig. 7-16. Finding contour lines with a hand level.

90-ft contour the rod target is set at 7.3 and the rod moved downhill perpendicular to the traverse line until the target is at the level of the instrument. The offset is then measured from the station. The target is then set at 10 ft, the level is moved to the position of the rod, and the process is repeated to find the 85-ft contour, etc.

In working uphill the rod is first held at the station with the target set at 7.7. The instrument is moved uphill until it is level with the target. The instrument is then on ground 2.7 ft above the station and therefore at the 95-ft contour. The offset is measured. The target is set at 10 ft, the rod moved to the instrument position, and the process repeated to find the 100-ft contour, etc.

Usually the results are plotted in the field book, and the contours between stations are sketched in while the ground is being viewed.

Problems

7-1, 7-2. Show the notes for the right-hand page of the field book for topography shown in Figs. 7-17 and 7-18, respectively. Note the length of 100 ft and give approximate distances and angles.

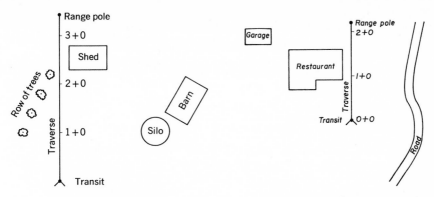

Fig. 7-17. Problem 7-1.   **Fig. 7-18.** Problem 7-2.

7-3. Draw a sketch illustrating an actual example of field conditions that make each of the following horizontal ties the best tie to use: Fig. 7-1, Ties 3, 5, and 7. Show the ties.

7-4. Same as Prob. 7-3, but Ties 4, 6, and 8.

7-5. Draw a sketch of the locus of a point that is exterior to a triangle and 10 ft from it.

7-6. Draw a sketch of the locus of a point that is equidistant from the two sides of an angle.

7-7. Name the key feature that makes the strongest horizontal tie.

7-8. Name the three rules for making a strong tie in the field.

Suggested Field Exercises

7-1. Establish a traverse of one or two courses and tie in several objects by various methods of making ties. Demonstrate weak ties and check the ties being made for strength.

7-2. Have the map drawn in the office.

7-3. Have two or more objects tied together and to the control with one or two weak ties and several strong ties.

7-4. Have these plotted to show the effect of weak ties.

8

Stadia and Photogrammetry

8-1. The Stadia Method. The stadia method provides a means of measuring a direction, a horizontal distance, and a difference in elevation, all three in one operation, with a transit and rod. It is used chiefly to measure ties from survey control stations and benchmarks to topographic features and to locate ties to photogrammetric control points (see Sec. 8-15). Short stadia traverses are often run to make these ties, and surveys for small topographic maps can be made entirely by stadia (see Sec. 8-7).

8-2. Stadia Hairs. For stadia measurements, a transit must be equipped with a vertical circle and stadia hairs. Stadia hairs are two supplementary horizontal cross hairs equally spaced above and below the center hair (Fig. 8-1). The stadia hairs are placed so that their lines of sight separate at a rate of 1 to 100 (Fig. 8-2) from a point at the center of modern, internal-focusing instruments. This point is called the **anallactic point.** The anallactic point moves slightly backward and forward with changes in focus, but the movement is negligible. In old-fashioned exterior-focusing instruments, the anallactic point is about 1 ft in front of the instrument center.

The Stadia Formulas. Figure 8-3 shows the geometry of a stadia shot. S' is the stadia intercept, i.e., the distance between the two points on the rod where the stadia hairs appear to fall. A is the vertical angle observed. Then, from Fig. 8-1,

$$H = D \cos A \qquad D = 100\,S \qquad S = S' \cos A \text{ (very nearly)}$$
$$V = D \sin A$$

hence, substituting,

$$H = 100S' \cos^2 A$$
$$V = 100S' \sin A \cos A$$

Stadia tables or **stadia slide rules** are used to compute H and V. Stadia tables are given in the back of this book (Tables IX and X) with directions for their use. The tabular values are multiplied by $100S'$ with an ordinary slide rule.

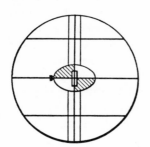

Fig. 8-1. Stadia hairs and rod.

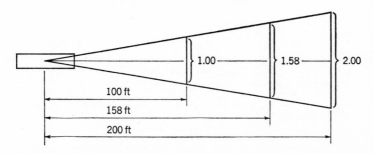

Fig. 8-2. The distance to the rod is equal to 100 times the stadia intercept.

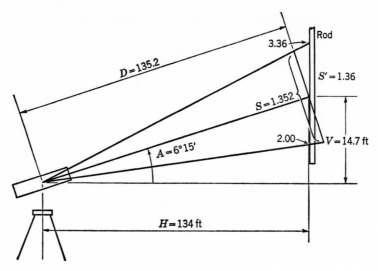

Fig. 8-3. Typical stadia shot.

Note: If the stadia survey is made with an old-type **exterior-focusing** transit, before computing by any method add 0.01 ft to the actual stadia intercept. In the example, this would give $S' = 1.37$. Then compute as before. To recognize an exterior-focusing instrument, turn the focusing screw. If the objective lens moves in and out, it is exterior-focusing.

8-3. The Stadia Slide Rule.   Figure 8-4 shows a stadia slide rule designed by the author. With this rule, except in extreme cases, the user never "runs off the rule" and thus is forced to reset the slide. In addition, the rule gives the position of the decimal point in the result. To use the rule, set the zero (which is red) at 100 times the stadia intercept on the R_1 scale. Set the hairline at the vertical angle on the red H scale and read

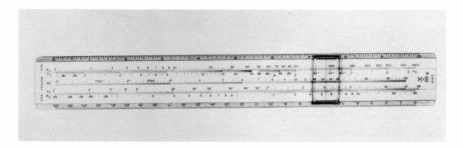

Fig. 8-4. The Kissam stadia slide rule. (*Keuffel & Esser Co.*)

the R_1 scale to the nearest foot. This is H.

Without moving the slide, set the hairline at the vertical angle on the V_1 or V_2 scale (whichever is convenient) and read the adjacent R scale to the nearest 0.01 ft. This is the V.

For very accurate values of H, set the hairline at the vertical angle on the HC scale. Subtract the reading on the R_2 scale from 100 times the stadia intercept. When the vertical angle is smaller than 1°, no correction is required.

Two extreme cases might possibly occur, viz.:

1. Intercept larger than 10.00 ft. Multiply the intercept by 10 (instead of 100) and multiply all results by 10.
2. Vertical angle less than 1 minute. To find V, set the hairline on the V_3 scale, read the R_1 scale, and divide by 10,000. If this runs off the scale, do both (1) and (2).

Example.   Stadia intercept 12.00 ft; vertical angle 30 seconds. Set the red zero at 120 on R_1. Set the hairline at 30 seconds on V_3. Read 175 on V_3.

$$\text{Divide by } 10,000 = 0.0175$$
$$\text{Multiply by } 10 = 0.175'$$

TRANSIT-STADIA FIELD METHOD

8-4. The Method Described.   There are several methods of taking transit-stadia observations. The method described here is thought to be the most rapid, the one which is most likely to be free from blunders, and capable of as accurate results as is possible with stadia.

Approximate Accuracy of Method

Distance measured once $= \frac{1}{300}$
Distance measured forward and back $= \frac{1}{500}$
Difference in elevation measured once $= 0.03' \sqrt{H}$
Difference in elevation measured forward and back $= 0.02' \sqrt{H}$

where H = number of hundreds of feet in length of the survey.

These accuracies are usually well within the limits required for mapping.

8-5. Equipment.   The transit should have a vertical circle with a vernier that reads to 1 minute or better. The rod can be an ordinary level rod graduated in units of 0.01 ft.

8-6. Azimuth.   All directions should be recorded in azimuths for ease in plotting. Figure 8-5 shows how this is accomplished. In the upper figure, the first stadia station is station 9 of the control traverse. The forward azimuth of the line station 8 to station 9 is known from the control survey. It is 69°15'.

When the instrument is set up at station 9, to orient the circle:

1. Compute the back azimuth, 69°15′ ± 180° = 249°15′.
2. Using the upper motion, set the vernier at this reading, clockwise.
3. Using the lower motion, aim at station 8.
4. Free the upper motion and now, in whatever direction the instrument is pointed, the clockwise reading will be the azimuth of the point.
5. Aim at *B*. The azimuth of *B* is read with the vernier, 312°18′
6. After all the topographic shots have been taken, aim at station 8 as a check. The vernier should still read 249°15′.

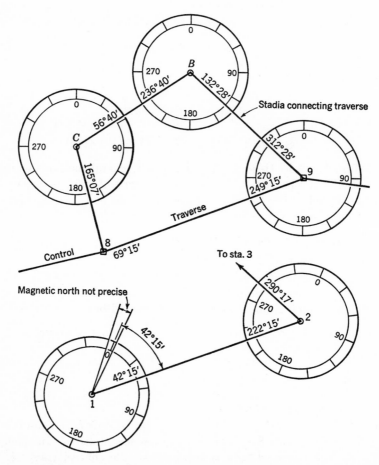

Fig. 8-5. Azimuth directions. How the circle is oriented so that it will give the azimuth for every direction at every station. Upper figure, oriented to a control traverse. Lower figure, oriented to a magnetic azimuth.

At B the back azimuth is computed, $312°28' \pm 180° = 132°28'$, and the process is repeated.

When the shot from the last stadia station is made, which in this case is the shot from C to 8, the azimuth of 8 ($165°07'$) will be measured. To check , set up 8 and orient by sighting C with the vernier set at $345°07'$. Turn to 9 and note how closely the azimuth checks the original value, $69°15'$. It should check within about $(0°1'30'')\sqrt{\text{number of stations}}$, in this case four stations, and therefore within $0°3'$.

In the lower figure, the first stadia station is 1. There are no control stations available, so either the azimuth 1 to 2 is assumed or the magnetic azimuth is used. As the magnetic azimuth cannot be determined more accurately than to ± 15 minutes without a large expenditure of time, it is used only at the first station. However, if it is used, any azimuth determined throughout the survey can be checked approximately by reading the compass azimuth.

To establish the magnetic azimuth:

1. Free the needle.
2. Set the vernier at zero with the upper motion.
3. With the lower motion, turn the alidade until the north point in the compass box is at the north end of the needle.
4. Then free the upper motion, aim at 2, and read the azimuth, $42°15'$.
5. At 2 set the vernier at $42°15' \pm 180° = 222°15'$ and aim at 1, using the lower motion.
   The process is continued and checked out as before.

PROCEDURE FOR A STADIA SURVEY

8-7. References.   Figure 8-6 shows a map resulting from a stadia survey. Figure 8-7 shows the field notes. Only the first four columns and the right-hand page are used in the field. The stadia procedure described here is based on this example. Note that all stadia operations are identical except the means of original orientation at the first station, as previously described. The steps in the procedure follow.

8-8. Planning.   First, plan the locations for the stadia stations and mark them. This is a very important operation, as a well-planned arrangement of stations reduces the field time and improves the accuracy. The stations must be intervisible and should preferably be located not more than 500 ft apart. In the example, several are 800 ft apart, as high accuracy is not required.

A long shot is difficult to observe, so that the transitman takes longer to make the observation. It usually saves time to err on the side of too many stations, as this reduces the lengths of the numerous side shots.

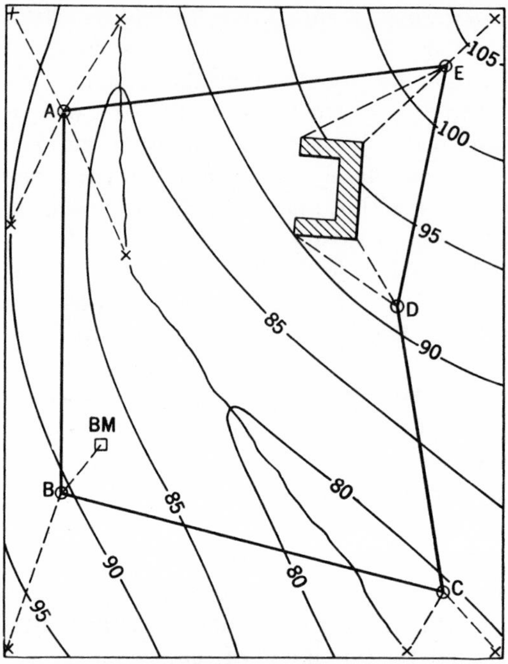

Plant #2 site survey

(a)

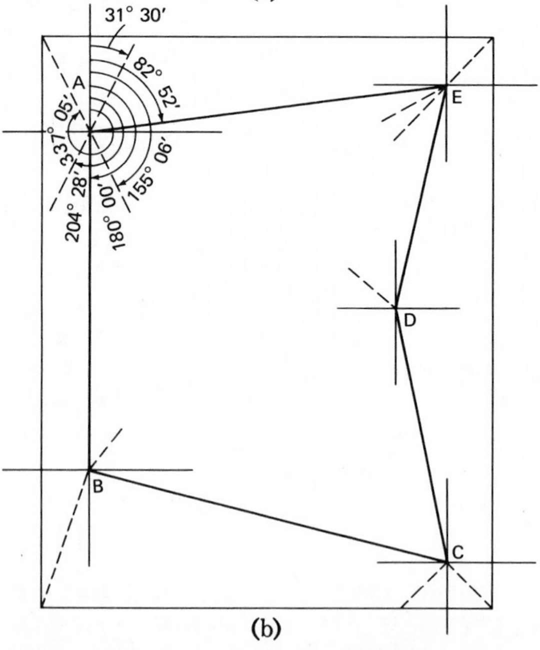

(b)

Fig. 8-6. a) Stadia loop traverse and map. b) Azimuth readings at A in a.

		PLANT #2 SITE SURVEY					Ch Smith	Date
							TT Jones	Windy
Sta	S	Azim	V∠	H	V	Elev	Rod Kole	
⊼ A	h.i.	4.65				87.72		
E	8.07	82° 52'	+1° 06'					
B	8.00	180° 00'	+0° 12'	800	+2.83			
	2.40	337° 05'	+1° 19'	240	+5.5	93.2	Boundary corner	
	2.28	31° 30'	−0° 15'	228	−1.0	86.7		
	3.33	155° 06'	−0° 52'	333	−5.0	82.7		
	2.65	204° 28'	+0° 30'	265	+2.3	90.0	Boundary line	
⊼ B	h.i.	4.26				90.55	87.06 + 7.75 − 4.26 = 90.55	
A	8.00	0° 00'	−0° 12'					
C	8.24	104° 02'	−0° 46'	824	−11.27			
BM	Level		7.75			87.06	Previously established BM	
	3.59	199° 22'	+1° 11'	360	+7.4	98.0	Boundary corner	
⊼ C	h.i.	4.85				79.28		
B	8.26	284° 02'	+0° 48'					
D	6.10	350° 32'	+1° 04'	610	−11.56			
	1.70	137° 35'	−0° 04'	170	−0.2	79.1	Boundary corner	
	1.50	208° 58'	−1° 27'	150	−3.8	75.5		
⊼ D	h.i.	4.36						
C	6.10	170° 32'	−1° 06'			90.84		
E	5.11	11° 18'	+1° 22'	512	−12.36			
I	1.64	239° 03'	+0° 40'	164	+1.9	92.7		
2	2.60	215° 20'	−0° 11'	260	−0.8	90.0		
⊼ E	h.i.	4.72						
D	5.13	191° 18'	−1° 24'			103.20		
A	8.09	262° 52'	−1° 06'	808	−15.48			
3	2.35	225° 55'	−1° 38'	235	−6.7	96.5		
4	3.34	242° 38'	−1° 44'	334	−10.1	93.1		
	1.42	44° 10'	+1° 57'	142	+4.8	108.0	Boundary	

Fig. 8-7. Field notes from stadia survey shown in Fig. 8-6a.

The stations should be placed so that all the topographic features, including the key points for locating contours (Sec. 8-10), can be readily observed and so that the traverse closes on itself or on the control systems.

Sometimes only part of the traverse is planned and the remainder filled in as the work proceeds. This method should be avoided if possible.

8-9. Steps in Procedure

1. Set up at station *A* (Fig. 8-6) and measure h.i. = 4.65. This is the height of the telescope axle above the top of the stake (or other object) which marks station *A* (see Figs. 8-8 and 8-9). Record in the first line of notes. Have the rod target set at this value.

2. Orient the circle, in this case by compass, as described in Sec. 8-6.

3. Take a shot on *E* (the last station). All stadia shots are taken in the same way except the orientation shot. Orientation shots differ only in that the lower motion is used. A stadia shot is described in the next section.

A Stadia Shot from A. With the upper motion and the vertical motion, aim at the target; i.e., place the vertical cross hair and the central horizontal hair on the target. With the vertical-motion tangent screw, lower or raise the aim until the lower stadia hair is on the nearest foot mark. Read the rod at the upper hair. From this reading, subtract the number of feet at the lower hair. The difference is the stadia intercept,

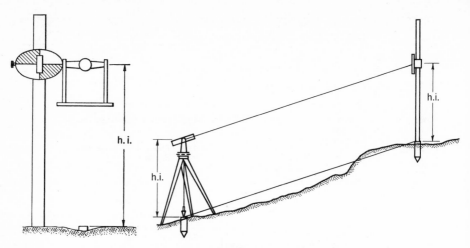

Fig., 8-8. Determining the h.i.

Fig. 8-9. The difference in height between the two stations is the same as the *V* computed from stadia data since the slope of the line of sight is parallel to the slope between the stations.

which on *E* is 8.07. Return the central hair to the target. Wave O.K. to the rodman, who moves to the next point to be observed, in this case station *B*.

While she or he is moving, record the stadia intercept 8.07 in the *S* column. Read and record the azimuth 82°52′ in the azimuth column, and read and record the vertical angle + 1°06′ in the vertical angle column. Check the sign of the vertical angle. If the telescope bubble is toward the front, the sign is plus.

The vertical angle is confusing to read, as the vernier is below the circle, contrary to the verniers on the horizontal circle. Be sure to estimate the position of the vernier zero on the scale before making a precise reading.

Note: If the shot is an orientation shot, read the azimuth, even if it has just been set on the vernier, as a precaution against using the upper motion in aiming for orientation.

Also, if the orientation shot is along a line previously determined by stadia, the stadia intercept should check within about ±0.02 ft and the vertical angle should have the opposite sign and check within about ±0°02 minutes.

4. Take stadia shots on the following, in the order given: station *B* and the four points marked *X*. The *X* points are for elevation only. They require no description.

5. When all shots are complete, take a check shot on *E*. All values should agree closely. The azimuth of *E* is most important because, if it does not agree exactly with the previous shot on *E*, the wrong motion has been used during the work.

Repeat the same procedure at each station, finally checking the shot from E to A against the original shot from A to E for S, azimuth, and vertical angle.

Note, at stations D and E, how the building is located. The dimensions of the building are measured with a woven tape.

The points not described in the notes (Fig. 8-7) are used for locating contour lines. In general, once the traverse shots are complete, the rodman should move from point to point in a continuous clockwise direction unless this increases the distances he or she must move. Often, however, this reduces the number of movements and simplifies plotting the map. Occasionally, points can be chosen on the same azimuth. This also reduces mapping time.

Difficult Shots. When mapping to a large scale, which is usually the case, to locate the center of a tree or the corner of a building, where the rod cannot be held, the rodman holds the rod close to the object and at the same distance from the transit as the object. The transitman reads the intercept, returns the center cross hair to the target, then aims at the object. All the readings made thereafter will then be correct for the true point he or she wishes to locate.

When a point in a gully is to be located which is so deep that the target is invisible, or the target is obscured by tree leaves, etc., the observation can be made on any convenient mark on the rod instead of the target. Of course, the value of mark chosen must be noted. The resulting elevation must be corrected by the difference between the mark used and the target position (the h.i.) as follows:

$$E = E' + \text{h.i.} - m$$

where E = true elevation

E' = computed elevation

m = value of rod mark used

8-10. To Choose Points for Side Shots

Locating Contours. The methods of establishing control and making ties have been covered in this and other chapters. The problem remains of how to locate contours. In area surveys the positions and elevations of certain key points are determined, and the contour lines are interpolated between them. When greater accuracy is necessary, elevations are also determined at grid positions as described in Sec. 6-16. The key points must be included even when the grid system is used.

Key Points for Contours. In general, key points are **those points between which the ground has a uniform slope.** Since the ground never slopes uniformly, the accuracy of the map depends on how small a change in slope is considered significant for the contour interval desired. The ability to select key points so that the desired map accuracy can be obtained with a minimum of field work is an art that develops with experience. However, if each of the following conformations is considered with a view to using it as a key point, there will be little chance for omissions:

1. Summits
2. Saddles (low points in ridges)
3. Depressions
4. Valley profiles
5. Ridge profiles
6. Boundary and building corners
7. Profiles along buildings and boundaries
8. Profiles along toes of slopes
9. Profiles along brows of hills (tops of slopes)
10. Profiles along shoulders

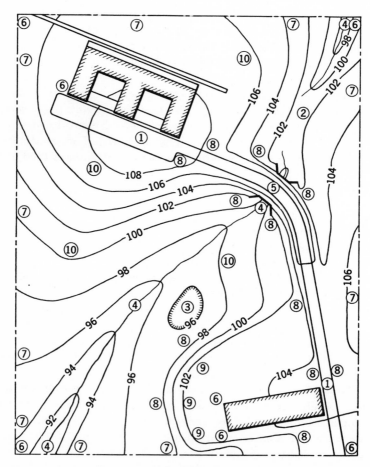

Fig. 8-10. Key points for contours.

Figure 8-10 illustrates the typical key points found on a plant site. The numbers refer to the list above. Although many points fall into more than one classification, only one classification is shown.

Time can usually be saved by taking the elevations of points that must be located horizontally, such as the corners of buildings and the bends in streams. These points are usually key points, and, of course, no further measurements for horizontal position are required.

TRANSIT–STADIA OFFICE PROCEDURE

8-11. Preliminary Computations. Extract the traverse data from the field notes as shown in Fig. 8-11. In the column "Course" are the names of each traverse course. Under S are the two stadia intercepts (forward and backward) for that course. Under $V \not\subset$ are the two vertical angles. The sign of each backward angle is changed.

The average of each pair is computed.

The values of H and V are carefully computed from these averages by one of the methods of computation.

Course	S	V ⊀	H	V	Cor.	Elev. V
A – B	8.00	+0° 12'				87.72
	8.00	+0° 12'				
	8.00	+0° 12'	800	+2.80	+0.03	+2.83
B – C	8.24	–0° 46'				90.55
	8.26	–0° 48'				
	8.25	–0° 47'	825	–11.30	+0.03	–11.27
C – D	6.10	+1° 04'				79.28
	6.10	+1° 06'				
	6.10	+1° 05'	610	+11.53	+0.02³	+11.56
D – E	5.11	+1° 22'				90.84
	5.13	+1° 24'				
	5.12	+1° 23'	512	+12.34	+0.02	+12.36
E – A	8.09	–1° 06'				103.20
	8.07	–1° 06'				
	8.08	–1° 06'	808	–15.51	+0.03	–15.48
			3555	– 0.14	+0.14	87.72

Estimated error $0.02\sqrt{35.55} = {}^{\pm}0.12$

Cor. $= \dfrac{0.14}{3555} \times H$

Fig. 8-11. Stadia traverse data and reduction. Taken from Fig. 8-7.

The H values are entered in the H column in the field book as shown in Fig. 8-7, on the line on which the data to the end of each course are recorded.

The V values should add to zero. If the sum approximates the estimated closure, as given in Sec. 8-4, they can be used unchanged or they can be adjusted. In Fig. 8-11 the adjustment is carried out. Like the compass traverse adjustment, the adjustment is proportional to the lengths of the courses. The final V's are entered in the field notes (Fig. 8-7) in the same line as the H's.

8-12. Elevations. Unless assumed elevations are used, at least one shot should be taken to a B.M. If this shot is taken in the normal way, the sign of the vertical angle is reversed and the computed V is added algebraically to the elevation of the B.M. This gives the elevation of the station from which the B.M. was observed.

In the example, a level shot was taken. In this case, the stadia intercept is unnecessary. The telescope level is centered, and the rod reading (7.75) is recorded. This is added to the known elevation of the B.M. (87.06) to give the H.I. of the instrument (94.81). The h.i. (4.26) is subtracted to give the elevation of the station (90.55).

The elevations of the remaining stations are computed from the final V's as shown in Fig. 8-11. These are placed in the elevation column in the field notes at the beginning of each course.

The H's and V's for side shots are computed usually by stadia slide rule and placed in the appropriate column (Fig. 8-7).

The elevations of the points observed are computed by adding the V's algebraically to the elevations of the stations from which they were observed. The data are now ready to be mapped.

TO PLOT THE STADIA MAP

8-13. To Plot Stations and Points Observed. Using a rough sketch, the map is placed on the sheet so that all the shots can be mapped. After the position of station A is selected, two lines at right angles are drawn through it to represent the 0° azimuth and the quadrants.

The remainder of the stations are plotted by protractor and scale and lines are drawn through each station parallel to the lines through A. The protractor should be used with great care, as this is the greatest source of error. The last line drawn should close on A. If there is no blunder in plotting, the residual error should be eliminated by shifting the stations, each parallel to the error and by amounts proportional to their distances along the traverse from the beginning point A, as shown in Fig. 8-12. The lengths of the corrections can usually be established by eye.

When the traverse has been corrected, the elevation of each station is marked and the side shots are plotted from their respective stations. As each side shot is plotted, its elevation is marked. All topographic details should be drawn in as the plotting progresses to avoid any questions as to which shots are used.

8-14. Plotting the Contours. To plot the contour lines, first inter-

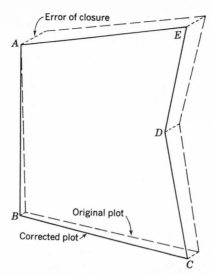

Fig. 8-12. The method of eliminating the errors of plotting. Corrections are parallel to the error and proportioned to the length of the traverse from A to the station where the correction is made.

polate the contours along streams and dry-stream or valley lines. Note the summits and the saddles. Interpolate along the shortest lines that connect adjacent elevations. Draw in the contour lines as described in Sec. 6-16.

8-15. Photogrammetry.[1] For large surveys, map details, including contour lines, are best located from vertical aerial photographs by **photogrammetry.** Photographs are taken in strips which overlap, as shown in Fig. 8-13. Clearly defined, well-distributed points (**control points**) are selected on the photographs and are often located for position and usually for elevation by ground surveys. Their positions are plotted on the map sheet.

Details from the photographs are placed on the map by means of **plotters.** The principle of the plotters is best understood in the case of the **multiplex plotter,** here described.

In the multiplex plotter, Fig. 8-14, a **tracing table,** Fig. 8-15, is placed over each plotted position in turn so that the tiny white light at the center of the circular white surface (the **platten**) is at the scale position and elevation of the control point.

A transparent positive (diapositive) made from each photograph is placed in an overhead projector on an adjustable mount. The projector is adjusted in height, position, and aim so that the projected image of

---

[1] Photogrammetry is more completely described in Philip Kissam, "Surveying for Civil Engineers," McGraw-Hill Book Company, 1956, and "Surveying: Instruments and Methods for Surveys of Limited Extent," 2d ed., McGraw-Hill Book Company, 1956.

every control point falls on the light on the platten positioned over that point. Usually a row of projectors are similarly placed and thus reconstruct to scale the geometric conditions that existed when the photographs were taken.

Stereo-Viewing. The chief means by which a person estimates the relative distances of nearby objects is based on the difference between the views seen by each eye (the **retinal disparity**), which occurs because of the difference of the two eye positions. Figure 8-16 shows an overlapping pair of diapositives projected on the platten of a tracing table set at the elevation of a desired contour line. Note also Fig. 8-17. At the left, the rays to the object *A* on the ground would intersect at the true scale position of *A*. The light from the left projector is colored green and that from the right is colored red. The operator wears glasses similarly colored. As a result, he or she sees the left image on the platten at *D* and the right image at *E*. This gives the illusion that the object *A* at *B* on the **stereo model** is far below the white light that is seen with both eyes. When the operator moves the tracing table toward the right, the stereo model appears to rise until it reaches the light at *C*. The light is then on the contour. A pencil under the light is lowered and the tracing table is moved (in this case forward or back) so that the light appears to remain on the ground, thus drawing the contour line for which the platten is set. All contour lines are drawn in this manner. Other features are drawn by raising or lowering the platten until the light appears on the ground at the location of the feature and tracing the feature. Since the diapositives together form a continuous series of pairs of diapositives that form stereo models, the whole strip can be completed.

Problems

8-1. Copy Fig. 8-18 approximately to scale, making a rectangle $5\frac{1}{4}$ by $7\frac{1}{2}$ in. Draw the 5-ft contour lines.

8-2. Copy Fig. 8-14 approximately to scale, making a rectangle $5\frac{1}{4}$ by $7\frac{1}{2}$ in. Draw the 5-ft contour lines.

8-3. *a.* Compute the required values from the stadia notes in Fig. 8-15. *b.* Draw the map to a scale of 1 in. = 200 ft.

8-4. *a.* Compute the required values from the stadia notes in Fig. 8-16. *b.* Draw the map to a scale of 1 in. = 100 ft. Use a sheet $8\frac{1}{2}$ by 11 in. Place station *A* $2\frac{1}{2}$ in. from the top and $2\frac{1}{4}$ in. from the side. Establish zero azimuth up the paper and parallel to the sides.

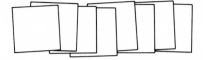

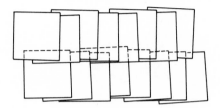

Fig. 8-13. Photographs overlap so that every object is on at least two, and sometimes as many as four, photographs.

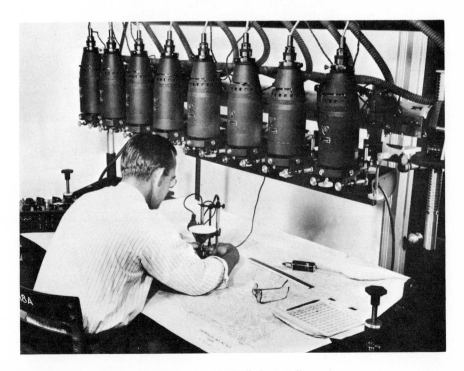

Fig. 8-14. A multiplex plotter in use. (*U.S. Geological Survey.*)

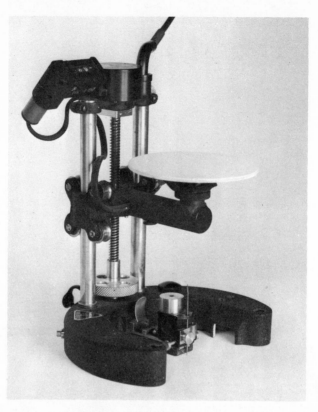

Fig. 8-15. A tracing table for the multiplex. Note that it is adjustable in height. A needle point or a pencil point marks its position on the map.

Fig. 8-16. The true model is formed as shown. It cannot be seen. The stereo model, with heights exaggerated, is seen instead. By moving the platten, on which the stereo model is seen, points on the true model are found. Their horizontal positions are transferred to the map by pencil. (*U.S. Geological Survey.*)

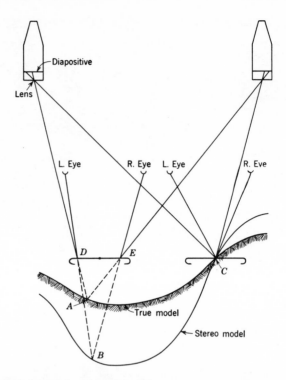

Fig. 8-17. Multiplex optics.

Fig. 8-18. A modern Kelsh plotter. In principle it is like two multiplex projectors.

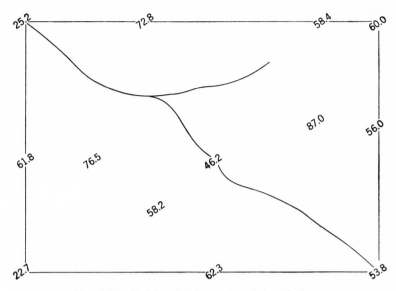

Fig. 8-19. Problem 8-1 for contour interpolation.

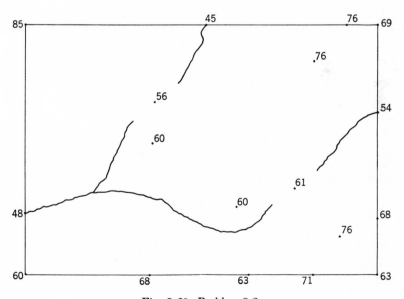

Fig. 8-20. Problem 8-2.

MILL SITE SURVEY						Ch. Smith Date		
						π Jones Hot, Calm		
Sta	S	Azim	V✗	H	V	Elev.	Rod	Cole
πA	h.i.	4.38						On Rock in Stream
E	6.08	260°32'	+1°41'					
B	9.00	0°00'	+2°10'					
	1.79	180°00'	+2°18'					Property Corner
πB	h.i.	4.63						
A	9.00	180°00'	-2°10'					
C	9.04	276°20'	-1°14'					
	5.67	223°25'	-2°24'					Saddle
	3.47	287°02'	-5°56'					₵ Stream
	.99	0°00'	-0°17'					Property Corner
πC	h.i.	4.71						Property Corner
B	9.06	96°20'	+1°14'					
D	7.07	171°52'	+1°04'					
πD	h.i.	4.22						
C	7.05	351°52'	-1°02'					
E	4.46	153°26'	-1°14'					
1	1.57	71°20'	-3°40'					
2	2.60	104°40'	-2°38'					
	1.06	290°15'	+1°48'					
πE	h.i.	4.68						
D	4.46	333°26'	+1°14'					
A	6.06	80°32'	-1°41'					
B.M.	3.11	255°17'	+0°19'			67.43		Mon at Property Cor.

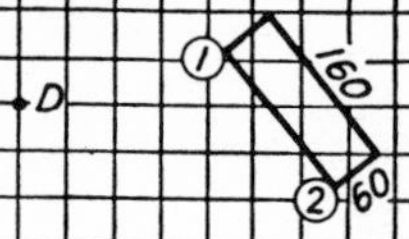

Fig. 8-21. Example of stadia notes for Prob. 8-3.

Sta	S	Azim	V⊀	H	V
△A	h.i.	4.73			
D	2.68	87° 30'	−4° 18'		
B	4.03	176° 50'	−3° 31'		
①	1.48	155° 30'	−3° 27'		
Mon BM	1.46	311° 32'	+3° 07'		
△B	h.i.	4.59			
A	4.03	356° 58'	+3° 33'		
C	3.82	70° 42'	+2° 14'		
Stream	1.47	245° 20'	−1° 41'		
Mon.	2.08	219° 50'	−0° 35'		
Stream	1.33	108° 15'	−0° 10'		
△C	h.i.	4.82			
B	3.82	250° 42'	−2° 14'		
D	3.11	338° 45'	−1° 55'		
Saddle	1.46	291° 25'	−3° 57'		
Mon.	3.02	159° 40'	+0° 06'		
Boundary	1.21	121° 50'	+3° 43'		
△D	h.i.	4.61			
C	3.13	158° 45'	+1° 53'		
A	2.70	267° 28'	+4° 18'		
②	1.19	249° 20'	+5° 08'		
Stream	1.43	55° 05'	−3° 08'		
Mon.	2.31	69° 20'	−0° 28'		

① 50 145 Bld.
Elev. 100.00

② Bld.

Fig. 8-22. Field notes for Prob. 8-4.

Suggested Field Exercises

The field exercises should, of course, consist of any of the various types of stadia surveys.

It is well to begin with a stadia traverse of three stations with one or two side shots at each station, including at least one benchmark. This usually eliminates future mistakes in more extensive stadia surveys.

9

Construction Surveys

9-1. Introduction. Once the map is available, detailed plans for the construction of a project can be completed. It is the function of the surveyor to mark out the exact locations and the important dimensions of the project according to these final plans. To accomplish this, the surveyor must mark the horizontal positions and the elevations of the construction planned—a process that is begun before the work is started and usually continues throughout the entire construction period. The surveyor must gauge her or his work so that the necessary marks are always available to the builder for each day's operation but never so far ahead of the work that the marks might be destroyed in the rough and tumble of the building process.

This function of the surveyor is called **construction surveying, location surveying,** or merely **giving line and grade.** It is a complicated process that requires many special techniques. This chapter covers the surveying procedures required for the usual type of construction. Certain surveys for highways, railways, and the like, called **route surveys,** are covered in Chaps. 11 through 13.

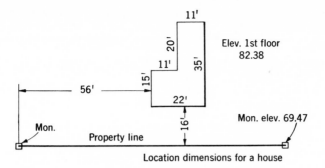

Location dimensions for a house

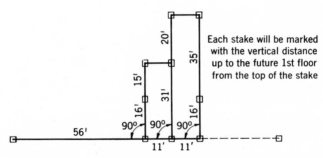

Fig. 9-1. One method of staking out a house, showing stakes set and angles and distances measured.

9-2. Methods. The plans for construction always give, either by scale or by actual dimensions, the positions and elevations of the new work relative to existing structures or to survey control marks. The dimensions of the construction shown on the plans complete the necessary data for giving line and grade.

For example, Fig. 9-1 shows the data for a house, and Fig. 9-2 shows the data for a curb. Indicated are the stakes and tacks that might be set in the two cases to mark position and elevation.

The stakes are sometimes set at the corners or other points required, later to be transferred to nearby marks which will not be disturbed by the construction and from which the construction can be located by short measurements with a carpenter's rule and level (see Figs. 9-1 and 9-2). Sometimes the marks are originally set clear of the work so that they will not be disturbed (see Figs. 9-3 and 9-4). Sometimes the positions are marked on the work itself as construction progresses.

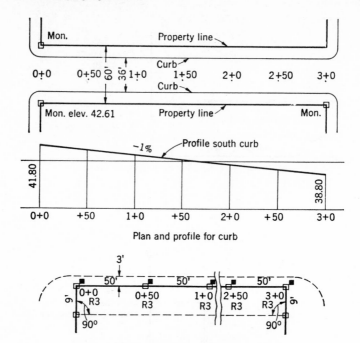

Plan and profile for curb

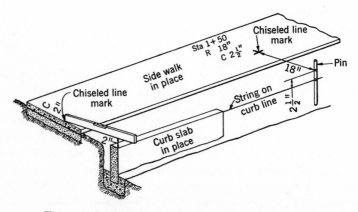

One method of placing line and grade stakes

The line stakes are set every 50 feet
3 feet back from future face of curb

Legend   Line stakes □
         Grade stakes ■

Fig. 9-2. Staking out a curb.

Fig. 9-3. Setting pins to give line and grade for curb.

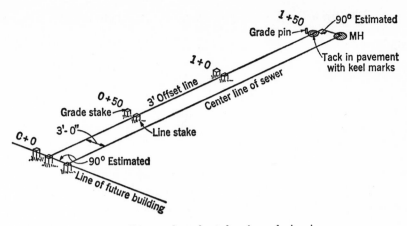

Fig. 9-4. Line and grade stakes for a drain pipe.

When it is necessary to stake out long lines for construction, the positions are numbered as for profile leveling. Each 100 ft is called a **station,** and points between are called **pluses.** Stations and pluses are marked as shown below:

Distance from beginning	Enumeration of stations and pluses
300.00	3 + 0
528.72	5 + 28.72
425.9	4 + 25.9

When a construction line must be marked so that it will not be disturbed, the marks are usually placed along a parallel line. Such a line is called an **offset** line. In looking along the true line from station 0, if the offset line is to the right, it is a right offset line. Thus a 4-ft right offset line means a line 4 ft to the right of the true line. Stakes on the offset line are marked with the station numbers of the points they are opposite and with the offset distance, thus:

$$4 + 73 \text{ R } 4$$

ESTABLISHING LINE IN THE FIELD

9-3. Marking Position. Without further examples, it is clear that the process of giving line consists in establishing predetermined angles and distances and placing a series of marks in line usually at given distances. The angles and alignment are almost invariably established with a transit, and the distances are measured with a steel tape.

9-4. Setting a Predetermined Angle.   An angle can be established by setting up the transit at the angle point, or vertex, and proceeding as follows:

1. Set the A vernier at zero, using the upper motion.
2. Point at the mark, using the lower motion.
3. Turn off the angle, using the upper motion, setting the A vernier accurately at the value of the angle.
4. Set a mark on the new line.

Obviously such an angle can be set only to the nearest half minute. When greater accuracy is desired, the angle thus established must be measured by repetition and the tack adjusted accordingly. The distance the tack must be shifted is computed by trigonometry (see Fig. 9-5). The following formula is useful and accurate if the error is not more than 3 minutes of arc:

$$D = 0.00000485SR$$

where D = distance tack is shifted
0.00000485 = number of radians in 1 second of arc
S = seconds of error
R = distance from transit to stake set

It is well to check by measuring the final angle by repetition.

9-5. Establishing Direction.   When the direction of a line is to be established either by turning an angle from a mark or by merely pointing at a mark on line, if more than one mark is available, the mark at the greatest distance from the transit should be used to establish the original direction of the line of sight. In general, the direction of a line should be established from a line longer than itself.

The transit can never be set up **exactly** over a point, nor can the signal be placed **exactly** over its mark. Obviously, the longer the line sighted, the less these errors will affect direction (see Fig. 9-6).

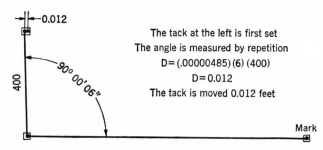

The tack at the left is first set
The angle is measured by repetition
D = (.00000485) (6) (400)
D = 0.012
The tack is moved 0.012 feet

Fig. 9-5. Establishing an accurate angle.

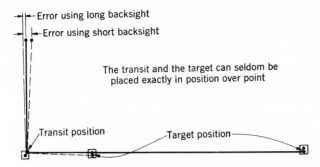

Fig. 9-6. Using a long backsight reduces error.

Movement of the Transit.   The transit is always subject to possible motion. Changes in temperature, settlement of the tripod, vibration, and readjustment of stresses in the tripod are contributing causes. Therefore, whenever a series of marks are to be set on a line, the direction of the line sight should be frequently checked by pointing at the original mark and always checked after the last mark is set.

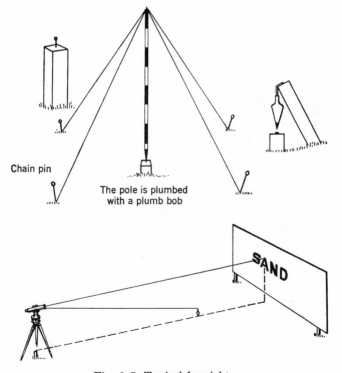

Fig. 9-7. Typical foresights.

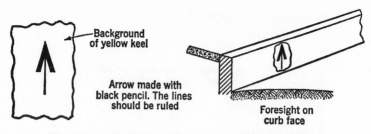

Fig. 9-8. A type of foresight that is easily established.

Use of Foresight.   It is clear that the transit must be pointed repeatedly at certain marks. When these marks cannot be seen, much time is wasted by sending someone with a plumb bob or a range pole to them whenever a sight is necessary. This can be avoided by establishing foresights for these points by one of the following methods:

1. Instead of tacks, use finishing nails driven so that the heads will remain $\frac{1}{4}$ in. above the top of the stakes.
2. Rig a plumb bob, range pole, or other device over the mark (see Fig. 9-7).
3. Choose or establish a special mark anywhere on line.

To Establish a Foresight.   After **taking line** by pointing on a plumb bob or range pole held at the mark, look for an object that happens to be anywhere on line. Letters on signboards are especially useful for this purpose. If nothing is available, choose any flat vertical surface on line. Set two pencil marks in line on this surface, one about 6 in. above the other. Using a pencil and yellow keel, construct a target that offers a precise line centered on these marks and one that is easily found and identified. See Fig. 9-8.

Permanent Construction Lines.   On large construction, important lines should be permanently marked with monuments, and permanent foresights should be built at each end.

9-6. Establishing Distance.   The procedure for taping to set marks at certain distances on line, i.e., marking line and distance, is given in Sec. 3-8.

ESTABLISHING GRADE IN THE FIELD

9-7. Marking Elevations.   Marking elevations is usually called **giving grade** or **grade staking.** It consists in setting marks such as tops of stakes, nails in vertical surfaces, and keel marks at required elevations or setting marks at random elevations and indicating the vertical heights at which the future construction is to be built above or below them. Marks for grade are usually placed near the work and transferred to the work by carpenter's levels and rules. Sometimes they can be placed on the work itself.

9-8. Definitions. The word "grade" is used loosely in surveying parlance. In this text, **rate of grade** is used to mean the steepness of slope, and **grade** is used as the equivalent of **elevation of future construction.** The use of the word "grade" alone to mean slope is avoided.

Rate of Grade. The rate of grade, often called gradient, is the rate of change of elevation expressed as a ratio of the change in elevation divided by the horizontal distance. For example, if a street sloped downward 1 ft in a horizontal distance of 100 ft, the rate of grade would be −0.01, or −1 per cent.

Cut and Fill. When the grade is above the grade mark, the notation **fill so many feet and inches** is written at the mark, thus, F 3′-6″; when below, **cut** is used, thus, C 1′-10″. The words "fill" and "cut" in this usage mean only up and down from a mark and have nothing to do with embankment or excavation.

9-9. Three Methods of Giving Grade. There are three methods of grade staking, here called **setting grade marks, shooting in grade,** and **indicating cuts and fills.**

Setting Grade Marks. The problem is to set a mark at a given grade. Starting at a benchmark, a line of levels is carried to the vicinity of the work. The instrument is thus brought into a position at a known H.I. from which the rod on the mark may be observed.

The **grade rod** (G.R.) is then determined. The grade rod is the reading on the rod that would be obtained from the present instrument position if the rod were placed on the required grade.

$$\text{G.R.} = \text{H.I.} - \text{grade}$$

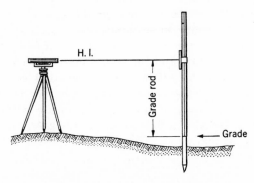

Fig. 9-9. Setting a stake at grade.

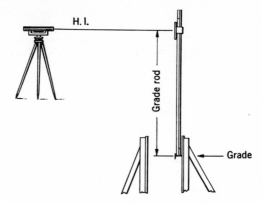

Fig. 9-10. Setting a nail at grade.

The target is set at this value. If the top of a stake is to be used for a mark, it is driven down until, when the rod is placed upon it, the target appears on the line of sight (see Fig. 9-9). The top of the stake is covered with keel and the station marked on the side. The letter G is often placed on the stake to indicate that the top is at grade.

When a grade mark is to be placed on a vertical surface, the rod is held against the surface and moved up and down until the target is on the line of sight. A mark or nail is then placed at the bottom of the rod (see Fig. 9-10).

Obviously, several grades can be set from one instrument position. The line of levels can then be carried to other locations and more grades set. Finally, the line of levels must be carried to the original or to another benchmark for a check.

Setting Grades When No Support Is Available at the Proper Elevation. Very often no support is available in the vicinity of the work on which the grade can be marked. For example, the actual grade for a pipeline or for a platform cannot be marked on the ground. Under these circumstances it is customary to set grade stakes at a certain number of half feet above or below grade, the number of half feet used being often different at different stakes, and the stakes marked accordingly.

This is accomplished by setting the target at a certain number of half feet above or below the value of the grade rod. **If the grade-rod value is larger than the rod setting, the grade will be below the top of the stake by the difference.** In this case, the stake will be marked cut, or C, so many feet. This may be stated as follows:

$$C = G.R. - rod$$

(where a negative value of C is taken as fill and marked F).

Thus, when the ground is not at the right height for setting a stake at grade, the problem is to determine how many half feet to add to or to subtract from the grade rod so that a stake may be set.

After the grade rod has been computed, a rod is read on the ground where the stake is to be driven. Obviously, when the stake has been driven and a rod placed on it, the reading must be equal to or less than this value. Therefore, the proper number of half feet is chosen such that, when applied to the grade rod, a value will be obtained that is as nearly as possible equal to, yet less than, the reading when the rod is held on the ground.

Example 1 (see Fig. 9-11). The H.I. is 55.28; the grade is 46.94.

$$\text{G.R.} = 55.28 - 46.94 = 8.34$$

Accordingly, values like 7.34, 7.84, 8.34, 8.84, 9.34, etc., can be used.
The rod on the ground is 3.5.
Choose 3.34 as the nearest to 3.5 and yet less than it. Set the target at 3.34.
Compute.

$$C = 8.34 - 3.34 = 5'0''$$

Example 2 (see Fig. 9-12). The H.I. is 48.52; the grade is 42.27.

$$\text{G.R.} = 48.52 - 42.27 = 6.25$$

Accordingly, values like 5.25, 5.75, 6.25, 6.75, 7.25, etc., can be used.
The rod on the ground is 9.9.
Choose 9.75 as the nearest to 9.9 and yet less than it. Set the target at 9.75.
Compute.

$$C = 6.25 - 9.75 = -3'6'' \qquad \text{or} \qquad F = 3'6''$$

Procedure for Setting Grade Stakes. The following procedure is recommended for the method of setting grade stakes described above (see Fig.

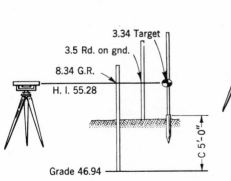

Fig. 9-11. Setting a grade stake when the ground is too high.

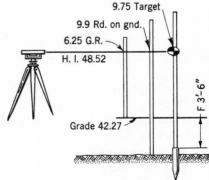

Fig. 9-12. Setting a grade stake when the ground is too low.

SMITH ST. GDS. FOR N. CURB					
Sta	*+*	*HI*	*−*	*Rod*	*Elev*
BM#5	5.02	62.27			57.25
TP#1	0.27	55.28	7.26		55.01
0+0				3.34	51.94
1+0				4.42	50.86
TP#2	2.66	48.52	9.42		45.86
2+0				9.75	38.77
BM#6			7.17		41.35
	7.95		23.85		

π Brown — Clear
Rec Jones — 60°F
Rod King Stks Hall — Date

Maple and William Sts., Fire Hydrant "R" in Corey

```
                    55.28 HI              Gnd 3.5
C 5'-0"            −46.94 Gd
                    8.34 Gd. Rod   55.28      4.5
C 6'-6"                          −44.36
                   48.52         10.92       9.9
                  −42.27
                    6.25
F 3'-6"
                                  7.95
                          Check −23.85
                                −15.90
                                −57.25
                                 41.35
```

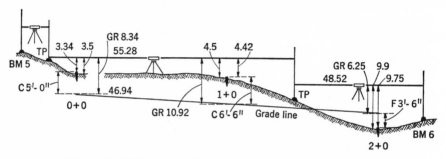

Fig. 9-13. Field notes and procedure for setting grade stakes.

9-13). Benchmark data, a list of required grades, and a sketch of the work must be taken into the field.

List of Grades

Station	*Grade*
0 + 0	46.94
1 + 0	44.36
2 + 0	42.27

1. By benchmark leveling, obtain an H.I. in the vicinity of the work, 55.28.
2. Compute G.R. on the right-hand page of the field notes, 8.34.
3. Take a rod on the ground at the stake location, and record on the right-hand page, 3.5.
4. Set the target at 3.34, and drive a stake until, when the rod is on stake, the target is on the line of sight.
5. Read the rod through the target to check the target setting, and record in the rod column, 3.34.

6. Compute the elevation of the stake, and record the elevation in the elevation column, 51.94.

7. Compute cut or fill by two methods as a check:

$$C = G.R. - rod \qquad 8.34 - 3.34 = 5.00$$
$$C = elev. - grade \qquad 51.94 - 46.94 = 5.00$$

8. Record cut or fill in feet and inches on the right-hand page, and mark it on the stake. Keel the top of the stake. If there is no cut or fill, mark the stake G.

9. Set any other grades possible from the present H.I., or carry levels to another H.I., finally checking on the B.M.

Shooting in Grade. When marks are to be set for a uniform rate of grade, computation and field work can be saved by a process known as shooting in grade.

This process is not independent. First it is necessary to set a grade stake or mark at each end of the uniform grade. A transit or level is then set up over the mark at one end (see Fig. 9-14). The difference in height between the instrument and the mark is measured (4.07), and the target on the rod is set at this value. The rod is held at the mark at the other end of the slope and the line of sight directed at it. This places the line of sight parallel to the grade line at a known height above it. With this arrangement, a grade mark can be set wherever desired by holding the rod at that point and raising or lowering the rod until the target is on the line of sight. The position of the foot of the rod is marked.

A foresight on the line of sight should be established if many grade marks are to be set. When only a few are necessary, the slope of the

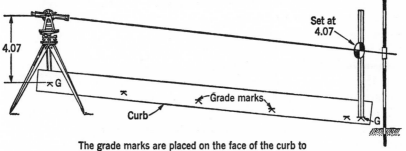

The grade marks are placed on the face of the curb to indicate the grade of the gutter.
The marks labeled G are first established at an established grade by the usual method of setting grade marks.
The foresight shown consists of a piece of paper wrapped around a range pole and held by an elastic

Fig. 9-14. Shooting in grade.

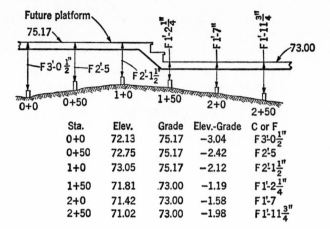

Fig. 9-15. Giving grade by indicating cut or fill.

Sta.	Elev.	Grade	Elev.-Grade	C or F
0+0	72.13	75.17	−3.04	F 3'-0½"
0+50	72.75	75.17	−2.42	F 2'-5
1+0	73.05	75.17	−2.12	F 2'-1½"
1+50	71.81	73.00	−1.19	F 1'-2¼"
2+0	71.42	73.00	−1.58	F 1'-7
2+50	71.02	73.00	−1.98	F 1'-11¾"

line of sight should be checked when the work is completed by holding the rod at the original mark and making sure the line of sight strikes the target.

Giving Grade by Indicating Cuts and Fills. The most rapid and in many ways the best method of giving grade is to indicate the cuts or fills measured from convenient objects near the work. Usually the tops of the line stakes or other line marks are used.

The elevations of the tops of the line stakes or the objects chosen are determined by profile leveling. The values of the cuts or fills are computed by comparing the elevation of each mark with the grade at that particular position. They are computed in hundredths of a foot, reduced to inches, and marked on the stakes or near the marks (see Fig. 9-15). The tops of the stakes or other objects are usually covered with keel to indicate that grade should be measured from those points.

Reducing Hundredths to Inches. One inch equals $8\frac{1}{3}$ hundredths of a foot. The quarters of a foot can be expressed exactly both in hundredths and in inches. By adding or subtracting 8 to or from the nearest quarter point, the inch values in hundredths of a foot can be computed within one-third of a hundredth. These values should be computed and memorized, as given in Table 9-1.

To reduce hundredths to inches, choose the nearest inch value and correct for the odd hundredths by calling them eighths of an inch. Thus,

$$0.89 \text{ ft} = 0.92 \text{ ft} - 0.03 \text{ ft} = 11 \text{ in.} - \tfrac{3}{8} \text{ in.} = 10\tfrac{5}{8} \text{ in.}$$
$$0.44 \text{ ft} = 0.42 \text{ ft} + 0.02 \text{ ft} = 5 \text{ in.} + \tfrac{2}{8} \text{ in.} = 5\tfrac{1}{4} \text{ in.}$$
$$0.71 \text{ ft} = 0.75 \text{ ft} - 0.04 \text{ ft} = 9 \text{ in.} - \tfrac{4}{8} \text{ in.} = 8\tfrac{1}{2} \text{ in.}$$

The error is never greater than 0.005 ft.

Table 9-1

In.	Quarter points	Computations	In. values, hundredths of ft
0	0		0
1		0 + 8	8
2		25 − 8	17
3	25		25
4		25 + 8	33
5		50 − 8	42
6	50		50
7		50 + 8	58
8		75 − 8	67
9	75		75
10		75 + 8	83
11		100 − 8	92
12	100		100

9-10. Signals for Giving Grade.   The only signals used for giving grade that are not used for profile leveling are "up" and "down." Up is signaled by moving the hand upward from shoulder height, usually with the index finger pointed up. Down is signaled by lowering the hand from waist height, with the index finger pointed down. Large, slow motions indicate large amounts, and vice versa. Usually the estimated distance is signaled immediately afterward in hundredths of a foot.

EXAMPLES OF FIELD METHODS

9-11. Batter Boards.   When the surveyor has set marks for line and grade, it is often necessary to use string or wire to guide the actual construction. These are usually supported on pins or **batter boards** (see Figs. 9-16 and 9-17).

The method of using batter boards for buildings has become standardized. They are usually so designed that they will support the string or wire so that it marks the exterior face of the building and also the elevation of the first floor. The exterior face of a building is a technical term. Figure 9-18 illustrates various building faces.

Sometimes the surveyor indicates the fill from each corner stake up to the first-floor elevation. In this case the contractor adjusts the wire, using a plumb bob to set the alignment and a rule to measure up from the stake. Sometimes line marks are transferred from the stakes to the batter boards with a transit, and the grade of the first floor is marked directly on the batter boards (see Fig. 9-16).

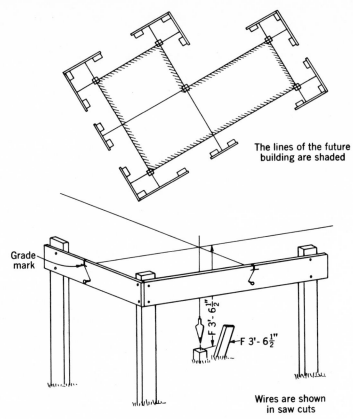

The lines of the future
building are shaded

Grade
mark

F 3'- 6½"

F 3'- 6½"

Wires are shown
in saw cuts

Fig. 9-16. Batter boards and wires in place over original stakes.

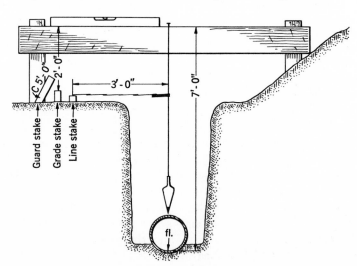

C 5'-0"

2'-0"

3'-0"

7'-0"

Guard stake

Grade stake

Line stake

fl.

Fig. 9-17. Batter board for a pipeline.

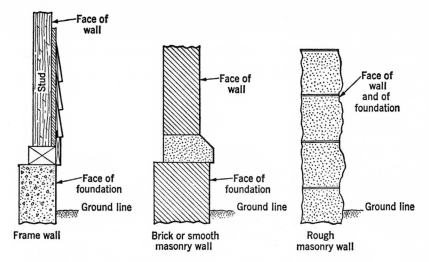

Fig. 9-18. Typical walls. Measurements to buildings are made to the face of a wall or to the face of the foundation and so noted.

9-12. Drainage Terms. Certain technical terms are used in connection with drainage facilities. **Flow line** is the bottom inside of a drainage pipe. **Invert** is the bottom inside of a drainage channel. Drainage manholes usually contain drainage channels (see Fig. 9-19). Inverts are sometimes also called flow lines. Flow lines and inverts are the lines always used in referring to elevations and alignment for drainage. They are often abbreviated to f.l. and Ivt.

9-13. Field Location. Often a project is so simple that the entire engineering process can be carried out in the field. The reconnaissance,

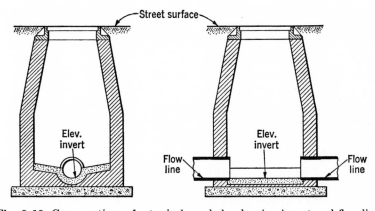

Fig. 9-19. Cross sections of a typical manhole, showing invert and flow line.

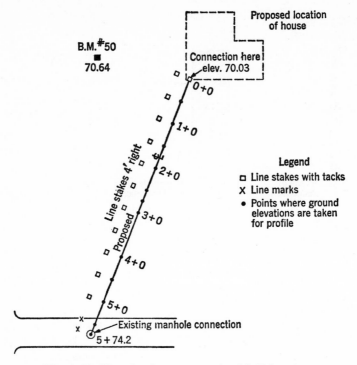

Fig. 9-20. Plan showing an example of field location.

preliminary survey, map, plan, and location survey can be executed in a few hours. The process is illustrated by the following problem.

Problem of Field Location.   Suppose that it is necessary to build a branch drain from a house to the main street drain. It is assumed that the f.l. must be at least 3 ft below the ground surface to prevent freezing. The minimum rate of grade of the f.l. should be 0.004. No breaks in rate of grade or direction should occur except at manholes, for sediment collects at such points.

Figure 9-20 illustrates the problem. The elevation of the flow line at the house as indicated on the architect's plan is 70.03, and a connection already exists at the street manhole as shown by the city records. The elevation of the connection must be determined during the preliminary leveling by opening the manhole cover and observing a rod held on the flow line.

Outline of Method.   It is evident at once from the reconnaissance that the line can be straight. An investigation by a preliminary survey must be made to determine whether or not a straight grade line can be used without bringing the flow line too near the surface. A profile of the ground

Table 9-2. Design of Grades for Field Location. (See Fig. 9-20)

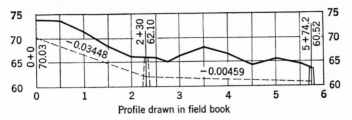

Profile drawn in field book

Computations of Grades

	Station	Grade	Station	Grade	Grade used
Start........	0 + 0	70.03	0 + 0	70.03	70.03
End.........	2 + 30	62.10		− 1.724	
Diff.........	230	− 7.93	+ 50	68.306	68.31
				− 1.724	
Rate $= \dfrac{-7.93}{230} = -.03448$			1 + 0	66.582	66.58
				− 1.724	
			+ 50	64.858	64.86
				− 1.724	
Change in grade = distance × rate			2 + 0	63.134	63.13
50(−0.03448) = −1.724				− 1.034	
30(−0.03448) = −1.034			+ 30	62.10	62.10
				− 0.092	
	Station	Grade	+ 50	62.008	62.01
				− 0.230	
Start........	2 + 30	62.10	3 + 0	61.778	61.78
End.........	5 + 74.2	60.52		− 0.230	
Diff.........	344.2	− 1.58	+ 50	61.548	61.55
				− 0.230	
Rate $= \dfrac{-1.58}{344.2} = -0.00459$			4 + 0	61.318	61.32
				− 0.230	
			+ 50	61.088	61.09
				− 0.230	
Change in grade = distance × rate			5 + 0	60.858	60.86
20(−0.00459) = −0.092				− 0.230	
50(−0.00459) = −0.230			+ 50	60.628	60.63
24.2(−0.00459) = −0.111				− 0.111	
			+ 74.2	60.52	60.52

is run and plotted (see Table 9-2). This constitutes the map. On this profile, a straight line representing a possible flow line is drawn from the known elevation (70.03) of the flow line at the house to any point not below the manhole connection (60.52). It is discovered that such a line comes too near the ground. Other flow lines are tried with various locations and elevations for breaks in rate of grade, the object being to find an arrangement that complies with the specifications and requires a minimum quantity of excavation and number of manholes. In this case a break in the rate of grade of the flow line located at about station 2 + 30 at an elevation of about 62.1 (as indicated by scaling) will solve the problem. It will require one new manhole (at 2 + 30). The existing connection at the street manhole can be used. Its f.l. elevation is 60.52.

It is now necessary to compute grades for the intervening points. The grades must be such that they will produce absolutely straight slopes for the flow line. For this purpose an **exact** position and elevation must be assumed for the invert of the new manhole. Accordingly, station 2 + 30 and elevation 62.10 are chosen, and the grades are computed by proportion. This completes the plan (see Table 9-2).

It is decided to give grade by indicating the cut from the top of the line stakes. It is to be remembered that cut is the distance from the top of the line stakes down to the flow line. It is **not** the excavation, which would be the distance from the ground down to the bottom of the trench.

To indicate cuts, the elevations of the tops of the line stakes must be found by leveling and the individual cuts computed by subtracting the required grades.

It is also decided to place the line stakes at a 4-ft offset to prevent disturbance when the trench is excavated.

With the above in mind, the procedure (the location survey) is planned to require a minimum of field work.

Field Procedure for Field-location Problem. The field steps are the following:

1. Stake out a 4-ft offset line, placing stake 0 + 0 beside the point in the house where the house connection is located and a stake every 50 ft thereafter. Carry the measurement to a point beside the manhole, and determine its plus.
2. Find the elevation of the ground at each 50-ft point along the true line and at all breaks in ground slope. The rod is held on the ground at an estimated 4 ft from and opposite to each offset stake. This places the rod at the true position on the construction line. The rod is read to tenths.
3. At the same time, determine the elevation of the tops of each of the offset-line stakes. On these the rod is read to hundredths.

Sta	+	HI	-	Rod	Elev	Grade	Cut	Mark Stk.								
HOUSE CONNECTION						Chief Smith / π Jones		H.C. Cole / R.C. Doe	Fair / 60° / Date							
BM#50	6.78	77.42			70.64	Nail in	Maple	Near House								
0+0 S				3.15	74.27	70.03	4.24	C 4'-2 7/8"								
G				3.2	74.2											
+50 S				4.00	73.42	68.31	5.11	C 5'-1 3/8"								
G				4.5	72.9											
1+0 S				5.41	72.01	66.58	5.43	C 5'-5 1/8"								
G				6.0	71.4											
+50 S				9.15	68.27	64.86	3.41	C 3'-4 7/8"								
G				9.3	68.1											
2+0 S				11.04	66.38	63.13	3.25	C 3'-3"								
G				11.1	66.3											
TP#1	4.03	70.50	10.95		66.47											
+50 S				4.39	66.11	62.01	4.10	C 4'-1 1/4"								
G				4.5	66.0											
+75 G				5.5	65.0											
3+0 S				4.07	66.43	61.78	4.65	C 4'-7 7/8"								
G				4.1	66.4											
+50 S				2.35	68.15	61.55	6.60	C 6'-7 1/4"								
G				2.5	68.0											
4+0 S				4.18	66.32	61.32	5.00	C 5'-0"								
G				4.1	66.4											
+50 S				6.13	64.37	61.09	3.28	C 3'-3 3/8"								
G				6.2	64.3											
5+0 S				5.22	65.28	60.86	4.42	C 4'-5"								
G				5.2	65.3											
+50 S				5.90	64.60	60.63	3.97	C 3'-11 5/8"								
G				5.9	64.6											
+74.2 S				6.90	63.60											
G				6.9	63.6											
Connect				9.98	60.52											
TP#2	5.89	72.92	3.47		67.03											
BM#50				2.29	70.63											
BM#50	7.42	78.06			70.64											
2+30 S				11.81	66.25	62.10	4.15	C 4'-1 3/4"								
BM#50			7.42		70.64											

Fig. 9-21. Field notes for field location.

4. Draw the profile of the ground elevations, and determine the grade profile for the flow line.
5. Compute the cuts, and mark the stakes.
6. Measuring along the offset line, place a stake for the new manhole, find the elevation of the top of the stake set, and mark the cut for the invert.

The form of notes is shown in Fig. 9-21.

SPECIAL ALIGNMENT OPERATIONS

9-14. Location-survey Operations. The five operations described in the following paragraphs are chiefly useful in establishing line, i.e., in the location survey.

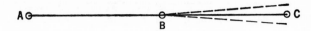

Fig. 9-22. Method of double centering.

Double Centering. When a line such as *AB* is to be prolonged from *B* to *C*, the transit may be set up at *A* and pointed at *B*, and *C* may be set on line (see Fig. 9-22).

This method is unsatisfactory for a long prolongation, for the point *C* may be too far away to be set accurately, or rolling ground may interfere.

The usual method, therefore, is to set up the transit at *B*, point at *A* with the telescope reversed, then transit the telescope to its direct position and set *C*. If the transit is in adjustment, this method will give correct results. If the transit is out of adjustment, particularly if the line of sight is not perpendicular to the horizontal axis, this method will not produce a straight line.

If the transit is not known to be in adjustment, the operation described must be repeated with the telescope in the opposite positions to those used before. *A* is pointed with the telescope **direct** and *C* set with it **reversed.** This will result in a second mark for *C* if the transit is out of adjustment. The final point *C* is then set halfway between the two marks.

The process is known as **double centering.** It is difficult in the field when the surveyors are out of earshot of one another. Each one must have a clear idea of exactly what is being accomplished during each operation.

To Buck in, between Two Points. Frequently it is necessary to establish a point on line between two marks when it is impossible to set up over either of them. This occurs often in shop alignment and in the field when a point must be set on a hill that intervenes between the two marks.

In Fig. 9-23 assume that it is required to set a point *C* between the marks *A* and *B*. Set up at *C'* approximately on line. Choosing the most distant mark, *A*, point on *A* reversed, transit, and set *B'*. Measure

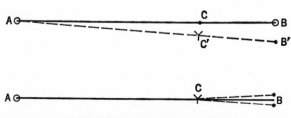

Fig. 9-23. Bucking in over a hill.

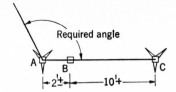

Fig. 9-24. To set a point close to the transit.

B' to B. Estimate the ratio AC/AB, and move the transit from C' to C, computing this distance as follows:

$$\frac{C'C}{B'B} = \frac{AC}{AB}$$

Repeat the procedure until B' falls on B. When $B'B$ becomes small, the position B' must be established each time by double centering. When the direct and reversed shots are equally spaced on each side of B, the transit is on line, and C can be set under the plumb bob.

To Set a Point near a Transit. The telescope cannot be lowered far enough or focused close enough to set a point on line nearer the transit than at about 4 ft. At less than 4 ft the following is necessary (see Fig. 9-24).

Set up at A, it being required to set B. Set a point C on the proper line. Set up on C. Point A, and set B on line.

To Set a Point of Intersection (PI). Frequently it is necessary to establish a point at the intersection of two lines, for example, the lines AB and CD (see Fig. 9-25).

Set up at C. Set E by double centering, and point on the final position of E. Set F on line. E and F should be as near together as possible and yet lie one on each side of the prolongation of AB.

Tie a string from E to F.

Set up at B, and set a stake G on the line AB prolonged and also under the string. The stake should be driven down until the top just touches

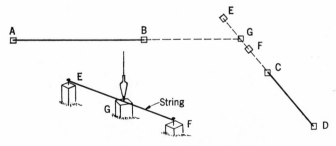

Fig. 9-25. To set a PI (point of intersection).

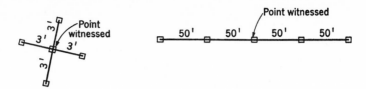

Fig. 9-26. Methods of witnessing a point.

the string. Draw a pencil line on the top of the stake just under the string. Find the exact point of intersection on the pencil line by double centering from *B*.

Witnessing a Mark.   Frequently it is necessary to make arrangements so that a mark can be easily replaced if disturbed. Two methods are indicated in Fig. 9-26. The supplementary marks are called **witnesses** or **witness marks,** and any ties are called **witness measurements.** It is well to use a method that will serve even if only two witness marks remain.

OBSTACLES TO LINE AND LEVELS

9-15. Obstacles to Measurement but Not to Line.   Water or other obstacles to measurement but not to sight are crossed by triangulation. The measured base should be about as long as the computed distance (see Fig. 9-27). All the angles should be measured, checked, and adjusted so that their sum is 180°.

9-16. Right-angle Offset.   In staking out a line, it is often necessary to carry distance and direction accurately beyond a small obstacle, as from *AB* to *CD* in Fig. 9-28.

At *B* turn 90°, and set *E* at a convenient number of feet. At *E* point a swing offset from *A*, and set *G* by double centering. Point *G*, and set *F* at a convenient number of feet. At *F* turn 90°, and set *C* so that *FC* = *BE*. At *C*, point on a swing offset from *G*. Set *D*. While this method is simple and accurate, it takes about 2 hr and should be avoided if a quicker method is available.

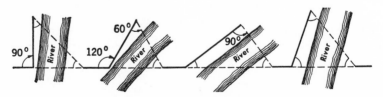

Fig. 9-27. Triangulation to cross a river.

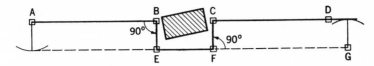

Fig. 9-28. A rectangular offset.

9-17. Parallel Offset to Obstructed Line. A property or construction line is often marked at both ends, but the entire length is obstructed. A parallel offset line is usually established and used instead (see Fig. 9-29).

Set C by estimating a position opposite A. Point a swing offset equal to AC at B, turn 90°, and measure a swing offset at A (usually a very small distance).

If the swing offset from A is large, move C back to C' and repeat the process. If it is small, add the value to measurements along the offset line from C.

9-18. Random Line. When a parallel offset line is impossible, a random line can be used to establish line points between the ends of an obstructed line (see Fig. 9-30).

Set C at random but visible from A. Measure the angle at C-AB. Compute any other desired positions as D, E, from the proportion:

$$\frac{FD}{AF} = \frac{GE}{AG} = \frac{CB}{AC}$$

Exaggerated error in positioning C

Fig. 9-29. To establish a line parallel to an obstructed line.

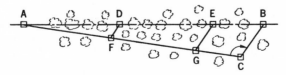

Fig. 9-30. To establish an obstructed line by a random line.

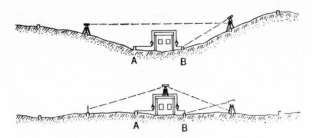

Fig. 9-31. To measure over an obstacle.

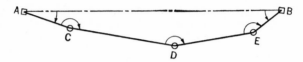

Fig. 9-32. Random traverse. Choose a coordinate system so that *CD* is due east.

and stake out using the same angle, i.e.,

$$F\text{-}AD = G\text{-}AE = C\text{-}AB$$

9-19. High Obstacle on Line. Frequently an obstacle can be avoided by setting a point on high ground from which a line may be established over it or by setting a station on it. Distance can be carried over the obstacle by long plumb bobs, slope measurements, or triangulation (see Fig. 9-31).

9-20. Random Traverse. Of course, any obstructed line can be replaced by a traverse. The length and direction of the obstructed line can then be computed by trigonometry, traverse-computation technique being used if desirable. Sometimes time can be saved by orienting the coordinate system so that it coincides with one of the lines of the random traverse (see Fig. 9-32).

9-21. Measuring Vertical Clearance. To determine vertical clearance or the elevation of a point above the H.I., the rod can be used upside down. The rod reading must be given the opposite sign to that ordinarily used. It is often called a **minus rod** (see Fig. 9-33).

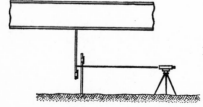

Fig. 9-33. To measure a vertical clearance.

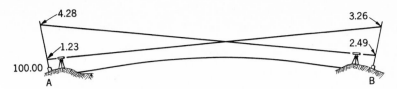

Fig. 9-34. Leveling across a wide river.

9-22. Reciprocal Leveling. When it is necessary to carry levels over a body of water, the plus sight and minus sight necessarily have different horizontal lengths. This introduces the instrument errors and, in long sights, the effect of earth curvature. To eliminate these errors, the levels should be carried from the mark on one side to the other by two instrument setups, one on each side of the body of water. This will result in two elevations for *B*. The average is used (see Fig. 9-34).

Sta.	+	H.I.	−	Rod elev.
B.M. *A*	1.23	101.23		100.00
B.M. *B*			3.26	97.97
B.M. *A*	4.28	104.28		100.00
B.M. *B*			2.49	101.79
		97.97		
		101.79		
		2 ⎸199.76		
	Average	99.88 adopted elev. of *B*		

9-23. Trigonometric Leveling. Often it is necessary to determine the elevation of an inaccessible point (see Fig. 9-35). Trigonometric leveling is used. An allowance for the refraction of the air and earth curvature is necessary when the distances are great.

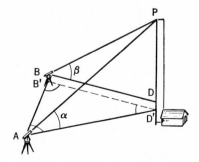

Fig. 9-35. To determine the height and position of a high point.

The length of the base AB and the angles shown are measured. The elevations of A and B are determined; then,

$$D' = 180° - (A + B')$$

$$AD' = \frac{AB'}{\sin D'} \sin B' \qquad D'P = AD' \tan \alpha$$

$$B'D' = \frac{AB'}{\sin D'} \sin A \qquad DP = B'D' \tan \beta$$

$$\text{Elev. } P = \text{elev. } A + D'P = \text{elev. } B + DP$$

9-24. Supplementary Uses of Laser.   A laser beam is a bright beam of red light. It spreads very little so that it retains its power for a long distance. It can be made so powerful that it will cut into iron or with so little power that it can be used safely. Like all other light beams, it cannot be seen unless it strikes some object like dust or a surface of some kind. On a flat surface it makes a bright red dot.

It can be used to give line. Once aimed at a foresight, the position of the line can be found without using another person (at the instrument). Laser is therefore especially useful when many points must be set on line, particularly when they are set at irregular location intervals.

This method is chiefly used to align pipelines for line and grade, to place dredges for bridge or tunnel construction, and to control the position of the concrete forms, pile clusters, and other pier features. On land it can be used for line and grade especially for large areas either level or at a required slope. In this case the beam is pointed upward, and a right-angle prism (a pentaprism) is rotated above it so that the beam generates a plane which can be level or sloped as desired. See Fig. 9-36.

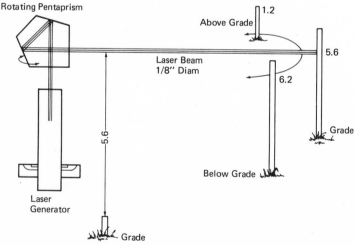

Fig. 9-36. Laser used to establish a horizontal plane.

Also, the whole device can be rotated so that the beam generates a vertical plane for aligning walls, columns, and the like.

Some devices indicate at a distance where the beam strikes in relation to a target if within an inch or two of the center.

In every case, one person is eliminated.

Problems

9-1. To set a mark distant 400 ft accurately at 90° from a line, the angle was measured by repetition and found to be 89°59'50". How far should the mark be moved?

9-2. Same as Prob. 9-1 except that the distance is 500 ft and the measured angle is 89°59'53".

9-3, 9-4. Compute the grades for each half station and station for a uniform rate of grade between the positions indicated.

	Station	*Grade*
9-3.	0 + 00	29.68
	6 + 73.41	34.25
9-4.	6 + 29.7	51.26
	12 + 16.5	72.49

9-5, 9-6. Convert the following feet and decimals into feet and inches:

9-5		9-6	
2.69	5.60	3.52	6.25
4.79	3.87	4.76	7.81
8.21	1.83	9.23	2.94
7.93	0.36	10.16	5.06
6.08	9.27	8.72	6.67

9-7, 9-8. Convert the following feet and inches to feet and hundredths of a foot:

9-7		9-8	
7' 2½"	3' 5¼"	2' 6¾"	4' 6⅛"
4' 9¾"	8' 8⅝"	1' 10⅛"	7' 2⅞"
5' 7⅝"	9' 4⅛"	3' 7⅝"	5' 4¾"
6' 4⅞"	2' 6½"	6' 3½"	8' 7⅜"
4' 3⅜"	10' 7¾"	9' 8½"	10' 5¼"

9-9, 9-10. Compute the cuts and fills to be written on the marks for the data given:

	9-9				9-10	
Station	Grade	Elev. mark		Station	Grade	Elev. mark
0 + 0	35.64	35.27		0 + 0	47.28	46.17
0 + 50		39.42		0 + 50		41.62
1 + 0		46.25		1 + 0		45.10
1 + 50	Uniform	47.31		1 + 50	Uniform	40.83
2 + 0	Slope	46.22		2 + 0	Slope	36.15
2 + 50		47.38		2 + 50		42.14
3 + 0		55.20		3 + 0		34.75
3 + 50		59.71		3 + 50		35.29
4 + 0		59.64		4 + 0		32.67
4 + 50	62.64	64.28		4 + 50	29.28	33.48

9-11, 9-12. Write out the form of notes with consistent numbers for setting grade stakes at a certain number of half feet above or below grade, as in Fig. 9-13, for the following:

	9-11				9-12		
H.I.	Station	Grade	Rod on ground	H.I.	Station	Grade	Rod on ground
37.28	0 + 0	32.61	8.2	81.29	0 + 0	80.32	1.4
	0 + 50	33.01	5.4		0 + 50	81.32	1.0
	1 + 0	33.41	2.3		1 + 0	82.32	0.6
	1 + 50	33.81	1.7		1 + 50	83.32	1.7
39.46	2 + 0	34.21	3.5		2 + 0	84.32	1.9
	2 + 50	35.61	4.7	92.42	2 + 50	85.32	2.8
	3 + 0	36.01	5.6		3 + 0	86.32	3.2
	3 + 50	36.41	7.2		3 + 50	87.32	4.2
	4 + 0	36.81	9.7		4 + 0	88.32	6.7
	4 + 50	37.21	10.6		4 + 50	89.32	7.8

Suggested Field Exercises

9-1. See Fig. 9-37. On a control line marked by a stake and tack (*A*) and a distant, sharply defined object, mark a 100-ft length *BC*. From each end of this length turn off 64°09′29″ by the precise method and establish marks 500 ft distant on these parallel lines (*D* and *E*).

From *E* set *F*.

Check *DE* equals 100 ft, *DF* equals 90 ft.

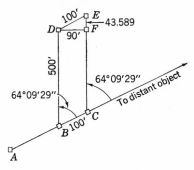

Fig. 9-37. Layout for suggested Field Exercise 9-1.

9-2. Stake out a six-sided building from a pair of stakes representing a monument line and determine the *F* to a given elevation representing the first-floor elevation.

9-3. Stake out 50-ft points on a traverse containing one angle point at an odd plus. Take a profile. Find three contours each side with a hand level.

Part 2

ADVANCED PROCEDURES

10

The Theodolite

10-1. Introduction. The American transit has one serious fault: it is difficult to read the circles precisely with the verniers. As a result, in order to measure or stake out an angle accurately, many repetitions must be used.

For some time many European instrument manufacturers have been turning out transits, usually spoken of as **theodolites,** having optical devices that make it possible to read the circles much more precisely than is possible with American instruments. In addition, they can be read more quickly and with less chance of a blunder. The **precision** varies usually from ± 1 second read directly on some instruments to ± 6 seconds read by estimation on others. Since no circle and no reading device can be perfect, the **accuracy** of measuring an angle by turning it once varies from instrument to instrument and from the products of one manufacturer to those of another. In general, measuring an angle once with one of these instruments will give an accuracy of ± 5 to 15 seconds.

When higher accuracy is desired, a program of observation must be set up that tends to eliminate the errors of the circle and the reading device. Usually such a program requires less field time than the system of repetition required to attain the same accuracy with an American transit.

Many of the reading devices give the average of the readings of two points on the circle 180° apart and thus automatically eliminate the errors due to eccentricity in the circle graduations.

In addition to an ordinary plumb bob, a theodolite has a so-called **optical plumb bob.** This is a small telescopic sight mounted in a vertical hole through the spindle and adjusted to coincide with the azimuth axis of the alidade. It is viewed through a horizontal eyepiece usually at the side of the alidade. After the theodolite is leveled, the optical plumb bob shows the position of the instrument with respect to the tack or other mark. Since the device does not swing, it is unaffected by wind. After the theodolite has been placed in position with the ordinary plumb bob and leveled, the position is checked with the optical plumb bob. However, an experienced transitman can set up the theodolite with the optical plumb bob alone.

In spite of the fact that theodolites have peculiarities of their own (described later), since these instruments usually save time and give higher accuracies than American transits, they are coming into general use in the United States and Canada.

Figure 10-1 shows a Wild T-2 theodolite with the names of the various parts. There are minor differences among theodolites. The following description gives general directions for their operation. The manufacturer's instructions should be consulted for any specific instrument. See Fig. 10-1 for the locations of the parts named.

10-2. To Set Up a Theodolite.   Place the instrument on the tripod. Underneath the tripod head is a screw with a knurled head called the **centering screw** that holds the instrument on the tripod. It must be screwed into the base of the instrument. If this is forgotten, the instrument will fall off the tripod when the instrument is picked up. When this screw is loosened, the instrument can be shifted laterally. Center the instrument on the tripod head and tighten the screw.

If the plumb bob is to be used, attach it and place the instrument over the point. Level the instrument approximately according to the circular level with the three leveling screws, as with a three-screw leveling instrument. See Sec. 6-11.

Turn the alidade so that the tubular (horizontal) level is parallel with a pair of leveling screws. Center the bubble with both screws, as with a four-screw leveling head. Turn the alidade 90°. Center the bubble with the one screw that was not used before. Repeat until the bubble centers in both positions.

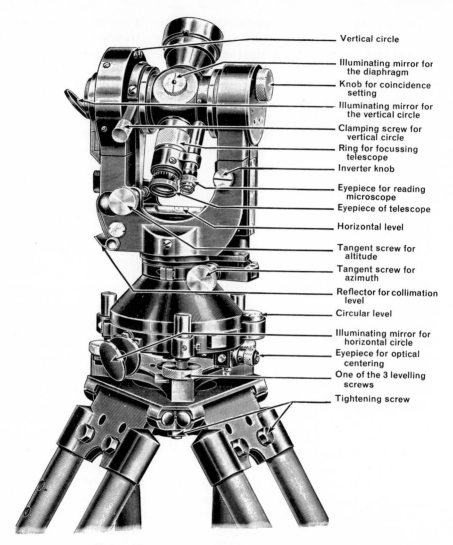

Vertical circle

Illuminating mirror for the diaphragm

Knob for coincidence setting

Illuminating mirror for the vertical circle

Clamping screw for vertical circle

Ring for focussing telescope

Inverter knob

Eyepiece for reading microscope

Eyepiece of telescope

Horizontal level

Tangent screw for altitude

Tangent screw for azimuth

Reflector for collimation level

Circular level

Illuminating mirror for horizontal circle

Eyepiece for optical centering

One of the 3 levelling screws

Tightening screw

Universal-Theodolite WILD T 2

Fig. 10-1. A Wild T-2. This is a second-order, optical-reading theodolite that reads directly to seconds.   (*Wild Heerbrugg Instruments, Inc.*)

Focus the optical plumb bob. If its cross hairs are not centered over the point, loosen the centering screw and shift the instrument over the point. Try to accomplish this without turning the leveling head in azimuth, as this throws the instrument out of level.

Tighten the centering screw and relevel if necessary. If releveling is necessary, again check the centering with the optical plumb bob. Repeat the centering and leveling process until both are satisfactory. Be sure to remember that the optical plumb bob is accurate **only when the instrument is level.**

10-3. To Set Up without the Ordinary Plumb Bob. Place the instrument over the point by eye. Focus the optical plumb bob. With the leveling screws, aim the optical plumb bob at the point. Adjust the lengths or positions of the tripod legs so that the circular bubble centers. Complete the centering and leveling as before.

10-4. To Illuminate the Circles. While looking through the reading microscope eyepiece, which is beside the telescope, adjust the mirrors for the horizontal and the vertical circles so that the field is bright. At night small electric lights are attached. Focus the reading microscope. On some instruments both the horizontal and the vertical circles are in the field of view. On others, like the one shown, an **inverter knob** is turned to select one or the other. Focus the eyepiece so that the graduations are clear.

10-5. The Motions. The movements of the alidade and the circle are quite different from those of a transit. Figure 10-2 shows the arrangement of the bearings of the azimuth axis. Note that both the circle and the alidade turn on their own separate bearings in the leveling head. A light friction between the circle and the leveling head keeps the circle from turning. The alidade turns freely.

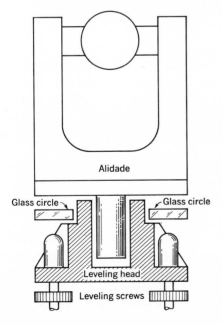

Fig. 10-2. Schematic drawing showing the arrangement of the vertical axis of a theodolite. The bearings have been opened up for clarity.

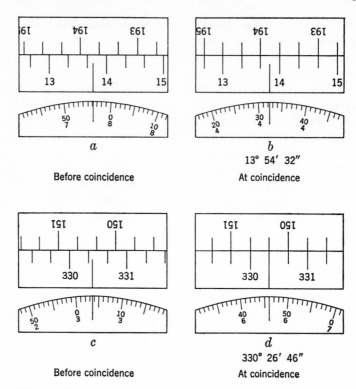

Fig. 10-3. Field of view in Wild T-2 optical micrometer for reading circles.

In most theodolites, like the one shown in Fig. 10-1, there is only one clamp and tangent screw on the azimuth axis. It connects the alidade with the leveling head. When necessary, if there is no other clamp, the circle can be rotated rapidly with respect to the leveling head by engaging a special finger-operated wheel.

The vertical motion is operated in the usual way by a clamp and tangent screw.

10-6. To Measure a Horizontal Angle. Using the clamp and tangent screw, aim at the left-hand point. Read the circle. Aim at the right-hand point and read the circle. Compute the angle from the difference in the readings.

10-7. To Read the Circle. Look through the eyepiece reading microscope and focus on the scales. In the Wild T-2 these will appear as in Fig. 10-3*a*. In this and most theodolites, the circle reads only clockwise angles. Turn the **knob for coincidence setting.** The scale in the lower part of the figure will rotate rapidly and the upper two scales will move slowly in opposite directions. Set them so the upper lines coincide as in

Fig. 10-3*b*. If some do not precisely coincide, make the coincidence of the two central ones perfect. There is a vertical stationary line at the bottom of the upper figure that marks the approximate center. The accuracy of the result depends almost entirely on the accuracy with which the coincidence is set.

Find the first right-side-up number to the left of the center line, in this case 13. This indicates 13°. Count, from the line at 13, the number of spaces to the inverted, symmetrically placed number on the right (193). Each of the spaces counted indicates 10 minutes, so this equals 50 minutes. On the small scale below, read the lower set of numbers. These are 4's, which indicates 4 minutes. The nearest mark to the index line is 32, or 32 seconds. The final reading then is 13°50′ plus 4′32″ or 13°54′32″.

Figure 10-3*c* shows another position of the circle, and Fig. 10-3*d* shows it after the coincidence has been set. The number to the left of the central line is 330, there are two spaces to the symmetrically spaced number (150), and the total reading is therefore 330°26′46″.

The vertical circle is read in the same way. However, the vertical circle reads in zenith angles (see Sec. 2-1). Subtract the reading from 90° to get the vertical angle if it is required. When the telescope is reversed, the angle will be over 180°. In this case subtract 270° from the reading to compute the vertical angle.

Example

	Zenith angle			Vertical angle
90° −	92°10′06″		=	− 2°10′16″
90 −	87 32 17		=	+ 2 27 43
	264 18 20	− 270°	=	− 5 41 40
	281 17 46	− 270	=	+11 17 46

10-8. Methods of Reading the Circle on Other Instruments. Figure 10-4*a* shows a reading system that is almost identical to the one shown in Fig. 10-3. Here the right-side-up numbers are above instead of below the horizontal line. As before, the first right-side-up number to the left of the central line is the number of degrees. The spaces are counted as before, and the final minutes and seconds are read on the small scale.

In Fig. 10-4*b*, when the coincidence-setting knob is turned, the sets of paired lines (*P*) and the single lines (*S*) in the smallest rectangle move in opposite directions. The observer places them so that the single lines bisect the spaces between the two lines of each pair as shown. The approximate reading is shown at the index line in the large rectangle, in this case, 25° and something over 20 minutes. In the lower rectangle, the exact value is given, 2′53″, so that the reading is 25°22′53″.

In Fig. 10-5*a*, turning the micrometer knob moves the degree lines in

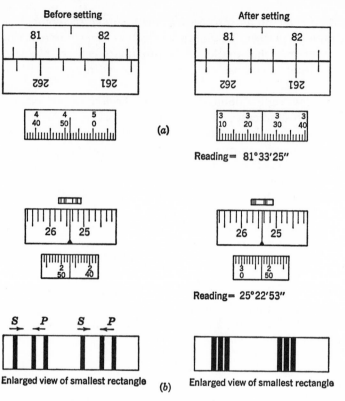

Fig. 10-4. Two types of reading devices which average the two sides of the circle, shown before and after coincidence.

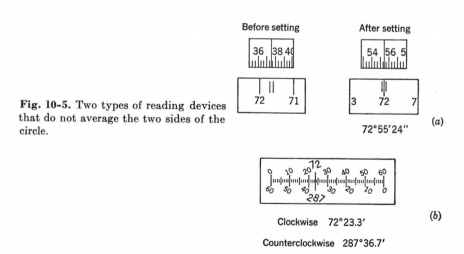

Fig. 10-5. Two types of reading devices that do not average the two sides of the circle.

the large rectangle. To read the angle, one of them is placed between the paired lines at the center of the rectangle. The numbers in the small rectangles are minutes. The seconds are estimated. In this case, the scale comes to a position that is 1.2 divisions beyond 55 minutes. As each division represents 20 seconds, each tenth of a division represents 2 seconds, so that the estimated reading is 72°55′20″ plus 4 seconds, or 72°55′24″.

For the system in Fig. 10-5*b* no micrometer setting is used. The scale is read as accurately as can be estimated. Note that the circle can be read both clockwise and counterclockwise. The large numbers are the degrees. The upper set of numbers, all of which slant downward to the right, is the set to use for clockwise angles. The clockwise reading is 72°23.3′, and the counterclockwise reading is 287°36.7′.

10-9. To Lay Out an Angle (with a Circle That Reads Only Clockwise). Aim at the reference point. Read the circle. Add the **clockwise** angle to be laid out. If 45° is to be laid out to the **right,** add 45°. If 45° is to be laid out to the **left,** subtract 45°. If the sum is greater than 360°, subtract 360°. If the difference is a minus value, add 360° to it.

Example. To lay out 30° to the right:

	First reading	*Set at*
	339°10′15″ + 30°	9°10′15″
	58 17 12 + 30	88 17 12

To lay out 30° to the left:

	First reading	*Set at*
	339°10′15″ − 30°	309°10′15″
	58 17 02 − 30	28 17 02

As an example, assume that the computed angle that must be set is 88°17′12″. With the knob for coincidence setting, set the lower scale at 7′12″. With the clamp and tangent screw turn the alidade to a coincidence that reads 88°10′. The line of sight is now pointed correctly.

When there is no coincidence-setting knob, as is the case in Fig. 10-5*b*, the desired angle can be set directly with the clamp and tangent screw.

10-10. Programs for Accurate Angle Measurement. It is impossible to repeat an angle with many types of theodolites by the method used with an American transit. Several types of theodolites have recently been equipped with **repetition clamps** which make this possible. They are described in Sec. 10-11.

As with a transit, almost any desired accuracy can be obtained with a theodolite. For measuring single angles as in a traverse, both instruments can attain the same accuracy in about the same length of time. On triangulation, when several angles must be measured at a single station, the theodolite is much faster.

In general, the procedure consists in taking the circle reading when the telescope is aimed at each point in turn, without moving the circle. The position of the circle is then changed and the process is repeated. The angles are computed from the differences in the readings and then averaged. The more circle positions used, the more accurate the results. Both the positions of the circle and the positions of the micrometer scale should be arranged to utilize all parts of each of them uniformly. Any one of the points sighted is chosen as the **initial point.** Circle settings are chosen to be used when pointed at the initial point which will accomplish this result.

The procedure is illustrated by the process of measuring the four angles formed by five lines extending from a triangulation station (see Fig. 10-6).

Type of Instrument. First determine whether the reading device for the horizontal circle gives the value of one point on the circle or automatically gives the average of two points 180° apart, as follows. While observing the reading, turn the alidade slowly with the tangent screw. If some lines where the coincidence is set move in one direction and others in the opposite direction, the instrument averages the two sides of the circle. Figure 10-4 shows reading devices of this type. If all the moving lines move in the same direction, this instrument may be one type or the other. However, very few whose lines move in the same direction average both sides of the circle, so if there is doubt, assume that only one side of the circle is used. Figure 10-5 shows this type.

Compute the Circle Settings. The more the circle settings, the higher the accuracy. Assume that four settings will give the accuracy required.

If the instrument is the averaging type, divide 180° into four parts thus: 0°, 45°, 90°, 135°. If the instrument reads a single point on the circle, divide 360° into four parts thus: 0°, 90°, 180°, 270°.

Determine the total range of the micrometer scale (assume 10 minutes) and divide this value into four parts thus: 0°, 2½ minutes, 5 minutes, 7½ minutes. For an averaging instrument, therefore, the four settings for the initial point will be:

$$0°0'$$
$$45\ 2\ 30''$$
$$90\ 5$$
$$135\ 7\ 30$$

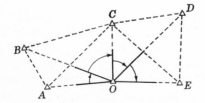

Fig. 10-6. The angles measured at station 0 in the example described.

Table 10-1. Form of Notes for Direction-theodolite Observations

Station	Station Occupied O				Av. angle
	Pos. 1 direct	Pos. 2 direct	Pos. 3 reversed	Pos. 4 reversed	
$A + 360°$	360–00–32	405–02–15	450–05–22	495–07–56	
E–A	172–25–51	172–25–42	172–25–45	172–25–43	172– 25– 45.2
E	187–34–41	232–36–33	277–39–37	322–42–13	
D–E	45–12–25	45–12–33	45–12–28	45–12–26	45– 12– 25.2
D	142–22–16	187–24–11	232–27–09	277–29–47	
C–D	47–11–54	47–12–00	47–12–01	47–11–58	47– 11– 58.2
C	95–10–22	140–12–11	185–15–08	230–17–49	
B–C	69–57–13	69–57–18	69–57–15	69–57–17	69– 57– 15.8
B	25–13–09	70–14–53	115–17–53	160–20–32	
A–B	25–12–37	25–12–38	25–12–31	25–12–36	25– 12– 35.5
A	0– 0–32	45–02–15	90–05–22	135–07–56	
				Check	358–117–179.9

Aim the instrument at the initial point, in this case A. Set the micrometer scale at 0. With the circle-setting knob, set the circle as nearly as possible to read zero. Perfect the coincidence with the micrometer. Read and record the angle, 0–0–32. See Table 10-1, at the bottom of the column, for position 1. Point at B, C, D, and E and record the readings for each, up the column, skipping spaces where the angles between the readings are computed.

After E has been observed, take a reading on A. This should check with the original reading on A, within the limits of reading the circle, to show that the circle has not moved. With the averaging instruments discussed, this should check to ± 2 seconds. If it checks, add 360° to the **first** reading on A and record. If it does not check, repeat the readings for this position.

For the second circle position, aim at the initial point and set the micrometer at 2′30″ and the circle at 45° as nearly as possible. Perfect the coincidence and record the reading 45–02–15 as before. Continue accordingly. When half the circle positions are completed, the telescope must be reversed before aiming at the initial point, and the work continued with the telescope in this position.

Compute the angles from the differences in the adjacent readings, average them, and place the results in the last column. The sum of the angles should add to exactly 360° except for rounding off. Distribute any rounding-off error to force the sum to equal 360°.

10-11. The Repetition or Zero-setting Clamp. In order to adapt theodolites to American practice, a special clamp has been added to some theodolites which connects the circle to the alidade. It has no tangent screw. Instruments equipped with this clamp usually have no circle-setting device. With this clamp, angles can be measured by repetition and laid out very much as with a transit. If desired, these theodolites can also be operated as previously described, using a set of circle positions.

To Set the Reading at Zero.   Turn the alidade until the circle reading is nearly zero. Set the repetition clamp. Turn the alidade so that it points about in the direction of the point to be first observed. The repetition clamp clamps the circle to the alidade so that, when the alidade is turned, the circle turns with the alidade, overcoming the slight friction between the circle and the leveling head, so that the circle reading remains the same. Set the micrometer to read zero. Free the repetition clamp. With the tangent screw, set the circle reading at exactly zero. Set the repetition clamp, and with the tangent screw aim at the first point. Free the repetition clamp. The instrument is now prepared either for measuring the angle by repetition or for laying out an angle to the right.

This procedure is necessary as the setting often changes slightly when the alidade is turned, after the repetition clamp has been locked, because of eccentricity between the bearing of the circle and the bearing of the main spindle.

To Measure an Angle by Repetition.   When the instrument has been prepared as described, aim at the second point using the tangent screw and clamp. Set the repetition clamp and aim at the first point. Free the repetition clamp and continue.

10-12. To Lay Off an Angle to the Right.   After preparation as in Sec. 10-11, set the micrometer screw for the minutes and seconds required and, with the clamp and tangent screw, set the degrees and tens of minutes.

10-13. To Lay Off an Angle to the Left.   If the scale and reading device read in both directions, as in Fig. 10-5*b*, the process is the same as that to the right except that the zero setting is made on the counterclockwise set of numbers (in this case the lower set) and the angle is laid off on this set. When there is no micrometer, as in this case, the instructions for setting the micrometer are omitted.

If the numbers read only clockwise, aim at the first point, set off the angle desired by the same procedure as for setting zero, and then turn the alidade to zero, handling the instrument as described.

10-14. Recapitulation.   To operate a theodolite the observer must thoroughly understand and keep the following constantly in mind.

1. The optical plumb bob operates only when the instrument is level.

2. For horizontal angles, there is only one clamp and tangent screw on most instruments, and this motion controls the position of the alidade with respect to the leveling head.

3. The position of the circle with respect to the leveling head can be changed rapidly with a special knob, if there is no repetition clamp. If there is a repetition clamp, this knob is omitted.

4. A repetition clamp, if the theodolite has one, clamps the circle to the alidade. It has no tangent screw.

With these four principles clearly in mind, the observer can think out how every operation can be performed.

10-15. To Measure Vertical Angles. Theodolites are designed to give quite accurate vertical angles. One of two devices not found on transits is used to reach this accuracy. These are, respectively, an **index level** and an **automatic index.**

When an index level is used, the procedure is the following:

Aim at the desired point. Adjust the reflector for the index (collimation) level so that plenty of light is available to show the bubble. If necessary, turn the inverter knob to bring the zenith angle scale into view. Center the bubble and read the zenith angle.

When an automatic index is used for accurate work, the zenith angle must be read direct and reversed. The average vertical angle is used.

Refer to the example in Sec. 10-7. Assume that the two zenith-angle readings are 85°26′10″ and 274°33′58″.

$$
\begin{array}{ll}
\qquad\qquad 89°59′60″ & \text{Reading} = \quad 274°33′58″ \\
\text{Subtract reading} = -85\ 26\ 10 & \qquad\qquad -270 \\
\hline
\text{Vertical angle} = +\ 4°33′50″ & \text{Vertical angle} = +\quad 4°33′58″
\end{array}
$$

$$
\text{Average vertical angle} = +4°33′54″
$$

10-16. Vertical Angles in Stadia. The degree of accuracy of stadia observations is not high enough to require direct and reversed zenith-angle readings.

For stadia computations, Tables IX and X (in the back of the book) read in zenith angles as well as vertical angles. Table XI provides a means of rapid conversion from zenith angles to vertical angles for use with a stadia slide rule.

10-17. Two-Screw Leveling. Some theodolites, such as the Zeiss Th 2 shown in Fig. 10-7, have only two leveling screws. The third screw is replaced by a pivot. To level the instrument, turn the alidade so that the plate level is in line with a leveling screw and the pivot. Center the bubble. Repeat, using the other screw. Check in both positions.

Fig. 10-7. A Zeiss Th 2. This is a second-order theodolite that reads directly to seconds. (*Keuffel & Esser Co.*)

Problems

10-1 to 10-4. Sketch the appearance of the reading device for a theodolite, as specified, after coincidence is complete.

Averages both sides of the circle. Reads to single seconds.

    10-1. Reading 97°37′22″

    10-2. Reading 95°12′37″

Does not average both sides of circle. Reads to 2 seconds by estimation.

    10-3. Reading 261°15′46″

    10-4. Reading 82°43′34″

10-5 to 10-8. List the settings for observing the initial point for six circle positions, for the reading devices shown in the figures stated.

10-5. Figure 10-4*a*.

10-6. Figure 10-4*b*.

10-7. Figure 10-5*a*.

10-8. Figure 10-5*b*.

Suggested Field Exercises

10-1. Set up theodolites with plumb bobs and check with optical centering.

10-2. Set up without plumb bobs.

10-3. Set various assigned readings.

10-4. Measure the directions of three, four, or five points by four circle positions and compute the angles.

10-5. Set out required angles left and right.

10-6. Measure the angles set out and correct the positions of the points.

10-7. Check the final positions by remeasuring.

10-8. If the theodolites have zero-setting clamps, measure the angles by repetition, preferably on angles previously measured by several circle positions.

10-9. Set out the angles by using the repetition clamp.

11

The Horizontal Curve

11-1. The Method of Staking Out a Simple Horizontal Curve.
Frequently a surveyor is required to stake out a horizontal curve. There
are several ways of accomplishing this. The procedure given here is used
more than any other and, once it is understood, the other systems can
be followed. With the exception of spirals or easement curves, which
are sometimes used, all horizontal curves are circular. A **simple** horizontal curve is composed of a single arc; a **compound** curve, of
two or more arcs of different radii; and a **reversed** curve, of two
arcs that curve in opposite directions. Usually the curve must be laid
out so as to join two straight lines called **tangents** which are marked
on the ground by **POT's** (points on tangent) (see Fig. 11-1a). The first or
back tangent has usually been marked off in stations. These tangents
are run to intersection, thus locating the **PI** (point of intersection). The
plus of the PI and the angle X are measured (see Fig. 11-1b). With these
values and any given value of R (the radius desired for the curve), the
data required for staking out the curve can be computed.

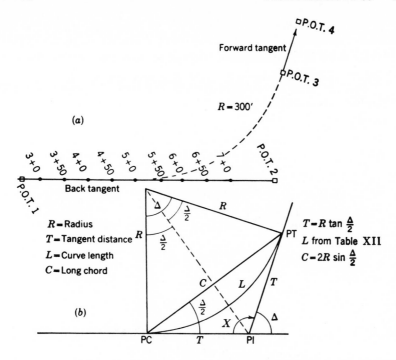

Fig. 11-1. Procedure for staking out a curve.

THE SIMPLE CURVE

The transit is set up at the **PC** (point of curve) at a distance T (tangent distance) from the PI and aimed at the PI with the vernier set at zero. Points on the curve, usually at 50-ft intervals, are then staked out by measuring the computed chord from each previous point and taking line from the transit set at the proper deflection angle, as shown in Fig. 11-3. The point where the curve ends at the forward tangent is the **PT** (point of tangency).

11-2. Principles of Chords and Deflection Angles. A deflection angle, in the sense applied to a curve, is the angle measured at the PC from the main tangent to a given point on the curve. A curve curving to the left as the station numbers increase, as in Fig. 11-3, is a left curve, and all the deflection angles are measured to the left. In general:

1. The angle between a tangent and a chord, measured at the point of tangency, is equal to one-half the central angle, and therefore one-half the angle of the arc subtended by the chord.

   In Fig. 11-2a, from geometry,

$$a = a' = a'' = \tfrac{1}{2}\underline{/MON} = \tfrac{1}{2} \text{ arc } MN$$

2. The angle between two chords that intersect on the circumference of a circle is equal to one-half the central angle subtended between them, and therefore one-half the angle of the arc subtended between the two chords.

In Fig. 11-2*b* the two chords are MN and MP, and the arc subtended between them is NP. Then

$$a = \tfrac{1}{2}\underline{/MON}$$
$$b = \tfrac{1}{2}\underline{/MOP}$$
$$c = b - a = \tfrac{1}{2}\underline{/NOP} = \tfrac{1}{2} \text{ arc } NP$$

3. The length of a chord is equal to twice the radius times the sine of half the angle subtended by the chord.

In Fig. 11-2*a*

$$\frac{1}{2}\frac{MN}{R} = \sin \underline{/MOC}$$
$$MN = 2R \sin \underline{/MOC}$$
$$= 2R \sin \tfrac{1}{2}\underline{/MON} \qquad (11\text{-}1)$$

11-3. Procedure for Staking Out a Simple Horizontal Curve. Follow Fig. 11-3 and the field notes in Fig. 11-4.

1. **Set the PI.** Intersect the two tangents. See Sec. 9-14, Fig. 9-25.
2. **Measure the plus of the PI.** Measure from station $7 + 0$, 47.64 ft. Plus PI $= 7 + 47.64$.
3. **Measure Δ.** Angle X should be measured, usually 2 D.R., at least 1 D.R., and Δ computed. $\Delta = 180° - X$:

$$\begin{array}{r} 179°59'60'' \\ -X = \underline{105\ 13\ 24} \\ \Delta = 74°46'36'' \end{array}$$

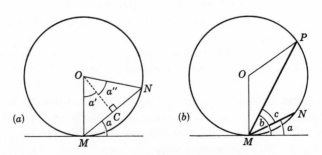

Fig. 11-2. Chords and deflection angles.

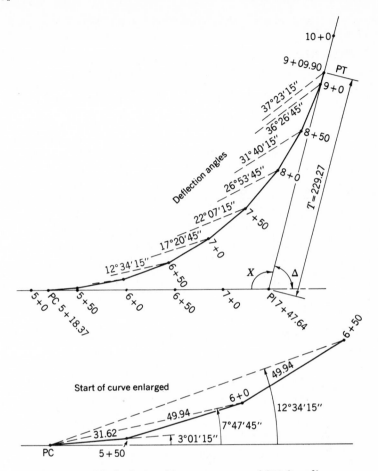

Fig. 11-3. Method of staking out a curve of 300 ft radius.

4. **Compute** T. From Fig. 11-1*b*

$$T = R \tan \frac{\Delta}{2} \qquad (11\text{-}2)$$

$$\tan 37°23'18'' = 0.76424$$
$$\times R = \underline{300}$$
$$229.27200$$
$$T = 229.27 \text{ ft}$$

5. **Compute** L. This is the actual circular length. Use Table XII (at the back of the book). Separate Δ as shown. To find values for single digits, divide the value for 10 times the digit by 10. The sum is the length of a curve having a $\Delta = 74°46'36''$ but with only a 1-ft radius. Multiply by the radius (300).

Example. For 4°. From Table XII, 40° = 0.69813. The value for 4° = $\frac{1}{10}$ × 0.69813 = 0.06981.

From Table XII:

$$\begin{aligned} 70° &= 1.22173 \\ 4° &= 0.06981 \\ 40' &= 0.01164 \\ 6' &= 0.00174 \\ 30'' &= 0.00015 \\ 6'' &= 0.00003 \\ \text{Sum} &= 1.30510 \end{aligned} \qquad \begin{aligned} & \ 1.30510 \\ & \times R = \quad 300 \\ & \ 391.53000 \end{aligned}$$

$$L = 391.53 \text{ ft}$$

Sta.	Chord	Deflec.	Curve Data	
+50				
⊙+09.90 PT	9.90	37° 23' 15"	R = 300 L	
9+0	49.94	36° 26' 45"		
			Δ= 74° 46' 36"	
+50	49.94	31° 40' 15"		
			Δ/2 =37° 23' 18"	
8+0	49.94	26° 53' 45"		
			=37° 23.30'	
+50	49.94	22° 07' 15"		
⊙+47.64 PI			T=229.27	
7+0	49.94	17° 20' 45"		
			L=391.53	
+50	49.94	12° 34' 15"		
6+0	49.94	7° 47' 45"		
+50	31.62	3° 01' 15"		
⊙+18.37 PC		0		
5+0				

Tan 37° 23' 18" = 0.76424
R = 300
T = 229.27200

Plus P.I. = 747.64
Less T = 229.27
Plus P.C. = 518.37
Add L = 391.53
Plus PT = 909.90

Plus 1st Sta = 550.00
Less Plus P.C. = 518.37
First arc = 31.63
17 19
284 67
3163
22141
3163
54371.97
181.24' = 3° 01.24'
3.00 ⟌543.71.97

Δ 70° 1.22173
4° .06981
40' .01164
6' .00174
30" .00015
6" .00003
Sum 1.30510
×R 300
L = 391.53000

1719
50
85950

286.50'=4° 46.50'
3.00 ⟌859.50
1719
Last arc = 9.90
15471
15471
17018.1
56.73 = 0° 56.73'
3.00 ⟌170.18.1

Fig. 11-4. Field notes for staking out a curve.

6. **Compute the pluses.** The stationing is usually continuous from the back tangent, around the curve, and continues on the forward tangent. See Fig. 11-3.

$$
\begin{aligned}
PI &= 7 + 47.64 \\
-T &= 2 \quad\ 29.27 \\
\hline
PC &= 5 + 18.37 \\
+L &= 3 \quad\ 91.53 \\
\hline
PT &= 9 + 09.90
\end{aligned}
$$

7. **Compute the deflection angles.** The first deflection angle subtends an arc from the PC to the first 50-ft point on the curve. Compute the arc length.

$$
\begin{aligned}
\text{First 50-ft point} &= 5 + 50.00 \\
\text{Minus the plus of the PC} &= 5 + 18.37 \\
\hline
\text{Length 1st arc} &= \quad\quad 31.63
\end{aligned}
$$

The formula for computing the number of minutes in an angle measured from a point on the curve which subtends a known arc is

$$\frac{\text{Length of arc}}{R} \times 1718.87 \tag{11-3}$$

Usually the number 1718.87 can be shortened to 1719 on all but long curves. On long curves 1718.87 should be used only for the regular (50-ft) intervals.

$$\frac{31.63}{300} \times 1719 = 181.24 \text{ minutes}$$
$$= 3°01.24'$$

which is the first deflection angle. Note where it is recorded in the field notes, Fig. 11-4. Note: 0.24 minutes = 15 seconds.

The second deflection angle equals the first deflection angle plus the angle which subtends 50 ft.

$$\frac{50}{300} \times 1719 = 286.50 \text{ minutes}$$
$$= 4°46.50'$$

This value is added for each 50-ft point until the point just previous to the PT, 9 + 00, is reached. The arc to the PT is 9.90 ft. The deflection angle for this arc is computed. Thus

$$\frac{9.9}{300} \times 1719 = 56.73 \text{ minutes}$$

Table 11-1

	Station	Deflection	Deflection to nearest 15″
PC	5 + 18.37	0	0
	5 + 50	3°01.24′	3°01′15″
		+4 46.50	
	6 + 0	7°47.74′	7 47 45
		+4 46.50	
	6 + 50	12°34.24′	12 34 15
		+4 46.50	
	7 + 0	17°20.74′	17 20 45
		+4 46.50	
	7 + 50	22°07.24′	22 07 15
		+4 46.50	
	8 + 0	26°53.74′	26 53 45
		+4 46.50	
	8 + 50	31°40.24′	31 40 15
		+4 46.50	
	9 + 0	36°26.74′	36 26 45
		+0 56.73	
PT	9 + 09.90	37°23.47′	37 23 30

With these values, the deflection angles can be computed as shown in Table 11-1.

The deflection angle computed for the PT should equal $\Delta/2$ or 37°23.30′ or 37°23′15″ (see Sec. 11-2). The small error is due to rounding off. A large error indicates a mistake in computation.

8. **Compute the chord lengths.** Since the length of each chord is slightly less than the length of the arc it subtends, the chord lengths must be computed. Equation (11-1) can be used or Table XIII, as shown.

Arc, ft	Correction, ft	Chord, ft
31.63	−0.01	31.62
50.00	−0.06	49.94
9.90	−0	9.90

Figure 11-4 shows how all these values are computed and recorded in the field notes.

9. **Set the PT.** Measure $T = 229.27$ from PI and set PT on line with the forward tangent.

10. **Set the PC.** Measure 18.37 from station 5 and line in.

11. **Set the stations.** Set up at the PC. Set the vernier at zero and aim at the PI. Turn left the deflection angle for 5 + 50 (3°01'15") and set 5 + 50 on line at the chord distance from the PC (31.62 ft).

Set the vernier at the next deflection angle (7°47'45") and set 6 + 0 on line and at the chord distance 49.94 from 5 + 50.

Continue in this way, setting off each successive deflection angle and measuring out the required chord length from the previous point until the last point is set previous to the PT (9 + 0).

12. **Measure to the PT.** Assume that from 9 + 0 to the PT measured 9.78 ft instead of the chord length 9.90, giving an error of 0.12 ft. Turn the transit to the deflection angle for the PT (37°23'15"). Measure the perpendicular distance from the line of sight to the PT. Assume this was 0.16 ft. Compute the total error.

$$\sqrt{12^2 + 16^2} = 20 \text{ hundredths of a foot}$$

13. **Compute the error ratio.** The total length of the survey is

$$2T + L = 850 \text{ ft}$$

Then $0.20/850 = 1$ part in 4250. The maximum ratio that should be accepted is 1 part in 3000.

11-4. Orientation on the Curve. It often occurs that some obstacle prevents sighting from the PC to distant points on a curve, as shown in Fig. 11-5.

The computed deflection angles for the stations to be measured at the PC are as follows.

Station	Defl. angles
PC	0
D	a
E	$a + b$
F	$a + b + c$

Note that all angles in the figure which have the same letter are equal, as they either are vertical angles or are equal to half the central angle of the same arc.

When the obstruction interferes, as in the line PC to E, the transit is moved to station D. The telescope is reversed, the vernier set at the deflection angle of the PC, which is zero, and the line of sight is aimed at the PC.

The telescope is then changed to direct so that it is sighting along the line PC to D prolonged. To set E, it must be turned through the angle

$a + b$. But note that $a + b$ is the deflection angle computed for E. This is, of course, true for all stations from D to E, or, for that matter, for all stations on the curve, as E represents any station. Thus, with this procedure the same list of deflection angles can be used as those originally computed. When this is the case, the transit is said to be **oriented to the curve.** It was oriented by sighting PC with the deflection angle of PC (zero) set on the vernier.

To set stations beyond E, the transit is moved to E. How can it be oriented to the curve? The deflection angle of D (angle a) is set on the vernier, and the line of sight aimed at D with telescope reversed.

The telescope is then changed to direct so that it now aims along the prolongation of the line DE. Remember that the vernier still reads the angle a.

To set F, the transit must be turned through the angle $b + c$ so that the total reading on the vernier will be $a + b + c$, which is the deflection angle of F. Evidently the transit is now oriented to the curve. Thus two rules can be stated.

To Orient to the Curve.   When the transit is on the curve, aim at any other station on the curve, with the telescope reversed for a point behind the transit station or direct for a point ahead of the transit station and with the vernier set at the deflection angle of the station at which it is aimed.

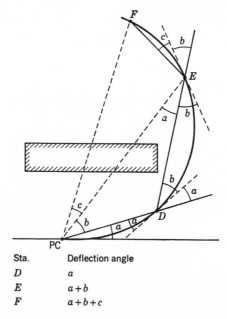

Sta.	Deflection angle
D	a
E	$a+b$
F	$a+b+c$

Fig. 11-5. To orient a transit to the curve.

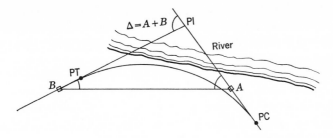

Fig. 11-6. Inaccessible PI.

When Oriented to the Curve.   When the transit is oriented to the curve, any station can be set on the curve by setting the transit at the deflection angle of the point to be set, with the telescope reversed for points behind and direct for points ahead. Also, after orientation, to establish a tangent to the curve at the transit station, turn to the deflection angle of the transit station.

11-5. Other Considerations.   When the curve is a left curve, all angles are set off to the left. For a right curve, all are set off to the right. If the transit is out of adjustment so that it will not prolong a line by reversing the telescope, instead of reversing the telescope merely add 180° to the deflection angle to be set on the vernier.

When the **degree of curvature arc definition** (D_a) is given instead of the radius, the radius can be found from this formula:

$$R = \frac{5729.578}{D_a} \tag{11-4}$$

where D_a is in degrees.

When the **degree of curvature chord definition** (D_c) is given,

$$R = \frac{50}{\sin \frac{1}{2} D_c} \tag{11-5}$$

11-6. When the PI Is Inaccessible.   Figure 11-6 shows what to do when the PI cannot be reached. Points A and B are set wherever convenient on the tangents. The distance AB and the angles A and B are measured. Then

$$\Delta = A + B$$
$$\text{PI to } A = \frac{AB}{\sin \Delta} \sin B$$
$$\text{PI to } B = \frac{AB}{\sin \Delta} \sin A$$

The distance to be measured for setting PT by measuring from B is computed from the above by using the value for T, and PC to A is computed similarly.

THE COMPOUND CURVE

11-7. The Compound Curve.   Often two curves of different radii are joined together as in Fig. 11-7. Point P is called the **point of compound curve** (PCC). GH is a common tangent. The subscript 1 refers to the curve of smaller radius.

The angle Δ is measured; R_1 and R_2, and either Δ_1 or Δ_2, are given. To find the curve data for the two curves, the following are computed:

From the figure:

$$\Delta_1 = \Delta - \Delta_2 \quad \text{or} \quad \Delta_2 = \Delta - \Delta_1$$
$$t_1 = R_1 \tan \tfrac{1}{2} \Delta_1 \qquad t_2 = R_2 \tan \tfrac{1}{2} \Delta_2$$
$$GH \text{ (the common tangent)} = t_1 + t_2$$
$$VG = \sin \Delta_2 \frac{GH}{\sin \Delta}$$
$$VH = \sin \Delta_1 \frac{GH}{\sin \Delta}$$
$$T_1 = AV = VG + t_1$$
$$T_2 = VB = VH + t_2$$

To stake out the curve, the deflection angles and the chords are computed for the two curves separately. When P is reached, the transit is

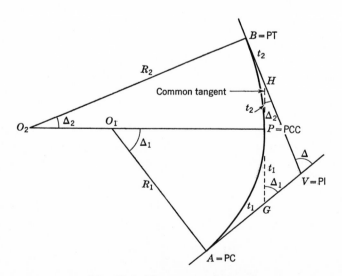

Fig. 11-7. The compound curve.

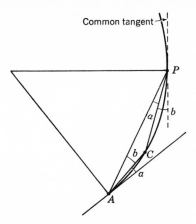

Fig. 11-8. To orient at the PCC.

oriented to the second curve by aiming it so that the vernier reads zero when pointed along the imaginary common tangent GH. To accomplish this, aim at any point on the first curve with the telescope reversed and the vernier set to the **right** (if the first curve is a left curve) at the deflection angle of the PCC on the first curve minus the deflection angle of the point sighted. To prove this, in Fig. 11-8, let C be any point on the first curve. On the first curve,

$$
\begin{aligned}
\text{Defl. angle of } P &= a + b \\
- \text{ Defl. angle of } C &= a \\
\hline
\text{Result} &= \qquad b
\end{aligned}
$$

Thus, if b is set off to the right and aimed at C, when the transit is then turned to zero the telescope will be on the common tangent and the vernier will read zero. Accordingly, once oriented in this way, the deflection angles computed for the second curve can be used.

THE REVERSED CURVE

11-8. The Reversed Curve. See Fig. 11-9. A reversed curve is used to connect a given point A on one tangent with a given point B on some other tangent, as shown in the figure. It is a means of moving the alignment partly sidewise.

In the field, the distance AB and the angles a and b are measured. As it is always an advantage to use the largest radius possible, the best method is to use equal radii.

$$
\cos c = \frac{SO_2}{O_2O_1} = \frac{R \cos a' + R \cos b'}{2R} \tag{11-6}
$$

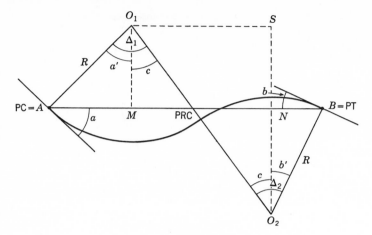

Fig. 11-9. The reversed curve.

But $a' = a$ and $b' = b$, sides perpendicular in the same order. Substituting and dividing both numerator and denominator by R,

$$\cos c = \tfrac{1}{2}\,(\cos a + \cos b)$$
$$AB = R\sin a + 2R\sin c + R\sin b$$
$$R = \frac{AB}{\sin a + 2\sin c + \sin b} \tag{11-7}$$
$$\Delta_1 = a + c \qquad \Delta_2 = b + c \tag{11-8}$$

The curves are computed separately. The first curve is staked out, and at the PRC (point of reversed curve) the transit is oriented to the second curve as in the compound curve (see Sec. 11-7).

References

For a more complete coverage of horizontal and vertical curves, the reader is referred to:

1. Ives, Howard Chapin, and Philip Kissam, "Highway Curves," 4th ed., John Wiley & Sons, Inc., New York, 1952.

For railroad curves, to:

2. Searles, William H., Howard Chapin Ives, and Philip Kissam, "Field Engineering," 22d ed., John Wiley & Sons, Inc., New York, 1949.

Problems

11-1. Given: $R = 350'$, $\Delta = 72°34'30''$, plus of PI $= 22 + 41.64$. Compute and set up the field notes.

11-2. Given: $R = 400'$, $\Delta = 66°18'24''$, plus of PI = 48 + 25.32. Compute and set up the field notes.

11-3. Given: $R = 500'$, $\Delta = 58°08'40''$, plus of PI = 38 + 17.25. Set up the field notes.

11-4. Given: $R = 600'$, $\Delta = 42°34'28''$, plus of PI = 28 + 37.42. Set up the field notes.

11-5. Given, for a compound curve: plus of PI = 14 + 29.31, $\Delta = 97°35'15''$; the first radius $R_1 = 400'$, $\Delta_1 = 63°22'18''$, $R_2 = 800'$. Compute the pluses of PC, PCC, and PT and the length T_2.

11-6. Given, for a compound curve: plus of PI = 12 + 87.93, $\Delta = 98°32'54''$, the first radius $R_1 = 300'$, $\Delta_1 = 62°18'34''$, $R_2 = 600'$. Compute the pluses of PC, PCC, and PT, and the length T_2.

11-7. Given, for a reversed curve: plus PC = A = 1729.38, $a = 47°29'14''$, $AB = 276.82$, $b = 22°34'16''$. Compute Δ_1, Δ_2, plus PRC, and plus PT.

11-8. Given, for a reversed curve: plus PC = A = 1532.71, $a = 44°32'10''$, $AB = 283.17'$, $b = 25°17'20''$. Compute $\Delta_1 = \Delta_2$, plus PRC, and plus PT.

Suggested Field Exercises

For the best results it is well to set out stakes and tacks to represent P.O.T.'s which establish two tangents. One stake should be given a definite station and plus.

These should be arranged so that the curves are not too long but, preferably, because of obstacles, require at least one setup on the curve. It is therefore necessary to make a rough survey to establish proper P.O.T.'s for each field party.

Short radii should be used and reasonably large Δ's so that the errors will be obvious.

It usually takes about 3 hr to complete a simple curve about 500 or 600 ft long, including the computations. The time necessary for compound and reversed curves can be estimated accordingly. It is often wise to run a compound curve with an inaccessible PI to avoid long measurements on the tangents. The possibilities are:

1. Simple curve
2. Simple curve, PI inaccessible
3. Compound curve
4. Reversed curve

12

The Vertical Curve

12-1. Definitions. A vertical curve is a curve in a vertical plane which is used to connect two grade lines. The two grade lines are called the back tangent and the forward tangent, respectively. Their intersection is called the PVI; the beginning of the curve, the PVC; and the end of the curve, the PVT.

In a vertical curve, the rate of grade starts at the rate of grade of the back tangent at the PVC and changes at a constant rate until it reaches the PVT, where it has the rate of grade of the forward tangent. It is therefore a true parabola with a vertical axis (see Fig. 12-1).

12-2. To Compute a Vertical Curve by the Stage Method. There are several methods of computing a vertical curve. The stage method described first can usually be performed on a slide rule, as the numbers required are not large. In the following example, the known values given are those usually given on the plans.

Example.   See Fig. 12-1. Given on plans:

$$\text{PVI station} = 11 + 02.43$$
$$\text{Elevation} = 43.32$$
$$\text{Back tangent grade } g_1 = +6\%$$
$$\text{Forward tangent grade } g_2 = -2\%$$
$$\text{Length of curve } L = 550'$$

The length of curve is usually chosen in units of 50 ft. Usually a grade stake must be set every 25 or 50 ft. Assume 50 ft; then 50 ft is called a **stage**.

$$L = 550 \text{ ft} = 11 \text{ stages}$$

Computation Form.   Set up the form shown in Table 12-1. Under x are the numbers of the stage points. Under x^2 are the squares of these numbers.

1. **Compute** the station and elevation of the PVC and the PVT. Note that the horizontal distances to each from the PVI are both equal to $L/2$.

	PVI $= 11 + 02.43$	PVI $= 11 + 02.43$
	$-L/2 = 2 + 75.00$	$+L/2 = 2 + 75$
Sta.:	PVC $= 8 + 27.43$	PVT $= 13 + 77.43$
	PVI $= 43.32$	PVI $= 43.32$
	$-6\% \cdot L/2 = -16.50$	$-2\% \cdot L/2 = 5.50$
Elev.:	PVC $= 26.82$	PVT $= 37.82$

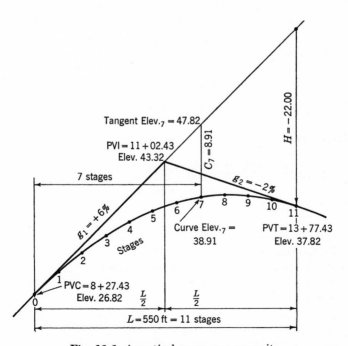

Fig. 12-1. A vertical curve on a summit.

Table 12-1. Form for Computing a Summit Vertical Curve by Stages

Station	x	x^2	$C = -0.1818x^2$	Tangent elevation	Curve elevation
PVC 8 + 27.43	0	0	0	26.82	26.82
8 + 77.43	1	1	−0.18	29.82	29.64
9 + 27.43	2	4	−0.73	32.82	32.09
9 + 77.43	3	9	−1.64	35.82	34.18
10 + 27.43	4	16	−2.91	38.82	35.91
10 + 77.43	5	25	−4.54	41.82	37.28
11 + 27.43	6	36	−6.54	44.82	38.28
11 + 77.43	7	49	−8.91	47.82	38.91
12 + 27.43	8	64	−11.64	50.82	39.18
12 + 77.43	9	81	−14.73	53.82	39.09
13 + 27.43	10	100	−18.18	56.82	38.64
PVT 13 + 77.43	11	121	−22.00	59.82	37.82

2. **Compute** *H.* *H* is the vertical distance to the PVT from the back tangent prolonged.

$$H = (g_2 - g_1)\frac{L}{2} = (-0.02 - 0.06)275 = -22' \qquad (12\text{-}1)$$

3. **Compute** *C.* *C* is the expression for the **tangent correction** for each stage point. When the stage method is used, the *L* and *x* in this formula are expressed in stages and *x* represents the distance from the PVC (measured in stages) to the stage point for which *C* is determined.

$$C = \frac{H}{L^2}x^2 = \frac{-22}{121}x^2 = -0.1818x^2 \qquad (12\text{-}2)$$

4. **Compute** the values in the *C* column by multiplying each x^2 by -0.1818 ft. Note that *C* for PVT = *H*.
5. **Compute** the elevations on the back tangent at the stage points, starting with the elevation of the PVC and adding 6 per cent · 50 = 3.00 ft successively.
6. **Compute** the curve elevations by applying the tangent corrections *C* to tangent elevations.

To Check. Compute, using the forward tangent prolonged.

12-3. To Compute High or Low Point. It is sometimes required to find the station and the elevation of the highest point on a summit curve (like the example) or the lowest point on a sag. The formulas are the same for both.

$$a = g_1\frac{L}{g_1 - g_2} = 0.06\frac{550}{0.06 + 0.02} = 412.50' \qquad (12\text{-}3)$$

where a = distance in feet from the PVC

L = curve length in feet

Find the tangent correction C, and apply it to the tangent elevation that occurs at 412.50 from the PVC.

$$C = \left(\frac{a}{L}\right)^2 H = \left(\frac{412.5}{550}\right)^2 (-22) = -12.38' \qquad (12\text{-}4)$$

	Station		*Elevation*
PVC =	8 + 27.43	PVC =	26.82
a =	4 + 12.50	+ 6% × 412.5 =	24.75
Highest point =	12 + 39.93	Tangent elev. =	51.57
		C = −	12.38
		Highest point =	39.19

12-4. To Compute a Vertical Curve by the Direct Method. The elevations found by the stage method described above must be staked out at the stage points, which are located at PVC 8 + 27.43, 8 + 77.43, 9 + 77.43, etc., and not at the regular stations. If the plus of the PVI and the length L had been chosen so that the PVC fell at a 50-ft point (as is often the case), the regular stations could be used. If this is not the case (as here) and if it is absolutely essential to set the grade stakes at the regular stations, the **direct method** is probably the best solution. Large numbers occur in this method, so that a computing machine should be employed for computation.

The example of the direct method described here is applied to the same curve that is used in the example of the stage method.

Example.   See Table 12-2. All values are in feet. The values under a are the distances from the PVC taken to the nearest 0.1 ft. To compute the a^2 values, start at the bottom of the column, write in the first four significant figures in a^2, and add zeros to show the position of the decimal point. Thereafter use the same number of zeros throughout the column.

Steps 1 and 2 are the same as before.

3. **Compute** C to four significant figures.

$$C = \frac{H}{L^2} a^2 = \frac{-22}{302,500} = -0.00007273a^2 \qquad (12\text{-}2a)$$

4. **Compute** the values in the C column by multiplying each a^2 by -0.00007273. This can be made easier by disregarding the zeros in the a^2 column and moving the decimal point in the multiplier accordingly. In this case the first computation is

$$5 \times -0.007273$$

5. **Compute** the elevations on the back tangent at the regular station points.

Table 12-2. Form for Computing a Summit Vertical Curve by the Direct Method*

	Station	a	a^2	$C =$ $-0.00007273a^2$	Tangent elevation	Curve elevation
PVC	8 + 27.43	0	0	0	26.82	26.82
	8 + 50	22.6	500	−0.04	28.17	28.13
	9 + 00	72.6	5300	−0.39	31.17	30.78
	9 + 50	122.6	15000	−1.09	34.17	33.08
	10 + 00	172.6	29800	−2.17	37.17	35.00
	10 + 50	222.6	49600	−3.61	40.17	36.56
	11 + 00	272.6	74300	−5.40	43.17	37.77
	11 + 50	322.6	104100	−7.57	46.17	38.60
	12 + 00	372.6	138800	−10.49	49.17	39.08
	12 + 50	422.6	178600	−12.99	52.17	39.18
	13 + 00	472.6	223400	−16.25	55.17	38.92
	13 + 50	522.6	273100	−19.86	58.17	38.31
PVT	13 + 77.43	550.0	302500	−22.00	59.82	37.82

* This is the same curve as in Table 12-1.

6. **Compute** the curve elevations by applying the tangent corrections C to the tangent elevations.

The highest point and the check are computed as before.

12-5. A Vertical Curve in a Sag Where the Stage Method Applies.
See Fig. 12-2 and Table 12-3.

Example

$$H = (0.03 + 0.05)225 = +18 \text{ ft}$$
$$C = \frac{18.00}{81} x^2 = +0.2222 \, x^2$$

Lowest point:

$$a = -0.05 \frac{450}{-0.05 - 0.03} = 281.25$$
$$C = \left(\frac{281.25}{450}\right)^2 18 = 7.03 \text{ ft}$$

	Station		Elevation
PVC =	19 + 00	PVC =	94.04
a =	2 + 81.25	(−0.05)(281.25) =	−14.06
Lowest point =	21 + 81.25 ft	Tangent elev. =	79.98
		C =	+ 7.03
		Lowest point =	87.01 ft

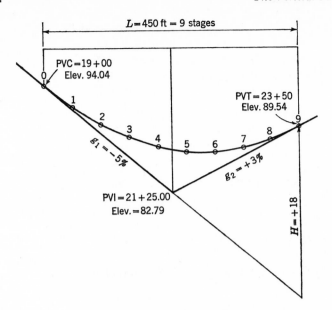

Fig. 12-2. A vertical curve in a sag.

Table 12-3. Computation of a Vertical Curve in a Sag by the Stage Method

Station		x	x^2	$C = +0.2222x^2$	Tangent elevation	Curve elevation
PVC	$19 + 00$	0	0	0	94.04	94.04
	$19 + 50$	1	1	$+0.22$	91.54	91.76
	$20 + 00$	2	4	$+0.89$	89.04	89.93
	$20 + 50$	3	9	$+2.00$	86.54	88.54
	$21 + 00$	4	16	$+3.56$	84.04	87.60
	$21 + 50$	5	25	$+5.56$	81.54	87.10
	$22 + 00$	6	36	$+8.00$	79.04	87.04
	$22 + 50$	7	49	$+10.89$	76.54	87.43
	$23 + 00$	8	64	$+14.22$	74.04	88.26
PVT	$23 + 50$	9	81	$+18.00$	71.54	89.54

12-6. A Vertical Curve in a Sag by the Direct Method. See Fig. 12-3 and Table 12-14. Given on the plans:

PVI station $= 32 + 11.61$
PVI elevation $= 54.18$
Back tangent grade $g_1 = -4\%$
Forward tangent grade $g_2 = +7\%$
Length of curve $L = 600$ ft

Sta. PVI $= 32 + 11.61$	Sta. PVI $= 32 + 11.61$
$-L/2 = \underline{\ 3 + 00.00}$	$+L/2 = \underline{\ 3 + 00.00}$
Sta. PVC $= 29 + 11.61$	Sta. PVT $= 35 + 11.61$
Elev. PVI $= 54.18$	Elev. PVI $= 54.18$
$+4\% \cdot L/2 = \underline{12.00}$	$+7\% \cdot L/2 = \underline{21.00}$
Elev. PVC $= 66.18$	Elev. PVT $= 75.18$

$$H = (g_2 - g_1)\frac{L}{2} = (0.07 + 0.04)300 = 33 \text{ ft}$$

$$C = \frac{H}{L^2}x^2 = \frac{33}{360,000}x^2 = 0.00009167x^2$$

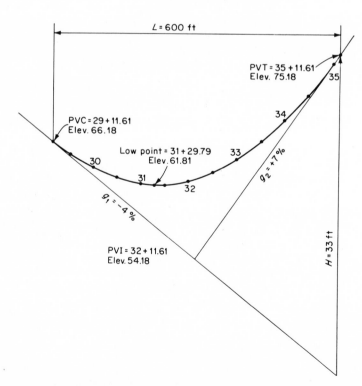

Fig. 12-3. A vertical curve in a sag. Direct method.

Table. 12-4. Form for Computing a Vertical Curve in a Sag by the Direct Method

Station	a	a^2	$c =$ 0.00009167x^2	Tangent elevation	Curve elevation
PVC 29 + 11.61	0	0	0	66.18	66.18
29 + 50	38.4	1,500	0.14	64.64	64.78
30 + 00	88.4	7,800	0.72	62.64	63.36
30 + 50	138.4	19,200	1.76	60.64	62.40
31 + 00	188.4	35,500	3.25	58.64	61.89
31 + 50	238.4	56,800	5.20	56.64	61.84
32 + 00	288.4	83,200	7.63	54.64	62.27
32 + 50	338.4	114,500	10.50	52.64	63.14
33 + 00	388.4	150,900	13.83	50.64	64.47
33 + 50	438.4	192,200	17.62	48.64	66.26
34 + 00	488.4	238,500	21.86	46.64	68.50
34 + 50	538.4	289,900	26.58	44.64	71.22
35 + 00	588.4	346,200	31.74	42.64	74.38
35 + 11.61	600.0	360,000	33.00	42.18	75.18

Lowest point:

$$a = -0.04 \frac{600}{0.04 - 0.07} = 218.18$$

$$c = \left(\frac{218.18}{600}\right)^2 33 = 4.36$$

Station	*Elevation*
PVC = 29 + 11.61	PVT = 66.18
2 + 18.18	(−0.04)(218.18) = − 8.73
31 + 29.79	57.45
	+ 4.36
	61.81

Problems

For all problems, grade stakes are to be set at 50-ft intervals. Fill in the form showing curve elevations.

12-1. Given: PVI station 29 + 25.00; elevation 87.52; $g_1 = +2.5\%$; $g_2 = -4.5\%$; $L = 550'$.

12-2. Given: PVI station 14 + 75.00; elevation 76.29; $g_1 = +3.4\%$; $g_2 = -4.8\%$; $L = 450'$.

Fill in the form showing curve elevations.

12-3. Given: PVI station $18 + 50.00$; elevation 69.32; $g_1 = -2.8\%$; $g_2 = +5.6\%$; $L = 600'$.

Fill in the form showing curve elevations, and give the position and elevation of the lowest point.

12-4. Given: PVI station $10 + 00.00$; elevation 54.71; $g_1 = -3.2\%$; $g_2 = +5.8\%$; $L = 500'$.

Fill in the form showing curve elevations and give the position and elevation of the lowest point.

13

Slope Staking

13-1. Definition. Slope staking is a procedure for giving line and grade for the construction of sloping surfaces when these surfaces meet uneven ground. It is used chiefly for laying out earth excavations and embankments and often for staking out retaining walls. The extent of its use can best be visualized when it is realized that slope stakes must be placed usually for every 50 ft of highway or railroad before construction can begin. The procedure for setting slope stakes is very nearly the same for all types of construction so that, once it is understood for one type of construction, there is no difficulty in applying it to other types. Since its greatest use today is for highway construction, slope staking is described here for marking out cuts and fills for highways.

13-2. Method. The method consists in placing a stake in the existing ground surface, on each side of every center-line stake, where the edge of a cut or the toe of a fill will come when the earthwork is completed (see Fig. 13-1). Each slope stake must be marked with its horizontal distance left or right from the center line and the vertical distance from

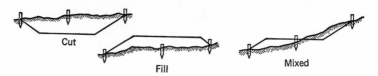

Fig. 13-1. Slope staking. The center-line stake is in place and the stakes on each side are the slope stakes. They are placed where the limits of the earthwork must be located.

the ground at the stake down to the elevation of the bottom of a cut or up to the elevation of the top of a fill, i.e., to the elevation of the **base** (see Fig. 13-2). All stakes should also be marked with the station number.

The position of the stake depends on:

1. Elevation and slope of the base
2. Width of the base
3. Slope of the sides
4. Elevation of the ground where the stake is placed

Evidently the position of the stake must be found by trial. In principle, the procedure is simple, but it is rather confusing until the surveyor becomes familiar with it.

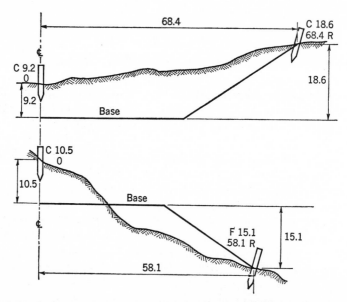

Fig. 13-2. Marks on stakes for slope staking. All stakes should also be marked with the station number. Note that C (cut) or F (fill) is not the cut or fill at the stake, but the vertical distance from the ground at the stake to the elevation of the base.

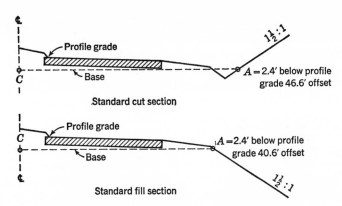

Fig. 13-3. Typical sections. These are the kinds of drawings shown in the plans. The line CA is an arbitrary line placed a certain number of feet below the profile grade, often where the surface of the cut or fill is first constructed.

13-3. Information Required. Figure 13-3 is a sample of the two **typical half sections** given on the plans, one for cut and one for fill. They are often called the **template** sections. They are used for each station throughout the project except when the cross section of the road is changed or unusual conditions necessitate special cross sections. The **base** (line CA) is usually the line to which the cut or fill is first constructed. It may be level or sloping and at different elevations below the profile grade. On curves, the whole half section is tilted up or down by different amounts. Whatever the conditions may be, the elevation and offset of the point A can be determined from the plans.

The procedure for slope staking is described by means of the examples given in the next three sections. These cover the conditions usually encountered.

In the examples given, it is assumed that the base is level, its elevation has been computed, and the offsets for A are 46.6 ft in cuts and 40.6 ft in fills. The slopes of the sides are always given in terms of the distance on the horizontal divided by the vertical height (see Fig. 13-4). A $1\frac{1}{2}$:1 slope is used in all the examples.

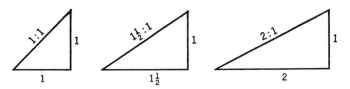

Fig. 13-4. Designation of side slopes.

13-4. Discussion of Fig. 13-5.   In Fig. 13-5, the first H.I. is located at E, and its elevation is found from previous leveling to be 95.72. The scratch work is then started. A rod shot will be first taken on the ground beside the center stake, hence the notation zero for the rod position. See scratch work below the drawing.

1. Compute the **grade rod** (G.R.). This is the theoretical rod reading that would occur if the rod were standing on the desired grade elevation when read from a given H.I. In this case the H. I. is 95.72 and the desired elevation is that of the base, as shown on the plans, 71.9. The formula is

$$\text{G.R.} = \text{H.I.} - \text{grade} \qquad (13.1)$$

$$\begin{aligned} \text{H.I.} &= 95.7 \\ \text{Less base} &= -71.9 \\ \hline \text{G.R.} &= 23.8 \end{aligned}$$

Follow the first scratch-work column.

2. Read the rod when held on the ground beside the center stake (reading, 5.2). Compute the cut at the centerline. The formula is

$$\text{Cut} = \text{G.R.} - \text{rod} \qquad (13.2)$$

$$\begin{aligned} \text{G.R.} &= 23.8 \\ -\text{Rod} &= -5.2 \\ \hline \text{Center cut} &= 18.6 \end{aligned}$$

This means that the base is 18.6 ft below the ground at the center stake C.

Note: **If this comes out minus, the value represents fill.**

When the grade rod, 23.8, and the centerline cut, 18.6, are known, the first slope stake can be set.

To Set the Left Grade Stake L.   Estimate the offset to the left grade stake (L). This may be a guess based on experience. If the volumes have been determined on an electronic computer or the cross sections have been plotted, a very close estimate will be available from these sources. A practical field method, shown in the scratch work, is as follows:

Compute the offset to L that would occur if the ground were level. This would be the offset to A_L (46.6) plus $1\frac{1}{2}$ times the center cut.

	For a $1\frac{1}{2}$:1 slope	For a 2:1 slope
Center cut	18.6	18.6
Plus $\frac{1}{2}$	9.3	18.6
A offset	46.6	46.6
Calculated offset to stake L	74.5	83.8

But the ground slopes downward, so that the cut at the stake L would be less than the center cut. Hence the offset would be somewhat less than 74.5 for the $1\frac{1}{2}:1$ slope. Try, for example, 55. The rod is held at offset 55 shown on the drawing at 1. The offsets are usually measured with a woven tape.

The offset for this rod reading is computed in the second column. Its value is 69.7.

The Key Procedure. It is now known that the cut measured at 55 should occur at 69.7. This indicates that the rod should be moved from its position at 55 toward 69.7. Should it be moved more or less than the whole distance? **To know which to do is the key.** Here are the rules:

1. When the slopes are opposite, move less.
2. When the slopes are the same, move more.

The two slopes are the slope of the ground and the side slope of the earthwork. They are opposite when one slopes down and the other up. They are the same when both slope up or both slope down. In the example, they are opposite, so move the rod less than called for.

For example, try 69 (2 in the drawing).

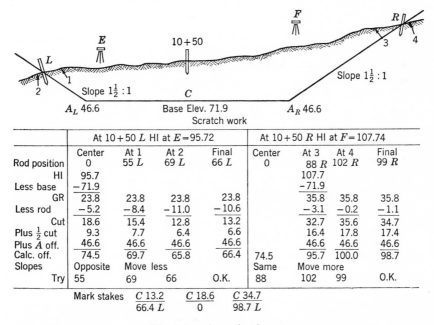

Scratch work

	At 10+50 L HI at $E=95.72$				At 10+50 R HI at $F=107.74$			
	Center	At 1	At 2	Final	Center	At 3	At 4	Final
Rod position	0	55 L	69 L	66 L	0	88 R	102 R	99 R
HI	95.7				107.7			
Less base	−71.9				−71.9			
GR	23.8	23.8	23.8	23.8	35.8	35.8	35.8	
Less rod	−5.2	−8.4	−11.0	−10.6	−3.1	−0.2	−1.1	
Cut	18.6	15.4	12.8	13.2	32.7	35.6	34.7	
Plus $\frac{1}{2}$ cut	9.3	7.7	6.4	6.6	16.4	17.8	17.4	
Plus A off.	46.6	46.6	46.6	46.6	46.6	46.6	46.6	
Calc. off.	74.5	69.7	65.8	66.4	74.5	95.7	100.0	98.7
Slopes	Opposite	Move less			Same	Move more		
Try	55	69	66	O.K.	88	102	99	O.K.

Mark stakes $\dfrac{C\ 13.2}{66.4\ L}$ $\dfrac{C\ 18.6}{0}$ $\dfrac{C\ 34.7}{98.7\ L}$

Fig. 13-5. A section in cut.

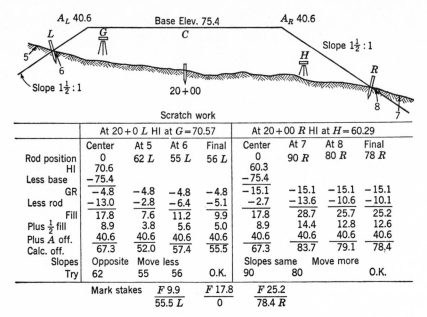

	At 20+0 L HI at G = 70.57				At 20+00 R HI at H = 60.29			
	Center	At 5	At 6	Final	Center	At 7	At 8	Final
Rod position	0	62 L	55 L	56 L	0	90 R	80 R	78 R
HI	70.6				60.3			
Less base	−75.4				−75.4			
GR	−4.8	−4.8	−4.8	−4.8	−15.1	−15.1	−15.1	−15.1
Less rod	−13.0	−2.8	−6.4	−5.1	−2.7	−13.6	−10.6	−10.1
Fill	17.8	7.6	11.2	9.9	17.8	28.7	25.7	25.2
Plus ½ fill	8.9	3.8	5.6	5.0	8.9	14.4	12.8	12.6
Plus A off.	40.6	40.6	40.6	40.6	40.6	40.6	40.6	40.6
Calc. off.	67.3	52.0	57.4	55.5	67.3	83.7	79.1	78.4
Slopes	Opposite	Move less			Slopes same	Move more		
Try	62	55	56	O.K.	90	80		O.K.

Mark stakes   F 9.9   F 17.8   F 25.2

55.5 L   0   78.4 R

Fig. 13-6. A section in fill.

At 69 the calculated offset turns out to be 65.8. The rod should be moved from 69 toward 65.8 but, as before, not all the way. Try 66. Here the calculated distance is 66.4. This is near enough to the actual rod position. A difference of 0.5 or less is near enough.

Set the stake at the **calculated offset** (66.4) and assume that the rod reading is the same as at 66. Therefore, mark the stake C 13.2/66.4 L, as shown below the scratch work.

To Set the Right Grade Stake R.   From the previous work the cut at the center is known to be 18.6; so with level ground the calculated offset is 74.5 as before. But here the slopes are the same; therefore move more.

For example, try 88, shown at 3. The calculated offset is 95.7. Move from 88 toward 95.7 and more.

Try 102. The calculated offset is 100.0. Move from 102 toward 100.0 and more.

Try 99. The calculated offset is 98.7, which is near enough.

13-5. Discussion of Fig. 13-6.   Note here that the grade rod from H.I. at G is minus (−4.8). The computed cut is −17.8. Call it a **plus fill.**

The trials are shown in Table 13-1.

Note that the center cut and calculated offset were taken from H.I. at H as well as from G as a check.

Table 13-1

Rod position	Calculated offset	Move	Try
0	67.3	Less	62 L
62 L	52.0	Less	55 L
55 L	57.4	Less	56 L
56 L	55.5	O.K.	
0	67.3	More	90 R
90 R	83.7	More	80 R
80 R	79.1	More	78 R
78 R	78.4	O.K.	

13-6. Discussion of Fig. 13-7. Here the center cut had to be taken from an H.I. not shown. It turned out to be -11.1 and therefore is a plus fill of 11.1, which gives a calculated offset of 57.3.

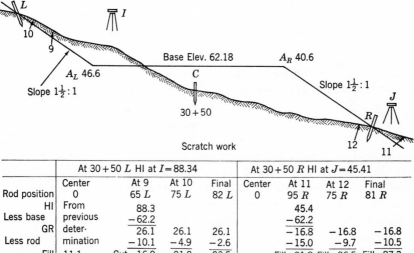

Scratch work

	At 30+50 L HI at $I=88.34$				At 30+50 R HI at $J=45.41$			
	Center	At 9	At 10	Final	Center	At 11	At 12	Final
Rod position	0	65 L	75 L	82 L	0	95 R	75 R	81 R
HI	From	88.3				45.4		
Less base	previous	−62.2				−62.2		
GR	deter-	26.1	26.1	26.1		−16.8	−16.8	−16.8
Less rod	mination	−10.1	−4.9	−2.6		−15.0	−9.7	−10.5
Fill	11.1	Cut 16.0	21.2	23.5	Fill 31.8	Fill 26.5	Fill 27.3
Plus $\frac{1}{2}$ fill	5.6	Cut 8.0	10.6	11.8	15.9	13.2	13.6
Plus A off.	40.6	46.6	46.6	46.6	40.6	40.6	40.6
Calc. off.	57.3	70.6	78.4	81.9	53.3	88.3	80.3	81.5
Slopes	Same Move more				Same Move more			
Try	65	75	82	O.K.	95	75	81	O.K.
Mark stakes		C 23.5	F 11.1		F 27.3			
		81.9 L	0		81.5 R			

Fig. 13-7. A mixed section.

Table 13-2

Rod position	Calculated offset	Move	Try
0	57.3	More	65 L
65 L	70.6	More	75 L
75 L	78.4	More	82 L
82 L	81.9	O.K.	
0	53.3	More	95 R
95 R	88.3	More	75 R
75 R	80.3	More	81 R
81 R	81.5	O.K.	

Here the surveyor must use judgment. By observing the ground, one must realize that, despite the fact that at the center there is fill, at the left side the ground is so high that cut will be required. Therefore the slopes are the same and the move is **more,** not less, and the cut offset for A (46.6) is used.

The trials are shown in Table 13-2.

13-7. To Sum Up.   Memorize these rules:

1. G.R. = H.I. − grade.
2. Cut = G.R. − rod.
3. Move **toward** calculated offset.
4. Slopes opposite, move less.
5. Slopes same, move more.

The first estimate is often a wild guess. Once its calculated offset is known, however, the next try is usually quite accurate. As the slope of the ground is usually fairly uniform from station to station, once the first slope stakes are worked out, the trials for the remainder become more and more accurate, as the surveyor soon has a feel for **how far** to move less or **how far** to move more.

The number of trials may vary considerably. Three trials are used only as illustrations. The level instrument can be moved as desired, and ordinary level notes are used between the required H.I.'s. Often several downhill or uphill slope stakes can be set from one setup. The leveling should of course, start and end at benchmarks.

13-8. Field Notes.   The field notes are exactly like ordinary level notes except when an H.I. is reached where one or more slope stakes are to be set. The form for the six H.I.'s used in the examples is shown in Fig. 13-8. As shown in these field notes, the last rod shot taken, i.e., where the

Sta.	+	HI	−	Rod	Elev.	Gds	C, F	L	¢	R
10+50 ¢		95.72		5.2	90.5	71.9	C 18.6	C 13.2 $\frac{}{66.4}$	$\frac{C18.6}{0}$	
66.4 L				10.6	85.1	71.9	C 13.2			
98.7 R		107.74		1.1	106.6	71.9	C 34.7			$\frac{C34.7}{98.7}$
20+00 ¢		70.57		13.0	57.6	75.4	F 17.8	F 9.9 $\frac{}{55.5 L}$	$\frac{F 17.8}{0}$	
55.5 L				5.1	65.5	75.4	F 9.9			
¢		60.29		2.7	57.6	75.4	F 17.8		$\frac{F 17.8}{0}$	
78.4 R				10.1	50.2	75.4	F 25.2			$\frac{F 25.2}{78.4 R}$
30+50 ¢		88.34		2.6	85.7	62.2	C 23.5	C 23.5 $\frac{}{81.9 L}$	$\frac{F 11.1}{0}$	
81.9 L										
81.5 R		45.41		10.5	34.9	62.2	F 27.3			$\frac{F 27.3}{81.5}$

Fig. 13-8. Suggested field notes for slope staking. Standard level notes are used except when any H.I. is reached from which one or more slope stakes are set.

slope stake is to be set, is recorded in the rod column. The elevation is computed, and the cut or fill is computed from this elevation and the required grade elevation. The formula is as follows:

$$\text{Cut} = \text{ground elevation} - \text{grade elevation}$$

(Minus indicates fill.)

These values should check with the cuts or fills computed by the grade-rod method.

On the right-hand side is the record of the marks placed on the slope stakes.

13-9. Volume Computation. Once the slope staking operations are complete, the quantities of the excavations and of the fills are separately computed. For this purpose, the **end-area formula** (although approximate) is specified in most contracts. This formula is based on the assumption that the area of the cut or the fill at any cross section is constant from a point halfway back toward the previous section to a point halfway forward toward the next section. When the sections are 50 ft apart, this distance is obviously 50 ft. See Fig. 13-9. The end-area formula can then be written:

$$\text{Volume in cubic yards} = \frac{1}{27} 50A \tag{13.3}$$

where A is the area in square feet of any section.

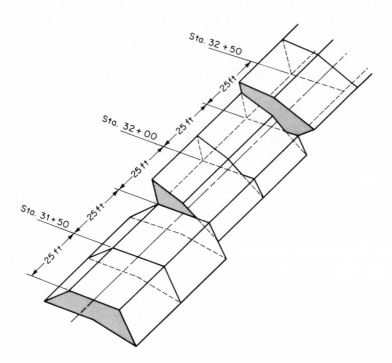

Fig. 13-9. Illustration of the assumption of the end-area formula.

13-10. Areas of Sections.   It is clear that the area of cut and the area of fill must be separately computed for each section. During grade staking, or previously, shots are taken at breaks in the ground at each section and recorded in the same manner as the grade stakes.

Usually each section is plotted on cross-section paper by measuring from a horizontal line drawn to represent the base. For wide, shallow sections, the vertical scale is usually exaggerated.

Figure 13-10, upper drawing, shows a section plotted, together with the *template section* (areas 1–8). The template section is the cross section of the pavement, which is the same for many stations. It is not drawn in (as here). The template is slightly different for cuts and fills. The areas of each are computed and applied as a correction to the area computed for each section.

Methods of Computation.   Usually the areas are computed by one of the following methods, although others are used: (1) **triangles,** (2) **stripping,** (3) **desk calculators,** and (4) **electronic computers.**

The first two are described here.

Figure 13-10 shows a **typical mixed section** (a section having both cut and fill). The base here used in the field is the **profile grade.** Slope stakes are at 57 ft left and 53.5 ft right. Breaks are shown. At 7 ft right, the ground is at the profile grade.

To Compute by Triangles.   This method gives exact areas. Draw in the dashed lines shown in the center drawing and scale the altitude of the first and last triangles as shown. Note triangles that have a common (dashed) base. Together the area equals one-half base times the sum of the altitudes. See also Appendix E.

Computation (start at left in figure):

Double Areas in Square Feet

	Cut		*Fill*
	$8.8(57 - 31) = 229$		$1.5(21 - 7) = 21$
	$5(45 - 12) = 165$		$9(40.5 - 11.5) = 261$
	$12(31 - 0) = 372$		$11.4(53 - 21) = 370$
	$3(12 - 0) = \underline{\ \ 36}$		
	$2/\overline{678}$		$2/\overline{652}$
Area:	339 sq ft	Area:	326 sq ft
Template correction:	$+ 67$		$- 55$
Special correction:	$+ \ 10$		$+ \ \ 6$
	$\overline{416}$ sq ft		$\overline{277}$ sq ft
	$\times 50$		$\times 50$
	$27/\overline{20,800}$ sq ft		$27/\overline{13,850}$ sq ft
	770 cu yd		513 cu yd

Note: The template correction increases the cut and decreases the fill. Also in a mixed section such as this there is an area requiring special corrections that are different for different sections. This is shown in black. The whole black area is added to the cut because it was not included in the cut computation. The part below the base is added to the fill because it was subtracted in the template correction.

To Compute by Stripping.   The lower drawing in Fig. 13-10 shows the method of stripping. Each area is divided into vertical strips; in this case each strip is 2 ft wide. The sum of the altitudes of these strips is measured by placing a strip of paper successively over each strip, as shown in Fig. 13-11.

Then:

$$\text{Vol. in cubic yards} = 50\,\frac{w\Sigma A}{27}$$

where w = width of each strip

ΣA = sum of the altitudes

A scale is prepared on cross-section paper to the same scale as the drawing of the cross section. The total length on the strip of paper, ΣA, is determined from the scale. This is multiplied by $50(w/27)$ to compute the number of cubic yards. The same template corrections are applied as before.

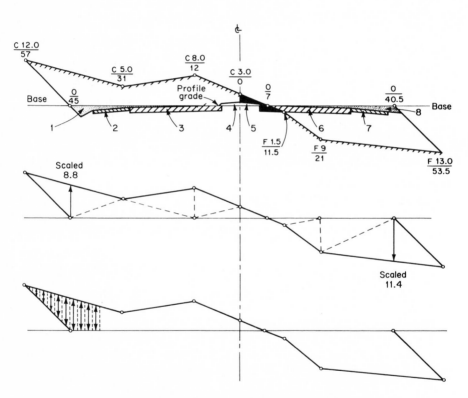

Fig. 13-10. A mixed cross section. See also Appendix E.

13-11. Borrow Pits. Figure 13-12 shows by small squares where permanent marks are established outside the borrow pit area, usually at intervals of 25 ft. Elevations are taken at the intersections of the grid, before and after excavation, and each change in elevation is computed. The changes are multiplied by the number of grid squares they touch,

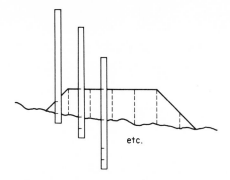

Fig. 13-11. The sum of the altitudes is the total length marked off on the strip.

as shown by numbers in the figure. The sum of these results is divided by 4 and multiplied by the area of one square.

Volumes outside the squares used are computed by the wedge or pyramid formula, whichever applies.

$$\text{Volume of a wedge } = \tfrac{1}{2} \text{ area of base } \times \text{ altitude}$$
$$\text{Volume of a pyramid } = \tfrac{1}{3} \text{ area of base } \times \text{ altitude}$$

The total quantity is divided by 27 to determine the number of cubic yards excavated.

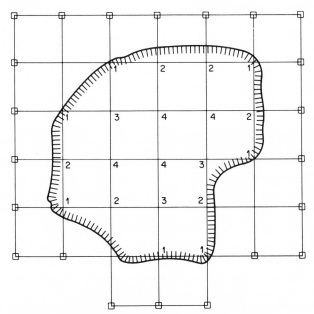

Fig. 13-12. A borrow-pit grid.

Problems

13-1 to 13-4. In the following, state whether the slopes are the same or opposite and whether to move the rod more or less.

13-1. Downhill cut

13-2. Downhill fill

13-3. Uphill cut

13-4. Uphill fill

13-5 to 13-12. In the following, fill in the "Try" column. Assume that the side slopes are $1\frac{1}{2}$:1.

Problem	Cut or fill	Ground slope	Rod position	Calculated offset	Try
13-5	C	Up steep	70	55	
13-6	F	Up steep	75	85	
13-7	C	Down steep	80	90	
13-8	F	Down steep	75	60	
13-9	F	Up medium	70	55	
13-10	C	Up medium	75	85	
13-11	F	Down medium	80	90	
13-12	C	Down medium	75	60	

13-13 to 13-14. Draw the section and compute the volume. Do not correct for template.

13-13. From data in Fig. 13-5.

13-14. From data in Fig. 13-7.

Suggested Field Exercises

A minimum field party consists of a chief, a levelman, and a rodman. The chief directs the work, does the scratch work, records the notes, and holds the zero end of the cross-section tape at the center stake. The levelman runs the level and also records the notes. The rodman holds the rod, reads the cross-section tape, and carries a chain pin to be used for long offsets.

Lay out a straight line of stakes, marked with their station numbers, at 50-ft intervals on a slight diagonal along a steep slope. Set at least three or four stakes more than the number of field parties.

The problem is to set slope stakes for $1\frac{1}{2}$:1 side slopes for a single-track railroad with an 18-ft base for fills and a 20-ft base for cuts.

Establish at least half as many benchmarks as field parties. Provide each field party with a list of the benchmarks and their elevations, and the base elevation for each station at about a 1 per cent rate of grade.

Each field party sets as many slope stakes as possible.

It is often necessary for the instructor to show each field party, in turn, how one cross section is slope-staked.

14

Trigonometry for Surveying

RIGHT TRIANGLES

14-1. Trigonometry. Trigonometry makes it possible to compute the relationships between the lengths of the sides and the sizes of the angles of a triangle. Since it is the only link between straight-line measurement and angular measurement, it has innumerable uses and it is essential in surveying computations.

Trigonometry is based on the relationships in right triangles. Every right triangle has two acute angles, such as A and B in the triangle (shown twice) in Fig. 14-1. Each of these acute angles is considered separately, and the sides of the triangle are named with respect to the angle considered. When, for example, angle B is considered, side CB is called the **side adjacent** (adj.) to angle B and CA is called the **side opposite** (opp.) angle B, since these are their positions with respect to B. The side opposite the right angle is always called the **hypotenuse** (hyp.). Note that the same system is used for angle A (see Fig. 14-1).

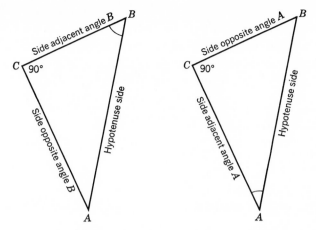

Fig. 14-1. Nomenclature of sides with respect to each acute angle.

14-2. Trigonometric Functions. From geometry, when two right triangles have an acute angle of one equal to an acute angle of the other, they are similar and their sides are proportional (see Fig. 14-2). It follows that, in such triangles, the ratio of any one side divided by another side is the same no matter how long the sides may be. There are six proportions or ratios,

$$\frac{\text{opp.}}{\text{hyp.}} \qquad \frac{\text{adj.}}{\text{hyp.}} \qquad \frac{\text{opp.}}{\text{adj.}} \qquad \frac{\text{adj.}}{\text{opp.}} \qquad \frac{\text{hyp.}}{\text{adj.}} \qquad \frac{\text{hyp.}}{\text{opp.}}$$

and the values of all these ratios will change if the acute angle changes.

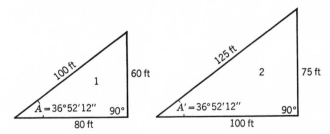

Fig. 14-2. If $A = A'$, the ratios among the sides of each triangle are the same.

Tri-angle	$\dfrac{\text{opp.}}{\text{hyp.}}$	$\dfrac{\text{adj.}}{\text{hyp.}}$	$\dfrac{\text{opp.}}{\text{adj.}}$	$\dfrac{\text{adj.}}{\text{opp.}}$	$\dfrac{\text{hyp.}}{\text{adj.}}$	$\dfrac{\text{hyp.}}{\text{opp.}}$
1	$\dfrac{60}{100} = .60000$	$\dfrac{80}{100} = .80000$	$\dfrac{60}{80} = .75000$	$\dfrac{80}{60} = 1.3333$	$\dfrac{100}{80} = 1.2500$	$\dfrac{100}{60} = 1.6667$
2	$\dfrac{75}{125} = .60000$	$\dfrac{100}{125} = .80000$	$\dfrac{75}{100} = .75000$	$\dfrac{100}{75} = 1.3333$	$\dfrac{125}{100} = 1.2500$	$\dfrac{125}{75} = 1.6667$

Each of these ratios therefore has a certain value for any given angle; and, for angles between 0 and 90°, once the value of any one of these ratios is known, the size of the angle is known, and vice versa.

Since the values of these ratios depend on the size of A, for example, they are called **trigonometric functions** of A. Table 14-1 lists the names of these functions and also two special functions which are useful in surveying.

Note that

$$\csc = 1/\sin \qquad \sec = 1/\cos \qquad \cot = 1/\tan$$

so that only the first three functions are absolutely necessary. The **reciprocal functions** are given to simplify computation.

Also note that, from geometry, $B = 90° - A$, that is, B is the com-

Table 14-1. Nomenclature of Trigonometric Functions

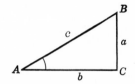

Ratios in triangle	Definition of function	Name of function	Abbreviation of name
$\dfrac{a}{c}$	$\dfrac{\text{side opposite } \angle A}{\text{hypotenuse}}$	sine A	sin A
$\dfrac{b}{c}$	$\dfrac{\text{side adjacent } \angle A}{\text{hypotenuse}}$	cosine A	cos A
$\dfrac{a}{b}$	$\dfrac{\text{side opposite } \angle A}{\text{side adjacent } \angle A}$	tangent A	tan A
$\dfrac{b}{a}$	$\dfrac{\text{side adjacent } \angle A}{\text{side opposite } \angle A}$	cotangent A	cot A or ctn A
$\dfrac{c}{b}$	$\dfrac{\text{hypotenuse}}{\text{side adjacent } \angle A}$	secant A	sec A
$\dfrac{c}{a}$	$\dfrac{\text{hypotenuse}}{\text{side opposite } \angle A}$	cosecant A	cosec A or csc A
$1 - \dfrac{b}{c}$	1 minus cosine $\angle A$	versine A	vers A
$\dfrac{c}{b} - 1$	secant $\angle A$ minus 1	exsecant A	exsec A

Table 14-2. Values of Trigonometric Functions

Angle, deg	Sine	Cosine	Tangent	Cotangent
0	.00000	1.00000	.00000	∞
10	.17365	.98481	.17633	5.67128
20	.34202	.93969	.36397	2.74748
30	.50000	.86603	.57735	1.73205
40	.64279	.76604	.83910	1.19175
50	.76604	.64279	1.19175	.83910
60	.86603	.50000	1.73205	.57735
70	.93969	.34202	2.74748	.36397
80	.98481	.17365	5.67128	.17633
90	1.00000	.00000	∞	.00000

plement of A. Every function of an angle is equal to the cofunction of its complement; thus

$$\sin A = \cos B \qquad \sin B = \cos A$$
$$\tan A = \cot B \qquad \tan B = \cot A$$
$$\sec A = \csc B \qquad \sec B = \csc A$$

14-3. Tables of Trigonometric Functions. It is possible to compute the values of all these trigonometric functions for every size of angle. Since the method of computation is complicated and not important to this text, it is omitted here. Table 14-2 gives the values, to five significant figures, of some of the functions of a few angles.

Examples of the Use of Table 14-2

Examples 1 through 4. Find the angle whose function is given.

1. cos = 0.76604   *Ans.* 40°
2. cot = 2.74748   *Ans.* 20°
3. tan = 5.67128   *Ans.* 80°
4. sin = 0.93969   *Ans.* 70°

Note: When a trigonometric function of an angle is known, but the value of the angle is yet to be found, as above, the **angle** is written thus:

$$\text{arc cos } 0.76604 \text{ or } \cos^{-1} 0.76604 \quad (=40°)$$
$$\text{arc cot } 2.74748 \text{ or } \cot^{-1} 2.74748 \quad (=20°)$$

For example, find arc tan 5.67128. *Ans.* 80°. Find $\sin^{-1} 0.93969$. *Ans.* 70°.

Examples 5 through 7. Solutions of right triangles. In Table 14-3, find the values of the unknown sides of the right triangles.

Table 14-3

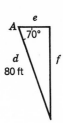
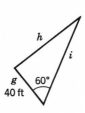
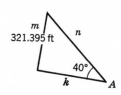

Example 5	Example 6	Example 7
$\dfrac{f}{d} = \sin 70°$	$\dfrac{h}{g} = \tan 60°$	$\dfrac{k}{m} = \cot 40°$
$f = d \sin 70°$	$h = g \tan 60°$	$k = m \cot 40°$
$f = 80(0.93969)$	$h = 40(1.73205)$	$k = (321.395)(1.19175)$
$f = 75.175$ ft	$h = 69.282$ ft	$k = 383.02$ ft
$\dfrac{e}{d} = \cos 70°$	$\dfrac{g}{i} = \cos 60°$	$\dfrac{m}{n} = \sin 40°$
$e = d \cos 70°$	$i = \dfrac{g}{\cos 60°}$	$n = \dfrac{m}{\sin 40°}$
$e = 80(0.34202)$	$i = \dfrac{40}{0.50000}$	$n = \dfrac{321.395}{0.64279}$
$e = 27.362$ ft	$i = 80.000$ ft	$n = 500.00$ ft
Check	*Check*	*Check*
$\tan A = \dfrac{f}{e}$	$\sin A = \dfrac{h}{i}$	$\cos A = \dfrac{k}{n}$
$\tan A = \dfrac{75.175}{27.362}$	$\sin A = \dfrac{69.282}{80.000}$	$\cos A = \dfrac{383.02}{500.00}$
$\tan A = 2.74742$	$\sin A = 0.86602$	$\cos A = 0.76604$
$A = 70°$	$A = 60°$	$A = 40°$

Explanation of Tables III and IV.   Tables III and IV[1] are an expansion of Table 14-2. In them are given the trigonometric functions of angles from 0 to 90°, for every minute of arc. With these tables, all the computations of trigonometry can be performed.

Note, in Table 14-2, that when the cos column is read upward it is identical to the sin column read downward, and the cot column is similarly identical to the tan column. To avoid repetition of numbers, trigonometric tables have a double system of labels, as explained in the next paragraphs.

To Use Tables III and IV from 0 to 44°.   At the top of each column is given the number of degrees. Down the left-hand side is the number of minutes. At the heads of the columns are the names of the functions.

[1] Tables numbered in Roman numerals are at the back of the book.

To Use Tables III and IV from 45 *to* 89°.   For these angles, the degrees are at the **bottom** of the pages, the minutes **up** the **right-hand** column, and the names of the functions at the **foot** of each column.

14-4. Interpolation.   When a function is desired of an angle in degrees, minutes, and **seconds,** it is found by proportion between the values for the nearest minutes. Assume that the tangent of 69°31′10″ is desired.

$$\tan 69°32' = 2.67937$$
$$\tan 69°31' = 2.67700$$

The tangent increases 237 in the last places when the angle increases 1 minute (60 seconds). This is the **tabular difference.** The increase for 10 seconds is

$$^{10}\!\!/_{60} \times 237 = 40$$

Thus $\tan 69°31'10'' = 2.67700 + 40 = 2.67740$

These small interpolation computations are usually best performed on the slide rule. Sometimes small extra tables are provided which aid interpolation.

Examples.   Find

1. sin 32°28′16″	*Ans.* 0.53687
2. cos 21°46′49″	*Ans.* 0.92861
3. tan 72°13′37″	*Ans.* 3.11967
4. cot 29°04′22″	*Ans.* 1.79866
5. cos 78°36′08″	*Ans.* 0.19762
6. sin 55°13′25″	*Ans.* 0.82139

The same principle applies in reverse. Assume that the angle whose cosine is 0.78064 is desired.

$$0.78079 = \cos 38°40'$$
$$0.78061 = \cos 38°41'$$

In the last two places, the cosine **decreases** 18 for an **increase** in the angle of 60 seconds. The tabular difference is therefore −18. The given cosine is 15 less than the cosine for 38°40′. This is **your** difference. Hence

$$^{15}\!\!/_{18} \times 60 = 50$$
$$38°40'50'' = \text{desired angle}$$

It is well to remember that functions whose names begin with co-, like cosine and cotangent, decrease as the angle becomes larger, between 0 and 90°.

Examples. Find:

1. arc tan 0.53291   *Ans.* 28°03'13"
2. arc sin 0.92318   *Ans.* 67°23'44"
3. $\sin^{-1}$ 0.43795   *Ans.* 25°58'23"
4. $\cot^{-1}$ 4.03620   *Ans.* 13°54'55"
5. arc cos 0.82174   *Ans.* 34°44'26"
6. $\tan^{-1}$ 1.84262   *Ans.* 61°30'40"

14-5. Solution of Right Triangles. By using the foregoing concepts, every right triangle can be solved if two of its parts are known, provided at least one side is known.

Example. Given the right triangle in Fig. 14-3, find the side c and the angles A and B.

$$\tan A = \frac{156.74}{240.38} = 0.65205$$

$$A = 33°06'23''$$

$$c = \frac{240.38}{\cos 33°06'23''} = \frac{240.38}{0.83766}$$

$$c = 286.97$$

$$\tan B = \frac{240.38}{156.74} = 1.53362$$

$$B = 56°53'37''$$

Check: $A = 33°06'23''$
           $+ B = 56°53'37''$
           $\overline{\ 89°59'60''}$
        or $90°00'00''$

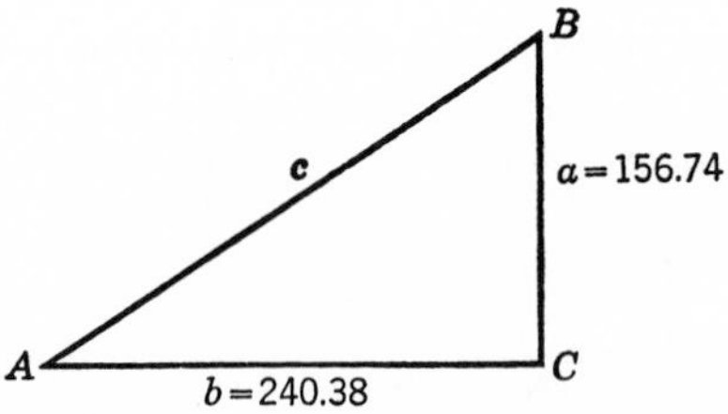

Fig. 14-3. Right triangle to be solved.

LOGARITHMS

14-6. Purpose of Logarithms. Accurate trigonometric computation obviously utilizes large numbers of digits. If possible, a desk or field calculating machine should be used. If no machine is available, logarithms are used, as they reduce the work of hand calculation. Logarithms are not necessary for an understanding of trigonometry. Sections 14-6 through 14-8, *Logarithms*, may be omitted if desired.

Principle. Note the following:

$$2 \times 2 \times 2 = (2)^3 = 8$$
$$2 \times 2 = (2)^2 = 4$$
$$2 \times 2 \times 2 \times 2 \times 2 = (2)^5 = 32$$

Therefore

$$(2)^2 \times (2)^3 = (2)^5 = 32$$
or
$$(2)^{2+3} = 2^5 = 32$$

In other words, adding exponents has the effect of multiplication.

Likewise,

$$\frac{2 \times 2 \times 2 \times 2 \times 2}{2 \times 2} = \frac{(2)^5}{(2)^2} = \frac{32}{4} = 8$$

Therefore,

$$\frac{(2)^5}{(2)^2} = (2)^{5-2} = 2^3 = 8$$

Subtracting exponents has the effect of division.

Definitions. A logarithm is an exponent (power) of a certain number called the base. In the above examples, the base is 2.

The logarithm of a number is the power to which the base must be raised to equal the number. For example, the logarithm of 32, to the base 2, is 5. This is written

$$\log_2 32 = 5$$

Since a logarithm is an exponent, multiplication can be performed by adding logarithms, and division by subtracting logarithms.

There is a great advantage in using 10 as a base. Note the following:

$$(10)^1 = 10 \quad \text{or} \quad \log_{10} 10 = 1$$
$$(10)^2 = 100 \quad \text{or} \quad \log_{10} 100 = 2$$
$$(10)^3 = 1000 \quad \text{or} \quad \log_{10} 1000 = 3$$
$$\cdots\cdots\cdots \qquad \cdots\cdots\cdots$$

and

or

$$100 \times 1000 = 100{,}000 \qquad \log_{10} 100 = 2$$
$$(10)^2 \times (10)^3 = (10)^5 \qquad + \log_{10} 1000 = 3$$
$$\overline{} \quad \overline{\log_{10} 100{,}000 = 5}$$

Consider the number 12. It is larger than $(10)^1$ and smaller than $(10)^2$. Therefore its $\log_{10}$ is somewhere between 1 and 2. The value of the logarithm of any number can be computed. A list of the logarithms of numbers is called a **table of logarithms.** From such a table the $\log_{10}$ of 12 is found to be 1.079181.

What is the $\log_{10}$ of 120?

$$120 = 10 \times 12$$
$$\log_{10} 10 = 1.000000$$
$$+ \log_{10} 12 = 1.079181$$
$$\overline{\log_{10} 120 = 2.079181}$$

Note that shifting the decimal point in the number merely changes the part of the logarithm in front of its decimal point. The rest of the logarithm remains unchanged. This is the great advantage of using the base 10. When no base is stated, the base is assumed to be 10; i.e., log 120 = log$_{10}$ 120. Also, logarithms to the base 10 are often called **common logarithms.**

The part of the logarithm in front of the decimal point is called the **characteristic,** and the part after the decimal point is called the **mantissa.** *Note:* When a logarithm is known, the corresponding number is called the **antilog.** Thus 120 is the antilog of 2.079181.

14-7. To Use Logarithms. Table I gives the mantissa, to the base 10, of numbers made up of four significant figures. By interpolation, five and even six significant figures can be found. A decimal point, not shown, is assumed to be in front of each mantissa. The numbers have no decimal point, as the mantissa is the same wherever the decimal point is located. For example, to find the mantissa of 2, look up the mantissa for 2000.

Example. Find the logarithm of 567.324. Since this number is between 100, or 10^2, and 1000, or 10^3, the characteristic is 2.

The rule for finding the characteristic is as follows:

Start to the right of the first significant figure and count the digits to the decimal point. Counting to the right indicates a plus characteristic. Counting to the left indicates a minus characteristic.

In this case, 5 is the first significant figure. According to the rule, +2 is the characteristic. Other examples are as follows.

Number	*Characteristic*
5673.24	+3
5.67324	0
0.567324	−1
0.00567324	−3
0.0000567324	−5

The mantissa of the number 567.324 is found as follows: Find the first three digits, 567, in the column headed "*N*." Find the next digit, 3, in the same line in the column headed "3." In this place appears 3813. Note that this is preceded by 75, which, to save space, is not repeated, so that the whole mantissa of 5673 is 753813. By the same process, the mantissa of 5674 is 753889.

The exact mantissa is found by interpolation. The difference between the mantissas for 5673 and 5674 is 76. This is known as the tabular difference. The number wanted is $^{24}/_{100}$ of this difference more than the mantissa for 5673:

$$^{24}/_{100} \times 76 = 18.24$$

Then
$$753813$$
$$+18.24$$
$$\overline{753831}$$

Applying the characteristic,

$$\log 567.324 = 2.753831$$

Use of Proportional Parts. Table I has additional tables which assist in interpolation. In the column "Diff." (which stands for **tabular differ-**

ence) are listed the differences between adjacent mantissas. They are placed so that most of the mantissas in the nearby lines differ by .the value stated, so that a glance will give the exact tabular difference.

On every pair of open pages are tables of **proportional parts** which give, for every tabular difference that appears on the two pages, the value of each difference multiplied by 0.1, 0.2, 0.3, etc., through 0.9.

To use these aids to interpolation in the example given, proceed as follows.

In the line where 3813 appears, the tabular difference listed is 77. A quick glance shows that the difference between 3813 and the next larger value 3889 ends in 6, so that the actual difference is 76. Find 76 under "Diff." in the proportional parts and write down the number given under 2, which is 15.2 (see below). This, added to the mantissa of 5673, gives the mantissa for 56732. In the same line in the proportional parts, divide the value under 4 by 10 (30.4/10 = 3.04) and write it down (see below). The value of the log of 567324 is found by adding; thus

Number	*Mantissa*
5673	753813
For the next digit (2)	+15.2
For the final digit (4)	+ 3.04
567324	753831

Applying the characteristic gives, as before,

$$\log 567.324 = 2.753831$$

When the number that precedes the listed mantissa changes, the point where it changes is shown by a horizontal line. Note that the complete mantissa for the number 5754 is 759970 and that for 5755 is 760045.

Examples of Computations by Logarithms

Example 1. Required: 56.935 × 487.67.

$$
\begin{array}{rl}
5693 = & 755341 \qquad \text{Diff. 76} \\
5 = & \underline{+38.0} \\
& 755379 \\
\log 56.935 = & 1.755379 \\
\\
4876 = & 688064 \qquad \text{Diff. 89} \\
7 = & \underline{+62.3} \\
& 688126 \\
\log 487.67 = & 2.688126
\end{array}
$$

The computation is set up as follows:

$$
\begin{aligned}
\log 56.935 &= 1.755379 \\
+ \log 487.67 &= 2.688126 \\
\hline
\text{Sum} = \log \text{ of product} &= 4.443505 \\
\text{Product (see below)} &= \quad 27765 \\
\text{or} \quad &\quad 27765.5
\end{aligned}
$$

To find a number from a logarithm, find the mantissa that is just less than the computed mantissa, 443505. This is 443419, corresponding to the number 2776. The mantissa wanted is 86 larger (your difference). The tabular difference is 157. In the proportional parts on the line for 157 find the number nearest[1] to your difference. It is 78.5 in the column for 5.

Thus, 27765 is the complete number. To find the decimal point, count, from the right of 2, four digits and place the decimal point after the 5. Then the product is 27765. If one more digit is wanted, proceed as follows:

$$
\begin{aligned}
\text{Your difference} &= \quad 86 \\
\text{Less tabular value in proportional parts} &= \underline{-78.5} \\
&\qquad 7.5
\end{aligned}
$$

Multiply this value by 10. Thus, $10 \times 7.5 = 75$. In the same line (157) find the number 75. It is 78.5 in the column headed "5." Then 5 is the digit placed at the end of the number wanted, giving 27765.5.

Example 2. Required: $738.26 \div 29.828$.

$$
\begin{aligned}
\log 738.26 &= 2.868209 \\
- \log 29.828 &= 1.474624 \\
\hline
\log \text{ quotient} &= 1.393585 \\
\text{Quotient} &= 24.751
\end{aligned}
$$

Logarithms of Negative Numbers. When a negative number is multiplied or divided by logarithms, a minus sign in parentheses $(-)$ or the notation **neg** is placed after the logarithm merely to keep track of the signs.

Example. Required: $(2.871) \times (-472.8)$.

$$
\begin{aligned}
\log 2.871 &= \quad 0.458033 \\
\log -472.8 &= \quad 2.674677 \ (-) \\
\hline
\log \text{ product} &= \quad 3.132710 \ (-) \\
\text{Product} &= -1357.41
\end{aligned}
$$

[1] If six digits are desired, choose the number just less than your difference.

Minus Characteristics. Going back to the log of 12 = 1.079181, what is the log of $\frac{1}{12}$?

$$\begin{array}{rcl} \log 1 &=& 0.000000 \\ -\log 12 &=& 1.079181 \\ \hline \log \tfrac{1}{12} &=& -1.079181 \end{array}$$

To use this log would require a whole table of minus mantissas. Instead, the log of $\frac{1}{12}$ is written 9.920819 − 10. This is the same log because, if 10 is subtracted from 9.920819, the result is −1.079181. The computation is as follows. Note that to clarify the computation, 10 is added to and subtracted from the log of 1.

$$\begin{array}{rcl} \log 1 &=& 10.000000 - 10 \\ -\log 12 &=& 1.079181 \\ \hline \log \tfrac{1}{12} &=& 8.920819 - 10 \end{array}$$

The mantissa 0.920819 gives the number 83333. The characteristic −2 places the decimal point. The result is 0.083333, which is the value of $\frac{1}{12}$ very nearly.

When no confusion would result, the −10 at the end is omitted but understood to be present.

Example 1. Log 0.0073826 = 7.868209 − 10.
Example 2. Antilog 6.329032 − 10 = 0.00021332.
Example 3. Find the value of 0.036241/0.12539.

$$\begin{array}{rcl} \log 0.036241 &=& 8.559200 - 10 \\ -\log 0.12539 &=& 9.098263 - 10 \end{array}$$

Since this subtraction cannot be performed without introducing a minus mantissa, 10 is added to and then subtracted from the first logarithm. Thus

$$\begin{array}{rcl} \log 0.036241 &=& 18.559200 - 20 \\ -\log 0.12539 &=& 9.098263 - 10 \\ \hline \log \text{result} &=& 9.460937 - 10 \\ \text{Result} &=& 0.28903 \end{array}$$

14-8. Tables of Logarithms of Trigonometric Functions. Tables II and IIa are much like tables of functions; but, instead of the functions, they give the logarithms of the functions. For the present, the values given in the table of angles larger than 90° need not be used. Note that the characteristics of the logarithms are given every 5 degrees. The −10 after the log is omitted but understood to be present.

This type of table is always used in logarithmic computations to avoid first looking up the natural functions and then looking up their logarithms.

To Use Table II.   Table II gives the logarithms of the sines, cosines, tangents, and cotangents for each minute of angle from 2 to 88°. In the column "D.1"." are listed the values of the tabular differences divided by 60 and therefore the tabular differences **per second.** Note that these differences are the same for the tangent and cotangent columns.

Example 1.   To find: log sin 23°19′42″.

$$\begin{aligned} \log \sin 23°19′ &= 9.597490 - 10 \qquad \text{D.1″.} = 4.88\\ +4.88 \times 42 &= \underline{+205}\\ \log \sin 23°19′42″ &= 9.597695 - 10 \end{aligned}$$

Example 2.   To find: arc log tan 9.656317 − 10.

$$\begin{aligned} \text{Given} \quad & 9.656317 - 10\\ \log \tan 24°22′ = & \underline{9.656020 - 10} \qquad \text{D.1″} = 5.60\\ \text{Your difference} = & \quad 297\\ 297/5.60 = & \qquad 53 \text{ seconds}\\ \text{Desired angle} = & \; 24°22′53″ \end{aligned}$$

Example 3.   To find: log cos 23°19′42″.

$$\begin{aligned} \log \cos 23°19′ &= 9.962999 - 10 \qquad \text{D.1″} = -0.90\\ -0.90 \times 42 &= \underline{-38}\\ \log \cos 23°19′42″ &= 9.962961 - 10 \end{aligned}$$

Example 4.   To find: arc log cot 0.345528.

$$\begin{aligned} \text{Given} \quad & 0.345528\\ \log \cot 24°17′ = & \underline{0.345663} \qquad \text{D.1″} = -5.62\\ \text{Your difference} = & \quad -135\\ -135/-5.62 = & \qquad 24 \text{ seconds}\\ \text{Desired angle} = & \; 24°17′24″ \end{aligned}$$

To Use Table IIa.   The tabular differences of the following logarithmic functions change so rapidly that interpolation by proportion is inaccurate:

> 0 to 2°:   log sines, log tangents, and log cotangents
> 88 to 90°: log cosines, log tangents, and log cotangents

Fortunately, the logarithm of the total number of seconds in an angle can be determined accurately. Also, to change such a logarithm into the logarithm of the desired function, a logarithm called a **conversion logarithm** is applied which changes so slowly, for the angles discussed, that it need be looked up only to the nearest minute. These conversion logs are named S, T, and C and are applied to the log of the number of seconds, as shown below:

For sines: log of no. seconds $+ S$
For tangents: log of no. seconds $+ T$
For cotangents: $C -$ log of no. seconds

For angles from 88 to 90°, find the log cofunction of the complement. This is equal to the desired value, as explained in Sec. 14-2.

In Table II*a*, beside each minute is the total number of seconds in the given angle, as well as the values for S, T, and C.

Example 1. To find: the logs of the functions of 0°27′14″.

From Table II*a*,

″	′	S	T	C
1620	27	4.685570	4.685584	15.314416

Hence

$$\text{Seconds in 27 minutes} = 1620$$
$$\underline{+14}$$
$$\text{Seconds in } 0°27′14″ = 1634$$

$$\log 1634 = 3.213252$$
$$\underline{+S = 4.685570 - 10}$$
$$\log \sin 0°27′14″ = 7.898822 - 10$$

$$\log 1634 = 3.213252$$
$$\underline{+T = 4.685584 - 10}$$
$$\log \tan 0°27′14″ = 7.898836 - 10$$

$$C = 15.314416 - 10$$
$$\underline{- \log 1634 = 3.213252}$$
$$\log \cot 0°27′14″ = 2.101164$$

$$\log \cos 0°27′14″ = 9.999987 \text{ directly from table}$$

Example 2. To find: the logs of the functions of 89°32′46″. The complement is $90 - 89°32′46″ = 0°27′14″$.

$$\log \cos 89°32′46″ = \log \sin 0°27′14″ = 7.898822 - 10$$
$$\log \tan 89°32′46″ = \log \cot 0°27′14″ = 2.101164$$
$$\log \cot 89°32′46″ = \log \tan 0°27′14″ = 7.898836 - 10$$
$$\log \sin 89°32′46″ = 9.999987 \text{ directly from table}$$

Example 3. To find: arc log tan 7.898836 − 10.
From Table II*a*, the nearest value is 7.895099 for 0°27′, for which $T = 4.685584$.
Then

$$\log \tan \text{ angle desired} = 7.898836 - 10$$
$$\underline{-T = 4.685584 - 10}$$
$$\log \text{ no. seconds in desired angle} = 3.213252$$
$$\text{No. seconds} = 1634$$
$$\underline{- \text{No. seconds in 27 minutes} = 1620}$$
$$14$$
$$\text{Desired angle} = 0°27′14″$$

Example 4.   To find: arc log cot 12.101164.

From Table IIa, the nearest value is 12.104901 for 0°27′, for which $C = 15.314416$. Then

$$C = 15.314416 - 10$$
$$- \text{ log cot angle desired} = 12.101164 - 10$$
$$\overline{\text{log no. seconds in desired angle} = 3.213252}$$
$$\text{No. seconds} = 1634$$
$$- \text{ No. seconds in 27 minutes} = 1620$$
$$\overline{14}$$
$$\text{Desired angle} = 0°27′14″$$

From these examples the remaining combinations can be worked out.

OBLIQUE TRIANGLES

14-9. Trigonometric Functions of Angles Greater Than 90°.

Figure 14-4 shows how the functions of angles from 0 to 360° are formed, as illustrated by angle B and the sides a, b, and c.

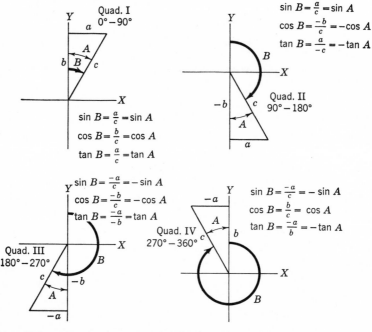

Fig. 14-4. Trigonometric functions of angles 0 to 360°. In Quadrant I, all functions of B are the same as those of A; in the other quadrants, they are as follows:

$$\sin B \text{ Quad. II} = \sin A$$
$$\cos B \text{ Quad. IV} = \cos A$$
$$\tan B \text{ Quad. III} = \tan A$$

In all other cases they are equal to the minus functions of A.

The functions are formed slightly differently in each of the four **quadrants.** The quadrants are named as follows: 0 to 90°, Quadrant I; 90 to 180°, Quadrant II; 180 to 270°, Quadrant III; and 270 to 360°, Quadrant IV.

Note the following:

1. All functions of B are numerically equal to the corresponding functions of A.
2. The signs of the functions of B are the same as signs of the functions of A, as follows:

> Quad. I: all functions
> Quad. II: sin
> Quad. III: tan
> Quad. IV: cos

Otherwise a function of B is equal to the function of A with the sign changed. A reciprocal function is handled in the same way as the function, viz., cosecants as sines, secants as cosines, cotangents as tangents.

3. To find a function of B: From the value B, compute the value of A by using Fig. 14-4, and look up the function of A. Then apply rules 1 and 2 above.

Example.   To find: cos 130°. This angle is in Quadrant II, where:

$$A = 180° - B = 50°, \text{ and } \cos B = - \cos A$$

$$\cos A = 0.64279$$

Therefore

$$\cos B = -0.64279$$

4. From Fig. 14-4, it is evident that every trigonometric function is the function of two different angles. Thus

> arc sin   0.42262 =   25° and 155°
> arc sin −0.42262 = 205° and 335°
> arc cos   0.90631 =   25° and 335°
> arc cos −0.90631 = 155° and 205°
> arc tan   0.46631 =   25° and 205°
> arc tan −0.46631 = 155° and 335°

14-10. Nomenclature.   The conventional system of naming the parts of an oblique triangle is used in this chapter. Capital letters are used for angles, and the side opposite each angle is given the same letter but in the lower case. See Fig. 14-6.

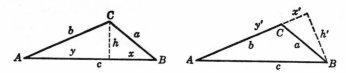

Fig. 14-5. Nomenclature of oblique triangles for derivations.

14-11. Solutions of Oblique Triangles. When three parts of an oblique triangle, one of which is a side, are given, the other parts can be computed. There are four possible cases, which are separately described in the following sections. Before following out the examples given, plot each triangle to the scale 1 in. = 100 ft, and check the results for gross errors by scaling.

Case I. Given One Side and Any Two Angles. In Fig. 14-5, first figure,

$$\frac{h}{a} = \sin B \qquad \frac{h}{b} = \sin A$$
$$h = a \sin B \qquad h = b \sin A$$

Equating,

$$a \sin B = b \sin A$$
$$\frac{a}{\sin A} = \frac{b}{\sin B}$$

In the second figure (which shows the same triangle),

$$\frac{h'}{a} = \sin (180° - C) = \sin C \qquad \frac{h'}{c} = \sin A$$
$$h' = a \sin C \qquad\qquad\qquad h' = c \sin A$$

Equating,

$$a \sin C = c \sin A$$
$$\frac{a}{\sin A} = \frac{c}{\sin C}$$

Hence

$$\frac{a}{\sin A} = \frac{b}{\sin B} = \frac{c}{\sin C} \tag{14-1}$$

(This is known as the law of sines.)

Note: To check the solution of any triangle, of any case, substitute the results in the formula $A + B + C = 180°$ and in the law of sines above. If the sum of the angles is 180° and the three parts, or the logs of the three parts of the law of sines, are equal, the solution is correct.

Example.   Given: $A = 38°54'37''$, $B = 78°12'42''$, $a = 326.39$ ft.

$$C = 180 - (A + B) \qquad b = \sin B \frac{a}{\sin A} \qquad c = \sin C \frac{a}{\sin A}$$

See the law of sines, Eq. (14-1).

$$
\begin{array}{ll}
A = \;\;38°54'37'' & 180°00'00'' \\
+B = \;\;78\;12\;42 & -117\;07\;19 \\
\hline
117°07'19'' \quad C = & 62°52'41''
\end{array}
$$

By Machine:

$$\frac{a}{\sin A} = \frac{326.39}{0.62810} = 519.647$$

$$b = (0.97891)(519.647) \qquad c = (0.89004)(519.647)$$
$$b = 508.69 \text{ ft} \qquad\qquad c = 462.51 \text{ ft}$$

By Logarithms:

$$\log a = 2.513737$$
$$-\log \sin A = 9.798031 - 10$$
$$\overline{\log \text{ quotient} = 2.715706}$$

$$
\begin{array}{ll}
\log \sin B = 9.990742 - 10 & \log \sin C = 9.949408 - 10 \\
+\log \text{ quotient} = 2.715706 & +\log \text{ quotient} = 2.715706 \\
\hline
\log b = 2.706448 & \log c = 2.665114 \\
b = 508.68 \text{ ft} & c = 462.50 \text{ ft}
\end{array}
$$

Results:

By machine:	$C = 62°52'41''$	$b = 508.69$ ft	$c = 462.51$ ft
By logs:	$C = 62°52'41''$	$b = 508.68$ ft	$c = 462.50$ ft

Case II. Given Two Sides and the Angle Opposite One of Them. If the side opposite the given angle is shorter than the other given side, there are two solutions (see Fig. 14-6). From the Eq. (14-1),

$$\sin B = b\,\frac{\sin A}{a} \qquad C = 180° - (A + B) \qquad c = \sin C\,\frac{a}{\sin A}$$

Example. Given: $A = 38°54'37''$, $a = 326.39$ ft, $b = 508.69$ ft.
By Machine:

$$\sin B = 508.69\,\frac{0.62810}{326.39} = 0.97891$$

$$B = 78°12'40'' \qquad \text{or} \qquad 180° - 78°12'40'' = 101°47'20''$$

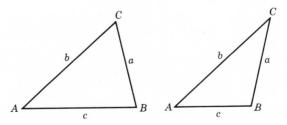

Fig. 14-6. The two solutions of Case II. Note that angle A and the sides a and b are each the same in both triangles.

See Sec. 14-9. Call the obtuse angle B'. Then

$$C = 62°52'43'' \quad \text{and} \quad C' = 39°18'03''$$

$$c = 0.89004 \frac{326.39}{0.62810} = 462.51$$

$$c' = 0.63339 \frac{326.39}{0.62810} = 329.14$$

By Logarithms:

$$\log \sin A = 9.798031 - 10$$
$$- \log a = 2.513737$$
$$\log \sin A/a = \overline{7.284294 - 10}$$
$$+ \log b = 2.706453$$
$$\log \sin B = \overline{9.990747 - 10}$$

$$B = 78°12'53'' \quad \text{or} \quad 180° - 78°12'53'' = 101°47'07''$$

See Sec. 14-9. Call the obtuse angle B'. Then

$$C = 62°52'30'' \quad \text{and} \quad C' = 39°18'16''$$

$$\log a = 12.513737 - 10$$
$$- \log \sin A = 9.798031$$
$$\log a/\sin A = \overline{2.715706}$$

$\log a/\sin A = 12.715706 - 10$	$\log a/\sin A = 12.715706 - 10$
$+ \log \sin C = 9.949396 - 10$	$+ \log \sin C' = 9.801706 - 10$
$\log c = \overline{2.665102}$	$\log c' = \overline{2.517412}$
$c = 462.49$	$c' = 329.16$

Results:

By machine:	$B = 78°12'40''$	$C = 62°52'43''$	$c = 462.51$
or	$B' = 101°47'20''$	$C' = 39°18'03''$	$c' = 329.14$
By logs:	$B = 78°12'53''$	$C = 62°52'30''$	$c = 462.49$
or	$B' = 101°47'07''$	$C' = 39°18'16''$	$c' = 329.16$

Case III. Given Two Sides and the Included Angle.   From Fig. 14-5,

$$\tan B = \frac{h}{x} = \frac{b \sin A}{c - b \cos A} \qquad a = \frac{x\dagger}{\cos B} \qquad C = 180° - (A + B)$$

$$a = \sin A \frac{b}{\sin B} = \frac{h}{\sin B}$$

Example.   Given: $b = 508.69$ ft, $c = 462.51$ ft, $A = 38°54'37''$.
By Machine:

$$\tan B = \frac{h}{x} = \frac{508.69(0.62810)}{462.51 - 508.69(0.77813)} = \frac{319.508}{66.683}$$

$$\tan B = 4.79145 \qquad B = 78°12'41''$$

$$a = \frac{66.683}{0.20450} = 326.40 \qquad a = \frac{319.508}{.97891} = 326.39$$

$$C = 62°52'42''$$

$\dagger$ If x is small, use the Eq. (14-1) for side a.

By Logarithms:

$$\begin{aligned}
\log b &= 2.706453 \\
+ \log \sin A &= 9.798031 - 10 \\
\hline
\log h &= 2.504484 \\
- \log x &= 1.823996 \\
\hline
\log \tan B &= 0.680497 \\
B &= 78°12'44''
\end{aligned}$$

$$\begin{aligned}
\log b &= 2.706453\ (-) \\
+ \log \cos A &= 9.891052 - 10 \\
\hline
\log \text{product} &= 2.597505\ (-) \\
\text{antilog} &= -395.83 \\
+c &= \quad 462.51 \\
\hline
x &= \quad 66.68
\end{aligned}$$

$$\begin{aligned}
\log x &= 11.823996 - 10 \\
- \log \cos B &= -9.310241 \\
\hline
\log a &= \quad 2.513755 \\
a &= \quad 326.40
\end{aligned}$$

$$\begin{aligned}
\log h &= 12.504484 \\
- \log \sin B &= \quad 9.990743 \\
\hline
\log a &= \quad 2.513741 \\
a &= \quad 326.39
\end{aligned}$$

$$C = 62°52'39''$$

Results:

By machine:	$B = 78°12'41''$	$C = 62°52'42''$	$a = 326.39$
By logs:	$B = 78°12'44''$	$C = 62°52'39''$	$a = 326.39$

Case IV. Given Three Sides.   From Fig. 14-5,

$$h^2 = a^2 - x^2 \qquad h^2 = b^2 - y^2$$
$$a^2 - x^2 = b^2 - y^2$$
$$y^2 - x^2 = b^2 - a^2$$
$$(y + x)(y - x) = (b + a)(b - a)$$
$$y - x = \frac{(b + a)(b - a)}{y + x}$$

From Fig. 14-5,

$$y + x = c$$
$$y - x = \frac{(b + a)(b - a)}{c} \tag{14-2}$$

$$\begin{array}{ll}
y = c - x & x = c - y \\
y = y & x = x \\
2y = c + (y - x) & 2x = c - (y - x)
\end{array}$$

Adding,

$$y = \frac{c + (y - x)}{2} \tag{14-3}$$

$$x = \frac{c - (y - x)}{2} \tag{14-4}$$

$$\cos A = \frac{y}{b} \tag{14-5}$$

$$\cos B = \frac{x}{a} \tag{14-6}$$

$$C = 180° - (A + B) \tag{14-7}$$

Example.   Given: $a = 326.39$ ft, $b = 508.69$ ft, $c = 462.51$ ft. Then

$$b + a = 835.08 \qquad b - a = 182.30$$

By Machine:
From Eq. (14-2),

$$y - x = \frac{(835.08)(182.30)}{462.51} = \frac{152235}{462.51} = 329.150$$

From Eqs. (14-3) and (14-4)

$c + (y - x) = 791.66$	$c - (y - x) = 133.36$
Divide by 2 $y = 395.83$	$x = 66.68$

From Eqs. (14-5) and (14-6),

$$\cos A = \frac{y}{b} = \frac{395.83}{508.69} = 0.77814 \qquad \cos B = \frac{x}{a} = \frac{66.68}{326.39} = 0.20430$$

$$A = 38°54'32'' \qquad\qquad B = 78°12'41''$$

From Eq. (14-7),

$$C = 62°52'47''$$

By Logarithms:

From Eq. (14-2),

$$\begin{aligned}
\log 835.08 &= 2.921728 \\
+ \log 182.30 &= 2.260787 \\
\hline
&\quad\, 5.182515 \\
- \log 462.51 &= 2.665121 \\
\hline
\log (y - x) &= 2.517394 \\
y - x &= \quad 329.15
\end{aligned}$$

From Eqs. (14-3) and (14-4)

$$c + (y - x) = \quad 791.66 \qquad\qquad c - (y - x) = 133.36$$

Dividing by 2 gives

$$y = \quad 395.83 \qquad\qquad x = 66.68$$

From Eqs. (14-5) and (14-6)

$$\begin{aligned}
\log y &= 12.597509 - 10 & \log x &= 11.823996 - 10 \\
- \log b &= 2.706453 & - \log b &= 2.513738 \\
\hline
\log \cos A &= 9.891056 - 10 & \log \cos B &= 9.310258 - 10 \\
A &= 38°54'35'' & B &= 78°12'42''
\end{aligned}$$

From Eq. (14-7),

$$C = 62°52'43''$$

Results:

By machine: $A = 38°54'32''$ $B = 78°12'41''$ $C = 62°52'47''$
By logs: $A = 38°54'35''$ $B = 78°12'42''$ $C = 62°52'43''$

TRIGONOMETRIC IDENTITIES

14-12. Most Useful Identities. A trigonometric identity is an equation that is true for all values of the angles expressed. Identities are very useful for solving many surveying problems.[1] Note the right triangle in Table 14-1.

$$\frac{a}{c} \times \frac{c}{b} = \frac{a}{b}$$

Hence

$$\frac{\sin A}{\cos A} = \tan A \tag{Id. 1}$$

Also, $$a^2 + b^2 = c^2$$

Dividing by c^2, $$\frac{a^2}{c^2} + \frac{b^2}{c^2} = 1$$

Hence, $$\sin^2 A + \cos^2 A = 1 \tag{Id. 2}$$

Again, $$a^2 + b^2 = c^2$$

Dividing by b^2, $$\frac{a^2}{b^2} + 1 = \frac{c^2}{b^2}$$

Hence, $$\tan^2 A + 1 = \sec^2 A \tag{Id. 3}$$

From Id. 1,

$$\sin A = \tan A \cos A$$

$$\sin A = \frac{\tan A}{\sec A}$$

$$\sin A = \frac{\tan A}{\sqrt{\sec^2 A}}$$

From Id. 3,

$$\sin A = \frac{\tan A}{\sqrt{1 + \tan^2 A}} \tag{Id. 4}$$

These are the most useful identities for surveying. A short list of others is given without showing their development.

14-13. Additional Identities. In these identities, A and B are independent, and either can have any value.

$$\cos A = \frac{1}{\sqrt{1 + \tan^2 A}}$$

$$\tan A = \frac{\sin A}{\sqrt{1 - \sin^2 A}} = \frac{\sqrt{1 - \cos^2 A}}{\cos A}$$

$$\sin(A \pm B) = \sin A \cos B \pm \cos A \sin B$$

$$\cos(A \pm B) = \cos A \cos B \mp \sin A \sin B$$

$$\tan(A \pm B) = \frac{\tan A \pm \tan B}{1 \mp \tan A \tan B}$$

---

[1] When a trigonometric function is squared, as $(\sin A)^2$, it is written $\sin^2 A$ to avoid the parentheses.

$$\cot (A \pm B) = \frac{\cot A \cot B \mp 1}{\cot B \pm \cot A}$$

$$\sin A + \sin B = 2 \sin \tfrac{1}{2} (A + B) \cos \tfrac{1}{2} (A - B)$$
$$\sin A - \sin B = 2 \cos \tfrac{1}{2} (A + B) \sin \tfrac{1}{2} (A - B)$$
$$\cos A + \cos B = 2 \cos \tfrac{1}{2} (A + B) \cos \tfrac{1}{2} (A - B)$$
$$\cos A - \cos B = - 2 \sin \tfrac{1}{2} (A + B) \sin \tfrac{1}{2} (A - B)$$

$$\sin 2A = 2 \sin A \cos A$$
$$\cos 2A = \cos^2 A - \sin^2 A = 1 - 2 \sin^2 A = 2 \cos^2 A - 1$$
$$\tan 2A = \frac{2 \tan A}{1 - \tan^2 A}$$
$$\cot 2A = \frac{\cot^2 A - 1}{2 \cot A}$$
$$\sin \tfrac{1}{2}A = \sqrt{\tfrac{1}{2}(1 - \cos A)}$$
$$\cos \tfrac{1}{2}A = \sqrt{\tfrac{1}{2}(1 + \cos A)}$$
$$\tan \tfrac{1}{2}A = \frac{\tan A}{1 + \sec A} = \csc A - \cot A = \frac{1 - \cos A}{\sin A} = \sqrt{\frac{1 - \cos A}{1 + \cos A}}$$
$$\cot \tfrac{1}{2}A = \frac{1 + \cos A}{\sin A} = \frac{1}{\csc A - \cot A} = \sqrt{\frac{1 + \cos A}{1 - \cos A}}$$

Problems

Natural Functions 0 to 90°

14-1 to 14-16. Find the values of the following between 0 and 90° to seconds:

14-1. tan 27°32′17″	**14-2.** tan 25°41′26″
14-3. cos 64°19′43″	**14-4.** cos 71°23′46″
14-5. sin 31°14′27″	**14-6.** sin 35°17′46″
14-7. cot 79°58′13″	**14-8.** cot 76°24′32″
14-9. arc sin 0.54786	**14-10.** arc sin 0.42634
14-11. $\cos^{-1}$ 0.81323	**14-12.** $\cos^{-1}$ 0.92138
14-13. $\cot^{-1}$ 3.82974	**14-14.** $\cot^{-1}$ 4.60729
14-15. arc tan 0.68329	**14-16.** arc tan 0.54726

Logarithms of Numbers

14-17 to 14-24. Find the values of the following:

14-17. log 431.82	**14-18.** log 239.45
14-19. log 0.054776	**14-20.** log 0.062314
14-21. log 2.8954	**14-22.** log 2.9528
41-23. log 37.7467	**14-24.** log 36.4825

14-25 to 14-32. Find the values
figures:

14-25. antilog 2.579264   **14-26.** antilog 3.742859
14-27. antilog 8.724381 − 10   **14-28.** antilog 7.428136 − 10
14-29. antilog 0.516437 (−)   **14-30.** antilog 0.463597 (−)
14-31. antilog 3.640102   **14-32.** antilog 2.650215

Logarithms of Functions

14-33 to 14-40. Find the values of the following to six decimal places:

14-33. log tan 23°13′54″   **14-34.** log tan 28°42′38″
14-35. log cos 58°42′25″   **14-36.** log cos 63°34′17″
14-37. log sin 42°55′14″   **14-38.** log sin 43°40′29″
14-39. log cot 67°32′28″   **14-40.** log cot 72°37′51″

14-41 to 14-60. Find the values of the following between 0 and 180°
to seconds:

14-41. arc log sin 9.712234   **14-42.** arc log sin 9.625532
14-43. arc log cos 9.257426   **14-44.** arc log cos 9.285741
14-45. log cot⁻¹ 0.465374   **14-46.** log cot⁻¹ 0.483742
14-47. log tan⁻¹ 9.775843   **14-48.** log tan⁻¹ 9.786743
14-49. log sin 1°23′18″   **14-50.** log sin 1°32′21″
14-51. log cot 88°15′34″   **14-52.** log cot 88°27′42″
14-53. log cot 0°17′52″   **14-54.** log cot 0°28′19″
14-55. arc log sin 8.427458   **14-56.** arc log sin 8.384792
14-57. arc log cot 7.983742   **14-58.** arc log cot 8.014368
14-59. arc log tan 1.795431   **14-60.** arc log tan 1.623847

Right Triangles

14-61 to 14-88. Plot all triangles to the scale 1 in. = 100 ft and then
compute the parts not given.

14-61. $A = 40°10′13″$, hyp. = 402.36 ft
14-62. $A = 42°23′12″$, hyp. = 437.25 ft
14-63. $A = 62°09′15″$, opp. = 338.74 ft
14-64. $A = 61°28′47″$, opp. = 345.51 ft
14-65. $A = 36°22′10″$, adj. = 360.41 ft
14-66. $A = 35°46′17″$, adj. = 358.17 ft
14-67. Hyp. = 428.29 ft, opp. = 397.06 ft
14-68. Hyp. = 432.89 ft, opp. = 398.24 ft
14-69. Hyp. = 409.31 ft, adj. = 274.82 ft

14-70. Hyp. = 471.65 ft, adj. = 270.46 ft
14-71. Opp. = 375.82 ft, adj. = 276.05 ft
14-72. Opp. = 368.47 ft, adj. = 274.61 ft

Oblique Triangles

Note: Use the convention that capital letters indicate angles; lowercase, sides. Each side is opposite the angle having the same letter.

14-73. $A = 63°29'10''$, $B = 58°42'07''$, $b = 458.24$ ft
14-74. $A = 74°22'53''$, $B = 34°15'45''$, $a = 287.46$ ft
14-75. $A = 27°38'14''$, $B = 32°18'25''$, $c = 348.27$ ft
14-76. $A = 48°17'35''$, $B = 64°26'41''$, $c = 396.41$ ft
14-77. $A = 35°21'54''$, $a = 315.46$ ft, $b = 478.28$ ft
14-78. $A = 25°04'16''$, $a = 228.71$ ft, $b = 517.09$ ft
14-79. $A = 64°27'13''$, $a = 357.46$ ft, $b = 295.87$ ft
14-80. $A = 59°17'23''$, $a = 451.14$ ft, $b = 398.36$ ft
14-81. $A = 51°10'13''$, $b = 358.15$ ft, $c = 307.01$ ft
14-82. $A = 55°42'35''$, $b = 426.82$ ft, $c = 411.28$ ft
14-83. $A = 61°50'29''$, $b = 451.63$ ft, $c = 197.17$ ft
14-84. $A = 67°04'41''$, $b = 475.74$ ft, $c = 162.27$ ft
14-85. $a = 289.95$ ft, $b = 363.75$ ft, $c = 497.38$ ft
14-86. $a = 305.13$ ft, $b = 485.27$ ft, $c = 572.16$ ft
14-87. $a = 626.14$ ft, $b = 286.21$ ft, $c = 512.69$ ft
14-88. $a = 582.31$ ft, $b = 250.28$ ft, $c = 436.14$ ft

15

Drawing Maps and Keeping Records

MAPPING

15-1. Maps. Maps have many uses and are made accordingly. When maps are to be used for design, it must be possible to determine distances, elevations, and angles from them by scaling. Usually the entire map must have a uniform standard of accuracy so that these data may be determined anywhere on it with equally accurate results. Since the survey on which a map is based can be readily made more accurate than can any drafting procedure, map accuracy is limited mainly by the accuracy of drafting. This imposes a high standard on the drafting technique required, and special methods must be employed. Steel straightedges must be used for important lines, important positions must be pricked with a needle, and important directions must be established by precise linear scaling. The T square and wooden-edged drawing board cannot be relied on for drawing parallel lines, and any parallel or angular drafting equipment must be tested thoroughly before it can be safely adopted.

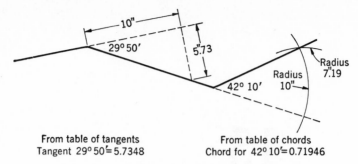

Fig. 15-1. Plotting angles by tangents or chords.

When less precision is desired, methods of plotting will occur to the reader. When coordinates are not computed, angles can be plotted by protractor or by setting out tangents or chords (see Fig. 15-1), and lengths can be scaled directly. Only precise methods are covered in the following paragraphs.

15-2. Steps in Making a Map.   The procedure for mapping is as follows:

1. Determine the types of features to be included
2. Draw an approximate small-scale sketch of the area to be included in the map
3. Establish the scale
4. Establish the size of the map sheets
5. Determine the arrangement of the map
6. Construct the grid (the graticule)
7. Plot the control
8. Plot features
9. Add details
10. Ink

15-3. Features to Be Included in the Map.   The features chosen to be included on the map, of course, depend on the purpose of the map and are usually stipulated before the survey is begun. Ordinarily all data obtained by the survey are included, for the cost of the survey is high, and maps are often used for purposes never considered when they were made. Data are sometimes omitted to avoid confusion.

Four items, independent of the survey, should **always** be included. They are

1. Statement of scale
2. Graphical indication of scale in case the sheet shrinks or expands or is photographically reproduced at an unknown scale

3. Title
4. North point, even if very approximate

Whenever it is necessary to give lengths or elevations to an accuracy greater than shown graphically, the values are lettered on the map near the features they represent.

15-4. The Approximate Sketch.   Before the map can be started, the arrangement of the map must be planned—if only to keep all the map on the paper. To plan the map, a small-scale sketch of the outline of the map must be obtained. Usually a sketch is made of the perimeter of the control system and the controlling external topographic observations. Sometimes the outline can be sketched by memory on an existing map (see Fig. 15-2).

15-5. Choice of Map Scale.   Three types of map scales are in common use. They are generally known as **engineer's scales, architect's scales** (usually in metric terms), and **ratio scales.** An engineer's scale gives the number of feet represented by 1 in. on the map and is written **1 in. = 20 ft** or **1 in. = 100 ft**, etc.

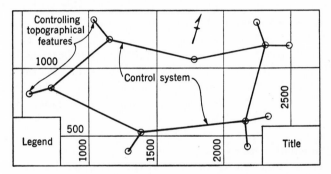

Fig. 15-2. Sketch map with sheet arrangement completed.

An architect's scale gives the fraction of an inch on the map that represents 1 ft and is written ⅛ **in. = 1 ft,** ¹⁄₁₆ **in. = 1 ft,** etc. A ratio scale gives the relative size of a distance on the map to the represented distance on the ground and is written 1:20,000, 1:62,500, etc.

The engineer's scale is the most convenient for mapping limited areas and is the most generally used. Rules for engineer's scales are usually made with the following number of spaces per inch: 10, 20, 30, 40, 50, 60. Eighty graduations per inch are sometimes found. Map scales are selected accordingly. The most common are 1 in. = 20, 40, 50, 100, 200, 500, 1000, and 2000 ft.

The map scale chosen should be the smallest at which the desired precision can be obtained. It is generally assumed that distances on a map can be measured to $\frac{1}{50}$ in. Thus, if distances are required to the nearest 0.4 ft, 0.4 ft should be represented by $\frac{1}{50}$ in. on the map, and the map scale would be 1 in. = 20 ft. In like manner, if distances are required to the nearest 10 ft, the scale should be 1 in. = 500 ft.

For large areas, the entire survey is planned according to the desired map accuracy. For example, if the largest distance in the area covered is 5000 ft and it is necessary to scale to the nearest foot, the maximum error allowed in the survey would be ±0.5 ft. Accordingly, the control system should have a minimum accuracy of 1:10,000, i.e., 0.5/5000, and the ties should be made to the nearest half foot. Stadia measurements (having an accuracy of 1:300) should not be made over 150 ft long. With such limits it might, of course, be theoretically possible to find distances on the map that were in error by 2 ft, but the limits described are usually accepted, for the adjustment of control and the laws of chance would practically eliminate the possibility of this occurrence.

Within small areas, it is difficult to make a well-planned survey that does not give position within the mapping accuracy desired.

When a map is to be traced for blueprinting, the scale must often be increased so that small details can be shown in the wider ink lines required.

15-6. The Size of Map Sheets. If a standard size of sheet is not required, the map should be drawn on one sheet if possible. The difficulty of using more than one sheet is so great that sheets as large as 40 by 60 in are used. They are hard to file, but they can be used on a drawing board. The sheets must be large enough to include a minimum border of $\frac{1}{2}$ in. to protect the map. Larger borders give a better appearance. Space must be reserved for the title and any required legends. Frequently, when small working sheets are required, the original map is drawn on a large sheet but traced in small sections.

15-7. The Arrangement of the Map. The border lines adopted should be drawn to scale on the sketch of the area (see Figs. 15-2 and 15-3). Sometimes a considerable overlap is used so that parts of the map appear on more than one sheet as in Fig. 15-3. When map sheets overlap, the coordinate grid shows how the sections are matched. Match lines must be used if no grid is drawn. Sometimes the borders are oriented to the north of the coordinate system, and sometimes they are placed at coordinate lines having round-number values.

15-8. The Grid. The coordinate system is shown on the map by a **graticule** of lines parallel to the axes, 5 to 10 in. apart (for ease in plotting) and representing round-number coordinate values. The approximate position of this grid system is found by scaling from the border lines on the sketch. The grid system can be laid out with a T square and triangle; but if an accurate grid is desired, an accurate right angle should be established and all lines thereafter laid out by scaling. This procedure

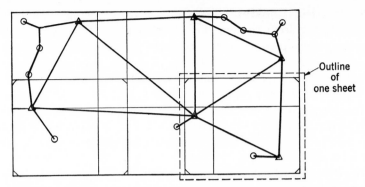

Fig. 15-3. Sketch map with overlapping sheet arrangement completed.

requires a steel straightedge but no precise drafting instrument other than the scale.

Figure 15-4 illustrates the method. Line 1 is placed by scaling the distances $S1$ and $S2$ on the sketch, and A is placed by scaling $S3$. The right angle at A is established by trial and error, the 3, 4, 5 method being used. Measuring from A, points are pricked along lines 1 and 2 according to the required spacing of the grid lines. Point D is located by trial measurements from B and C. Lines 3 and 4 are divided by prick marks, and finally opposite prick points are connected to form the grid.

All lines must be drawn with a very hard pencil sharpened to a round point and held against the bottom of the steel straightedge. Points must be pricked with a needle. After the grid is constructed, the spacing should be thoroughly checked with dividers.

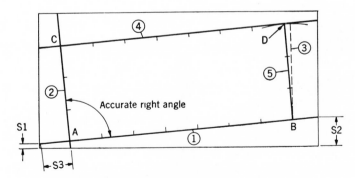

Fig. 15-4. Establishing an accurate grid. $S1$ and $S2$ are scaled from the sketch. Line 1 is drawn. $S3$ is scaled from the sketch and line 2 erected at exactly 90°. Point D is found by trial measurements from B and C.

15-9. Plotting Control. The control should be plotted by coordinates, measuring from the grid lines. The stations thus plotted are connected by lines representing the traverse or triangulation system, and their lengths and the angles between them are checked by protractor and scale against the **original field notes.** This check will disclose nearly all blunders and should never be omitted.

When the coordinates of control stations are not computed, at least the directions of the lines should be computed. Lines plotted by directions are not affected by the accumulated accidental errors caused by plotting successive angles.

15-10. Plotting Features. Stadia traverses and the positions of topographical features are plotted by protractor and scale, measuring from the control system. A drafting machine (Fig. 15-5) or a large paper protractor used as shown in Fig. 15-6 will be found to be a timesaver. Unless the drafting machine is quite accurate, the angular scale should be oriented at each station.

Plotters are available that can be programmed to plot maps, profiles, or cross sections from field notes.

15-11. The Title. In the past maps were often embellished with considerable artistry. Too frequently, however, the ornamentation covered lack of data. Today, some survival of this custom is found in ornate titles and other details. There may be some justification for them when the map is made for a private individual. Otherwise, the title and other details should be designed to give the maximum information at a glance. The

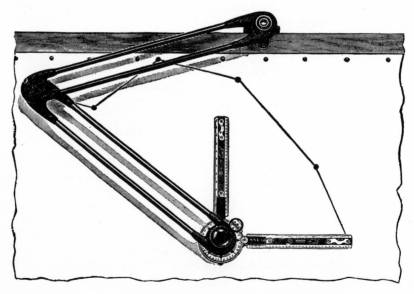

Fig. 15-5. A Keuffel & Esser drafting machine.   (*Keuffel & Esser Co.*)

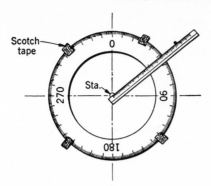

Fig. 15-6. Plotting topography by scale and cutout paper protractor.

JONES AND JONES CONSULTANTS

MAP

OF THE

SITE OF PROPOSED PLANT B

OF THE

SMITH MANUFACTURING CO.

LAKEVILLE, NEW YORK

MAY 3, 1947   SCALE 1 INCH=200 FEET

Fig. 15-7.

THE SMITH MANUFACTURING CO.		
Map of Site of Plant B *Lakeville, New York*		
Survey by: Thomas Smith		*May 3, 1947*
Scale: 1-Inch = 200 feet		*D'w'g. 2222*

Fig. 15-8.

title should contain the following items unless a good reason exists for their omission. They are stated in the usual order found in titles.

1. Organization making the map
2. Technical name (map, chart, or plan)
3. Name of the area mapped

4. Name of purchaser
5. Where area is located
6. Name of engineer responsible
7. Date of survey
8. Scale
9. Identification numbers

Frequently this complete list will result in duplication, and some items should be omitted.

Certain items should be emphasized by larger and heavier letters so that the map can be quickly selected from others. The order of emphasis should be the following:

Greatest emphasis, items 2, 3, 4
Medium emphasis, items 1, 5, 6
Least emphasis, items 7, 8, 9

ABCDEFGHIJKLM
NOPQRSTUVWXYZ
abcdefghijklmnopqrs
tuvwxyz&1234567890

ABCDEFGHIJKLM
NOPQRSTUVWXYZ
abcdefghijklmnopqrs
tuvwxyz&1234567890

Fig. 15-9.

A B C D E F G H I J

K L M N O P Q R S T

U V W X Y Z a b c d e f

g h i j k l m n o p q r s t u v

w x y z & 1 2 3 4 5 6 7 8 9 0

A B C D E F G H I J

K L M N O P Q R S T

U V W X Y Z a b c d e f

g h i j k l m n o p q r s t u v

w x y z & 1 2 3 4 5 6 7 8 9 0

Fig. 15-9. *(Continued)*

Examples of the same title used for different purposes are shown in Figs. 15-7 and 15-8. Figure 15-7 shows the kind of title that would be used by a consulting firm preparing the map for a manufacturing company. The simple form of title shown in Fig. 15-8 would be used by the manufacturing company when the map was prepared by their own personnel.

It will be found that vertical letters are more quickly read than slant letters. Roman type offers the best method of emphasis. Figure 15-9 gives useful alphabets. Special lettering devices that guide the pen, such as LeRoy,[1] Wrico,[2] or Varigraph,[3] are used to a large extent. They are available in various alphabets and symbols.

15-12. Other Details.   Figure 15-10 gives a good form of graphical scale. A graphical scale should be as long as possible, within reason. It can be omitted if a coordinate grid is used. A typical north point is shown

Fig. 15-10. A graphical scale.

in Fig. 15-11. Beside it should be a legend stating exactly whether it is true, magnetic, or assumed. Spot elevations and dimensions should be printed at the features to which they refer. Descriptive legends are often valuable aids to clarity. They are usually placed at the lower left-hand corner of the map.

15-13. Topographic Symbols.   Figure 15-12 shows the most useful typical topographic symbols. They are essential for small-scale maps where the features are small and detailed. A legend giving their meaning should accompany the map. On large-scale maps the features can usually be recognized and are easily labeled. The symbols are useful for outlining the limits of woods, swamps, etc.

15-14. Inking the Original Map.   The entire original map should be inked, including the coordinate grid and the survey control system. The identification numbers or letters and the coordinates of control stations, the identifications and elevations of benchmarks, the azimuths, and sometimes the lengths of lines should be printed in appropriate places. This information will prove of great assistance when additions to the map are required and when location surveys are necessary.

Fig. 15-11. A north point.

[1] ® Keuffel & Esser Co., Hoboken, N.J.

[2] ® Wood-Regan Instrument Co., Inc., Nutley, N.J.

[3] ® The Varigraph Company, Inc., Madison, Wis.

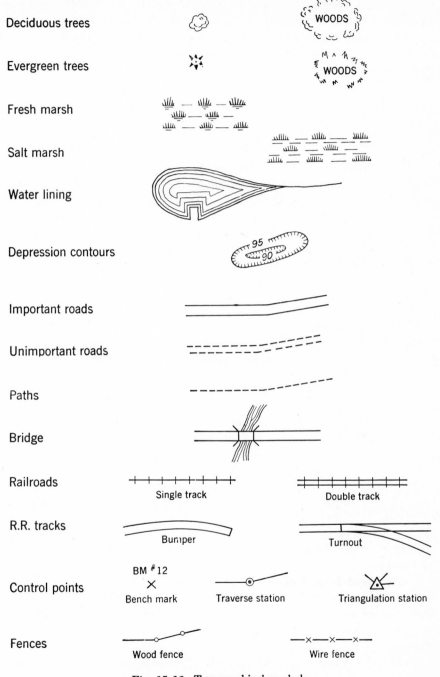

Fig. 15-12. Topographical symbols.

All control lines on the original map should be inked with very fine lines for accuracy and clarity. In general, fine lines are preferable except when emphasis is necessary. Often colored inks improve the appearance and legibility. The following schedule is representative of standard practice in the use of colors:

1. Culture (works of man).................... Black
2. Water................................... Blue
3. Relief (including contour lines)............. Brown
4. Survey points and lines.................... Red
5. Vegetation.............................. Various shades of green

15-15. Tracing. Tracings can never reach the accuracy of the original map and are almost always made for a special purpose. Usually, therefore, only the items and those parts of the map which are necessary to the purpose are traced, and heavy lines are used freely for emphasis. A graphical scale should be included, or the coordinate intersections should be shown by crosses to give scale, for both the tracing cloth and the blueprint change considerably in size from day to day.

RECORDS

15-16. Importance of Survey Records. Since survey data are invariably used for many purposes in addition to the specific purpose for which they were obtained, survey records are very valuable and should be preserved with great care.

15-17. Job Names. Every job must be given a name at the outset of the work. Methods of naming differ with the size of the organization and its functions. Usually the name of the purchaser or a **job number** is used. The job name must appear on all records except the records of the survey control points, for which it has no significance.

15-18. Records of Control Points. Control points nearly always have a continuing value. Each should be given a number for identification. A card index should be kept, with a card for each control point containing a description of the point, a sketch showing its position, including **witness** measurements to various nearby objects, and the coordinates or the elevations it represents. The cards, of course, are filed by number.

15-19. Field-book Records. The field books should be used chronologically and numbered and paged as they are issued to the field parties. The name or number of each job, the date, the members of the field party, and the weather should be recorded daily. Field books are filed by number.

15-20. The Computation Record. The computations should show the name or number of the job, the field-book number and page, and the date of computation. They are usually filed by job name or number.

15-21. The Map References.   The map shows the name of the job, the date of the survey, and the field-book number and page. Maps are filed by job name or number.

15-22. Survey Index Cards and the Index.   Since maps are hard to handle and often no map results from a survey, a card for each survey should be made that duplicates the references on the map or carries the same type of information when no map exists. Like maps, they are filed by job name or number.

A good index should originate with the item in the record most easily brought to mind. Accordingly, a survey index **must** originate with the location of the survey. By far the best index is an outline map of the areas where surveys are performed. On the outline map are placed, in their proper positions, the control points with their numbers, the job names and dates, and the field-book numbers and pages.

15-23. Method of Finding the Records.   With the system outlined above the survey records can be found according to the table below:

Known	Index	References to
Location...............	Outline map	Job name or number, field book, date, control points
Job name or number.....	Survey cards or map	Location, field book, data, computations
Date....................	Field book	Location, job name

Appendix A

Error Theory

A-1. Accidental Errors. In Chap. 2, a method of determining the 90 per cent error, the 50 per cent error, or any other percentage error was described. In each test a certain quantity was measured a number of times, the accidental error of each measurement determined, and the error equal to or larger than the desired percentage error selected from the list. Thus the result depended chiefly on a single measurement or the average measurement in each test. A more accurate result, which depends on all the measurements in each test, can be determined by applying the laws of chance. It is first necessary to determine the **standard error,** which is usually called lower-case sigma (σ).

A-2. To Determine the Standard Error. A measure of the accidental error of any procedure or single observation can be determined by the proper analysis of the results of previous surveys or by making special tests. The theory is more clearly demonstrated by using a test as an example. To make the test, proceed as follows:

Repeat the same procedure or single observation a number of times. Arrange the work so that the true value of the result is known, as it is when running a loop of levels beginning and ending at the same benchmark, or measuring the sum of all the angles in a complete circle around a point. Thirty repetitions are assumed to give a very accurate result. A few will give a useful result. Find the error of each run.

Continuing the example of a number of measurements of the same level loop, suppose that, instead of measuring the same loop a number of times, the same runs had been strung out, one after the other, in a series so that a single loop resulted; the error of such an arrangement would be equal to the sum of the errors of the individual loops. Since there is an equal chance that the error of any one of the runs had a plus or minus sign, the sum of the errors (E) would be expressed as

$$E = \pm v_1 \pm v_2 \pm v_3 \cdots \pm v_n$$

where v_i are the errors of the individual runs. How can this sum be found?

Let $E = \pm v_1 \pm v_2 \pm v_3$, the sums of the respective errors of three runs (any other number of runs can be used for demonstration). Then

$$E^2 = (\pm v_1 \pm v_2 \pm v_3)^2 = v_1{}^2 + v_2{}^2 + v_3{}^2 \pm 2v_1 v_2 \pm 2v_1 v_3 \pm 2v_2 v_3$$

Since the plus-or-minus terms tend to equal zero, they may be dropped; thus

$$E^2 = v_1{}^2 + v_2{}^2 + v_3{}^2$$

This can be written

$$E^2 = \Sigma v_i{}^2$$
$$E = \pm \sqrt{\Sigma v_i{}^2} \qquad \text{(A-1)}$$

328

where $\Sigma v_i{}^2$ is the sum of the squares of each of the errors.

When all the errors have the same value, this equation becomes

$$E = \pm \sqrt{nv^2} = \pm v \sqrt{n}$$

as in Eq. (2-2).

Let σ (sigma) be the value of an error such that, if each of the runs had the same error σ, E would be the same as in Eq. (A-1). σ is called the standard error of a procedure. It is sometimes called the **mean square error**, or the M.S.E.

To compute σ, let n equal the number of runs. Then, by Eq. (A-1),

$$E = \pm \sqrt{n\sigma^2} = \pm \sigma \sqrt{n} \qquad (A\text{-}2)$$

Equating the two values of E from Eqs. (A-2) and (A-1),

$$\pm \sigma \sqrt{n} = \pm \sqrt{\Sigma v_i{}^2}$$

Dividing by $\sqrt{n}$,

$$\sigma = \pm \sqrt{\frac{\Sigma v_i{}^2}{n}} \qquad (A\text{-}3)$$

If the runs cannot be arranged so that the true value is known, as, for example, runs between two benchmarks, the average of the results can be used as a basis for computing the various errors. In this case, $(1/n)$th part of each error is used in computing the average, and in n errors a total of one error is used, so that only $n - 1$ errors are actually contained in $\Sigma v_i{}^2$. In this case,

$$\sigma = \pm \sqrt{\frac{\Sigma v_i{}^2}{n - 1}} \qquad (A\text{-}4)$$

The rule is: The denominator under the radical must be the number of **redundant** measurements. When the true value is known, all the measurements are redundant. When the average is used, all measurements but one are redundant.

A-3. To Compute the Error of an Average.   Often, in order to obtain a very accurate value of a quantity, the quantity is measured several times and the average of the measurements is used. Let σ_0 = the standard error of an average. An average is found by dividing the sum of the measurements by n. The error of a sum is E. From Eq. (A-2),

$$E = \pm \sigma \sqrt{n}$$

Dividing by n gives the standard error of the average, when the true value is known:

$$\sigma_0 = \frac{E}{n} = \pm \sigma \frac{\sqrt{n}}{n}$$

$$= \pm \frac{\sigma}{\sqrt{n}}$$

From Eq. (A-4), when the true value is not known,

$$\sigma_0 = \pm \sqrt{\frac{\Sigma v_i{}^2}{n(n - 1)}} \qquad (A\text{-}5)$$

where the v's were computed from the average.

A-4. To Find Per Cent Errors from σ.   When σ is multiplied by various coefficients, per cent errors can be found, as follows:

$$0.6745\sigma = 50 \% \text{ error or the probable error (P.E.)} = E_{50} \qquad (A\text{-}6)$$

This is the error, exceeded in 50 per cent of the observations.

$$1.6449\sigma = 90\% \text{ error} = E_{90} \qquad \text{(A-7)}$$

which is the error exceeded in 10 per cent of the observations. This is the error used in designing a survey.

$$1.9599\sigma = 95\% \text{ error} = E_{95} \qquad \text{(A-8)}$$

This is the error exceeded in 5 per cent of the observations.

Note: Equations (A-1) and (A-2) apply to any of these percentage errors as well as to σ itself. For example,

E_{90} when n observations are involved $= (E_{90}$ for one observation$) \sqrt{n}$
E_{50} when n observations are involved $= (E_{50}$ for one observation$) \sqrt{n}$

A-5. Examples of Computational Procedures

Example 1.   Assume that a distance was measured five times, with the results shown in the first column of Table A-1. The probable error would be computed as shown.

Table A-1

Results	Average	v	v^2
$800.91 -$	$800.81 =$	$+0.10$	0.0100
$800.65 -$	$800.81 =$	-0.16	0.0256
$800.72 -$	$800.81 =$	-0.09	0.0081
$800.87 -$	$800.81 =$	$+0.06$	0.0036
$800.90 -$	$800.81 =$	$+0.09$	0.0081
Sums.............. 4004.05		0	0.0554
Average........... 800.81			

By Eq. (A-6)

$$\text{P.E. of one measurement} = 0.6745 \sqrt{\frac{0.0554}{4}} = \pm 0.079$$

This means that, if a single measurement of this distance is made by the same method, there is a 50 per cent chance that it will fall within the range 800.81 ± 0.079; that is to say, it will fall between 800.731 and 800.889, inclusive, provided, of course, that no blunders or systematic errors are introduced.

By Eqs. (A-5) and (A-6),

$$\text{P.E. of av.} = 0.6745 \sqrt{\frac{0.0554}{(5)(4)}} = \pm 0.034$$

Example 2.   Suppose that five traverses, all of which had been run by the same method, gave the results shown in Table A-2. What are the error data for 36,000 ft? The expected error in 36,000 ft, by Eq. (2-2), is

$$E = 0.45 \sqrt{36} = \pm 2.70 \text{ ft}$$

The standard error in 36,000 ft, by Eqs. (A-3), and (2-2), is

$$\sigma = \sqrt{0.2160} \sqrt{36} = \pm 2.79 \text{ ft}$$

Table A-2

Length, ft	Error, ft	Expected error in 1000 ft, by Eq. (2-2), v_i ft	v_i^2, sq ft
8,000	1.16	$\div \sqrt{8}\ = 0.41$	0.1681
10,000	0.95	$\div \sqrt{10} = 0.30$	0.0900
7,000	1.52	$\div \sqrt{7}\ = 0.57$	0.3249
16,000	2.40	$\div \sqrt{16} = 0.60$	0.3600
4,000	0.74	$\div \sqrt{4}\ = 0.37$	0.1369
		Sums 2.25	1.0799
		$\div 5$ 0.45	0.2160

The 90 per cent error in 36,000 ft, by Eq. (A-7), is

$$E_{90} = (1.645)(2.79) = \pm 4.59 \text{ ft}$$

Design accuracy (Sec. 2-17) is

$$4.59/36,000 = 1/7843$$

Thus, this method will have a greater accuracy than 1/7843 in a 36,000-ft survey in 90 per cent of such surveys.

Example 3. Assume the level runs shown in Table A-3. What will be the results of a level run 15,000 ft long which closes on itself?

Table A-3

Length, ft	Error, ft	Expected error in 1000 ft by Eq. (2-2), v_i ft	v_i^2, sq ft
2,000	0.018	$\div \sqrt{2}\ = 0.013$	0.000169
10,000	0.030	$\div \sqrt{10} = 0.009$	0.000081
7,000	0.035	$\div \sqrt{7}\ = 0.013$	0.000169
8,000	0.020	$\div \sqrt{8}\ = 0.007$	0.000049
12,000	0.038	$\div \sqrt{12} = 0.011$	0.000121
4,000	0.015	$\div \sqrt{4}\ = 0.008$	0.000064
9,000	0.040	$\div \sqrt{9}\ = 0.013$	0.000169
		Sums 0.074	0.000822
		$\div 7$ 0.011	0.000117

The expected error in 15,000′, by Eq. (2-2), is

$$0.011 \sqrt{15} = \pm 0.043′$$

The standard error in 15,000′, by Eqs. (A-3) and (2-2), is

$$\sqrt{0.000117} \sqrt{15} = \pm 0.042′$$

The 90 per cent error in 15,000′, by Eq. (A-7), is

$$(1.645)(0.042) = \pm 0.069′$$

Design accuracy, by Eq. (2-4), is

$$V = \pm 0.069 \sqrt{\frac{5,280}{15,000}} = \pm 0.041′$$

which is third order or better in 90 per cent of the runs.

Problems

In the following problems, find the expected error, σ, E_{90}, and design accuracy in a distance of 12,000 ft. Slide-rule accuracy is sufficient.

A-1. Traverses

Approx. length, ft	Total error, ft
5,000	0.92
8,000	2.78
10,000	2.81
7,000	1.09
20,000	5.13

A-2. Traverses

Approx. length, ft	Total error, ft
6,000	0.52
14,000	1.69
8,000	0.85
7,000	0.71
16,000	1.04

A-3. Level Runs

Approx. length, ft	Total error, ft
5,000	0.020
4,000	0.016
6,000	0.028
8,000	0.042
10,000	0.030

A-4. Level Runs

Approx. length, ft	Total error, ft
4,000	0.018
7,000	0.050
6,000	0.017
11,000	0.059
3,000	0.012

Appendix B

Physics of Tapes

B-1. To Find a New Tape Correction. When the tape is used with a type of support and/or a tension for which it has not been standardized, a new tape correction must be computed.

It is first necessary to find the weight per foot of the tape. Weigh the tape on the reel and weigh the reel separately.

Example

$$\text{Weight of tape and reel} = 1.98 \text{ lb}$$
$$\text{Weight of reel} = 0.27$$
$$\text{Weight of tape} = \overline{1.71} \text{ lb}$$
$$\text{Length of ribbon plus allowance for loops at end} = 100.86'$$
$$\text{Weight per foot} = \frac{1.71}{100.86} = 0.017 \text{ lb}$$

Since a steel bar 1 in. square weighs 3.40 lb per ft, the cross-sectional area of the tape is

$$\frac{0.017}{3.4} = 0.005 \text{ sq in.}$$

Correction for Sag. When the whole or any part of the tape spans a distance unsupported, the zero and 100-ft marks are brought nearer to each other than they would have been if the tape had been fully supported throughout. This shortening from the length supported throughout is computed by the following formula:

$$C_s = -\frac{w^2 l^3}{24 t^2}$$

where C_s = correction for sag, ft
w = weight per foot of tape, lb
l = unsupported span, ft
t = tension on tape at the time, lb

Correction for Tension. The steel ribbon stretches when increased tension is applied. The correction for tension is computed by the following formula:

$$C_p = \frac{L(t - t_0)}{ES}$$

where C_p = correction for a new tension, ft
  L = nominal tape length, ft
  t = new tension, lb
  t_0 = original tension, lb
  E = 28,000,000 psi
  S = cross-sectional area of tape, sq in.

Example. If the standardized length of the tape were 99.996 when supported at the zero and 100-ft points at a tension of 20 lb, what would be its length when supported for 60 ft and unsupported for a span of 40 ft at a tension of 10 lb? Assume that w = 0.017 and S = 0.005 sq in.

Solution:

$$\text{Standardized length of tape} = \quad 99.9960'$$

$$-C_s = +\frac{w^2l^3}{24t^2} = +\frac{(0.017)^2(100)^3}{24(20)^2} = +0.0301$$

$$+C_p = +\frac{L(t-t_0)}{ES} = +\frac{100(10-20)}{28,000,000(0.005)} = -0.0071$$

$$+C_s = -\frac{w^2l^3}{24t^2} = -\frac{(0.017)^2(40)^3}{24(10)^2} = -0.0077$$

$$\text{New tape length} = \text{algebraic sum} = \overline{100.011'}$$

Explanation. In these problems, unless it is known, first find the length supported throughout at the tension at which the tape was standardized. This is accomplished by applying the sag correction with a minus sign.

Second, apply the correction for change in tension.

Third, apply the sag correction for sag under the new conditions.

It is well to check the sign of each correction by deciding whether the tape is made longer (hence plus) or shorter (hence minus) by the new condition.

Problems

B-1, B-2. Find the tape length to be used for correcting field records when the following conditions existed.

	When standardized				When used	
Problem	Std.	w	t_0	Support	t	Support
B-1	99.982	0.020	20	At 0, 100	10	Throughout
B-2	100.006	0.024	10	At 0, 100	20	At 0, 100

Appendix C

Adjustments of Instruments

C-1. Introduction. The importance of instrument adjustment cannot be over-emphasized. While the errors resulting from lack of adjustment can be eliminated by the principle of reversal in the operation of the instrument, the time consumed for this procedure usually limits it to primary measurements. In locating topographic features, establishing a large number of gauge points in a jig, or, in fact, working the great bulk of instrument operations in the field or shop, the time necessary for reversals prohibits their use.

The instrumentmen must be trained so thoroughly in the theory and methods of instrument adjustment that they can be relied upon to perform the five following functions without fail:

1. They must recognize the operations that depend on the accuracy of instrument adjustment.
2. They must be able to test the instrument in the course of the work.
3. They must be able to adjust it with the minimum of delay.
4. They must be able to judge when the accuracy of measurement requires reversal even when the instrument is in adjustment.
5. When necessary, they must be able to operate the instrument so that the reversal principles will neutralize instrument errors.

Modern surveying instruments seldom get out of adjustment. However, they should be tested frequently. Before it is assumed that any part of an instrument is out of adjustment, the test should be repeated at least three times. If the error indicated is not the same each time the test is made, the test itself is faulty.

If possible, a cloudy day should be chosen to test and adjust the instrument. Otherwise the instrument must be set up in the shade. The tripod shoes must be tight, and the instrument should be set up exactly in accordance with the method given in Chap. 4.

Instrument makers supply instructions for the adjustments of their instruments. The basic principles are very nearly the same in all surveying instruments, but the methods of making the adjustments differ in detail. Given here are the adjustments of the transit and the dumpy level, since these are the instruments chiefly used in the United States and their adjustments clearly illustrate the basic principles involved.

C-2. Method of Presentation. To simplify the text without omitting details, schematic sketches are used to demonstrate the adjustment operations. Beside each screw shown in the sketches is an arrow that indicates the direction of movement of the side of the screw nearest the observer. The numbers in circles show the order of procedure. The fractions beside the arrows give very approximately the fractions of a turn that should be applied to the screws.

The operation **to transit** means to turn the telescope from its direct position to its reversed position or vice versa, whichever applies.

The operation **to traverse** means to turn the instrument in azimuth.

The sketches show typical adjustment parts. There are slight variations from instrument to instrument, but the reader will have no difficulty in discovering how to handle types not shown.

Order of Adjustment. Obviously, certain adjustments will upset others. It is necessary, therefore, to make adjustments according to the order in which they are given in the text.

Final Adjustment. Opposing screws or nuts are used on most adjustable parts. Forcing them in the wrong direction will strip the threads. However, when the final adjustment is complete, the screws must be firm. The last increment of adjustment must therefore be made by tightening one screw of the pair.

Neutralization of Residual Errors of Adjustment. After each adjustment there is described, under the heading Neutralization, the field procedure that will neutralize errors caused by lack of the adjustment. When highly accurate results are required, the instrument must be operated so that any residual errors are neutralized. When less accurate results are required and it becomes apparent that the instrument is not in adjustment, these methods can be employed until adjustment can be accomplished.

ADJUSTMENT OF TRANSIT

Adjustment A

C-3. Object. To make the vertical cross hair lie in a plane perpendicular to the horizontal axis.

Test. Point on some well-defined mark. Rotate the telescope slightly about the horizontal axis, using the vertical tangent screw. The point should remain on the cross hair.

Adjustment. Loosen two adjacent cross-hair adjusting screws, and by moving the screws rotate the cross hairs in the desired direction. Tighten the screws (see Fig. C-1).

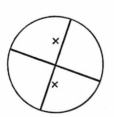

Appearance of field in both
erecting and inverting instruments

As the telescope is rotated about
the horizontal axis the point x
apparently moves along the line connecting the two positions shown

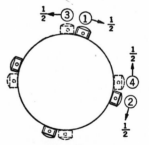

Rotate the cross hairs
counterclockwise until
the point x follows the
vertical cross hair

Fig. C-1. Transit adjustment A.

Repeat Test.

Neutralization.   Use only that part of the vertical hair which is close to the horizontal cross hair.

Geometry.   A lens forms an image that is rotated 180° around the lens axis with respect to the object. Direction of rotation therefore is not changed by a lens as it is by a mirror.

Adjustment 1

C-4. Object.   To make the plate bubbles center when the vertical axis is vertical.

Test.   Level the instrument. Rotate 180° in azimuth. The bubbles should remain centered.

Adjustment.   Assume that the bubble moves away from the adjustment end. Follow the indications on the sketch until the bubble moves halfway back (Fig. C-2).

Repeat Test.

Neutralization.   Level the instrument. Rotate 180° in azimuth. If the bubbles fail to center, bring them halfway back with the leveling screws.

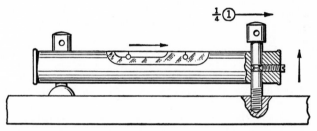

Fig. C-2. Transit adjustment 1, the adjustment of the plate levels. Continue until the bubble moves halfway back.

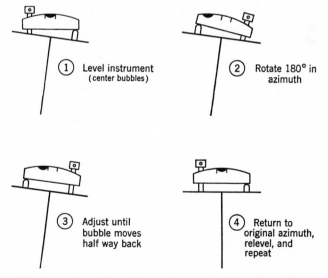

1. Level instrument (center bubbles)
2. Rotate 180° in azimuth
3. Adjust until bubble moves half way back
4. Return to original azimuth, relevel, and repeat

Fig. C-3. Principles of the adjustment of the plate levels.

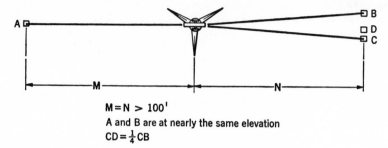

M = N > 100'
A and B are at nearly the same elevation
CD = ¼ CB

Fig. C-4. Transit adjustment 2.

Geometry. Figure C-3 shows that the error of adjustment creates an error of setup. When the instrument is turned 180°, the error of adjustment combines with the error of setup to move the bubble twice the amount caused by the error of adjustment alone.

Adjustment 2

C-5. Object. To make the line of sight perpendicular to the horizontal axis.

Test. Set up near the center of a level stretch. Point on a well-defined mark at *A*, 100 ft or more distant. Transit the telescope, and set *B* on line, at nearly the same elevation and distance as *A*. Traverse, and point on *A*. Transit, and the line of sight should fall on *B* (see Fig. C-4).

Adjustment. Assume that the line of sight falls south of the point. Set point *C* on line beside *B*. Follow the indications on the sketches until the cross hairs appear to move to *D*, one-fourth of the way from *C* toward *B* (see Figs. C-5 and C-6).

Note that in both erecting and inverting instruments, for the **same error,** the cross hairs are moved physically in the **same direction.**

Repeat Test.

Neutralization. Point on *A*. Transit, and set *B*. Traverse, and point on *A*. Transit, and set *C*. Set a line point halfway from *C* to *B*. This is called **double centering**.

Geometry. Since, when the cross hairs are adjusted, the line of sight pivots about the center of the objective, to move the exterior portion of the line of sight toward

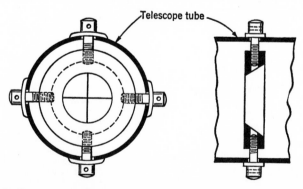

Telescope tube

Fig. C-5. Cross hairs and reticle. The four adjusting screws are in tension.

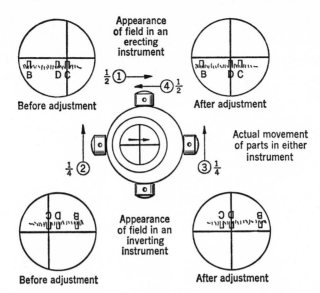

Fig. C-6. Transit adjustment 2. The appearance of the field of view before and after the adjustment.

the north the cross hairs must be physically moved south (see Fig. C-7). Therefore, in adjusting an erecting instrument, the cross hairs must be moved apparently in the wrong direction. In adjusting an inverting instrument, since in the inverted view north and south are apparently interchanged, the adjustment is made in apparently the right direction.

Before the cross-hair ring can be moved sideways, an upper or lower adjusting screw must be loosened.

The procedure outlined collects four times the error in adjustment between B and C, as shown in Fig. C-8. Hence D is set one-fourth of the distance from C toward B.

Adjustment 3

C-6. Object. To make the horizontal axis perpendicular to the vertical axis.

Test. Set up near a high, well-defined point such as a church steeple. Point on the high point A. Lower the telescope, and set point B on the ground in line about 20 ft

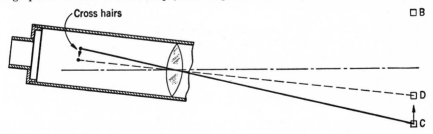

Fig. C-7. Direction to move cross hairs. When D is north of C, the cross hairs must be moved south physically.

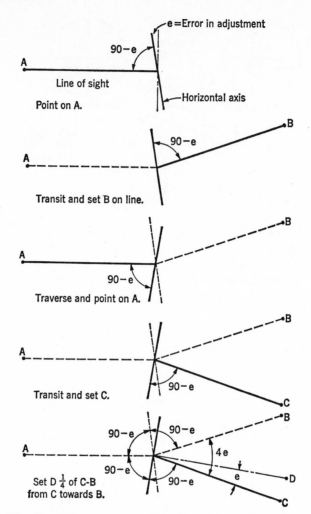

Fig. C-8. Principles of the adjustment of the line of sight.

in front of the instrument. Transit and traverse, and point on *B*. Raise the telescope, and the line of sight should fall on the high point.

Adjustment. Assume that the line of sight falls on the opposite side of the high point from the adjustable end of the horizontal axis. Follow the indications on the sketch (Fig. C-9) until the line of sight moves one-fourth to one-half of the way back.[1] On most transits, as shown in the sketch, the adjusting screws also control the bearing pressure. The pressure should prevent the telescope from plunging under its own weight but not create any noticeable friction.

[1] The exact fraction of *D* (the total divergence) $= \dfrac{1}{2}\dfrac{a}{a+b}$ where a = vertical angle to *A* and where b = vertical angle to *B*.

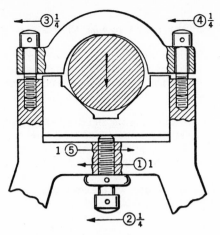

Fig. C-9. Transit adjustment 3. Procedure for lowering the adjustable end of the horizontal axis. On many transits, as shown here, the adjusting screws also control the bearing pressure. Play or excessive friction in the bearings must be avoided.

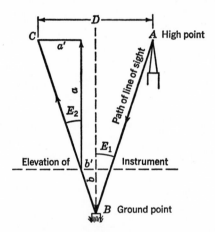

Fig. C-10. Geometry of transit adjustment 3 as seen from the instrument:

C = point where line of sight falls after test
D = divergence of line of sight from A
E_1, E_2 = angular error in horizontal axis to be corrected
a' = movement of line of sight necessary to correct angular error
a = vertical angle to A and C
b = vertical angle to B

$$a' + b' = D/2$$

$$\frac{a'}{a' + b'} = \frac{a}{a + b}$$

$$a' = \frac{D}{2}\frac{a + b}{a}$$

Repeat Test.

Neutralization.   Repeat any procedure with the telescope reversed, and use the average of the results.

Geometry.   The line of sight is first directed at A (see Fig. C-10). The left end of the horizontal axis is high; therefore, when the line of sight is moved downward, it also moves to the left, generating a plane that inclines from the vertical an amount equal to the error in the adjustment of the horizontal axis, viz., angle E_1. Point B is set accordingly.

The transit is now reversed, making the right end of the horizontal axis high. The line of sight is now directed at B and raised to C, level with A. As it is raised, it moves to the left, generating another plane inclined to the vertical (on the opposite side) an amount equal to the error in the adjustment of the horizontal axis, viz., E_2.

The problem now arises in toward what point should the line of sight be directed by adjusting the horizontal axis to eliminate the error? See Fig. C-10. It must be brought back the distance a'. If $b = 0$, $a' = D/2$; if $b = a$, $a' = D/4$.

Adjustment 4

C-7. Object.   To make the telescope-level bubble center when the line of sight is horizontal. The procedure is often known as the **peg adjustment**.

Test.   Set stakes at D, A, C, and B, as shown in Fig. C-11. Set up at C. **Center the bubble** and read the rod on A, turn to B, and **center the bubble** and read the rod on B.

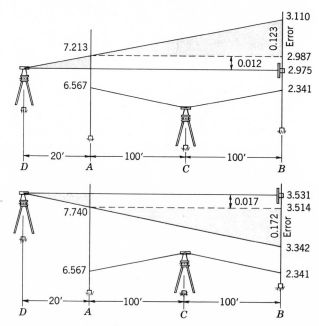

Fig. C-11. Principles of the peg adjustment. The two possibilities are shown.

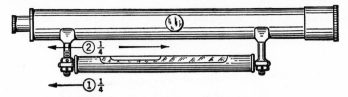

Fig. C-12. Transit adjustment 4. Adjustment of the telescope level. After pointing on the target, adjust until the bubble moves all the way to the center.

If the line of sight slopes, it will slope exactly the same amount when each reading is taken since the bubble is centered both times. Since the instrument is exactly halfway between A and B, the **difference** in the two readings will be the same as it would be if the line of sight were level. This difference always expresses the difference in elevation of stakes A and B.

Transit at C

Rod A = 6.567
Less Rod B = 2.341
Difference = 4.226

Thus the rod B reading should be 4.226 less than the rod A reading wherever the instrument is placed if the instrument is in adjustment.

Set up at D, center the bubble, and take readings on rods A and B.

	Top figure	Bottom figure
A	7.213	7.740
B is less; less diff.	4.226	4.226
B should read	2.987	3.514
B does read	3.110	3.342
Error	0.123 too high	0.172 too low

The target must now be set so that it is at the same elevation as the instrument. Note that the two shaded triangles in each figure are similar and the smaller has a base of 20 ft or one-tenth the 200-ft base of the larger. It follows that the target must be set a distance from the value that B should read equal to one-tenth the error.

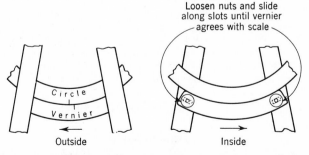

Fig. C-13. Transit adjustment B. The adjustment of the vertical vernier.

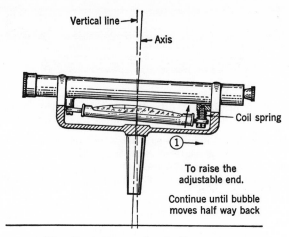

Fig. C-14. Dumpy-level adjustment 1.

	Top figure	*Bottom figure*
B should read	2.987	3.514
Apply ⅒ error	−0.012	+0.017
Set target at	2.975	3.531

Using the vertical tangent screw, point on the target. Follow the indications on the sketch (Fig. C-12) until the bubble moves to the center.

Repeat Test.

Neutralization. Set up the instrument at equal distances from the points observed.

Geometry. The geometry is illustrated in Fig. C-11.

Adjustment B

C-8. Object. To make the vertical circle indicate zero when the line of sight is perpendicular to the vertical axis.

Test. Level the instrument, using the telescope bubble (see Sec. 4-18). Using the lower motion, turn the instrument until the telescope is on line with a pair of opposite leveling screws. Center the telescope bubble, using the vertical motion.

The vertical circle should now read zero.

Adjustment. Loosen the nuts behind the vernier. Tap the vernier into position. Tighten the nuts (Fig. C-13).

Neutralization. Read the vertical angles direct and reversed, and use the average.

Geometry. The vertical axis has been made vertical by the method of leveling. The line of sight is horizontal as the telescope bubble is centered. Therefore the line of sight and the vertical axis are perpendicular, and the vernier should read zero.

ADJUSTMENT OF DUMPY LEVEL

Adjustment 1

C-9. Object. To make the bubble center when the vertical axis is vertical.

Test. Level approximately over both pairs of opposite leveling screws. Level carefully over one pair. Turn the instrument 180° around the vertical axis. The bubble should remain centered.

Adjustment. Assume that the bubble moves away from the adjustment end. Adjust as shown in the sketch (Fig. C-14) until the bubble moves halfway back.

Repeat Test.

Neutralization. Relevel for any pointing of the instrument.

Adjustment 2

C-10. Object. To make the line of sight horizontal when the bubble is centered.

Test. The same test is used as that for the level attached to the transit telescope. When the target has been properly set, the test is complete (see Sec. C-7).

Adjustment. Assume that the line of sight is too high on the rod. Level carefully. Adjust the reticule until the line of sight is brought on the target.

Repeat Test.

Neutralization. Balance the horizontal lengths of the plus and minus sights.

Problems

C-1. Draw sketches of the field of view as seen by an observer when, at the end of the test in transit adjustment 2, the line of sight actually falls to the left of point B.

C-2. Assume that at the end of the test in transit adjustment 3 the line of sight apparently missed the high point by the distance CA. If the vertical angles were $+40°$ and $-20°$, respectively, how far should the line of sight be moved by adjustment?

C-3, C-4. Compute the target setting for transit adjustment 4 if the data are the following:

Problem	*Readings at C*		*Readings at D*	
	Rod A	Rod B	Rod A	Rod B
C-3	2.341	1.268	4.632	3.548
C-4	2.341	1.268	4.632	3.583

C-5. Using sketches, demonstrate why, in dumpy-level adjustment A, the crosshair ring is rotated in the direction apparently required for correction in both erecting and inverting instruments.

C-6, C-7. What indications in dumpy-level operation would indicate that the following instrument adjustments were not accurate?

C-6. Adjustment 1

C-7. Adjustment 2

C-8. What precautions in the work would you take if you thought that a level was out of adjustment and no time was available to test it?

Appendix D

Property Surveys

D-1. Property Surveys. Often called **boundary surveys** or **land surveys,** they require surveying skill, a thorough understanding of related law, a knowledge of surveys in the locality, and considerable experience. No matter how skillful a surveyor might be, he should never attempt a property survey unless he has had considerable experience working under the direction of a land surveyor.

The simplest property survey occurs when a lot is to be marked out in a real-estate development, because it requires only surveying skill. Figure D-1 shows, for such a lot, an example of the final map that should be filed with the description in the deed.

Enough monuments and dimensions usually appear in the development plan to establish at least one property corner and the **bearing base** to be used. A bearing base is the slightly arbitrary bearing chosen for one line in the development from which all bearings are computed through appropriate angles. If these exist, the field operation depends entirely on good surveying procedures.

Once the locations of the property corners are established, dimensions must be measured from the property lines to show the positions of all permanent buildings, rights-of-way, and encroachments.

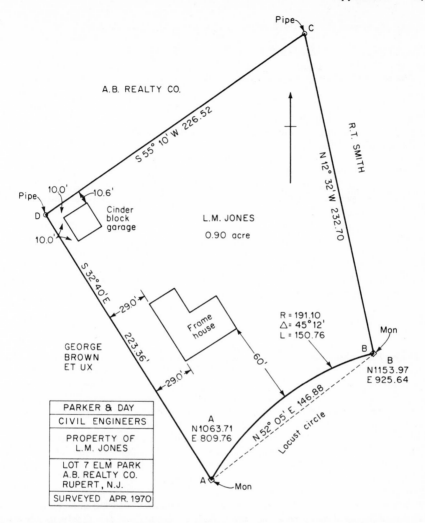

Fig. D-1. Example of a property survey. Note: Subtract from the area of the traverse the area of the segment. See Appendix E.

The actual property lines are frequently obscured, so that parallel offset lines must be used or a closed traverse must be run inside the property with traverse corners placed as near as possible to the property corners. The traverse must then be computed in the office and from this the angles and distances required to set the property corners from the traverse marks.

Finally, the lengths and bearings of the actual property lines must be determined and the area must be computed.

Any experienced field chief could handle this job. Unfortunately, many property surveys introduce complications that cannot be solved by mere surveying ability. These include lost or questionable markers, inadequate or erroneous descriptions, overlapping or separated claims, public rights-of-way, riparian rights, etc.

The following is an example of a description of the property in Fig. D-1 that would be entered in the deed.

The property of L. M. Jones being Lot 7 situated in Elm Park in the city of Blankville, County of Blank, State of Blank.

Beginning at a concrete monument in the northwesterly section of the line of Locust Circle and the southerly corner of the lot hereby conveyed:

1. Thence, along the said northwesterly line on a circular arc curving to the right at a radius of 191.10 feet, a distance of 150.76 feet, and a central angle of 45°12′, to a concrete monument at the easterly corner of the lot hereby conveyed. The chord of said arc running N52°05′E 146.88 feet;
2. Thence, along the line now or formerly of R. T. Smith N12°32′W 232.70 feet to an iron pipe at the northerly corner of the lot hereby conveyed (bearing base);
3. Thence, along the line of A. B. Realty Company S55°10′W 226.52 feet to an iron pipe at the westerly corner of the lot hereby conveyed;
4. Thence, along the line now or formerly of George Brown et Ux S32°40′E 223.36′ feet to the point of beginning.

As surveyed by Parker and Day Civil Engineers, in April 1970.

Office Procedure for Typical Properties. Figure D-2 shows the field notes.

Step 1. Check the Angles Measured. In this case the interior angles were measured. The sum of the interior angles should be equal to $(n - 2)$ 180° where n is the number of angles. If the exterior angles were measured their sum should be 360°, irrespective of the number of angles. See Figs. D-3 and D-4. If the sum is incorrect by enough to show an important error, the angles must be remeasured.

Step 2. Plot the Traverse. Use the measured angles and lengths. The plotted traverse should end closely at the starting point. If it misses by a short distance obviously due to slight errors in plotting, adjust the lines by eye to close the traverse.

Step 3. Compute the Bearings. First choose a *bearing base* (in this case, AB N80°20′W). This should be the basis for all bearings generally recognized for lines in the area. It is seldom magnetic or true. Instead, it is often the magnetic bearing of some important old line determined many years ago at an unknown declination. If a satisfactory bearing base cannot be reached easily, the present approximate magnetic or true bearing can be used. An indication of the orientation is often useful to the architect and others.

Starting with the bearing base, compute the bearing of each line successively, using the angle at each station.

Section 5-4, page 110 indicates the method. See Fig. D-5

Step 4. Compute the Coordinates. Use the latitudes and the departures. See Fig. D-6.

Step 5. Plot the Traverse by Coordinates. See Fig. D-7.

Step 6. Write the Legal Description. Example. This is a minimum description, but entirely legal and adequate.

Beginning at a concrete monument on the southerly side of Birk Street and running.

1. Thence: along the southerly side of Birk Street N80°20′W 232.46 feet to a concrete monument.

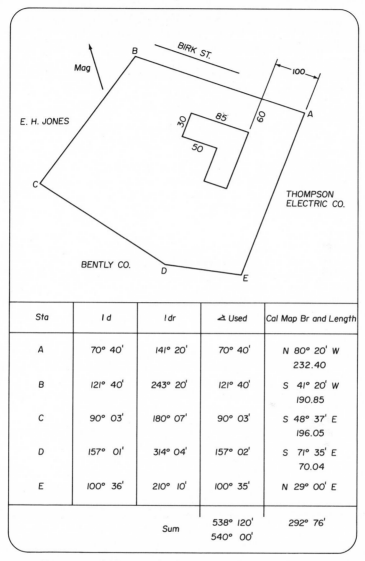

Fig. D-2. Field notes. Clockwise interior angles measured.

Sta	∠ d	∠ dr	∠ Used	Cal Map Br and Length
A	70° 40'	141° 20'	70° 40'	N 80° 20' W 232.40
B	121° 40'	243° 20'	121° 40'	S 41° 20' W 190.85
C	90° 03'	180° 07'	90° 03'	S 48° 37' E 196.05
D	157° 01'	314° 04'	157° 02'	S 71° 35' E 70.04
E	100° 36'	210° 10'	100° 35'	N 29° 00' E
		Sum	538° 120' 540° 00'	292° 76'

2. Thence: along the lands now or formerly of E. H. Jones S41°20′W 190.85 feet to an iron pipe.
3. Thence: along the lands of the Bently Company S48°37′E 196.05 feet to an iron pipe.
4. Thence: along the lands of the aforesaid Bently Company 71°35′E 70.04 feet to an iron pipe.
5. Thence: along the right-of-way of the South Thompson Electric Company N29°30′E 292.76 feet to point and place of beginning.

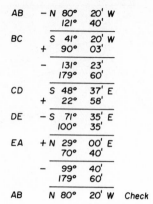

AB		− N 80°	20′ W
		121°	40′
BC		S 41°	20′ W
	+	90°	03′
	−	131°	23′
		179°	60′
CD		S 48°	37′ E
	+	22°	58′
DE		− S 71°	35′ E
		100°	35′
EA	+	N 29°	00′ E
		70°	40′
	−	99°	40′
		179°	60′
AB		N 80°	20′ W Check

Fig. D-3. Computation of bearings. See Sec. 5-3.

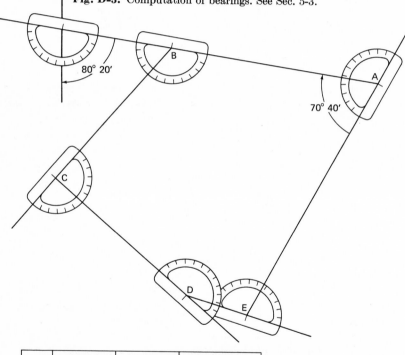

	Exterior Angles	Interior Angles	Bearings	
A	109° 20′	70° 40′	AB	N 80° 20′ W
B	58° 20′	212° 40′	BC	S 41° 20′ W
C	89° 57′	90° 03′	CD	S 48° 37′ E
D	22° 58′	157° 02′	DE	S 31° 35′ E
E	79° 25′	100° 35′	EA	N 29° 88′ E
Sums	360° 00′	540° 00′		

$(n-2)\ 180° = 540°$

Fig. D-4. Relationships of the exterior and interior angles with respect to the bearing base and the length of the sides.

Sta	Bearing	Length	Cos	Lat	Sin	Dept.
AB	N 80 20 W	232.46	.16792	N 39.03	.98580	W 229.16
BC	S 41 20 W	190.85	.75088	S 143.30	.66044	W 126.04
CD	S 48 37 E	196.05	.66109	S 129.61	.75030	E 147.10
DE	S 71 35 E	70.04	.31593	S 22.13	.94878	E 66.45
EA	N 29 30 E	292.76	.87462	N 256.65	.48481	E 141.93

Sum   982.16

	N Lat	E Dept
A	500.00	500.00
	+ 39.03	− 229.16
B	$\overline{539.03}$	$\overline{270.84}$
	− 143.30	− 126.04
C	$\overline{395.73}$	$\overline{144.80}$
	− 129.61	+ 147.10
D	$\overline{266.12}$	$\overline{291.90}$
	− 22.13	+ 66.45
E	$\overline{243.99}$	$\overline{358.35}$
	+ 256.05	+ 141.93
A	$\overline{500.04}$	$\overline{500.28}$
Error	N 0.04	E 0.28

$.04^2 + .28^2 = .0700$

Total Error

$\sqrt{.0700} = 0.283$

$$\frac{.263}{982.16} = \frac{1}{3469}$$

Fig. D-5. To lay out the traverse using measured angles and lengths.

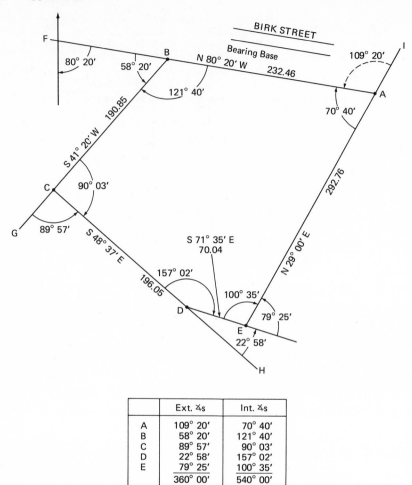

	Ext. ∢s	Int. ∢s
A	109° 20′	70° 40′
B	58° 20′	121° 40′
C	89° 57′	90° 03′
D	22° 58′	157° 02′
E	79° 25′	100° 35′
	360° 00′	540° 00′

Fig. D-6. Computation of latitudes, departures, and coordinates.

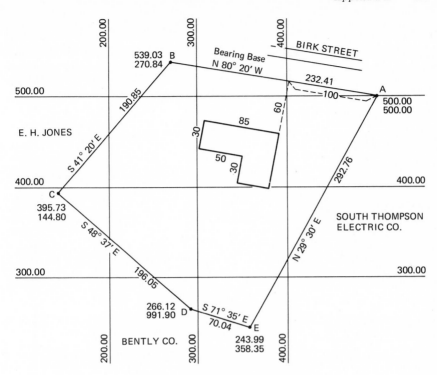

Fig. D-7. Plotting by coordinates.

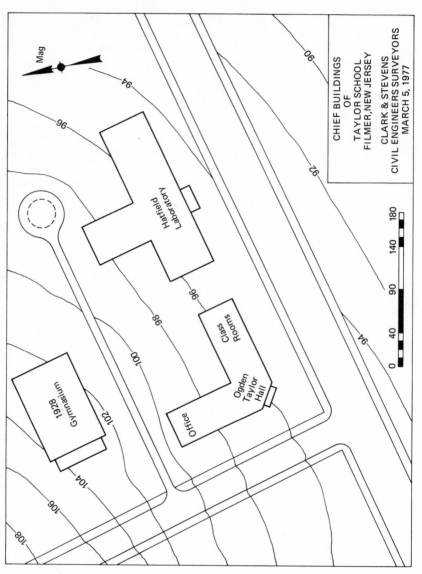

Fig. D-8. A typical map.

Units of Measure

There is a definite movement to change from English units of measurement to metric units. Although metric units simplify computation, they are difficult for the surveyor to use as a 100-m tape is too long for general use and a 10-m tape is too short. Furthermore, all records of land ownership or transfer are written in terms of the English system almost exclusively. It follows that the surveyor will have to be familiar with both systems.

Conversion Factors

(All values are exact except when followed by + or −.)

1 mile = 1.609344 kilometers (km)

1 foot = 0.3048 meter (m)

1 square yard = 0.83612736 square meter (m³)

1 cubic yard = 0.76455486− cubic meter [m²]

1 Gunter's chain = 66 ft = 20.1168 meters

1 acre = 10 square Gunter's chains

1 acre = 43560 square feet

1 acre = 0.40468564− hectare

1 acre = 40.468564− ares

1 kilometer = 0.62137119+ feet

1 meter = 3.2808399− feet

1 square meter = 1.1959900+ square yard

1 cubic meter = 1.3079505+ cubic yard

1 are = 100 square meters

1 hectare = 100 ares

1 square kilometer = 100 hectares

1 hectare = 10,000 square meters

1 hectare = 2.4716645+ acres

1 are = 0.024716645+ acres

The sizes of the various units of each system are fundamentally based on the meter, the size of which is defined very precisely in terms of certain light waves. The English units are established by the following ratio: one yard = 0.9144 m exactly.

See also Chap. 3.

Problems

D-1. Draw the boundaries of the lot from the following description. Insert the dimensions and bearings.

The property of Dan Bray located at the southeasterly corner of Roe Street and Marcus Avenue in Blank Town, Blank County, State of Blank, more particularly described as follows:

Beginning at a stone bound in the southerly line of Roe Street and the westerly line of Marcus Avenue distant westerly 1081.66 feet from the westerly line of Jones Avenue measured along the southerly line of Roe Street and running:

1. Thence along the southerly line of Roe Street S82°41′W 425.31 feet to an iron pipe at the corner of Jacob Wrenn.
2. Thence along the line of the property of Jacob Wrenn S12°31′W 426.05 feet to a corner marked with an iron pipe.
3. Thence still along Jacob Wrenn's line S65°05′E 345.28 feet to a corner of the property of John Jones marked by an iron pipe pin.
4. Thence along the line of John Jones N76°59′ 322.21 feet to a corner marked by an iron pipe on the easterly line of Marcus Avenue.
5. Thence along the easterly line of Marcus Avenue N11°45′W 554.09 feet to the point and place of beginning.

Surveyed by George Kane, Civil Engineers, in October 1907.

D-2. Draw the boundaries of the lot from the following description.

The property of H. A. Smith known as 22 Elm Street being Lot 27 in the Green Hill development in Blank City, County of Blank, State of Blank.

Beginning at the stone bound on the northerly line of Elm Street distant easterly 561.82 feet from the intersection of the northerly line of Elm Street and the easterly line of Johnson Avenue marking the southwesterly corner of the property hereby conveyed and running:

1. Thence S85°03'E 161.04 feet along the northerly line of Elm Street to an iron pipe at the southeasterly corner of the lot hereby conveyed.
2. Thence along the westerly line of Elmer Jones N7°08' East 260.68 feet to an iron pipe at the northeasterly corner of the lot hereby conveyed.

Thence on three courses along the southerly line now or formerly of J. M. Parker as follows:

3. S60°36'W 67.09 feet to an iron pipe
4. N58°18'W 95.42 feet to an iron pipe
5. S67°25'W 80.00 feet to an iron pipe at the westerly corner of the property hereby conveyed.
6. Thence along the easterly line of John Acker S3°44'E 231.80 feet to the place of beginning.

Bearing base; the northerly line of Elm Street.

From a survey of Parker and Day registered land surveyors in January 1963.

Appendix E

Areas of Regular Plane Figures

Areas of Plane Figures. Figure E-1 shows the areas chiefly used in surveying together with the formulas used to compute them. In general, the formulas can be derived by the manipulations shown. Triangles can be moved or equal triangles can be added to give a figure whose area can be computed by a known formula; or the desired area can be broken down into parts that are figures whose areas can be computed by known formulas.

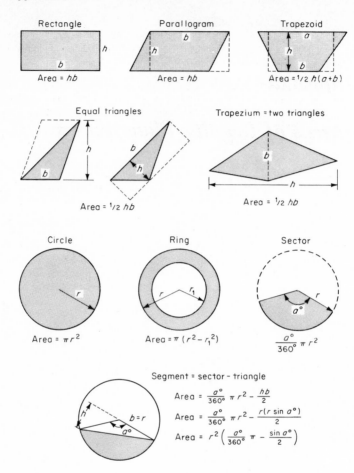

Fig. E-1. Areas of plane figures.

Table I. Logarithms of Numbers

No. 100
Log. 000

No. 109
Log. 040

N.	0	1	2	3	4	5	6	7	8	9	Diff.
100	00 0000	0434	0868	1301	1734	2166	2598	3029	3461	3891	432
1	4321	4751	5181	5609	6038	6466	6894	7321	7748	8174	428
2	8600	9026	9451	9876	0300	0724	1147	1570	1993	2415	424
3	01 2837	3259	3680	4100	4521	4940	5360	5779	6197	6616	420
4	7033	7451	7868	8284	8700	9116	9532	9947	0361	0775	416
105	02 1189	1603	2016	2428	2841	3252	3664	4075	4486	4896	412
6	5306	5715	6125	6533	6942	7350	7757	8164	8571	8978	408
7	9384	9789	0195	0600	1004	1408	1812	2216	2619	3021	404
8	03 3424	·3826	4227	4628	5029	5430	5830	6230	6629	7028	400
9	7426	7825	8223	8620	9017	9414	9811	0207	0602	0998	397
04											

PROPORTIONAL PARTS

Diff.	1	2	3	4	5	6	7	8	9
434	43.4	86.8	130.2	173.6	217.0	260.4	303.8	347.2	390.6
433	43.3	86.6	129.9	173.2	216.5	259.8	303.1	346.4	389.7
432	43.2	86.4	129.6	172.8	216.0	259.2	302.4	345.6	388.8
431	43.1	86.2	129.3	172.4	215.5	258.6	301.7	344.8	387.9
430	43.0	86.0	129.0	172.0	215.0	258.0	301.0	344.0	387.0
429	42.9	85.8	128.7	171.6	214.5	257.4	300.3	343.2	386.1
428	42.8	85.6	128.4	171.2	214.0	256.8	299.6	342.4	385.2
427	42.7	85.4	128.1	170.8	213.5	256.2	298.9	341.6	384.3
426	42.6	85.2	127.8	170.4	213.0	255.6	298.2	340.8	383.4
425	42.5	85.0	127.5	170.0	212.5	255.0	297.5	340.0	382.5
424	42.4	84.8	127.2	169.6	212.0	254.4	296.8	339.2	381.6
423	42.3	84.6	126.9	169.2	211.5	253.8	296.1	338.4	380.7
422	42.2	84.4	126.6	168.8	211.0	253.2	295.4	337.6	379.8
421	42.1	84.2	126.3	168.4	210.5	252.6	294.7	336.8	378.9
420	42.0	84.0	126.0	168.0	210.0	252.0	294.0	336.0	378.0
419	41.9	83.8	125.7	167.6	209.5	251.4	293.3	335.2	377.1
418	41.8	83.6	125.4	167.2	209.0	250.8	292.6	334.4	376.2
417	41.7	83.4	125.1	166.8	208.5	250.2	291.9	333.6	375.3
416	41.6	83.2	124.8	166.4	208.0	249.6	291.2	332.8	374.4
415	41.5	83.0	124.5	166.0	207.5	249.0	290.5	332.0	373.5
414	41.4	82.8	124.2	165.6	207.0	248.4	289.8	331.2	372.6
413	41.3	82.6	123.9	165.2	206.5	247.8	289.1	330.4	371.7
412	41.2	82.4	123.6	164.8	206.0	247.2	288.4	329.6	370.8
411	41.1	82.2	123.3	164.4	205.5	246.6	287.7	328.8	369.9
410	41.0	82.0	123.0	164.0	205.0	246.0	287.0	328.0	369.0
409	40.9	81.8	122.7	163.6	204.5	245.4	286.3	327.2	368.1
408	40.8	81.6	122.4	163.2	204.0	244.8	285.6	326.4	367.2
407	40.7	81.4	122.1	162.8	203.5	244.2	284.9	325.6	366.3
406	40.6	81.2	121.8	162.4	203.0	243.6	284.2	324.8	365.4
405	40.5	81.0	121.5	162.0	202.5	243.0	283.5	324.0	364.5
404	40.4	80.8	121.2	161.6	202.0	242.4	282.8	323.2	363.6
403	40.3	80.6	120.9	161.2	201.5	241.8	282.1	322.4	362.7
402	40.2	80.4	120.6	160.8	201.0	241.2	281.4	321.6	361.8
401	40.1	80.2	120.3	160.4	200.5	240.6	280.7	320.8	360.9
400	40.0	80.0	120.0	160.0	200.0	240.0	280.0	320.0	360.0
399	39.9	79.8	119.7	159.6	199.5	239.4	279.3	319.2	359.1
398	39.8	79.6	119.4	159.2	199.0	238.8	278.6	318.4	358.2
397	39.7	79.4	119.1	158.8	198.5	238.2	277.9	317.6	357.3
396	39.6	79.2	118.8	158.4	198.0	237.6	277.2	316.8	356.4
395	39.5	79.0	118.5	158.0	197.5	237.0	276.5	316.0	351.5

Table I. (Continued)

No. 110
Log. 041

No. 119
Log. 078

N.	0	1	2	3	4	5	6	7	8	9	Diff.
110	04 1393	1787	2182	2576	2969	3362	3755	4148	4540	4932	393
1	5323	5714	6105	6495	6885	7275	7664	8053	8442	8830	390
2	9218	9606	9993	0380	0766	1153	1538	1924	2309	2694	386
3	05 3078	3463	3846	4230	4613	4996	5378	5760	6142	6524	383
4	6905	7286	7666	8046	8426	8805	9185	9563	9942	0320	379
115	06 0698	1075	1452	1829	2206	2582	2958	3333	3709	4083	376
6	4458	4832	5206	5580	5953	6326	6699	7071	7443	7815	373
7	8186	8557	8928	9298	9668	0038	0407	0776	1145	1514	370
8	07 1882	2250	2617	2985	3352	3718	4085	4451	4816	5182	366
9	5547	5912	6276	6640	7004	7368	7731	8094	8457	8819	363

PROPORTIONAL PARTS

Diff.	1	2	3	4	5	6	7	8	9
395	39.5	79.0	118.5	158.0	197.5	237.0	276.5	316.0	355.5
394	39.4	78.8	118.2	157.6	197.0	236.4	275.8	315.2	354.6
393	39.3	78.6	117.9	157.2	196.5	235.8	275.1	314.4	353.7
392	39.2	78.4	117.6	156.8	196.0	235.2	274.4	313.6	352.8
391	39.1	78.2	117.3	156.4	195.5	234.6	273.7	312.8	351.9
390	39.0	78.0	117.0	156.0	195.0	234.0	273.0	312.0	351.0
389	38.9	77.8	116.7	155.6	194.5	233.4	272.3	311.2	350.1
388	38.8	77.6	116.4	155.2	194.0	232.8	271.6	310.4	349.2
387	38.7	77.4	116.1	154.8	193.5	232.2	270.9	309.6	348.3
386	38.6	77.2	115.8	154.4	193.0	231.6	270.2	308.8	347.4
385	38.5	77.0	115.5	154.0	192.5	231.0	269.5	308.0	346.5
384	38.4	76.8	115.2	153.6	192.0	230.4	268.8	307.2	345.6
383	38.3	76.6	114.9	153.2	191.5	229.8	268.1	306.4	344.7
382	38.2	76.4	114.6	152.8	191.0	229.2	267.4	305.6	343.8
381	38.1	76.2	114.3	152.4	190.5	228.6	266.7	304.8	342.9
380	38.0	76.0	114.0	152.0	190.0	228.0	266.0	304.0	342.0
379	37.9	75.8	113.7	151.6	189.5	227.4	265.3	303.2	341.1
378	37.8	75.6	113.4	151.2	189.0	226.8	264.6	302.4	340.2
377	37.7	75.4	113.1	150.8	188.5	226.2	263.9	301.6	339.3
376	37.6	75.2	112.8	150.4	188.0	225.6	263.2	300.8	338.4
375	37.5	75.0	112.5	150.0	187.5	225.0	262.5	300.0	337.5
374	37.4	74.8	112.2	149.6	187.0	224.4	261.8	299.2	336.6
373	37.3	74.6	111.9	149.2	186.5	223.8	261.1	298.4	335.7
372	37.2	74.4	111.6	148.8	186.0	223.2	260.4	297.6	334.8
371	37.1	74.2	111.3	148.4	185.5	222.6	259.7	296.8	333.9
370	37.0	74.0	111.0	148.0	185.0	222.0	259.0	296.0	333.0
369	36.9	73.8	110.7	147.6	184.5	221.4	258.3	295.2	332.1
368	36.8	73.6	110.4	147.2	184.0	220.8	257.6	294.4	331.2
367	36.7	73.4	110.1	146.8	183.5	220.2	256.9	293.6	330.3
366	36.6	73.2	109.8	146.4	183.0	219.6	256.2	292.8	329.4
365	36.5	73.0	109.5	146.0	182.5	219.0	255.7	292.0	328.5
364	36.4	72.8	109.2	145.6	182.0	218.4	254.8	291.2	327.6
363	36.3	72.6	108.9	145.2	181.5	217.8	254.1	290.4	326.7
362	36.2	72.4	108.6	144.8	181.0	217.2	253.4	289.6	325.8
361	36.1	72.2	108.3	144.4	180.5	216.6	252.7	288.8	324.9
360	36.0	72.0	108.0	144.0	180.0	216.0	252.0	288.0	324.0
359	35.9	71.8	107.7	143.6	179.5	215.4	251.3	287.2	323.1
358	35.8	71.6	107.4	143.2	179.0	214.8	250.6	286.4	322.2
357	35.7	71.4	107.1	142.8	178.5	214.2	249.9	285.6	321.3
356	35.6	71.2	106.8	142.4	178.0	213.6	249.2	284.8	320.4

Table I. (Continued)

No. 120   No. 134
Log. 079   Log. 130

N.	0	1	2	3	4	5	6	7	8	9	Diff.
120	07 9181	9543	9904	0266	0626	0987	1347	1707	2067	2426	360
1	08 2785	3144	3503	3861	4219	4576	4934	5291	5647	6004	357
2	6360	6716	7071	7426	7781	8136	8490	8845	9198	9552	355
3	9905	0258	0611	0963	1315	1667	2018	2370	2721	3071	352
4	09 3422	3772	4122	4471	4820	5169	5518	5866	6215	6562	349
125	6910	7257	7604	7951	8298	8644	8990	9335	9681	0026	346
6	10 0371	0715	1059	1403	1747	2091	2434	2777	3119	3462	343
7	3804	4146	4487	4828	5169	5510	5851	6191	6531	6871	341
8	7210	7549	7888	8227	8565	8903	9241	9579	9916	0253	338
9	11 0590	0926	1263	1599	1934	2270	2605	2940	3275	3609	335
130	3943	4277	4611	4944	5278	5611	5943	6276	6608	6940	333
1	7271	7603	7934	8265	8595	8926	9256	9586	9915	0245	330
2	12 0574	0903	1231	1560	1888	2216	2544	2871	3198	3525	328
3	3852	4178	4504	4830	5156	5481	5806	6131	6456	6781	325
4	7105	7429	7753	8076	8399	8722	9045	9368	9690	0012	323
	13										

PROPORTIONAL PARTS

Diff.	1	2	3	4	5	6	7	8	9
355	35.5	71.0	106.5	142.0	177.5	213.0	248.5	284.0	319.5
354	35.4	70.8	106.2	141.6	177.0	212.4	247.8	283.2	318.6
353	35.3	70.6	105.9	141.2	176.5	211.8	247.1	282.4	317.7
352	35.2	70.4	105.6	140.8	176.0	211.2	246.4	281.6	316.8
351	35.1	70.2	105.3	140.4	175.5	210.6	245.7	280.8	315.9
350	35.0	70.0	105.0	140.0	175.0	210.0	245.0	280.0	315.0
349	34.9	69.8	104.7	139.6	174.5	209.4	244.3	279.2	314.1
348	34.8	69.6	104.4	139.2	174.0	208.8	243.6	278.4	313.2
347	34.7	69.4	104.1	138.8	173.5	208.2	242.9	277.6	312.3
346	34.6	69.2	103.8	138.4	173.0	207.6	242.2	276.8	311.4
345	34.5	69.0	103.5	138.0	172.5	207.0	241.5	276.0	310.5
344	34.4	68.8	103.2	137.6	172.0	206.4	240.8	275.2	309.6
343	34.3	68.6	102.9	137.2	171.5	205.8	240.1	274.4	308.7
342	34.2	68.4	102.6	136.8	171.0	205.2	239.4	273.6	307.8
341	34.1	68.2	102.3	136.4	170.5	204.6	238.7	272.8	306.9
340	34.0	68.0	102.0	136.0	170.0	204.0	238.0	272.0	306.0
339	33.9	67.8	101.7	135.6	169.5	203.4	237.3	271.2	305.1
338	33.8	67.6	101.4	135.2	169.0	202.8	236.6	270.4	304.2
337	33.7	67.4	101.1	134.8	168.5	202.2	235.9	269.6	303.3
336	33.6	67.2	100.8	134.4	168.0	201.6	235.2	268.8	302.4
335	33.5	67.0	100.5	134.0	167.5	201.0	234.5	268.0	301.5
334	33.4	66.8	100.2	133.6	167.0	200.4	233.8	267.2	300.6
333	33.3	66.6	99.9	133.2	166.5	199.8	233.1	266.4	299.7
332	33.2	66.4	99.6	132.8	166.0	199.2	232.4	265.6	298.8
331	33.1	66.2	99.3	132.4	165.5	198.6	231.7	264.8	297.9
330	33.0	66.0	99.0	132.0	165.0	198.0	231.0	264.0	297.0
329	32.9	65.8	98.7	131.6	164.5	197.4	230.3	263.2	296.1
328	32.8	65.6	98.4	131.2	164.0	196.8	229.6	262.4	295.2
327	32.7	65.4	98.1	130.8	163.5	196.2	228.9	261.6	294.3
326	32.6	65.2	97.8	130.4	163.0	195.6	228.2	260.8	293.4
325	32.5	65.0	97.5	130.0	162.5	195.0	227.5	260.0	292.5
324	32.4	64.8	97.2	129.6	162.0	194.4	226.8	259.2	291.6
323	32.3	64.6	96.9	129.2	161.5	193.8	226.1	258.4	290.7
322	32.2	64.4	96.6	128.8	161.0	193.2	225.4	257.6	289.8

Table I. (Continued)

No. 135
Log. 130

No. 149
Log. 175

N.	0	1	2	3	4	5	6	7	8	9	Diff.
135	13 0334	0655	0977	1298	1619	1939	2260	2580	2900	3219	321
6	3539	3858	4177	4496	4814	5133	5451	5769	6086	6403	318
7	6721	7037	7354	7671	7987	8303	8618	8934	9249	9564	316
8	9879	0194	0508	0822	1136	1450	1763	2076	2389	2702	314
9	14 3015	3327	3639	3951	4263	4574	4885	5196	5507	5818	311
140	6128	6438	6748	7058	7367	7676	7985	8294	8603	8911	309
·1	9219	9527	9835	0142	0449	0756	1063	1370	1676	1982	307
2	15 2288	2594	2900	3205	3510	3815	4120	4424	4728	5032	305
3	5336	5640	5943	6246	6549	6852	7154	7457	7759	8061	303
4	8362	8664	8965	9266	9567	9868	0168	0469	0769	1068	301
145	16 1368	1667	1967	2266	2564	2863	3161	3460	3758	4055	299
6	4353	4650	4947	5244	5541	5838	6134	6430	6726	7022	297
7	7317	7613	7908	8203	8497	8792	9086	9380	9674	9968	295
8	17 0262	0555	0848	1141	1434	1726	2019	2311	2603	2895	293
9	3186	3478	3769	4060	4351	4641	4932	5222	5512	5802	291

PROPORTIONAL PARTS

Diff.	1	2	3	4	5	6	7	8	9
321	32.1	64.2	96.3	128.4	160.5	192.6	224.7	256.8	288.9
320	32.0	64.0	96.0	128.0	160.0	192.0	224.0	256.0	288.0
319	31.9	63.8	95.7	127.6	159.5	191.4	223.3	255.2	287.1
318	31.8	63.6	95.4	127.2	159.0	190.8	222.6	254.4	286.2
317	31.7	63.4	95.1	126.8	158.5	190.2	221.9	253.6	285.3
316	31.6	63.2	94.8	126.4	158.0	189.6	221.2	252.8	284.4
315	31.5	63.0	94.5	126.0	157.5	189.0	220.5	252.0	283.5
314	31.4	62.8	94.2	125.6	157.0	188.4	219.8	251.2	282.6
313	31.3	62.6	93.9	125.2	156.5	187.8	219.1	250.4	281.7
312	31.2	62.4	93.6	124.8	156.0	187.2	218.4	249.6	280.8
311	31.1	62.2	93.3	124.4	155.5	186.6	217.7	248.8	279.9
310	31.0	62.0	93.0	124.0	155.0	186.0	217.0	248.0	279.0
309	30.9	61.8	92.7	123.6	154.5	185.4	216.3	247.2	278.1
308	30.8	61.6	92.4	123.2	154.0	184.8	215.6	246.4	277.2
307	30.7	61.4	92.1	122.8	153.5	184.2	214.9	245.6	276.3
306	30.6	61.2	91.8	122.4	153.0	183.6	214.2	244.8	275.4
305	30.5	61.0	91.5	122.0	152.5	183.0	213.5	244.0	274.5
304	30.4	60.8	91.2	121.6	152.0	182.4	212.8	243.2	273.6
303	30.3	60.6	90.9	121.2	151.5	181.8	212.1	242.4	272.7
302	30.2	60.4	90.6	120.8	151.0	181.2	211.4	241.6	271.8
301	30.1	60.2	90.3	120.4	150.5	180.6	210.7	240.8	270.9
300	30.0	60.0	90.0	120.0	150.0	180.0	210.0	240.0	270.0
299	29.9	59.8	89.7	119.6	149.5	179.4	209.3	239.2	269.1
298	29.8	59.6	89.4	119.2	149.0	178.8	208.6	238.4	268.2
297	29.7	59.4	89.1	118.8	148.5	178.2	207.9	237.6	267.3
296	29.6	59.2	88.8	118.4	148.0	177.6	207.2	236.8	266.4
295	29.5	59.0	88.5	118.0	147.5	177.0	206.5	236.0	265.5
294	29.4	58.8	88.2	117.6	147.0	176.4	205.8	235.2	264.6
293	29.3	58.6	87.9	117.2	146.5	175.8	205.1	234.4	263.7
292	29.2	58.4	87.6	116.8	146.0	175.2	204.4	233.6	262.8
291	29.1	58.2	87.3	116.4	145.5	174.6	203.7	232.8	261.9
290	29.0	58.0	87.0	116.0	145.0	174.0	203.0	232.0	261.0
289	28.9	57.8	86.7	115.6	144.5	173.4	202.3	231.2	260.1
288	28.8	57.6	86.4	115.2	144.0	172.8	201.6	230.4	259.2
287	28.7	57.4	86.1	114.8	143.5	172.2	200.9	229.6	258.3
286	28.6	57.2	85.8	114.4	143.0	171.6	200.2	228.8	257.4

Table I. (Continued)

No. 150
Log. 176

No. 169
Log. 230

N.	0	1	2	3	4	5	6	7	8	9	Diff.
150	17 6091	6381	6670	6959	7248	7536	7825	8113	8401	8689	289
1	8977	9264	9552	9839	0126	0413	0699	0986	1272	1558	287
2	18 1844	2129	2415	2700	2985	3270	3555	3839	4123	4407	285
3	4691	4975	5259	5542	5825	6108	6391	6674	6956	7239	283
4	7521	7803	8084	8366	8647	8928	9209	9490	9771	0051	281
155	19 0332	0612	0892	1171	1451	1730	2010	2289	2567	2846	279
6	3125	3403	3681	3959	4237	4514	4792	5069	5346	5623	278
7	5900	6176	6453	6729	7005	7281	7556	7832	8107	8382	276
8	8657	8932	9206	9481	9755	0029	0303	0577	0850	1124	274
9	20 1397	1670	1943	2216	2488	2761	3033	3305	3577	3848	272
160	4120	4391	4663	4934	5204	5475	5746	6016	6286	6556	271
1	6826	7096	7365	7634	7904	8173	8441	8710	8979	9247	269
2	9515	9783	0051	0319	0586	0853	1121	1388	1654	1921	267
3	21 2188	2454	2720	2986	3252	3518	3783	4049	4314	4579	266
4	4844	5109	5373	5638	5902	6166	6430	6694	6957	7221	264
165	7484	7747	8010	8273	8536	8798	9060	9323	9585	9846	262
6	22 0108	0370	0631	0892	1153	1414	1675	1936	2196	2456	261
7	2716	2976	3236	3496	3755	4015	4274	4533	4792	5051	259
8	5309	5568	5826	6084	6342	6600	6858	7115	7372	7630	258
9	7887	8144	8400	8657	8913	9170	9426	9682	9938	0193	256
23											

PROPORTIONAL PARTS

Diff.	1	2	3	4	5	6	7	8	9
285	28.5	57.0	85.5	114.0	142.5	171.0	199.5	228.0	256.5
284	28.4	56.8	85.2	113.6	142.0	170.4	198.8	227.2	255.6
283	28.3	56.6	84.9	113.2	141.5	169.8	198.1	226.4	254.7
282	28.2	56.4	84.6	112.8	141.0	169.2	197.4	225.6	253.8
281	28.1	56.2	84.3	112.4	140.5	168.6	196.7	224.8	252.9
280	28.0	56.0	84.0	112.0	140.0	168.0	196.0	224.0	252.0
279	27.9	55.8	83.7	111.6	139.5	167.4	195.3	223.2	251.1
278	27.8	55.6	83.4	111.2	139.0	166.8	194.6	222.4	250.2
277	27.7	55.4	83.1	110.8	138.5	166.2	193.9	221.6	249.3
276	27.6	55.2	82.8	110.4	138.0	165.6	193.2	220.8	248.4
275	27.5	55.0	82.5	110.0	137.5	165.0	192.5	220.0	247.5
274	27.4	54.8	82.2	109.6	137.0	164.4	191.8	219.2	246.6
273	27.3	54.6	81.9	109.2	136.5	163.8	191.1	218.4	245.7
272	27.2	54.4	81.6	108.8	136.0	163.2	190.4	217.6	244.8
271	27.1	54.2	81.3	108.4	135.5	162.6	189.7	216.8	243.9
270	27.0	54.0	81.0	108.0	135.0	162.0	189.0	216.0	243.0
269	26.9	53.8	80.7	107.6	134.5	161.4	188.3	215.2	242.1
268	26.8	53.6	80.4	107.2	134.0	160.8	187.6	214.4	241.2
267	26.7	53.4	80.1	106.8	133.5	160.2	186.9	213.6	240.3
266	26.6	53.2	79.8	106.4	133.0	159.0	186.2	212.8	239.4
265	26.5	53.0	79.5	106.0	132.5	159.0	185.5	212.0	238.5
264	26.4	52.8	79.2	105.6	132.0	158.4	184.8	211.2	237.6
263	26.3	52.6	78.9	105.2	131.5	157.8	184.1	210.4	236.7
262	26.2	52.4	78.6	104.8	131.0	157.2	183.4	209.6	235.8
261	26.1	52.2	78.3	104.4	130.5	156.6	182.7	208.8	234.9
260	26.0	52.0	78.0	104.0	130.0	156.0	182.0	208.0	234.0
259	25.9	51.8	77.7	103.6	129.5	155.4	181.3	207.2	233.1
258	25.8	51.6	77.4	103.2	129.0	154.8	180.6	206.4	232.2
257	25.7	51.4	77.1	102.8	128.5	154.2	179.9	205.6	231.3
256	25.6	51.2	76.8	102.4	128.0	153.6	179.2	204.8	230.4
255	25.5	51.0	76.5	102.0	127.5	153.0	178.5	204.0	229.5

Table I. (Continued)

No. 170
Log. 230

No. 189
Log. 278

N.	0	1	2	3	4	5	6	7	8	9	Diff.
170	23 0449	0704	0960	1215	1470	1724	1979	2234	2488	2742	255
1	2996	3250	3504	3757	4011	4264	4517	4770	5023	5276	253
2	5528	5781	6033	6285	6537	6789	7041	7292	7544	7795	252
3	8046	8297	8548	8799	9049	9299	9550	9800	0050	0300	250
4	24 0549	0799	1048	1297	1546	1795	2044	2293	2541	2790	249
175	3038	3286	3534	3782	4030	4277	4525	4772	5019	5266	248
6	5513	5759	6006	6252	6499	6745	6991	7237	7482	7728	246
7	7973	8219	8464	8709	8954	9198	9443	9687	9932	0176	245
8	25 0420	0664	0908	1151	1395	1638	1881	2125	2368	2610	243
9	2853	3096	3338	3580	3822	4064	4306	4548	4790	5031	242
180	5273	5514	5755	5996	6237	6477	6718	6958	7198	7439	241
1	7679	7918	8158	8398	8637	8877	9116	9355	9594	9833	239
2	26 0071	0310	0548	0787	1025	1263	1501	1739	1976	2214	238
3	2451	2688	2925	3162	3399	3636	3873	4109	4346	4582	237
4	4818	5054	5290	5525	5761	5996	6232	6467	6702	6937	235
185	7172	7406	7641	7875	8110	8344	8578	8812	9046	9279	234
6	9513	9746	9980	0213	0446	0679	0912	1144	1377	1609	233
7	27 1842	2074	2306	2538	2770	3001	3233	3464	3696	3927	232
8	4158	4389	4620	4850	5081	5311	5542	5772	6002	6232	230
9	6462	6692	6921	7151	7380	7609	7838	8067	8296	8525	229

PROPORTIONAL PARTS

Diff.	1	2	3	4	5	6	7	8	9
255	25.5	51.0	76.5	102.0	127.5	153.0	178.5	204.0	229.5
254	25.4	50.8	76.2	101.6	127.0	152.4	177.8	203.2	228.6
253	25.3	50.6	75.9	101.2	126.5	151.8	177.1	202.4	227.7
252	25.2	50.4	75.6	100.8	126.0	151.2	176.4	201.6	226.8
251	25.1	50.2	75.3	100.4	125.5	150.6	175.7	200.8	225.9
250	25.0	50.0	75.0	100.0	125.0	150.0	175.0	200.0	225.0
249	24.9	49.8	74.7	99.6	124.5	149.4	174.3	199.2	224.1
248	24.8	49.6	74.4	99.2	124.0	148.8	173.6	198.4	223.2
247	24.7	49.4	74.1	98.8	123.5	148.2	172.9	197.6	222.3
246	24.6	49.2	73.8	98.4	123.0	147.6	172.2	196.8	221.4
245	24.5	49.0	73.5	98.0	122.5	147.0	171.5	196.0	220.5
244	24.4	48.8	73.2	97.6	122.0	146.4	170.8	195.2	219.6
243	24.3	48.6	72.9	97.2	121.5	145.8	170.1	194.4	218.7
242	24.2	48.4	72.6	96.8	121.0	145.2	169.4	193.6	217.8
241	24.1	48.2	72.3	96.4	120.5	144.6	168.7	192.8	216.9
240	24.0	48.0	72.0	96.0	120.0	144.0	168.0	192.0	216.0
239	23.9	47.8	71.7	95.6	119.5	143.4	167.3	191.2	215.1
238	23.8	47.6	71.4	95.2	119.0	142.8	166.6	190.4	214.2
237	23.7	47.4	71.1	94.8	118.5	142.2	165.9	189.6	213.3
236	23.6	47.2	70.8	94.4	118.0	141.6	165.2	188.8	212.4
235	23.5	47.0	70.5	94.0	117.5	141.0	164.5	188.0	211.5
234	23.4	46.8	70.2	93.6	117.0	140.4	163.8	187.2	210.6
233	23.3	46.6	69.9	93.2	116.5	139.8	163.1	186.4	209.7
232	23.2	46.4	69.6	92.8	116.0	139.2	162.4	185.6	208.8
231	23.1	46.2	69.3	92.4	115.5	138.6	161.7	184.8	207.9
230	23.0	46.0	69.0	92.0	115.0	138.0	161.0	184.0	207.0
229	22.9	45.8	68.7	91.6	114.5	137.4	160.3	183.2	206.1
228	22.8	45.6	68.4	91.2	114.0	136.8	159.6	182.4	205.2
227	22.7	45.4	68.1	90.8	113.5	136.2	158.9	181.6	204.3
226	22.6	45.2	67.8	90.4	113.0	135.6	158.2	180.8	203.4

Table I. (Continued)

No. 190
Log. 278

No. 214
Log. 332

N.	0	1	2	3	4	5	6	7	8	9	Diff.
190	27 8754	8982	9211	9439	9667	9895	0123	0351	0578	0806	228
1	28 1033	1261	1488	1715	1942	2169	2396	2622	2849	3075	227
2	3301	3527	3753	3979	4205	4431	4656	4882	5107	5332	226
3	5557	5782	6007	6232	6456	6681	6905	7130	7354	7578	225
4	7802	8026	8249	8473	8696	8920	9143	9366	9589	9812	223
195	29 0035	0257	0480	0702	0925	1147	1369	1591	1813	2034	222
6	2256	2478	2699	2920	3141	3363	3584	3804	4025	4246	221
7	4466	4687	4907	5127	5347	5567	5787	6007	6226	6446	220
8	6665	6884	7104	7323	7542	7761	7979	8198	8416	8635	219
9	8853	9071	9289	9507	9725	9943	0161	0378	0595	0813	218
200	30 1030	1247	1464	1681	1898	2114	2331	2547	2764	2980	217
1	3196	3412	3628	3844	4059	4275	4491	4706	4921	5136	216
2	5351	5566	5781	5996	6211	6425	6639	6854	7068	7282	215
3	7496	7710	7924	8137	8351	8564	8778	8991	9204	9417	213
4	9630	9843	0056	0268	0481	0693	0906	1118	1330	1542	212
205	31 1754	1966	2177	2389	2600	2812	3023	3234	3445	3656	211
6	3867	4078	4289	4499	4710	4920	5130	5340	5551	5760	210
7	5970	6180	6390	6599	6809	7018	7227	7436	7646	7854	209
8	8063	8272	8481	8689	8898	9106	9314	9522	9730	9938	208
9	32 0146	0354	0562	0769	0977	1184	1391	1598	1805	2012	207
210	2219	2426	2633	2839	3046	3252	3458	3665	3871	4077	206
1	4282	4488	4694	4899	5105	5310	5516	5721	5926	6131	205
2	6336	6541	6745	6950	7155	7359	7563	7767	7972	8176	204
3	8380	8583	8787	8991	9194	9398	9601	9805	0008	0211	203
4	33 0414	0617	0819	1022	1225	1427	1630	1832	2034	2236	202

PROPORTIONAL PARTS

Diff.	1	2	3	4	5	6	7	8	9
225	22.5	45.0	67.5	90.0	112.5	135.0	157.5	180.0	202.5
224	22.4	44.8	67.2	89.6	112.0	134.4	156.8	179.2	201.6
223	22.3	44.6	66.9	89.2	111.5	133.8	156.1	178.4	200.7
222	22.2	44.4	66.6	88.8	111.0	133.2	155.4	177.6	199.8
221	22.1	44.2	66.3	88.4	110.5	132.6	154.7	176.8	198.9
220	22.0	44.0	66.0	88.0	110.0	132.0	154.0	176.0	198.0
219	21.9	43.8	65.7	87.6	109.5	131.4	153.3	175.2	197.1
218	21.8	43.6	65.4	87.2	109.0	130.8	152.6	174.4	196.2
217	21.7	43.4	65.1	86.8	108.5	130.2	151.9	173.6	195.3
216	21.6	43.2	64.8	86.4	108.0	129.6	151.2	172.8	194.4
215	21.5	43.0	64.5	86.0	107.5	129.0	150.5	172.0	193.5
214	21.4	42.8	64.2	85.6	107.0	128.4	149.8	171.2	192.6
213	21.3	42.6	63.9	85.2	106.5	127.8	149.1	170.4	191.7
212	21.2	42.4	63.6	84.8	106.0	127.2	148.4	169.6	190.8
211	21.1	42.2	63.3	84.4	105.5	126.6	147.7	168.8	189.9
210	21.0	42.0	63.0	84.0	105.0	126.0	147.0	168.0	189.0
209	20.9	41.8	62.7	83.6	104.5	125.4	146.3	167.2	188.1
208	20.8	41.6	62.4	83.2	104.0	124.8	145.6	166.4	187.2
207	20.7	41.4	62.1	82.8	103.5	124.2	144.9	165.6	186.3
206	20.6	41.2	61.8	82.4	103.0	123.6	144.2	164.8	185.4
205	20.5	41.0	61.5	82.0	102.5	123.0	143.5	164.0	184.5
204	20.4	40.8	61.2	81.6	102.0	122.4	142.8	163.2	183.6
203	20.3	40.6	60.9	81.2	101.5	121.8	142.1	162.4	182.7
202	20.2	40.4	60.6	80.8	101.0	121.2	141.4	161.6	181.8

Table I. (Continued)

No. 215
Log. 332

No. 239
Log. 380

N.	0	1	2	3	4	5	6	7	8	9	Diff.
215	33 2438	2640	2842	3044	3246	3447	3649	3850	4051	4253	202
6	4454	4655	4856	5057	5257	5458	5658	5859	6059	6260	201
7	6460	6660	6860	7060	7260	7459	7659	7858	8058	8257	200
8	8456	8656	8855	9054	9253	9451	9650	9849	0047	0246	199
9	34 0444	0642	0841	1039	1237	1435	1632	1830	2028	2225	198
220	2423	2620	2817	3014	3212	3409	3606	3802	3999	4196	197
1	4392	4589	4785	4981	5178	5374	5570	5766	5962	6157	196
2	6353	6549	6744	6939	7135	7330	7525	7720	7915	8110	195
3	8305	8500	8694	8889	9083	9278	9472	9666	9860	0054	194
4	35 0248	0442	0636	0829	1023	1216	1410	1603	1796	1989	193
225	2183	2375	2568	2761	2954	3147	3339	3532	3724	3916	193
6	4108	4301	4493	4685	4876	5068	5260	5452	5643	5834	192
7	6026	6217	6408	6599	6790	6981	7172	7363	7554	7744	191
8	7935	8125	8316	8506	8696	8886	9076	9266	9456	9646	190
9	9835	0025	0215	0404	0593	0783	0972	1161	1350	1539	189
230	36 1728	1917	2105	2294	2482	2671	2859	3048	3236	3424	188
1	3612	3800	3988	4176	4363	4551	4739	4926	5113	5301	188
2	5488	5675	5862	6049	6236	6423	6610	6796	6983	7169	187
3	7356	7542	7729	7915	8101	8287	8473	8659	8845	9030	186
4	9216	9401	9587	9772	9958	0143	0328	0513	0698	0883	185
235	37 1068	1253	1437	1622	1806	1991	2175	2360	2544	2728	184
6	2912	3096	3280	3464	3647	3831	4015	4198	4382	4565	184
7	4748	4932	5115	5298	5481	5664	5846	6029	6212	6394	183
8	6577	6759	6942	7124	7306	7488	7670	7852	8034	8216	182
9	8398	8580	8761	8943	9124	9306	9487	9668	9849	0030	181
	38										

PROPORTIONAL PARTS

Diff.	1	2	3	4	5	6	7	8	9
202	20.2	40.4	60.6	80.8	101.0	121.2	141.4	161.6	181.8
201	20.1	40.2	60.3	80.4	100.5	120.6	140.7	160.8	180.9
200	20.0	40.0	60.0	80.0	100.0	120.0	140.0	160.0	180.0
199	19.9	39.8	59.7	79.6	99.5	119.4	139.3	159.2	179.1
198	19.8	39.6	59.4	79.2	99.0	118.8	138.6	158.4	178.2
197	19.7	39.4	59.1	78.8	98.5	118.2	137.9	157.6	177.3
196	19.6	39.2	58.8	78.4	98.0	117.6	137.2	156.8	176.4
195	19.5	39.0	58.5	78.0	97.5	117.0	136.5	156.0	175.5
194	19.4	38.8	58.2	77.6	97.0	116.4	135.8	155.2	174.6
193	19.3	38.6	57.9	77.2	96.5	115.8	135.1	154.4	173.7
192	19.2	38.4	57.6	76.8	96.0	115.2	134.4	153.6	172.8
191	19.1	38.2	57.3	76.4	95.5	114.6	133.7	152.8	171.9
190	19.0	38.0	57.0	76.0	95.0	114.0	133.0	152.0	171.0
189	18.9	37.8	56.7	75.6	94.5	113.4	132.3	151.2	170.1
188	18.8	37.6	56.4	75.2	94.0	112.8	131.6	150.4	169.2
187	18.7	37.4	56.1	74.8	93.5	112.2	130.9	149.6	168.3
186	18.6	37.2	55.8	74.4	93.0	111.6	130.2	148.8	167.4
185	18.5	37.0	55.5	74.0	92.5	111.0	129.5	148.0	166.5
184	18.4	36.8	55.2	73.6	92.0	110.4	128.8	147.2	165.6
183	18.3	36.6	54.9	73.2	91.5	109.8	128.1	146.4	164.7
182	18.2	36.4	54.6	72.8	91.0	109.2	127.4	145.6	163.8
181	18.1	36.2	54.3	72.4	90.5	108.6	126.7	144.8	162.9
180	18.0	36.0	54.0	72.0	90.0	108.0	126.0	144.0	162.0
179	17.9	35.8	53.7	71.6	89.5	107.4	125.3	143.2	161.1

Table I. (Continued)

No. 240
Log. 380

No. 269
Log. 431

N.	0	1	2	3	4	5	6	7	8	9	Diff.
240	38 0211	0392	0573	0754	0934	1115	1296	1476	1656	1837	181
1	2017	2197	2377	2557	2737	2917	3097	3277	3456	3636	180
2	3815	3995	4174	4353	4533	4712	4891	5070	5249	5428	179
3	5606	5785	5964	6142	6321	6499	6677	6856	7034	7212	178
4	7390	7568	7746	7924	8101	8279	8456	8634	8811	8989	178
245	9166	9343	9520	9698	9875	0051	0228	0405	0582	0759	177
6	39 0935	1112	1288	1464	1641	1817	1993	2169	2345	2521	176
7	2697	2873	3048	3224	3400	3575	3751	3926	4101	4277	176
8	4452	4627	4802	4977	5152	5326	5501	5676	5850	6025	175
9	6199	6374	6548	6722	6896	7071	7245	7419	7592	7766	174
250	7940	8114	8287	8461	8634	8808	8981	9154	9328	9501	173
1	9674	9847	0020	0192	0365	0538	0711	0883	1056	1228	173
2	40 1401	1573	1745	1917	2089	2261	2433	2605	2777	2949	172
3	3121	3292	3464	3635	3807	3978	4149	4320	4492	4663	171
4	4834	5005	5176	5346	5517	5688	5858	6029	6199	6370	171
255	6540	6710	6881	7051	7221	7391	7561	7731	7901	8070	170
6	8240	8410	8579	8749	8918	9087	9257	9426	9595	9764	169
7	9933	0102	0271	0440	0609	0777	0946	1114	1283	1451	169
8	41 1620	1788	1956	2124	2293	2461	2629	2796	2964	3132	168
9	3300	3467	3635	3803	3970	4137	4305	4472	4639	4806	167
260	4973	5140	5307	5474	5641	5808	5974	6141	6308	6474	167
1	6641	6807	6973	7139	7306	7472	7638	7804	7970	8135	166
2	8301	8467	8633	8798	8964	9129	9295	9460	9625	9791	165
3	9956	0121	0286	0451	0616	0781	0945	1110	1275	1439	165
4	42 1604	1768	1933	2097	2261	2426	2590	2754	2918	3082	164
265	3246	3410	3574	3737	3901	4065	4228	4392	4555	4718	164
6	4882	5045	5208	5371	5534	5697	5860	6023	6186	6349	163
7	6511	6674	6836	6999	7161	7324	7486	7648	7811	7973	162
8	8135	8297	8459	8621	8783	8944	9106	9268	9429	9591	162
9	9752	9914	0075	0236	0398	0559	0720	0881	1042	1203	161
	43										

PROPORTIONAL PARTS

Diff.	1	2	3	4	5	6	7	8	9
178	17.8	35.6	53.4	71.2	89.0	106.8	124.6	142.4	160.2
177	17.7	35.4	53.1	70.8	88.5	106.2	123.9	141.6	159.3
176	17.6	35.2	52.8	70.4	88.0	105.6	123.2	140.8	158.4
175	17.5	35.0	52.5	70.0	87.5	105.0	122.5	140.0	157.5
174	17.4	34.8	52.2	69.6	87.0	104.4	121.8	139.2	156.6
173	17.3	34.6	51.9	69.2	86.5	103.8	121.1	138.4	155.7
172	17.2	34.4	51.6	68.8	86.0	103.2	120.4	137.6	154.8
171	17.1	34.2	51.3	68.4	85.5	102.6	119.7	136.8	153.9
170	17.0	34.0	51.0	68.0	85.0	102.0	119.0	136.0	153.0
169	16.9	33.8	50.7	67.6	84.5	101.4	118.3	135.2	152.1
168	16.8	33.6	50.4	67.2	84.0	100.8	117.6	134.4	151.2
167	16.7	33.4	50.1	66.8	83.5	100.2	116.9	133.6	150.3
166	16.6	33.2	49.8	66.4	83.0	99.6	116.2	132.8	149.4
165	16.5	33.0	49.5	66.0	82.5	99.0	115.5	132.0	148.5
164	16.4	32.8	49.2	65.6	82.0	98.4	114.8	131.2	147.6
163	16.3	32.6	48.9	65.2	81.5	97.8	114.1	130.4	146.7
162	16.2	32.4	48.5	64.8	81.0	97.2	113.4	129.6	145.8
161	16.1	32.2	48.3	64.4	80.5	96.6	112.7	128.8	144.9

Table I. (Continued)

No. 270
Log. 431

No. 299
Log. 476

N.	0	1	2	3	4	5	6	7	8	9	Diff.
270	43 1364	1525	1685	1846	2007	2167	2328	2488	2649	2809	161
1	2969	3130	3290	3450	3610	3770	3930	4090	4249	4409	160
2	4569	4729	4888	5048	5207	5367	5526	5685	5844	6004	159
3	6163	6322	6481	6640	6799	6957	7116	7275	7433	7592	159
4	7751	7909	8067	8226	8384	8542	8701	8859	9017	9175	158
275	9333	9491	9648	9806	9964	0122	0279	0437	0594	0752	158
6	44 0909	1066	1224	1381	1538	1695	1852	2009	2166	2323	157
7	2480	2637	2793	2950	3106	3263	3419	3576	3732	3889	157
8	4045	4201	4357	4513	4669	4825	4981	5137	5293	5449	156
9	5604	5760	5915	6071	6226	6382	6537	6692	6848	7003	155
230	7158	7313	7468	7623	7778	7933	8088	8242	8397	8552	155
1	8706	8861	9015	9170	9324	9478	9633	9787	9941	0095	154
2	45 0249	0403	0557	0711	0865	1018	1172	1326	1479	1633	154
3	1786	1940	2093	2247	2400	2553	2706	2859	3012	3165	153
4	3318	3471	3624	3777	3930	4082	4235	4387	4540	4692	153
285	4845	4997	5150	5302	5454	5606	5758	5910	6062	6214	152
6	6366	6518	6670	6821	6973	7125	7276	7428	7579	7731	152
7	7882	8033	8184	8336	8487	8638	8789	8940	9091	9242	151
8	9392	9543	9694	9845	9995	0146	0296	0447	0597	0748	151
9	46 0898	1048	1198	1348	1499	1649	1799	1948	2098	2248	150
290	2398	2548	2697	2847	2997	3146	3296	3445	3594	3744	150
1	3893	4042	4191	4340	4490	4639	4788	4936	5085	5234	149
2	5383	5532	5680	5829	5977	6126	6274	6423	6571	6719	149
3	6868	7016	7164	7312	7460	7608	7756	7904	8052	8200	148
4	8347	8495	8643	8790	8938	9085	9233	9380	9527	9675	148
295	9822	9969	0116	0263	0410	0557	0704	0851	0998	1145	147
6	47 1292	1438	1585	1732	1878	2025	2171	2318	2464	2610	146
7	2756	2903	3049	3195	3341	3487	3633	3779	3925	4071	146
8	4216	4362	4508	4653	4799	4944	5090	5235	5381	5526	146
9	5671	5816	5962	6107	6252	6397	6542	6687	6832	6976	145

PROPORTIONAL PARTS

Diff.	1	2	3	4	5	6	7	8	9
161	16.1	32.2	48.3	64.4	80.5	96.6	112.7	128.8	144.9
160	16.0	32.0	48.0	64.0	80.0	96.0	112.0	128.0	144.0
159	15.9	31.8	47.7	63.6	79.5	95.4	111.3	127.2	143.1
158	15.8	31.6	47.4	63.2	79.0	94.8	110.6	126.4	142.2
157	15.7	31.4	47.1	62.8	78.5	94.2	109.9	125.6	141.3
156	15.6	31.2	46.8	62.4	78.0	93.6	109.2	124.8	140.4
155	15.5	31.0	46.5	62.0	77.5	93.0	108.5	124.0	139.5
154	15.4	30.8	46.2	61.6	77.0	92.4	107.8	123.2	138.6
153	15.3	30.6	45.9	61.2	76.5	91.8	107.1	122.4	137.7
152	15.2	30.4	45.6	60.8	76.0	91.2	106.4	121.6	136.8
151	15.1	30.2	45.3	60.4	75.5	90.6	105.7	120.8	135.9
150	15.0	30.0	45.0	60.0	75.0	90.0	105.0	120.0	135.0
149	14.9	29.8	44.7	59.6	74.5	89.4	104.3	119.2	134.1
148	14.8	29.6	44.4	59.2	74.0	88.8	103.6	118.4	133.2
147	14.7	29.4	44.1	58.8	73.5	88.2	102.9	117.6	132.3
146	14.6	29.2	43.8	58.4	73.0	87.6	102.2	116.8	131.4
145	14.5	29.0	43.5	58.0	72.5	87.0	101.5	116.0	130.5
144	14.4	28.8	43.2	57.6	72.0	86.4	100.8	115.2	129.6
143	14.3	28.6	42.9	57.2	71.5	85.8	100.1	114.4	128.7
142	14.2	28.4	42.6	56.8	71.0	85.2	99.4	113.6	127.8
141	14.1	28.2	42.3	56.4	70.5	84.6	98.7	112.8	126.9
140	14.0	28.0	42.0	56.0	70.0	84.0	98.0	112.0	126.0

Table I. (Continued)

No. 300
Log. 477

No. 339
Log. 531

N.	0	1	2	3	4	5	6	7	8	9	Diff.
300	47 7121	7266	7411	7555	7700	7844	7989	8133	8278	8422	145
1	8566	8711	8855	8999	9143	9287	9431	9575	9719	9863	144
2	48 0007	0151	0294	0438	0582	0725	0869	1012	1156	1299	144
3	1443	1586	1729	1872	2016	2159	2302	2445	2588	2731	143
4	2874	3016	3159	3302	3445	3587	3730	3872	4015	4157	143
305	4300	4442	4585	4727	4869	5011	5153	5295	5437	5579	142
6	5721	5863	6005	6147	6289	6430	6572	6714	6855	6997	142
7	7138	7280	7421	7563	7704	7845	7986	8127	8269	8410	141
8	8551	8692	8833	8974	9114	9255	9396	9537	9677	9818	141
9	9958	0099	0239	0380	0520	0661	0801	0941	1081	1222	140
310	49 1362	1502	1642	1782	1922	2062	2201	2341	2481	2621	140
1	2760	2900	3040	3179	3319	3458	3597	3737	3876	4015	139
2	4155	4294	4433	4572	4711	4850	4989	5128	5267	5406	139
3	5544	5683	5822	5960	6099	6238	6376	6515	6653	6791	139
4	6930	7068	7206	7344	7483	7621	7759	7897	8035	8173	138
315	8311	8448	8586	8724	8862	8999	9137	9275	9412	9550	138
6	9687	9824	9962	0099	0236	0374	0511	0648	0785	0922	137
7	50 1059	1196	1333	1470	1607	1744	1880	2017	2154	2291	137
8	2427	2564	2700	2837	2973	3109	3246	3382	3518	3655	136
9	3791	3927	4063	4199	4335	4471	4607	4743	4878	5014	136
320	5150	5286	5421	5557	5693	5828	5964	6099	6234	6370	136
1	6505	6640	6776	6911	7046	7181	7316	7451	7586	7721	135
2	7856	7991	8126	8260	8395	8530	8664	8799	8934	9068	135
3	9203	9337	9471	9606	9740	9874	0009	0143	0277	0411	134
4	51 0545	0679	0813	0947	1081	1215	1349	1482	1616	1750	134
325	1883	2017	2151	2284	2418	2551	2684	2818	2951	3084	133
6	3218	3351	3484	3617	3750	3883	4016	4149	4282	4415	133
7	4548	4681	4813	4946	5079	5211	5344	5476	5609	5741	133
8	5874	6006	6139	6271	6403	6535	6668	6800	6932	7064	132
9	7196	7328	7460	7592	7724	7855	7987	8119	8251	8382	132
330	8514	8646	8777	8909	9040	9171	9303	9434	9566	9697	131
1	9828	9959	0090	0221	0353	0484	0615	0745	0876	1007	131
2	52 1138	1269	1400	1530	1661	1792	1922	2053	2183	2314	131
3	2444	2575	2705	2835	2966	3096	3226	3356	3486	3616	130
4	3746	3876	4006	4136	4266	4396	4526	4656	4785	4915	130
335	5045	5174	5304	5434	5563	5693	5822	5951	6081	6210	129
6	6339	6469	6598	6727	6856	6985	7114	7243	7372	7501	129
7	7630	7759	7888	8016	8145	8274	8402	8531	8660	8788	129
8	8917	9045	9174	9302	9430	9559	9687	9815	9943	0072	128
9	53 0200	0328	0456	0584	0712	0840	0968	1096	1223	1351	128

PROPORTIONAL PARTS

Diff.	1	2	3	4	5	6	7	8	9
139	13.9	27.8	41.7	55.6	69.5	83.4	97.3	111.2	125.1
138	13.8	27.6	41.4	55.2	69.0	82.8	96.6	110.4	124.2
137	13.7	27.4	41.1	54.8	68.5	82.2	95.9	109.6	123.3
136	13.6	27.2	40.8	54.4	68.0	81.6	95.2	108.8	122.4
135	13.5	27.0	40.5	54.0	67.5	81.0	94.5	108.0	121.5
134	13.4	26.8	40.2	53.6	67.0	80.4	93.8	107.2	120.6
133	13.3	26.6	39.9	53.2	66.5	79.8	93.1	106.4	119.7
132	13.2	26.4	39.6	52.8	66.0	79.2	92.4	105.6	118.8
131	13.1	26.2	39.3	52.4	65.5	78.6	91.7	104.8	117.9
130	13.0	26.0	39.0	52.0	65.0	78.0	91.0	104.0	117.0
129	12.9	25.8	38.7	51.6	64.5	77.4	90.3	103.2	116.1
128	12.8	25.6	38.4	51.2	64.0	76.8	89.6	102.4	115.2
127	12.7	25.4	38.1	50.8	63.5	76.2	88.9	101.6	114.3

Table I. (Continued)

No. 340
Log. 531

No. 379
Log. 579

N.	0	1	2	3	4	5	6	7	8	9	Diff.
340	53 1479	1607	1734	1862	1990	2117	2245	2372	2500	2627	128
1	2754	2882	3009	3136	3264	3391	3518	3645	3772	3899	127
2	4026	4153	4280	4407	4534	4661	4787	4914	5041	5167	127
3	5294	5421	5547	5674	5800	5927	6053	6180	6306	6432	126
4	6558	6685	6811	6937	7063	7189	7315	7441	7567	7693	126
345	7819	7945	8071	8197	8322	8448	8574	8699	8825	8951	126
6	9076	9202	9327	9452	9578	9703	9829	9954	0079	0204	125
7	54 0329	0455	0580	0705	0830	0955	1080	1205	1330	1454	125
8	1579	1704	1829	1953	2078	2203	2327	2452	2576	2701	125
9	2825	2950	3074	3199	3323	3447	3571	3696	3820	3944	124
350	4068	4192	4316	4440	4564	4688	4812	4936	5060	5183	124
1	5307	5431	5555	5678	5802	5925	6049	6172	6296	6419	124
2	6543	6666	6789	6913	7036	7159	7282	7405	7529	7652	123
3	7775	7898	8021	8144	8267	8389	8512	8635	8758	8881	123
4	9003	9126	9249	9371	9494	9616	9739	9861	9984	0106	123
355	55 0228	0351	0473	0595	0717	0840	0962	1084	1206	1328	122
6	1450	1572	1694	1816	1938	2060	2181	2303	2425	2547	122
7	2668	2790	2911	3033	3155	3276	3398	3519	3640	3762	122
8	3883	4004	4126	4247	4368	4489	4610	4731	4852	4973	121
9	5094	5215	5336	5457	5578	5699	5820	5940	6061	6182	121
360	6303	6423	6544	6664	6785	6905	7026	7146	7267	7387	120
1	7507	7627	7748	7868	7988	8108	8228	8349	8469	8589	120
2	8709	8829	8948	9068	9188	9308	9428	9548	9667	9787	120
3	9907	0026	0146	0265	0385	0504	0624	0743	0863	0982	119
4	56 1101	1221	1340	1459	1578	1698	1817	1936	2055	2174	119
365	2293	2412	2531	2650	2769	2887	3006	3125	3244	3362	119
6	3481	3600	3718	3837	3955	4074	4192	4311	4429	4548	119
7	4666	4784	4903	5021	5139	5257	5376	5494	5612	5730	118
8	5848	5966	6084	6202	6320	6437	6555	6673	6791	6909	118
9	7026	7144	7262	7379	7497	7614	7732	7849	7967	8084	118
370	8202	8319	8436	8554	8671	8788	8905	9023	9140	9257	117
1	9374	9491	9608	9725	9842	9959	0076	0193	0309	0426	117
2	57 0543	0660	0776	0893	1010	1126	1243	1359	1476	1592	117
3	1709	1825	1942	2058	2174	2291	2407	2523	2639	2755	116
4	2872	2988	3104	3220	3336	3452	3568	3684	3800	3915	116
375	4031	4147	4263	4379	4494	4610	4726	4841	4957	5072	116
6	5188	5303	5419	5534	5650	5765	5880	5996	6111	6226	115
7	6341	6457	6572	6687	6802	6917	7032	7147	7262	7377	115
8	7492	7607	7722	7836	7951	8066	8181	8295	8410	8525	115
9	8639	8754	8868	8983	9097	9212	9326	9441	9555	9669	114

PROPORTIONAL PARTS

Diff.	1	2	3	4	5	6	7	8	9
128	12.8	25.6	38.4	51.2	64.0	76.8	89.6	102.4	115.2
127	12.7	25.4	38.1	50.8	63.5	76.2	88.9	101.6	114.3
126	12.6	25.2	37.8	50.4	63.0	75.6	88.2	100.8	113.4
125	12.5	25.0	37.5	50.0	62.5	75.0	87.5	100.0	112.5
124	12.4	24.8	37.2	49.6	62.0	74.4	86.8	99.2	111.6
123	12.3	24.6	36.9	49.2	61.5	73.8	86.1	98.4	110.7
122	12.2	24.4	36.6	48.8	61.0	73.2	85.4	97.6	109.8
121	12.1	24.2	36.3	48.4	60.5	72.6	84.7	96.8	108.9
120	12.0	24.0	36.0	48.0	60.0	72.0	84.0	96.0	108.0
119	11.9	23.8	35.7	47.6	59.5	71.4	83.3	95.2	107.1

Table I. (Continued)

No. 380
Log. 579

No. 414
Log. 617

N.	0	1	2	3	4	5	6	7	8	9	Diff.
380	57 9784	9898	0012	0126	0241	0355	0469	0583	0697	0811	114
1	58 0925	1039	1153	1267	1381	1495	1608	1722	1836	1950	
2	2063	2177	2291	2404	2518	2631	2745	2858	2972	3085	
3	3199	3312	3426	3539	3652	3765	3879	3992	4105	4218	113
4	4331	4444	4557	4670	4783	4896	5009	5122	5235	5348	
385	5461	5574	5686	5799	5912	6024	6137	6250	6362	6475	
6	6587	6700	6812	6925	7037	7149	7262	7374	7486	7599	
7	7711	7823	7935	8047	8160	8272	8384	8496	8608	8720	112
8	8832	8944	9056	9167	9279	9391	9503	9615	9726	9838	
9	9950	0061	0173	0284	0396	0507	0619	0730	0842	0953	
390	59 1065	1176	1287	1399	1510	1621	1732	1843	1955	2066	
1	2177	2288	2399	2510	2621	2732	2843	2954	3064	3175	111
2	3286	3397	3508	3618	3729	3840	3950	4061	4171	4282	
3	4393	4503	4614	4724	4834	4945	5055	5165	5276	5386	
4	5496	5606	5717	5827	5937	6047	6157	6267	6377	6487	
395	6597	6707	6817	6927	7037	7146	7256	7366	7476	7586	110
6	7695	7805	7914	8024	8134	8243	8353	8462	8572	8681	
7	8791	8900	9009	9119	9228	9337	9446	9556	9665	9774	
8	9883	9992	0101	0210	0319	0428	0537	0646	0755	0864	109
9	60 0973	1082	1191	1299	1408	1517	1625	1734	1843	1951	
400	2060	2169	2277	2386	2494	2603	2711	2819	2928	3036	108
1	3144	3253	3361	3469	3577	3686	3794	3902	4010	4118	
2	4226	4334	4442	4550	4658	4766	4874	4982	5089	5197	
3	5305	5413	5521	5628	5736	5844	5951	6059	6166	6274	
4	6381	6489	6596	6704	6811	6919	7026	7133	7241	7348	107
405	7455	7562	7669	7777	7884	7991	8098	8205	8312	8419	
6	8526	8633	8740	8847	8954	9061	9167	9274	9381	9488	
7	9594	9701	9808	9914	0021	0128	0234	0341	0447	0554	
8	61 0660	0767	0873	0979	1086	1192	1298	1405	1511	1617	106
9	1723	1829	1936	2042	2148	2254	2360	2466	2572	2678	
410	2784	2890	2996	3102	3207	3313	3419	3525	3630	3736	
1	3842	3947	4053	4159	4264	4370	4475	4581	4686	4792	
2	4897	5003	5108	5213	5319	5424	5529	5634	5740	5845	105
3	5950	6055	6160	6265	6370	6476	6581	6686	6790	6895	
4	7000	7105	7210	7315	7420	7525	7629	7734	7839	7943	

PROPORTIONAL PARTS

Diff.	1	2	3	4	5	6	7	8	9
118	11.8	23.6	35.4	47.2	59.0	70.8	82.6	94.4	106.2
117	11.7	23.4	35.1	46.8	58.5	70.2	81.9	93.6	105.3
116	11.6	23.2	34.8	46.4	58.0	69.6	81.2	92.8	104.4
115	11.5	23.0	34.5	46.0	57.5	69.0	80.5	92.0	103.5
114	11.4	22.8	34.2	45.6	57.0	68.4	79.8	91.2	102.6
113	11.3	22.6	33.9	45.2	56.5	67.8	79.1	90.4	101.7
112	11.2	22.4	33.6	44.8	56.0	67.2	78.4	89.6	100.8
111	11.1	22.2	33.3	44.4	55.5	66.6	77.7	88.8	99.9
110	11.0	22.0	33.0	44.0	55.0	66.0	77.0	88.0	99.0
109	10.9	21.8	32.7	43.6	54.5	65.4	76.3	87.2	98.1
108	10.8	21.6	32.4	43.2	54.0	64.8	75.6	86.4	97.2
107	10.7	21.4	32.1	42.8	53.5	64.2	74.9	85.6	96.3
106	10.6	21.2	31.8	42.4	53.0	63.6	74.2	84.8	95.4
105	10.5	21.0	31.5	42.0	52.5	63.0	73.5	84.0	94.5
104	10.4	20.8	31.2	41.6	52.0	62.4	72.8	83.2	93.6

Table I. (Continued)

No. 415
Log. 618

No. 459
Log. 662

N.	0	1	2	3	4	5	6	7	8	9	Diff.
415	61 8048	8153	8257	8362	8466	8571	8676	8780	8884	8989	105
6	9093	9198	9302	9406	9511	9615	9719	9824	9928	0032	
7	62 0136	0240	0344	0448	0552	0656	0760	0864	0968	1072	104
8	1176	1280	1384	1488	1592	1695	1799	1903	2007	2110	
9	2214	2318	2421	2525	2628	2732	2835	2939	3042	3146	
420	3249	3353	3456	3559	3663	3766	3869	3973	4076	4179	
1	4282	4385	4488	4591	4695	4798	4901	5004	5107	5210	103
2	5312	5415	5518	5621	5724	5827	5929	6032	6135	6238	
3	6340	6443	6546	6648	6751	6853	6956	7058	7161	7263	
4	7366	7468	7571	7673	7775	7878	7980	8082	8185	8287	
425	8389	8491	8593	8695	8797	8900	9002	9104	9206	9308	102
6	9410	9512	9613	9715	9817	9919	0021	0123	0224	0326	
7	63 0428	0530	0631	0733	0835	0936	1038	1139	1241	1342	
8	1444	1545	1647	1748	1849	1951	2052	2153	2255	2356	
9	2457	2559	2660	2761	2862	2963	3064	3165	3266	3367	
430	3468	3569	3670	3771	3872	3973	4074	4175	4276	4376	101
1	4477	4578	4679	4779	4880	4981	5081	5182	5283	5383	
2	5484	5584	5685	5785	5886	5986	6087	6187	6287	6388	
3	6488	6588	6688	6789	6889	6989	7089	7189	7290	7390	
4	7490	7590	7690	7790	7890	7990	8090	8190	8290	8389	100
435	8489	8589	8689	8789	8888	8988	9088	9188	9287	9387	
6	9486	9586	9686	9785	9885	9984	0084	0183	0283	0382	
7	64 0481	0581	0680	0779	0879	0978	1077	1177	1276	1375	
8	1474	1573	1672	1771	1871	1970	2069	2168	2267	2366	
9	2465	2563	2662	2761	2860	2959	3058	3156	3255	3354	99
440	3453	3551	3650	3749	3847	3946	4044	4143	4242	4340	
1	4439	4537	4636	4734	4832	4931	5029	5127	5226	5324	
2	5422	5521	5619	5717	5815	5913	6011	6110	6208	6306	
3	6404	6502	6600	6698	6796	6894	6992	7089	7187	7285	98
4	7383	7481	7579	7676	7774	7872	7969	8067	8165	8262	
445	8360	8458	8555	8653	8750	8848	8945	9043	9140	9237	
6	9335	9432	9530	9627	9724	9821	9919	0016	0113	0210	
7	65 0308	0405	0502	0599	0696	0793	0890	0987	1084	1181	
8	1278	1375	1472	1569	1666	1762	1859	1956	2053	2150	97
9	2246	2343	2440	2536	2633	2730	2826	2923	3019	3116	
450	3213	3309	3405	3502	3598	3695	3791	3888	3984	4080	
1	4177	4273	4369	4465	4562	4658	4754	4850	4946	5042	
2	5138	5235	5331	5427	5523	5619	5715	5810	5906	6002	96
3	6098	6194	6290	6386	6482	6577	6673	6769	6864	6960	
4	7056	7152	7247	7343	7438	7534	7629	7725	7820	7916	
455	8011	8107	8202	8298	8393	8488	8584	8679	8774	8870	
6	8965	9060	9155	9250	9346	9441	9536	9631	9726	9821	
7	9916	0011	0106	0201	0296	0391	0486	0581	0676	0771	95
8	66 0865	0960	1055	1150	1245	1339	1434	1529	1623	1718	
9	1813	1907	2002	2096	2191	2286	2380	2475	2569	2663	

PROPORTIONAL PARTS

Diff.	1	2	3	4	5	6	7	8	9
105	10.5	21.0	31.5	42.0	52.5	63.0	73.5	84.0	94.5
104	10.4	20.8	31.2	41.6	52.0	62.4	72.8	83.2	93.6
103	10.3	20.6	30.9	41.2	51.5	61.8	72.1	82.4	92.7
102	10.2	20.4	30.6	40.8	51.0	61.2	71.4	81.6	91.8
101	10.1	20.2	30.3	40.4	50.5	60.6	70.7	80.8	90.9
100	10.0	20.0	30.0	40.0	50.0	60.0	70.0	80.0	90.0
99	9.9	19.8	29.7	39.6	49.5	59.4	69.3	79.2	89.1

Table I. (Continued)

No. 460
Log. 662

No. 499
Log. 698

N.	0	1	2	3	4	5	6	7	8	9	Diff.
460	66 2758	2852	2947	3041	3135	3230	3324	3418	3512	3607	
1	3701	3795	3889	3983	4078	4172	4266	4360	4454	4548	
2	4642	4736	4830	4924	5018	5112	5206	5299	5393	5487	94
3	5581	5675	5769	5862	5956	6050	6143	6237	6331	6424	
4	6518	6612	6705	6799	6892	6986	7079	7173	7266	7360	
465	7453	7546	7640	7733	7826	7920	8013	8106	8199	8293	
6	8386	8479	8572	8665	8759	8852	8945	9038	9131	9224	
7	9317	9410	9503	9596	9689	9782	9875	9967	0060	0153	93
8	67 0246	0339	0431	0524	0617	0710	0802	0895	0988	1080	
9	1173	1265	1358	1451	1543	1636	1728	1821	1913	2005	
470	2098	2190	2283	2375	2467	2560	2652	2744	2836	2929	
1	3021	3113	3205	3297	3390	3482	3574	3666	3758	3850	
2	3942	4034	4126	4218	4310	4402	4494	4586	4677	4769	92
3	4861	4953	5045	5137	5228	5320	5412	5503	5595	5687	
4	5778	5870	5962	6053	6145	6236	6328	6419	6511	6602	
475	6694	6785	6876	6968	7059	7151	7242	7333	7424	7516	
6	7607	7698	7789	7881	7972	8063	8154	8245	8336	8427	
7	8518	8609	8700	8791	8882	8973	9064	9155	9246	9337	91
8	9428	9519	9610	9700	9791	9882	9973	0063	0154	0245	
9	68 0336	0426	0517	0607	0698	0789	0879	0970	1060	1151	
480	1241	1332	1422	1513	1603	1693	1784	1874	1964	2055	
1	2145	2235	2326	2416	2506	2596	2686	2777	2867	2957	
2	3047	3137	3227	3317	3407	3497	3587	3677	3767	3857	90
3	3947	4037	4127	4217	4307	4396	4486	4576	4666	4756	
4	4845	4935	5025	5114	5204	5294	5383	5473	5563	5652	
485	5742	5831	5921	6010	6100	6189	6279	6368	6458	6547	
6	6636	6726	6815	6904	6994	7083	7172	7261	7351	7440	
7	7529	7618	7707	7796	7886	7975	8064	8153	8242	8331	89
8	8420	8509	8598	8687	8776	8865	8953	9042	9131	9220	
9	9309	9398	9486	9575	9664	9753	9841	9930	0019	0107	
490	69 0196	0285	0373	0462	0550	0639	0728	0816	0905	0993	
1	1081	1170	1258	1347	1435	1524	1612	1700	1789	1877	
2	1965	2053	2142	2230	2318	2406	2494	2583	2671	2759	
3	2847	2935	3023	3111	3199	3287	3375	3463	3551	3639	88
4	3727	3815	3903	3991	4078	4166	4254	4342	4430	4517	
495	4605	4693	4781	4868	4956	5044	5131	5219	5307	5394	
6	5482	5569	5657	5744	5832	5919	6007	6094	6182	6269	
7	6356	6444	6531	6618	6706	6793	6880	6968	7055	7142	
8	7229	7317	7404	7491	7578	7665	7752	7839	7926	8014	
9	8100	8188	8275	8362	8449	8535	8622	8709	8796	8883	87

PROPORTIONAL PARTS

Diff.	1	2	3	4	5	6	7	8	9
98	9.8	19.6	29.4	39.2	49.0	58.8	68.6	78.4	88.2
97	9.7	19.4	29.1	38.8	48.5	58.2	67.9	77.6	87.3
96	9.6	19.2	28.8	38.4	48.0	57.6	67.2	76.8	86.4
95	9.5	19.0	28.5	38.0	47.5	57.0	66.5	76.0	85.5
94	9.4	18.8	28.2	37.6	47.0	56.4	65.8	75.2	84.6
93	9.3	18.6	27.9	37.2	46.5	55.8	65.1	74.4	83.7
92	9.2	18.4	27.6	36.8	46.0	55.2	64.4	73.6	82.8
91	9.1	18.2	27.3	36.4	45.5	54.6	63.7	72.8	81.9
90	9.0	18.0	27.0	36.0	45.0	54.0	63.0	72.0	81.0
89	8.9	17.8	26.7	35.6	44.5	53.4	62.3	71.2	80.1
88	8.8	17.6	26.4	35.2	44.0	52.8	61.6	70.4	79.2
87	8.7	17.4	26.1	34.8	43.5	52.2	60.9	69.6	78.3
86	8.6	17.2	25.8	34.4	43.0	51.6	60.2	68.8	77.4

Table I. (Continued)

No. 500
Log. 698

No. 544
Log. 736

N.	0	1	2	3	4	5	6	7	8	9	Diff.
500	69 8970	9057	9144	9231	9317	9404	9491	9578	9664	9751	
1	9838	9924	0011	0098	0184	0271	0358	0444	0531	0617	
2	70 0704	0790	0877	0963	1050	1136	1222	1309	1395	1482	
3	1568	1654	1741	1827	1913	1999	2086	2172	2258	2344	
4	2431	2517	2603	2689	2775	2861	2947	3033	3119	3205	
505	3291	3377	3463	3549	3635	3721	3807	3893	3979	4065	86
6	4151	4236	4322	4408	4494	4579	4665	4751	4837	4922	
7	5008	5094	5179	5265	5350	5436	5522	5607	5693	5778	
8	5864	5949	6035	6120	6206	6291	6376	6462	6547	6632	
9	6718	6803	6888	6974	7059	7144	7229	7315	7400	7485	
510	7570	7655	7740	7826	7911	7996	8081	8166	8251	8336	85
1	8421	8506	8591	8676	8761	8846	8931	9015	9100	9185	
2	9270	9355	9440	9524	9609	9694	9779	9863	9948	0033	
3	71 0117	0202	0287	0371	0456	0540	0625	0710	0794	0879	
4	0963	1048	1132	1217	1301	1385	1470	1554	1639	1723	
515	1807	1892	1976	2060	2144	2229	2313	2397	2481	2566	84
6	2650	2734	2818	2902	2986	3070	3154	3238	3323	3407	
7	3491	3575	3659	3742	3826	3910	3994	4078	4162	4246	
8	4330	4414	4497	4581	4665	4749	4833	4916	5000	5084	
9	5167	5251	5335	5418	5502	5586	5669	5753	5836	5920	
520	6003	6087	6170	6254	6337	6421	6504	6588	6671	6754	
1	6838	6921	7004	7088	7171	7254	7338	7421	7504	7587	
2	7671	7754	7837	7920	8003	8086	8169	8253	8336	8419	
3	8502	8585	8668	8751	8834	8917	9000	9083	9165	9248	83
4	9331	9414	9497	9580	9663	9745	9828	9911	9994	0077	
525	72 0159	0242	0325	0407	0490	0573	0655	0738	0821	0903	
6	0986	1068	1151	1233	1316	1398	1481	1563	1646	1728	
7	1811	1893	1975	2058	2140	2222	2305	2387	2469	2552	
8	2634	2716	2798	2881	2963	3045	3127	3209	3291	3374	82
9	3456	3538	3620	3702	3784	3866	3948	4030	4112	4194	
530	4276	4358	4440	4522	4604	4685	4767	4849	4931	5013	
1	5095	5176	5258	5340	5422	5503	5585	5667	5748	5830	
2	5912	5993	6075	6156	6238	6320	6401	6483	6564	6646	
3	6727	6809	6890	6972	7053	7134	7216	7297	7379	7460	
4	7541	7623	7704	7785	7866	7948	8029	8110	8191	8273	81
535	8354	8435	8516	8597	8678	8759	8841	8922	9003	9084	
6	9165	9246	9327	9408	9489	9570	9651	9732	9813	9893	
7	9974	0055	0136	0217	0298	0378	0459	0540	0621	0702	
8	73 0782	0863	0944	1024	1105	1186	1266	1347	1428	1508	
9	1589	1669	1750	1830	1911	1991	2072	2152	2233	2313	
540	2394	2474	2555	2635	2715	2796	2876	2956	3037	3117	
1	3197	3278	3358	3438	3518	3598	3679	3759	3839	3919	
2	3999	4079	4160	4240	4320	4400	4480	4560	4640	4720	80
3	4800	4880	4960	5040	5120	5200	5279	5359	5439	5519	
4	5599	5679	5759	5838	5918	5998	6078	6157	6237	6317	

PROPORTIONAL PARTS

Diff.	1	2	3	4	5	6	7	8	9
87	8.7	17.4	26.1	34.8	43.5	52.2	60.9	69.6	78.3
86	8.6	17.2	25.8	34.4	43.0	51.6	60.2	68.8	77.4
85	8.5	17.0	25.5	34.0	42.5	51.0	59.5	68.0	76.5
84	8.4	16.8	25.2	33.6	42.0	50.4	58.8	67.2	75.6

Table I. (Continued)

No. 545
Log. 736

No. 584
Log. 767

N.	0	1	2	3	4	5	6	7	8	9	Diff.
545	73 6397	6476	6556	6635	6715	6795	6874	6954	7034	7113	
6	7193	7272	7352	7431	7511	7590	7670	7749	7829	7908	
7	7987	8067	8146	8225	8305	8384	8463	8543	8622	8701	
8	8781	8860	8939	9018	9097	9177	9256	9335	9414	9493	
9	9572	9651	9731	9810	9889	9968	0047	0126	0205	0284	79
550	74 0363	0442	0521	0600	0678	0757	0836	0915	0994	1073	
1	1152	1230	1309	1388	1467	1546	1624	1703	1782	1860	
2	1939	2018	2096	2175	2254	2332	2411	2489	2568	2647	
3	2725	2804	2882	2961	3039	3118	3196	3275	3353	3431	
4	3510	3588	3667	3745	3823	3902	3980	4058	4136	4215	
555	4293	4371	4449	4528	4606	4684	4762	4840	4919	4997	
6	5075	5153	5231	5309	5387	5465	5543	5621	5699	5777	78
7	5855	5933	6011	6089	6167	6245	6323	6401	6479	6556	
8	6634	6712	6790	6868	6945	7023	7101	7179	7256	7334	
9	7412	7489	7567	7645	7722	7800	7878	7955	8033	8110	
560	8188	8266	8343	8421	8498	8576	8653	8731	8808	8885	
1	8963	9040	9118	9195	9272	9350	9427	9504	9582	9659	
2	9736	9814	9891	9968	0045	0123	0200	0277	0354	0431	
3	75 0508	0586	0663	0740	0817	0894	0971	1048	1125	1202	
4	1279	1356	1433	1510	1587	1664	1741	1818	1895	1972	
565	2048	2125	2202	2279	2356	2433	2509	2586	2663	2740	77
6	2816	2893	2970	3047	3123	3200	3277	3353	3430	3506	
7	3583	3660	3736	3813	3889	3966	4042	4119	4195	4272	
8	4348	4425	4501	4578	4654	4730	4807	4883	4960	5036	
9	5112	5189	5265	5341	5417	5494	5570	5646	5722	5799	
570	5875	5951	6027	6103	6180	6256	6332	6408	6484	6560	
1	6636	6712	6788	6864	6940	7016	7092	7168	7244	7320	76
2	7396	7472	7548	7624	7700	7775	7851	7927	8003	8079	
3	8155	8230	8306	8382	8458	8533	8609	8685	8761	8836	
4	8912	8988	9063	9139	9214	9290	9366	9441	9517	9592	
575	9668	9743	9819	9894	9970	0045	0121	0196	0272	0347	
6	76 0422	0498	0573	0649	0724	0799	0875	0950	1025	1101	
7	1176	1251	1326	1402	1477	1552	1627	1702	1778	1853	
8	1928	2003	2078	2153	2228	2303	2378	2453	2529	2604	
9	2679	2754	2829	2904	2978	3053	3128	3203	3278	3353	75
580	3428	3503	3578	3653	3727	3802	3877	3952	4027	4101	
1	4176	4251	4326	4400	4475	4550	4624	4699	4774	4848	
2	4923	4998	5072	5147	5221	5296	5370	5445	5520	5594	
3	5669	5743	5818	5892	5966	6041	6115	6190	6264	6338	
4	6413	6487	6562	6636	6710	6785	6859	6933	7007	7082	

PROPORTIONAL PARTS

Diff.	1	2	3	4	5	6	7	8	9
83	8.3	16.6	24.9	33.2	41.5	49.8	58.1	66.4	74.7
82	8.2	16.4	24.6	32.8	41.0	49.2	57.4	65.6	73.8
81	8.1	16.2	24.3	32.4	40.5	48.6	56.7	64.8	72.9
80	8.0	16.0	24.0	32.0	40.0	48.0	56.0	64.0	72.0
79	7.9	15.8	23.7	31.6	39.5	47.4	55.3	63.2	71.1
78	7.8	15.6	23.4	31.2	39.0	46.8	54.6	62.4	70.2
77	7.7	15.4	23.1	30.8	38.5	46.2	53.9	61.6	69.3
76	7.6	15.2	22.8	30.4	38.0	45.6	53.2	60.8	68.4
75	7.5	15.0	22.5	30.0	37.5	45.0	52.5	60.0	67.5
74	7.4	14.8	22.2	29.6	37.0	44.4	51.8	59.2	66.6

Table I. (Continued)

No. 585
Log. 767

No. 629
Log. 799

N.	0	1	2	3	4	5	6	7	8	9	Diff.
585	76 7156	7230	7304	7379	7453	7527	7601	7675	7749	7823	
6	7898	7972	8046	8120	8194	8268	8342	8416	8490	8564	74
7	8638	8712	8786	8860	8934	9008	9082	9156	9230	9303	
8	9377	9451	9525	9599	9673	9746	9820	9894	9968	0042	
9	77 0115	0189	0263	0336	0410	0484	0557	0631	0705	0778	
590	0852	0926	0999	1073	1146	1220	1293	1367	1440	1514	
1	1587	1661	1734	1808	1881	1955	2028	2102	2175	2248	
2	2322	2395	2468	2542	2615	2688	2762	2835	2908	2981	
3	3055	3128	3201	3274	3348	3421	3494	3567	3640	3713	
4	3786	3860	3933	4006	4079	4152	4225	4298	4371	4444	73
595	4517	4590	4663	4736	4809	4882	4955	5028	5100	5173	
6	5246	5319	5392	5465	5538	5610	5683	5756	5829	5902	
7	5974	6047	6120	6193	6265	6338	6411	6483	6556	6629	
8	6701	6774	6846	6919	6992	7064	7137	7209	7282	7354	
9	7427	7499	7572	7644	7717	7789	7862	7934	8006	8079	
600	8151	8224	8296	8368	8441	8513	8585	8658	8730	8802	
1	8874	8947	9019	9091	9163	9236	9308	9380	9452	9524	
2	9596	9669	9741	9813	9885	9957	0029	0101	0173	0245	72
3	78 0317	0389	0461	0533	0605	0677	0749	0821	0893	0965	
4	1037	1109	1181	1253	1324	1396	1468	1540	1612	1684	
605	1755	1827	1899	1971	2042	2114	2186	2258	2329	2401	
6	2473	2544	2616	2688	2759	2831	2902	2974	3046	3117	
7	3189	3260	3332	3403	3475	3546	3618	3689	3761	3832	
8	3904	3975	4046	4118	4189	4261	4332	4403	4475	4546	
9	4617	4689	4760	4831	4902	4974	5045	5116	5187	5259	
610	5330	5401	5472	5543	5615	5686	5757	5828	5899	5970	
1	6041	6112	6183	6254	6325	6396	6467	6538	6609	6680	71
2	6751	6822	6893	6964	7035	7106	7177	7248	7319	7390	
3	7460	7531	7602	7673	7744	7815	7885	7956	8027	8098	
4	8168	8239	8310	8381	8451	8522	8593	8663	8734	8804	
615	8875	8946	9016	9087	9157	9228	9299	9369	9440	9510	
6	9581	9651	9722	9792	9863	9933	0004	0074	0144	0215	
7	79 0285	0356	0426	0496	0567	0637	0707	0778	0848	0918	
8	0988	1059	1129	1199	1269	1340	1410	1480	1550	1620	
9	1691	1761	1831	1901	1971	2041	2111	2181	2252	2322	
620	2392	2462	2532	2602	2672	2742	2812	2882	2952	3022	70
1	3092	3162	3231	3301	3371	3441	3511	3581	3651	3721	
2	3790	3860	3930	4000	4070	4139	4209	4279	4349	4418	
3	4488	4558	4627	4697	4767	4836	4906	4976	5045	5115	
4	5185	5254	5324	5393	5463	5532	5602	5672	5741	5811	
625	5880	5949	6019	6088	6158	6227	6297	6366	6436	6505	
6	6574	6644	6713	6782	6852	6921	6990	7060	7129	7198	
7	7268	7337	7406	7475	7545	7614	7683	7752	7821	7890	
8	7960	8029	8098	8167	8236	8305	8374	8443	8513	8582	
9	8651	8720	8789	8858	8927	8996	9065	9134	9203	9272	69

PROPORTIONAL PARTS

Diff.	1	2	3	4	5	6	7	8	9
75	7.5	15.0	22.5	30.0	37.5	45.0	52.5	60.0	67.5
74	7.4	14.8	22.2	29.6	37.0	44.4	51.8	59.2	66.6
73	7.3	14.6	21.9	29.2	36.5	43.8	51.1	58.4	65.7
72	7.2	14.4	21.6	28.8	36.0	43.2	50.4	57.6	64.8
71	7.1	14.2	21.3	28.4	35.5	42.6	49.7	56.8	63.9
70	7.0	14.0	21.0	28.0	35.0	42.0	49.0	56.0	63.0
69	6.9	13.8	20.7	27.6	34.5	41.4	48.3	55.2	62.1

Table I. (Continued)

No. 630
Log. 799

No. 674
Log. 829

N.	0	1	2	3	4	5	6	7	8	9	Diff.
630	79 9341	9409	9478	9547	9616	9685	9754	9823	9892	9961	
1	80 0029	0098	0167	0236	0305	0373	0442	0511	0580	0648	
2	0717	0786	0854	0923	0992	1061	1129	1198	1266	1335	
3	1404	1472	1541	1609	1678	1747	1815	1884	1952	2021	
4	2089	2158	2226	2295	2363	2432	2500	2568	2637	2705	
635	2774	2842	2910	2979	3047	3116	3184	3252	3321	3389	
6	3457	3525	3594	3662	3730	3798	3867	3935	4003	4071	
7	4139	4208	4276	4344	4412	4480	4548	4616	4685	4753	
8	4821	4889	4957	5025	5093	5161	5229	5297	5365	5433	68
9	5501	5569	5637	5705	5773	5841	5908	5976	6044	6112	
640	80 6180	6248	6316	6384	6451	6519	6587	6655	6723	6790	
1	6858	6926	6994	7061	7129	7197	7264	7332	7400	7467	
2	7535	7603	7670	7738	7806	7873	7941	8008	8076	8143	
3	8211	8279	8346	8414	8481	8549	8616	8684	8751	8818	
4	8886	8953	9021	9088	9156	9223	9290	9358	9425	9492	
645	9560	9627	9694	9762	9829	9896	9964	0031	0098	0165	
6	81 0233	0300	0367	0434	0501	0569	0636	0703	0770	0837	
7	0904	0971	1039	1106	1173	1240	1307	1374	1441	1508	67
8	1575	1642	1709	1776	1843	1910	1977	2044	2111	2178	
9	2245	2312	2379	2445	2512	2579	2646	2713	2780	2847	
650	2913	2980	3047	3114	3181	3247	3314	3381	3448	3514	
1	3581	3648	3714	3781	3848	3914	3981	4048	4114	4181	
2	4248	4314	4381	4447	4514	4581	4647	4714	4780	4847	
3	4913	4980	5046	5113	5179	5246	5312	5378	5445	5511	
4	5578	5644	5711	5777	5843	5910	5976	6042	6109	6175	
655	6241	6308	6374	6440	6506	6573	6639	6705	6771	6838	
6	6904	6970	7036	7102	7169	7235	7301	7367	7433	7499	
7	7565	7631	7698	7764	7830	7896	7962	8028	8094	8160	
8	8226	8292	8358	8424	8490	8556	8622	8688	8754	8820	
9	8885	8951	9017	9083	9149	9215	9281	9346	9412	9478	66
660	9544	9610	9676	9741	9807	9873	9939	0004	0070	0136	
1	82 0201	0267	0333	0399	0464	0530	0595	0661	0727	0792	
2	0858	0924	0989	1055	1120	1186	1251	1317	1382	1448	
3	1514	1579	1645	1710	1775	1841	1906	1972	2037	2103	
4	2168	2233	2299	2364	2430	2495	2560	2626	2691	2756	
665	2822	2887	2952	3018	3083	3148	3213	3279	3344	3409	
6	3474	3539	3605	3670	3735	3800	3865	3930	3996	4061	
7	4126	4191	4256	4321	4386	4451	4516	4581	4646	4711	
8	4776	4841	4906	4971	5036	5101	5166	5231	5296	5361	65
9	5426	5491	5556	5621	5686	5751	5815	5880	5945	6010	
670	6075	6140	6204	6269	6334	6399	6464	6528	6593	6658	
1	6723	6787	6852	6917	6981	7046	7111	7175	7240	7305	
2	7369	7434	7499	7563	7628	7692	7757	7821	7886	7951	
3	8015	8080	8144	8209	8273	8338	8402	8467	8531	8595	
4	8660	8724	8789	8853	8918	8982	9046	9111	9175	9239	

PROPORTIONAL PARTS

Diff.	1	2	3	4	5	6	7	8	9
68	6.8	13.6	20.4	27.2	34.0	40.8	47.6	54.4	61.2
67	6.7	13.4	20.1	26.8	33.5	40.2	46.9	53.6	60.3
66	6.6	13.2	19.8	26.4	33.0	39.6	46.2	52.8	59.4
65	6.5	13.0	19.5	26.0	32.5	39.0	45.5	52.0	58.5
64	6.4	12.8	19.2	25.6	32.0	38.4	44.8	51.2	57.6

Table I. (Continued)

No. 675
Log. 829

No. 719
Log. 857

N.	0	1	2	3	4	5	6	7	8	9	Diff.
675	82 9304	9368	9432	9497	9561	9625	9690	9754	9818	9882	
6	9947	0011	0075	0139	0204	0268	0332	0396	0460	0525	
7	83 0589	0653	0717	0781	0845	0909	0973	1037	1102	1166	
8	1230	1294	1358	1422	1486	1550	1614	1678	1742	1806	64
9	1870	1934	1998	2062	2126	2189	2253	2317	2381	2445	
680	2509	2573	2637	2700	2764	2828	2892	2956	3020	3083	
1	3147	3211	3275	3338	3402	3466	3530	3593	3657	3721	
2	3784	3848	3912	3975	4039	4103	4166	4230	4294	4357	
3	4421	4484	4548	4611	4675	4739	4802	4866	4929	4993	
4	5056	5120	5183	5247	5310	5373	5437	5500	5564	5627	
685	5691	5754	5817	5881	5944	6007	6071	6134	6197	6261	
6	6324	6387	6451	6514	6577	6641	6704	6767	6830	6894	
7	6957	7020	7083	7146	7210	7273	7336	7399	7462	7525	
8	7588	7652	7715	7778	7841	7904	7967	8030	8093	8156	
9	8219	8282	8345	8408	8471	8534	8597	8660	8723	8786	63
690	8849	8912	8975	9038	9101	9164	9227	9289	9352	9415	
1	9478	9541	9604	9667	9729	9792	9855	9918	9981	0043	
2	84 0106	0169	0232	0294	0357	0420	0482	0545	0608	0671	
3	0733	0796	0859	0921	0984	1046	1109	1172	1234	1297	
4	1359	1422	1485	1547	1610	1672	1735	1797	1860	1922	
695	1985	2047	2110	2172	2235	2297	2360	2422	2484	2547	
6	2609	2672	2734	2796	2859	2921	2983	3046	3108	3170	
7	3233	3295	3357	3420	3482	3544	3606	3669	3731	3793	
8	3855	3918	3980	4042	4104	4166	4229	4291	4353	4415	
9	4477	4539	4601	4664	4726	4788	4850	4912	4974	5036	
700	5098	5160	5222	5284	5346	5408	5470	5532	5594	5656	62
1	5718	5780	5842	5904	5966	6028	6090	6151	6213	6275	
2	6337	6399	6461	6523	6585	6646	6708	6770	6832	6894	
3	6955	7017	7079	7141	7202	7264	7326	7388	7449	7511	
4	7573	7634	7696	7758	7819	7881	7943	8004	8066	8128	
705	8189	8251	8312	8374	8435	8497	8559	8620	8682	8743	
6	8805	8866	8928	8989	9051	9112	9174	9235	9297	9358	
7	9419	9481	9542	9604	9665	9726	9788	9849	9911	9972	
8	85 0033	0095	0156	0217	0279	0340	0401	0462	0524	0585	
9	0646	0707	0769	0830	0891	0952	1014	1075	1136	1197	
710	1258	1320	1381	1442	1503	1564	1625	1686	1747	1809	
1	1870	1931	1992	2053	2114	2175	2236	2297	2358	2419	
2	2480	2541	2602	2663	2724	2785	2846	2907	2968	3029	61
3	3090	3150	3211	3272	3333	3394	3455	3516	3577	3637	
4	3698	3759	3820	3881	3941	4002	4063	4124	4185	4245	
715	4306	4367	4428	4488	4549	4610	4670	4731	4792	4852	
6	4913	4974	5034	5095	5156	5216	5277	5337	5398	5459	
7	5519	5580	5640	5701	5761	5822	5882	5943	6003	6064	
8	6124	6185	6245	6306	6366	6427	6487	6548	6608	6668	
9	6729	6789	6850	6910	6970	7031	7091	7152	7212	7272	

PROPORTIONAL PARTS

Diff.	1	2	3	4	5	6	7	8	9
65	6.5	13.0	19.5	26.0	32.5	39.0	45.5	52.0	58.5
64	6.4	12.8	19.2	25.6	32.0	38.4	44.8	51.2	57.6
63	6.3	12.6	18.9	25.2	31.5	37.8	44.1	50.4	56.7
62	6.2	12.4	18.6	24.8	31.0	37.2	43.4	49.6	55.8
61	6.1	12.2	18.3	24.4	30.5	36.6	42.7	48.8	54.9
60	6.0	12.0	18.0	24.0	30.0	36.0	42.0	48.0	54.0

Table I. (Continued)

No. 720
Log. 857

No. 764
Log. 883

N.	0	1	2	3	4	5	6	7	8	9	Diff.
720	85 7332	7393	7453	7513	7574	7634	7694	7755	7815	7875	
1	7935	7995	8056	8116	8176	8236	8297	8357	8417	8477	
2	8537	8597	8657	8718	8778	8838	8898	8958	9018	9078	
3	9138	9198	9258	9318	9379	9439	9499	9559	9619	9679	60
4	9739	9799	9859	9918	9978	0038	0098	0158	0218	0278	
725	86 0338	0398	0458	0518	0578	0637	0697	0757	0817	0877	
6	0937	0996	1056	1116	1176	1236	1295	1355	1415	1475	
7	1534	1594	1654	1714	1773	1833	1893	1952	2012	2072	
8	2131	2191	2251	2310	2370	2430	2489	2549	2608	2668	
9	2728	2787	2847	2906	2966	3025	3085	3144	3204	3263	
730	3323	3382	3442	3501	3561	3620	3680	3739	3799	3858	
1	3917	3977	4036	4096	4155	4214	4274	4333	4392	4452	
2	4511	4570	4630	4689	4748	4808	4867	4926	4985	5045	
3	5104	5163	5222	5282	5341	5400	5459	5519	5578	5637	
4	5696	5755	5814	5874	5933	5992	6051	6110	6169	6228	
735	6287	6346	6405	6465	6524	6583	6642	6701	6760	6819	
6	6878	6937	6996	7055	7114	7173	7232	7291	7350	7409	59
7	7467	7526	7585	7644	7703	7762	7821	7880	7939	7998	
8	8056	8115	8174	8233	8292	8350	8409	8468	8527	8586	
9	8644	8703	8762	8821	8879	8938	8997	9056	9114	9173	
740	9232	9290	9349	9408	9466	9525	9584	9642	9701	9760	
1	9818	9877	9935	9994	0053	0111	0170	0228	0287	0345	
2	87 0404	0462	0521	0579	0638	0696	0755	0813	0872	0930	
3	0989	1047	1106	1164	1223	1281	1339	1398	1456	1515	
4	1573	1631	1690	1748	1806	1865	1923	1981	2040	2098	
745	2156	2215	2273	2331	2389	2448	2506	2564	2622	2681	
6	2739	2797	2855	2913	2972	3030	3088	3146	3204	3262	
7	3321	3379	3437	3495	3553	3611	3669	3727	3785	3844	
8	3902	3960	4018	4076	4134	4192	4250	4308	4366	4424	58
9	4482	4540	4598	4656	4714	4772	4830	4888	4945	5003	
750	5061	5119	5177	5235	5293	5351	5409	5466	5524	5582	
1	5640	5698	5756	5813	5871	5929	5987	6045	6102	6160	
2	6218	6276	6333	6391	6449	6507	6564	6622	6680	6737	
3	6795	6853	6910	6968	7026	7083	7141	7199	7256	7314	
4	7371	7429	7487	7544	7602	7659	7717	7774	7832	7889	
755	7947	8004	8062	8119	8177	8234	8292	8349	8407	8464	
6	8522	8579	8637	8694	8752	8809	8866	8924	8981	9039	
7	9096	9153	9211	9268	9325	9383	9440	9497	9555	9612	
8	9669	9726	9784	9841	9898	9956	0013	0070	0127	0185	
9	88 0242	0299	0356	0413	0471	0528	0585	0642	0699	0756	
760	0814	0871	0928	0985	1042	1099	1156	1213	1271	1328	
1	1385	1442	1499	1556	1613	1670	1727	1784	1841	1898	
2	1955	2012	2069	2126	2183	2240	2297	2354	2411	2468	57
3	2525	2581	2638	2695	2752	2809	2866	2923	2980	3037	
4	3093	3150	3207	3264	3321	3377	3434	3491	3548	3605	

PROPORTIONAL PARTS

Diff.	1	2	3	4	5	6	7	8	9
59	5.9	11.8	17.7	23.6	29.5	35.4	41.3	47.2	53.1
58	5.8	11.6	17.4	23.2	29.0	34.8	40.6	46.4	52.2
57	5.7	11.4	17.1	22.8	28.5	34.2	39.9	45.6	51.3
56	5.6	11.2	16.8	22.4	28.0	33.6	39.2	44.8	50.4

Table I. (Continued)

No. 765
Log. 883

No. 809
Log. 908

N.	0	1	2	3	4	5	6	7	8	9	Diff.
765	88 3661	3718	3775	3832	3888	3945	4002	4059	4115	4172	
6	4229	4285	4342	4399	4455	4512	4569	4625	4682	4739	
7	4795	4852	4909	4965	5022	5078	5135	5192	5248	5305	
8	5361	5418	5474	5531	5587	5644	5700	5757	5813	5870	
9	5926	5983	6039	6096	6152	6209	6265	6321	6378	6434	
770	6491	6547	6604	6660	6716	6773	6829	6885	6942	6998	
1	7054	7111	7167	7223	7280	7336	7392	7449	7505	7561	
2	7617	7674	7730	7786	7842	7898	7955	8011	8067	8123	
3	8179	8236	8292	8348	8404	8460	8516	8573	8629	8685	
4	8741	8797	8853	8909	8965	9021	9077	9134	9190	9246	
775	9302	9358	9414	9470	9526	9582	9638	9694	9750	9806	56
6	9862	9918	9974	0030	0086	0141	0197	0253	0309	0365	
7	89 0421	0477	0533	0589	0645	0700	0756	0812	0868	0924	
8	0980	1035	1091	1147	1203	1259	1314	1370	1426	1482	
9	1537	1593	1649	1705	1760	1816	1872	1928	1983	2039	
780	2095	2150	2206	2262	2317	2373	2429	2484	2540	2595	
1	2651	2707	2762	2818	2873	2929	2985	3040	3096	3151	
2	3207	3262	3318	3373	3429	3484	3540	3595	3651	3706	
3	3762	3817	3873	3928	3984	4039	4094	4150	4205	4261	
4	4316	4371	4427	4482	4538	4593	4648	4704	4759	4814	
785	4870	4925	4980	5036	5091	5146	5201	5257	5312	5367	
6	5423	5478	5533	5588	5644	5699	5754	5809	5864	5920	
7	5975	6030	6085	6140	6195	6251	6306	6361	6416	6471	
8	6526	6581	6636	6692	6747	6802	6857	6912	6967	7022	
9	7077	7132	7187	7242	7297	7352	7407	7462	7517	7572	
790	7627	7682	7737	7792	7847	7902	7957	8012	8067	8122	55
1	8176	8231	8286	8341	8396	8451	8506	8561	8615	8670	
2	8725	8780	8835	8890	8944	8999	9054	9109	9164	9218	
3	9273	9328	9383	9437	9492	9547	9602	9656	9711	9766	
4	9821	9875	9930	9985	0039	0094	0149	0203	0258	0312	
795	90 0367	0422	0476	0531	0586	0640	0695	0749	0804	0859	
6	0913	0968	1022	1077	1131	1186	1240	1295	1349	1404	
7	1458	1513	1567	1622	1676	1731	1785	1840	1894	1948	
8	2003	2057	2112	2166	2221	2275	2329	2384	2438	2492	
9	2547	2601	2655	2710	2764	2818	2873	2927	2981	3036	
800	3090	3144	3199	3253	3307	3361	3416	3470	3524	3578	
1	3633	3687	3741	3795	3849	3904	3958	4012	4066	4120	
2	4174	4229	4283	4337	4391	4445	4499	4553	4607	4661	
3	4716	4770	4824	4878	4932	4986	5040	5094	5148	5202	
4	5256	5310	5364	5418	5472	5526	5580	5634	5688	5742	54
805	5796	5850	5904	5958	6012	6066	6119	6173	6227	6281	
6	6335	6389	6443	6497	6551	6604	6658	6712	6766	6820	
7	6874	6927	6981	7035	7089	7143	7196	7250	7304	7358	
8	7411	7465	7519	7573	7626	7680	7734	7787	7841	7895	
9	7949	8002	8056	8110	8163	8217	8270	8324	8378	8431	

PROPORTIONAL PARTS

Diff.	1	2	3	4	5	6	7	8	9
57	5.7	11.4	17.1	22.8	28.5	34.2	39.9	45.6	51.3
56	5.6	11.2	16.8	22.4	28.0	33.6	39.2	44.8	50.4
55	5.5	11.0	16.5	22.0	27.5	33.0	38.5	44.0	49.5
54	5.4	10.8	16.2	21.6	27.0	32.4	37.8	43.2	48.6

Table I. (Continued)

No. 810
Log. 908

No. 854
Log. 931

N.	0	1	2	3	4	5	6	7	8	9	Diff.
810	90 8485	8539	8592	8646	8699	8753	8807	8860	8914	8967	
1	9021	9074	9128	9181	9235	9289	9342	9396	9449	9503	
2	9556	9610	9663	9716	9770	9823	9877	9930	9984	0037	
3	91 0091	0144	0197	0251	0304	0358	0411	0464	0518	0571	
4	0624	0678	0731	0784	0838	0891	0944	0998	1051	1104	
815	1158	1211	1264	1317	1371	1424	1477	1530	1584	1637	
6	1690	1743	1797	1850	1903	1956	2009	2063	2116	2169	
7	2222	2275	2328	2381	2435	2488	2541	2594	2647	2700	
8	2753	2806	2859	2913	2966	3019	3072	3125	3178	3231	
9	3284	3337	3390	3443	3496	3549	3602	3655	3708	3761	53
820	3814	3867	3920	3973	4026	4079	4132	4184	4237	4290	
1	4343	4396	4449	4502	4555	4608	4660	4713	4766	4819	
2	4872	4925	4977	5030	5083	5136	5189	5241	5294	5347	
3	5400	5453	5505	5558	5611	5664	5716	5769	5822	5875	
4	5927	5980	6033	6085	6138	6191	6243	6296	6349	6401	
825	6454	6507	6559	6612	6664	6717	6770	6822	6875	6927	
6	6980	7033	7085	7138	7190	7243	7295	7348	7400	7453	
7	7506	7558	7611	7663	7716	7768	7820	7873	7925	7978	
8	8030	8083	8135	8188	8240	8293	8345	8397	8450	8502	
9	8555	8607	8659	8712	8764	8816	8869	8921	8973	9026	
830	9078	9130	9183	9235	9287	9340	9392	9444	9496	9549	
1	9601	9653	9706	9758	9810	9862	9914	9967	0019	0071	
2	92 0123	0176	0228	0280	0332	0384	0436	0489	0541	0593	
3	0645	0697	0749	0801	0853	0906	0958	1010	1062	1114	
4	1166	1218	1270	1322	1374	1426	1478	1530	1582	1634	52
835	1686	1738	1790	1842	1894	1946	1998	2050	2102	2154	
6	2206	2258	2310	2362	2414	2466	2518	2570	2622	2674	
7	2725	2777	2829	2881	2933	2985	3037	3089	3140	3192	
8	3244	3296	3348	3399	3451	3503	3555	3607	3658	3710	
9	3762	3814	3865	3917	3969	4021	4072	4124	4176	4228	
840	4279	4331	4383	4434	4486	4538	4589	4641	4693	4744	
1	4796	4848	4899	4951	5003	5054	5106	5157	5209	5261	
2	5312	5364	5415	5467	5518	5570	5621	5673	5725	5776	
3	5828	5879	5931	5982	6034	6085	6137	6188	6240	6291	
4	6342	6394	6445	6497	6548	6600	6651	6702	6754	6805	
845	6857	6908	6959	7011	7062	7114	7165	7216	7268	7319	
6	7370	7422	7473	7524	7576	7627	7678	7730	7781	7832	
7	7883	7935	7986	8037	8088	8140	8191	8242	8293	8345	
8	8396	8447	8498	8549	8601	8652	8703	8754	8805	8857	
9	8908	8959	9010	9061	9112	9163	9215	9266	9317	9368	
850	9419	9470	9521	9572	9623	9674	9725	9776	9827	9879	51
1	9930	9981	0032	0083	0134	0185	0236	0287	0338	0389	
2	93 0440	0491	0542	0592	0643	0694	0745	0796	0847	0898	
3	0949	1000	1051	1102	1153	1204	1254	1305	1356	1407	
4	1458	1509	1560	1610	1661	1712	1763	1814	1865	1915	

PROPORTIONAL PARTS

Diff.	1	2	3	4	5	6	7	8	9
53	5.3	10.6	15.9	21.2	26.5	31.8	37.1	42.4	47.7
52	5.2	10.4	15.6	20.8	26.0	31.2	36.4	41.6	46.8
51	5.1	10.2	15.3	20.4	25.5	30.6	35.7	40.8	45.9
50	5.0	10.0	15.0	20.0	25.0	30.0	35.0	40.0	45.0

ntinued)

No. 855
Log. 931

No. 899
Log. 954

N.	0	1	2	3	4	5	6	7	8	9	Diff.
855	93 1966	2017	2068	2118	2169	2220	2271	2322	2372	2423	
6	2474	2524	2575	2626	2677	2727	2778	2829	2879	2930	
7	2981	3031	3082	3133	3183	3234	3285	3335	3386	3437	
8	3487	3538	3589	3639	3690	3740	3791	3841	3892	3943	
9	3993	4044	4094	4145	4195	4246	4296	4347	4397	4448	
860	4498	4549	4599	4650	4700	4751	4801	4852	4902	4953	
1	5003	5054	5104	5154	5205	5255	5306	5356	5406	5457	
2	5507	5558	5608	5709	5759	5759	5809	5860	5910	5960	
3	6011	6061	6111	6162	6212	6262	6313	6363	6413	6463	
4	6514	6564	6614	6665	6715	6765	6815	6865	6916	6966	
865	7016	7066	7116	7167	7217	7267	7317	7367	7418	7468	
6	7518	7568	7618	7668	7718	7769	7819	7869	7919	7969	
7	8019	8069	8119	8169	8219	8269	8320	8370	8420	8470	50
8	8520	8570	8620	8670	8720	8770	8820	8870	8920	8970	
9	9020	9070	9120	9170	9220	9270	9320	9369	9419	9469	
870	9519	9569	9619	9669	9719	9769	9819	9869	9918	9968	
1	94 0018	0068	0118	0168	0218	0267	0317	0367	0417	0467	
2	0516	0566	0616	0666	0716	0765	0815	0865	0915	0964	
3	1014	1064	1114	1163	1213	1263	1313	1362	1412	1462	
4	1511	1561	1611	1660	1710	1760	1809	1859	1909	1958	
875	2008	2058	2107	2157	2207	2256	2306	2355	2405	2455	
6	2504	2554	2603	2653	2702	2752	2801	2851	2901	2950	
7	3000	3049	3099	3148	3198	3247	3297	3346	3396	3445	
8	3495	3544	3593	3643	3692	3742	3791	3841	3890	3939	
9	3989	4038	4088	4137	4186	4236	4285	4335	4384	4433	
880	4483	4532	4581	4631	4680	4729	4779	4828	4877	4927	
1	4976	5025	5074	5124	5173	5222	5272	5321	5370	5419	
2	5469	5518	5567	5616	5665	5715	5764	5813	5862	5912	
3	5961	6010	6059	6108	6157	6207	6256	6305	6354	6403	
4	6452	6501	6551	6600	6649	6698	6747	6796	6845	6894	
885	6943	6992	7041	7090	7139	7189	7238	7287	7336	7385	
6	7434	7483	7532	7581	7630	7679	7728	7777	7826	7875	49
7	7924	7973	8022	8070	8119	8168	8217	8266	8315	8364	
8	8413	8462	8511	8560	8608	8657	8706	8755	8804	8853	
9	8902	8951	8999	9048	9097	9146	9195	9244	9292	9341	
890	9390	9439	9488	9536	9585	9634	9683	9731	9780	9829	
1	9878	9926	9975	0024	0073	0121	0170	0219	0267	0316	
2	95 0365	0414	0462	0511	0560	0608	0657	0706	0754	0803	
3	0851	0900	0949	0997	1046	1095	1143	1192	1240	1289	
4	1338	1386	1435	1483	1532	1580	1629	1677	1726	1775	
895	1823	1872	1920	1969	2017	2066	2114	2163	2211	2260	
6	2308	2356	2405	2453	2502	2550	2599	2647	2696	2744	
7	2792	2841	2889	2938	2986	3034	3083	3131	3180	3228	
8	3276	3325	3373	3421	3470	3518	3566	3615	3663	3711	
9	3760	3808	3856	3905	3953	4001	4049	4098	4146	4194	

PROPORTIONAL PARTS

Diff.	1	2	3	4	5	6	7	8	9
51	5.1	10.2	15.3	20.4	25.5	30.6	35.7	40.8	45.9
50	5.0	10.0	15.0	20.0	25.0	30.0	35.0	40.0	45.0
49	4.9	9.8	14.7	19.6	24.5	29.4	34.3	39.2	44.1
48	4.8	9.6	14.4	19.2	24.0	28.8	33.6	38.4	43.2

Table I. (Continued)

No. 900
Log. 954

No. 944
Log. 975

N.	0	1	2	3	4	5	6	7	8	9	Diff.
900	95 4243	4291	4339	4387	4435	4484	4532	4580	4628	4677	
1	4725	4773	4821	4869	4918	4966	5014	5062	5110	5158	
2	5207	5255	5303	5351	5399	5447	5495	5543	5592	5640	
3	5688	5736	5784	5832	5880	5928	5976	6024	6072	6120	
4	6168	6216	6265	6313	6361	6409	6457	6505	6553	6601	48
905	6649	6697	6745	6793	6840	6888	6936	6984	7032	7080	
6	7128	7176	7224	7272	7320	7368	7416	7464	7512	7559	
7	7607	7655	7703	7751	7799	7847	7894	7942	7990	8038	
8	8086	8134	8181	8229	8277	8325	8373	8421	8468	8516	
9	8564	8612	8659	8707	8755	8803	8850	8898	8946	8994	
910	9041	9089	9137	9185	9232	9280	9328	9375	9423	9471	
1	9518	9566	9614	9661	9709	9757	9804	9852	9900	9947	
2	9995	0042	0090	0138	0185	0233	0280	0328	0376	0423	
3	96 0471	0518	0566	0613	0661	0709	0756	0804	0851	0899	
4	0946	0994	1041	1089	1136	1184	1231	1279	1326	1374	
915	1421	1469	1516	1563	1611	1658	1706	1753	1801	1848	
6	1895	1943	1990	2038	2085	2132	2180	2227	2275	2322	
7	2369	2417	2464	2511	2559	2606	2653	2701	2748	2795	
8	2843	2890	2937	2985	3032	3079	3126	3174	3221	3268	
9	3316	3363	3410	3457	3504	3552	3599	3646	3693	3741	
920	3788	3835	3882	3929	3977	4024	4071	4118	4165	4212	
1	4260	4307	4354	4401	4448	4495	4542	4590	4637	4684	
2	4731	4778	4825	4872	4919	4966	5013	5061	5108	5155	
3	5202	5249	5296	5343	5390	5437	5484	5531	5578	5625	
4	5672	5719	5766	5813	5860	5907	5954	6001	6048	6095	47
925	6142	6189	6236	6283	6329	6376	6423	6470	6517	6564	
6	6611	6658	6705	6752	6799	6845	6892	6939	6986	7033	
7	7080	7127	7173	7220	7267	7314	7361	7408	7454	7501	
8	7548	7595	7642	7688	7735	7782	7829	7875	7922	7969	
9	8016	8062	8109	8156	8203	8249	8296	8343	8390	8436	
930	8483	8530	8576	8623	8670	8716	8763	8810	8856	8903	
1	8950	8996	9043	9090	9136	9183	9229	9276	9323	9369	
2	9416	9463	9509	9556	9602	9649	9695	9742	9789	9835	
3	9882	9928	9975	0021	0068	0114	0161	0207	0254	0300	
4	97 0347	0393	0440	0486	0533	0579	0626	0672	0719	0765	
935	0812	0858	0904	0951	0997	1044	1090	1137	1183	1229	
6	1276	1322	1369	1415	1461	1508	1554	1601	1647	1693	
7	1740	1786	1832	1879	1925	1971	2018	2064	2110	2157	
8	2203	2249	2295	2342	2388	2434	2481	2527	2573	2619	
9	2666	2712	2758	2804	2851	2897	2943	2989	3035	3082	
940	3128	3174	3220	3266	3313	3359	3405	3451	3497	3543	
1	3590	3636	3682	3728	3774	3820	3866	3913	3959	4005	
2	4051	4097	4143	4189	4235	4281	4327	4374	4420	4466	
3	4512	4558	4604	4650	4696	4742	4788	4834	4880	4926	
4	4972	5018	5064	5110	5156	5202	5248	5294	5340	5386	46

PROPORTIONAL PARTS

Diff.	1	2	3	4	5	6	7	8	9
47	4.7	9.4	14.1	18.8	23.5	28.2	32.9	37.6	42.3
46	4.6	9.2	13.8	18.4	23.0	27.6	32.2	36.8	41.4

Table I. (Continued)

No. 945
Log. 975

No. 989
Log. 995

N.	0	1	2	3	4	5	6	7	8	9	Diff.
945	97 5432	5478	5524	5570	5616	5662	5707	5753	5799	5845	
6	5891	5937	5983	6029	6075	6121	6167	6212	6258	6304	
7	6350	6396	6442	6488	6533	6579	6625	6671	6717	6763	
8	6808	6854	6900	6946	6992	7037	7083	7129	7175	7220	
9	7266	7312	7358	7403	7449	7495	7541	7586	7632	7678	
950	7724	7769	7815	7861	7906	7952	7998	8043	8089	8135	
1	8181	8226	8272	8317	8363	8409	8454	8500	8546	8591	
2	8637	8683	8728	8774	8819	8865	8911	8956	9002	9047	
3	9093	9138	9184	9230	9275	9321	9366	9412	9457	9503	
4	9548	9594	9639	9685	9730	9776	9821	9867	9912	9958	
955	98 0003	0049	0094	0140	0185	0231	0276	0322	0367	0412	
6	0458	0503	0549	0594	0640	0685	0730	0776	0821	0867	
7	0912	0957	1003	1048	1093	1139	1184	1229	1275	1320	
8	1366	1411	1456	1501	1547	1592	1637	1683	1728	1773	
9	1819	1864	1909	1954	2000	2045	2090	2135	2181	2226	
960	2271	2316	2362	2407	2452	2497	2543	2588	2633	2678	
1	2723	2769	2814	2859	2904	2949	2994	3040	3085	3130	
2	3175	3220	3265	3310	3356	3401	3446	3491	3536	3581	
3	3626	3671	3716	3762	3807	3852	3897	3942	3987	4032	
4	4077	4122	4167	4212	4257	4302	4347	4392	4437	4482	45
965	4527	4572	4617	4662	4707	4752	4797	4842	4887	4932	
6	4977	5022	5067	5112	5157	5202	5247	5292	5337	5382	
7	5426	5471	5516	5561	5606	5651	5696	5741	5786	5830	
8	5875	5920	5965	6010	6055	6100	6144	6189	6234	6279	
9	6324	6369	6413	6458	6503	6548	6593	6637	6682	6727	
970	6772	6817	6861	6906	6951	6996	7040	7085	7130	7175	
1	7219	7264	7309	7353	7398	7443	7488	7532	7577	7622	
2	7666	7711	7756	7800	7845	7890	7934	7979	8024	8068	
3	8113	8157	8202	8247	8291	8336	8381	8425	8470	8514	
4	8559	8604	8648	8693	8737	8782	8826	8871	8916	8960	
975	9005	9049	9094	9138	9183	9227	9272	9316	9361	9405	
6	9450	9494	9539	9583	9628	9672	9717	9761	9806	9850	
7	9895	9939	9983	0028·	0072	0117	0161	0206	0250	0294	
8	99 0339	0383	0428	0472	0516	0561	0605	0650	0694	0738	
9	0783	0827	0871	0916	0960	1004	1049	1093	1137	1182	
980	1226	1270	1315	1359	1403	1448	1492	1536	1580	1625	
1	1669	1713	1758	1802	1846	1890	1935	1979	2023	2067	
2	2111	2156	2200	2244	2288	2333	2377	2421	2465	2509	
3	2554	2598	2642	2686	2730	2774	2819	2863	2907	2951	
4	2995	3039	3083	3127	3172	3216	3260	3304	3348	3392	
985	3436	3480	3524	3568	3613	3657	3701	3745	3789	3833	
6	3877	3921	3965	4009	4053	4097	4141	4185	4229	4273	
7	4317	4361	4405	4449	4493	4537	4581	4625	4669	4713	44
8	4757	4801	4845	4889	4933	4977	5021	5065	5108	5152	
9	5196	5240	5284	5328	5372	5416	5460	5504	5547	5591	

PROPORTIONAL PARTS

Diff.	1	2	3	4	5	6	7	8	9
46	4.6	9.2	13.8	18.4	23.0	27.6	32.2	36.8	41.4
45	4.5	9.0	13.5	18.0	22.5	27.0	31.5	36.0	40.5
44	4.4	8.8	13.2	17.6	22.0	26.4	30.8	35.2	39.6
43	4.3	8.6	12.9	17.2	21.5	25.8	30.1	34.4	38.7

Table I. (Concluded)

No. 990
Log. 995

No. 999
Log. 999

N.	0	1	2	3	4	5	6	7	8	9	Diff.
990	99 5635	5679	5723	5767	5811	5854	5898	5942	5986	6030	
1	6074	6117	6161	6205	6249	6293	6337	6380	6424	6468	44
2	6512	6555	6599	6643	6687	6731	6774	6818	6862	6906	
3	6949	6993	7037	7080	7124	7168	7212	7255	7299	7343	
4	7386	7430	7474	7517	7561	7605	7648	7692	7736	7779	
995	7823	7867	7910	7954	7998	8041	8085	8129	8172	8216	
6	8259	8303	8347	8390	8434	8477	8521	8564	8608	8652	
7	8695	8739	8782	8826	8869	8913	8956	9000	9043	9087	
8	9131	9174	9218	9261	9305	9348	9392	9435	9479	9522	
9	9565	9609	9652	9696	9739	9783	9826	9870	9913	9957	43

TABLE IIa. VALUES OF *S*, *T*, AND *C* IN TABLE II, FOR ANGLES BETWEEN 0° AND 2° AND BETWEEN 88° AND 90°

If we were to plot the values of the logarithmic functions given in Table II as ordinates and corresponding minutes as abscissas, it would be found that the points for each function were on a curve with variable radius. It would be noted further that the curves for sines, tangents, and cotangents were of comparatively small radii when the angles were small; that the curves for cosines, cotangents, and tangents of angles near 90°, respectively, had the same shape as the curves for sines, tangents, and cotangents of the complements of the angles; and that other than the portions of the curves just mentioned were nearly straight lines for short distances.

When seconds are involved, it will be sufficiently accurate to interpolate in the ordinary manner between adjacent values in the tables —in other words, to assume that the curve joining two adjacent points is a straight line—for all functions between 2° and 88°, and also for sines of angles between 88° and 90° and for cosines of angles between 0° and 2°. The values in the columns headed *S*, *T*, and *C* provide a means (1) of accurately determining for any given angle between 0° and 2° the logarithmic sine, tangent, or cotangent, and for any given angle between 88° and 90°, the logarithmic cosine, cotangent, or tangent; or (2) for a given value of the logarithmic sine, tangent, or cotangent of accurately determining the angle when it lies between 0° and 2°, and for any given value of the cosine, cotangent, or tangent, the angle when it lies between 88° and 90°.

See also page 282.

Table IIa. Logarithmic Sines, Cosines, Tangents, and Cotangents

0°   179°

"	'	Sine.	S.•	T.•	Tang.	Cotang.	C.•	D. 1".	Cosine.	'
			4.685				15.314			
0	0	Inf. neg.	575	575	Inf. neg.	Inf. pos.	425		10.00 0000	60
60	1	6.46 3726	575	575	6.46 3726	13.53 6274	425		0000	59
120	2	.76 4756	575	575	.76 4756	.23 5244	425		0000	58
180	3	6.94 0847	575	575	6.94 0847	13.05 9153	425		0000	57
240	4	7.06 5786	575	575	7.06 5786	12.93 4214	425		0000	56
300	5	7.16 2696	575	575	7.16 2696	12.83 7304	425	.02	10.00 0000	55
360	6	.24 1877	575	575	.24 1878	.75 8122	425	.00	9.99 9999	54
420	7	.30 8824	575	575	.30 8825	.69 1175	425	.00	9999	53
480	8	.36 6816	574	576	.36 6817	.63 3183	424	.00	9999	52
540	9	.41 7968	574	576	.41 7970	.58 2030	424	.00	9999	51
								.02		
600	10	7.46 3726	574	576	7.46 3727	12.53 6273	424	.00	9.99 9998	50
660	11	.50 5118	574	576	.50 5120	.49 4880	424	.02	9998	49
720	12	.54 2906	574	577	.54 2909	.45 7091	423	.00	9997	48
780	13	.57 7668	574	577	.57 7672	.42 2328	423	.02	9997	47
840	14	.60 9853	574	577	.60 9857	.39 0143	423	.00	9996	46
900	15	7.63 9816	573	578	7.63 9820	12.36 0180	422	.02	9.99 9996	45
960	16	.66 7845	573	578	.66 7849	.33 2151	422	.00	9995	44
1020	17	.69 4173	573	578	.69 4179	.30 5821	422	.02	9995	43
1080	18	.71 8997	573	579	.71 9003	.28 0997	421	.02	9994	42
1140	19	.74 2478	573	579	.74 2484	.25 7516	421	.00	9993	41
1200	20	7.76 4754	572	580	7.76 4761	12.23 5239	420	.02	9.99 9993	40
1260	21	.78 5943	572	580	.78 5951	.21 4049	420	.02	9992	39
1320	22	.80 6146	572	581	.80 6155	.19 3845	419	.02	9991	38
1380	23	.82 5451	572	581	.82 5460	.17 4540	419	.02	9990	37
1440	24	.84 3934	571	582	.84 3944	.15 6056	418	.00	9989	36
1500	25	7.86 1662	571	583	7.86 1674	12.13 8326	417	.02	9.99 9989	35
1560	26	.87 8695	571	583	.87 8708	.12 1292	417	.02	9988	34
1620	27	.89 5085	570	584	.89 5099	.10 4901	416	.02	9987	33
1680	28	.91 0879	570	584	.91 0894	.08 9106	416	.02	9986	32
1740	29	.92 6119	570	585	.92 6134	.07 3866	415	.03	9985	31
1800	30	7.94 0842	569	586	7.94 0858	12.05 9142	414	.02	9.99 9983	30
1860	31	.95 5082	569	587	.95 5100	.04 4900	413	.02	9982	29
1920	32	.96 8870	569	587	.96 8889	.03 1111	413	.02	9981,	28
1980	33	.98 2233	568	588	.98 2253	.01 7747	412	.02	9980	27
2040	34	7.99 5198	568	589	7.99 5219	12.00 4781	411	.03	9979	26
2100	35	8.00 7787	567	590	8.00 7809	11.99 2191	410	.02	9.99 9977	25
2160	36	.02 0021	567	591	.02 0044	.97 9956	409	.02	9976	24
2220	37	.03 1919	566	592	.03 1945	.96 8055	408	.03	9975	23
2280	38	.04 3501	566	593	.04 3527	.95 6473	407	.02	9973	22
2340	39	.05 4781	566	593	.05 4809	.94 5191	407	.02	9972	21
2400	40	8.06 5776	565	594	8.06 5806	11.93 4194	406	.03	9.99 9971	20
2460	41	.07 6500	565	595	.07 6531	.92 3469	405	.02	9969	19
2520	42	.08 6965	564	596	.08 6997	.91 3003	404	.03	9968	18
2580	43	.09 7183	564	598	.09 7217	.90 2783	402	.03	9966	17
2640	44	.10 7167	563	599	.10 7203	.89 2797	401	.02	9964	16
2700	45	8.11 6926	562	600	8.11 6963	11.88 3037	400	.03	9.99 9963	15
2760	46	.12 6471	562	601	.12 6510	.87 3490	399	.03	9961	14
2820	47	.13 5810	561	602	.13 5851	.86 4149	398	.02	9959	13
2880	48	.14 4953	561	603	.14 4996	.85 5004	397	.03	9958	12
2940	49	.15 3907	560	604	.15 3952	.84 6048	396	.03	9956	11
3000	50	8.16 2681	560	605	8.16 2727	11.83 7273	395	.03	9.99 9954	10
3060	51	.17 1280	559	607	.17 1328	.82 8672	393	.03	9952	9
3120	52	.17 9713	558	608	.17 9763	.82 0237	392	.03	9950	8
3180	53	.18 7985	558	609	.18 8036	.81 1964	391	.03	9948	7
3240	54	.19 6102	557	611	.19 6156	.80 3844	389	.03	9946	6
3300	55	8.20 4070	556	612	8.20 4126	11.79 5874	388	.03	9.99 9944	5
3360	56	.21 1895	556	613	.21 1953	.78 8047	387	.03	9942	4
3420	57	.21 9581	555	615	.21 9641	.78 0359	385	.03	9940	3
3480	58	.22 7134	554	616	.22 7195	.77 2805	384	.03	9938	2
3540	59	.23 4557	554	618	.23 4621	.76 5379	382	.03	9936	1
3600	60	8.24 1855	553	619	8.24 1921	11.75 8079	381	.03	9.99 9934	0
			4.685				15.314			
"	'	Cosine.	S.•	T.•	Cotang.	Tang.	C.•	D. 1".	Sine.	'

90°   89°

* For use of S, T, and C see note, Sec. 14-8. Let D = log of number of seconds in angle; then,

$$\log \sin = S + D \qquad \log \tan = I + D \qquad \log \cot = C - D$$

Table IIa. (Concluded)

″	′	Sine.	S.*	T.*	Tang.	Cotang.	C.*	D. 1″.	Cosine.	′
			4.685				15.314			
3600	**0**	**8.24 1855**	553	619	**8.24 1921**	**11.75 8079**	381	.03	**9.99 9934**	**60**
3660	1	.24 9033	552	620	.24 9102	.75 0898	380	.05	9932	59
3720	2	.25 6094	551	622	.25 6165	.74 3835	378	.03	9929	58
3780	3	.26 3042	551	623	.26 3115	.73 6885	377	.03	9927	57
3840	4	.26 9881	550	625	.26 9956	.73 0044	375	.05	9925	56
3900	**5**	**8.27 6614**	549	627	**8.27 6691**	**11.72 3309**	373	.03	**9.99 9922**	**55**
3960	6	.28 3243	548	628	.28 3323	.71 6677	372	.03	9920	54
4020	7	.28 9773	547	630	.28 9856	.71 0144	370	.05	9918	53
4080	8	.29 6207	546	632	.29 6292	.70 3708	368	.03	9915	52
4140	9	.30 2546	546	633	.30 2634	.69 7366	367	.05	9913	51
4200	**10**	**8.30 8794**	545	635	**8.30 8884**	**11.69 1116**	365	.05	**9.99 9910**	**50**
4260	11	.31 4954	544	637	.31 5046	.68 4954	363	.03	9907	49
4320	12	.32 1027	543	638	.32 1122	.67 8878	362	.05	9905	48
4380	13	.32 7016	542	640	.32 7114	.67 2886	360	.05	9902	47
4440	14	.33 2924	541	642	.33 3025	.66 6975	358	.03	9899	46
4500	**15**	**8.33 8753**	540	644	**8.33 8856**	**11.66 1144**	356	.05	**9.99 9897**	**45**
4560	16	.34 4504	539	646	.34 4610	.65 5390	354	.05	9894	44
4620	17	.35 0181	539	648	.35 0289	.64 9711	352	.05	9891	43
4680	18	.35 5783	538	649	.35 5895	.64 4105	351	.05	9888	42
4740	19	.36 1315	537	651	.36 1430	.63 8570	349	.05	9885	41
4800	**20**	**8.36 6777**	536	653	**8.36 6895**	**11.63 3105**	347	.05	**9.99 9882**	**40**
4860	21	.37 2171	535	655	.37 2292	.62 7708	345	.05	9879	39
4920	22	.37 7499	534	657	.37 7622	.62 2378	343	.05	9876	38
4980	23	.38 2762	533	659	.38 2889	.61 7111	341	.05	9873	37
5040	24	.38 7962	532	661	.38 8092	.61 1908	339	.05	9870	36
5100	**25**	**8.39 3101**	531	663	**8.39 3234**	**11.60 6766**	337	.05	**9.99 9867**	**35**
5160	26	.39 8179	530	666	.39 8315	.60 1685	334	.05	9864	34
5220	27	.40 3199	529	668	.40 3338	.59 6662	332	.05	9861	33
5280	28	.40 8161	527	670	.40 8304	.59 1696	330	.07	9858	32
5340	29	.41 3068	526	672	.41 3213	.58 6787	328	.05	9854	31
5400	**30**	**8.41 7919**	525	674	**8.41 8068**	**11.58 1932**	326	.05	**9.99 9851**	**30**
5460	31	.42 2717	524	676	.42 2869	.57 7131	324	.07	9848	29
5520	32	.42 7462	523	679	.42 7618	.57 2382	321	.05	9844	28
5580	33	.43 2156	522	681	.43 2315	.56 7685	319	.05	9841	27
5640	34	.43 6800	521	683	.43 6962	.56 3038	317	.07	9838	26
5700	**35**	**8.44 1394**	520	685	**8.44 1560**	**11.55 8440**	315	.05	**9.99 9834**	**25**
5760	36	.44 5941	518	688	.44 6110	.55 3890	312	.07	9831	24
5820	37	.45 0440	517	690	.45 0613	.54 9387	310	.05	9827	23
5880	38	4893	516	693	5070	4930	307	.07	9824	22
5940	39	.45 9301	515	695	.45 9481	.54 0519	305	.07	9820	21
6000	**40**	**8.46 3665**	514	697	**8.46 3849**	**11.53 6151**	303	.05	**9.99 9816**	**20**
6060	41	.46 7985	512	700	.46 8172	.53 1828	300	.07	9813	19
6120	42	.47 2263	511	702	.47 2454	.52 7546	298	.07	9809	18
6180	43	.47 6498	510	705	.47 6693	.52 3307	295	.07	9805	17
6240	44	.48 0693	509	707	.48 0892	.51 9108	293	.07	9801	16
6300	**45**	**8.48 4848**	507	710	**8.48 5050**	**11.51 4950**	290	.05	**9.99 9797**	**15**
6360	46	.48 8963	506	713	.48 9170	.51 0830	287	.07	9794	14
6420	47	.49 3040	505	715	.49 3250	.50 6750	285	.07	9790	13
6480	48	.49 7078	503	718	.49 7293	.50 2707	282	.07	9786	12
6540	49	.50 1080	502	720	.50 1298	.49 8702	280	.07	9782	11
6600	**50**	**8.50 5045**	501	723	**8.50 5267**	**11.49 4733**	277	.07	**9.99 9778**	**10**
6660	51	.50 8974	499	726	.50 9200	.49 0800	274	.08	9774	9
6720	52	.51 2867	498	729	.51 3098	.48 6902	271	.07	9769	8
6780	53	.51 6726	497	731	.51 6961	.48 3039	269	.07	9765	7
6840	54	.52 0551	495	734	.52 0790	.47 9210	266	.07	9761	6
6900	**55**	**8.52 4343**	494	737	**8.52 4586**	**11.47 5414**	263	.07	**9.99 9757**	**5**
6960	56	.52 8102	492	740	.52 8349	.47 1651	260	.08	9753	4
7020	57	.53 1828	491	743	.53 2080	.46 7920	257	.07	9748	3
7080	58	5523	490	745	5779	4221	255	.07	9744	2
7140	59	.53 9186	488	748	.53 9447	.46 0553	252	.08	9740	1
7200	**60**	**8.54 2819**	487	751	**8.54 3084**	**11.45 6916**	249		**9.99 9735**	**0**
			4.685				15.314			
″	′	Cosine.	S.*	T.*	Cotang.	Tang.	C.*	D. 1″.	Sine.	′

* For use of S, T, and C see note, Sec. 14-8. Let D = log of number of seconds in angle; then,

$$\log \sin = S + D \qquad \log \tan = T + D \qquad \log \cot = C - D$$

Table II. Logarithmic Sines, Cosines, Tangents, and Cotangents

2°   177°

′	Sine.	D. 1″.	Cosine.	D.1″.	Tang.	D.1″.	Cotang.	′
0	8.54 2819	60.05	9.99 9735	.07	8.54 3084	60.12	11.45 6916	60
1	6422	59.55	9731	.08	.54 6691	59.62	.45 3309	59
2	.54 9995	59.07	9726	.07	.55 0268	59.15	.44 9732	58
3	.55 3539	58.58	9722	.08	3817	58.65	6183	57
4	.55 7054	58.10	9717	.07	.55 7336	58.20	.44 2664	56
5	8.56 0540	57.65	9.99 9713	.08	8.56 0828	57.72	11.43 9172	55
6	3999	57.20	9708	.07	4291	57.27	5709	54
7	.56 7431	56.75	9704	.08	.56 7727	56.83	.43 2273	53
8	.57 0836	56.30	9699	.08	.57 1137	56.38	.42 8863	52
9	4214	55.87	9694	.08	4520	55.95	5480	51
10	8.57 7566	55.43	9.99 9689	.07	8.57 7877	55.52	11.42 2123	50
11	.58 0892	55.02	9685	.08	.58 1208	55.10	.41 8792	49
12	4193	54.60	9680	.08	4514	54.68	5486	48
13	.58 7469	54.20	9675	.08	.58 7795	54.27	.41 2205	47
14	.59 0721	53.78	9670	.08	.59 1051	53.87	.40 8949	46
15	8.59 3948	53.40	9.99 9665	.08	8.59 4283	53.48	11.40 5717	45
16	.59 7152	53.00	9660	.08	.59 7492	53.08	.40 2508	44
17	.60 0332	52.62	9655	.08	.60 0677	52.70	.39 9323	43
18	3489	52.23	9650	.08	3839	52.32	6161	42
19	6623	51.85	9645	.08	.60 6978	51.93	.39 3022	41
20	8.60 9734	51.48	9.99 9640	.08	8.61 0094	51.58	11.38 9906	40
21	.61 2823	51.13	9635	.10	3189	51.22	6811	39
22	5891	50.77	9629	.08	6262	50.85	3738	38
23	.61 8937	50.42	9624	.08	.61 9313	50.50	.38 0687	37
24	.62 1962	50.05	9619	.08	.62 2343	50.15	.37 7657	36
25	8.62 4965	49.72	9.99 9614	.10	8.62 5352	49.80	11.37 4648	35
26	.62 7948	49.38	9608	.08	.62 8340	49.47	.37 1660	34
27	.63 0911	49.05	9603	.10	.63 1308	49.13	.36 8692	33
28	3854	48.70	9597	.08	4256	48.80	5744	32
29	6776	48.40	9592	.10	.63 7184	48.48	.36 2816	31
30	8.63 9680	48.05	9.99 9586	.08	8.64 0093	48.15	11.35 9907	30
31	.64 2563	47.75	9581	.10	2982	47.85	7018	29
32	5428	47.43	9575	.08	5853	47.52	4147	28
33	.64 8274	47.13	9570	.10	.64 8704	47.22	.35 1296	27
34	.65 1102	46.82	9564	.10	.65 1537	46.92	.34 8463	26
35	8.65 3911	46.52	9.99 9558	.08	8.65 4352	46.62	11.34 5648	25
36	6702	46.22	9553	.10	7149	46.32	2851	24
37	.65 9475	45.92	9547	.10	.65 9928	46.02	.34 0072	23
38	.66 2230	45.63	9541	.10	.66 2689	45.73	.33 7311	22
39	4968	45.35	9535	.10	5433	45.45	4567	21
40	8.66 7689	45.07	9.99 9529	.08	8.66 8160	45.17	11.33 1840	20
41	.67 0393	44.78	9524	.10	.67 0870	44.88	.32 9130	19
42	3080	44.52	9518	.10	3563	44.60	6437	18
43	5751	44.23	9512	.10	6239	44.35	3761	17
44	.67 8405	43.97	9506	.10	.67 8900	44.07	.32 1100	16
45	8.68 1043	43.70	9.99 9500	.12	8.68 1544	43.80	11.31 8456	15
46	3665	43.45	9493	.10	4172	43.53	5828	14
47	6272	43.18	9487	.10	6784	43.28	3216	13
48	.68 8863	42.92	9481	.10	.68 9381	43.03	.31 0619	12
49	.69 1438	42.67	9475	.10	.69 1963	42.77	.30 8037	11
50	8.69 3998	42.42	9.99 9469	.10	8.69 4529	42.53	11.30 5471	10
51	6543	42.17	9463	.12	7081	42.27	2919	9
52	.69 9073	41.93	9456	.10	.69 9617	42.03	.30 0383	8
53	.70 1589	41.68	9450	.12	.70 2139	41.78	.29 7861	7
54	4090	41.45	9443	.10	4646	41.57	5354	6
55	8.70 6577	41.20	9.99 9437	.10	8.70 7140	41.30	11.29 2860	5
56	.70 9049	40.97	9431	.12	.70 9618	41.08	.29 0382	4
57	.71 1507	40.75	9424	.10	.71 2083	40.85	.28 7917	3
58	3952	40.52	9418	.12	4534	40.63	5466	2
59	6383	40.28	9411	.12	6972	40.40	3028	1
60	8.71 8800		9.99 9404		8.71 9396		11.28 0604	0
′	Cosine.	D.1″.	Sine.	D.1″.	Cotang.	D.1″.	Tang.	′

92°   87°

Table II. (Continued)

3°   176°

′	Sine.	D.1″.	Cosine.	D.1″.	Tang.	D.1″.	Cotang.	′
0	8.71 8800	40.07	9.99 9404	.10	8.71 9396	40.17	11.28 0604	60
1	.72 1204	39.85	9398	.12	.72 1806	39.97	.27 8194	59
2	3595	39.62	9391	.12	4204	39.73	5796	58
3	5972	39.42	9384	.10	6588	39.52	3412	57
4	.72 8337	39.18	9378	.12	.72 8959	39.30	.27 1041	56
5	8.73 0688	38.98	9.99 9371	.12	8.73 1317	39.10	11.26 8683	55
6	3027	38.78	9364	.12	3663	38.88	6337	54
7	5354	38.55	9357	.12	5996	38.68	4004	53
8	7667	38.37	9350	.12	.73 8317	38.48	.26 1683	52
9	.73 9969	38.17	9343	.12	.74 0626	38.27	.25 9374	51
10	8.74 2259	37.95	9.99 9336	.12	8.74 2922	38.08	11.25 7078	50
11	4536	37.77	9329	.12	5207	37.87	4793	49
12	6802	37.55	9322	.12	7479	37.68	2521	48
13	.74 9055	37.37	9315	.12	.74 9740	37.48	.25 0260	47
14	.75 1297	37.18	9308	.12	.75 1989	37.30	.24 8011	46
15	8.75 3528	36.98	9.99 9301	.12	8.75 4227	37.10	11.24 5773	45
16	5747	36.80	9294	.12	6453	36.92	3547	44
17	.75 7955	36.60	9287	.13	.75 8668	36.73	.24 1332	43
18	.76 0151	36.43	9279	.12	.76 0872	36.55	.23 9128	42
19	2337	36.23	9272	.12	3065	36.35	6935	41
20	8.76 4511	36.07	9.99 9265	.13	8.76 5246	36.18	11.23 4754	40
21	6675	35.88	9257	.12	7417	36.02	2583	39
22	.76 8828	35.70	9250	.13	.76 9578	35.82	.23 0422	38
23	.77 0970	35.52	9242	.13	.77 1727	35.65	.22 8273	37
24	3101	35.37	9235	.13	3866	35.48	6134	36
25	8.77 5223	35.17	9.99 9227	.12	8.77 59˙5	35.32	11.22 4005	35
26	7333	35.02	9220	.13	.77 8114	35.13	.22 1886	34
27	.77 9434	34.83	9212	.12	.78 0222	34.97	.21 9778	33
28	.78 1524	34.68	9205	.13	2320	34.80	7680	32
29	3605	34.50	9197	.13	4408	34.63	5592	31
30	8.78 5675	34.35	9.99 9189	.13	8.78 6486	34.47	11.21 3514	30
31	7736	34.18	9181	.12	.78 8554	34.32	.21 1446	29
32	.78 9787	34.02	9174	.13	.79 0613	34.15	.20 9387	28
33	.79 1828	33.85	9166	.13	2662	33.98	7338	27
34	3859	33.70	9158	.13	4701	33.83	5299	26
35	8.79 5881	33.55	9.99 9150	.13	8.79 6731	33.68	11.20 3269	25
36	7894	33.38	9142	.13	.79 8752	33.52	.20 1248	24
37	.79 9897	33.25	9134	.13	.80 0763	33.37	.19 9237	23
38	.80 1892	33.07	9126	.13	2765	33.22	7235	22
39	3876	32.93	9118	.13	4758	33.07	5242	21
40	8.80 5852	32.78	9.99 9110	.13	8.80 6742	32.92	11.19 3258	20
41	7819	32.63	9102	.13	.80 8717	32.77	.19 1283	19
42	.80 9777	32.48	9094	.13	.81 0683	32.63	.18 9317	18
43	.81 1726	32.35	9086	.15	2641	32.47	7359	17
44	3667	32.20	9077	.13	4589	32.33	5411	16
45	8.81 5599	32.05	9.99 9069	.13	8.81 6529	32.20	11.18 3471	15
46	7522	31.90	9061	.13	.81 8461	32.05	.18 1539	14
47	.81 9436	31.78	9053	.15	.82 0384	31.90	.17 9616	13
48	.82 1343	31.62	9044	.13	2298	31.78	7702	12
49	3240	31.50	9036	.15	4205	31.63	5795	11
50	8.82 5130	31.35	9.99 9027	.13	8.82 6103	31.48	11.17 3897	10
51	7011	31.22	9019	.15	7992	31.37	2008	9
52	.82 8884	31.08	9010	.13	.82 9874	31.23	.17 0126	8
53	.83 0749	30.97	9002	.15	.83 1748	31.08	.16 8252	7
54	2607	30.82	8993	.15	3613	30.97	6387	6
55	8.83 4456	30.68	9.99 8984	.13	8.83 5471	30.83	11.16 4529	5
56	6297	30.55	8976	.15	7321	30.70	2679	4
57	8130	30.43	8967	.15	.83 9163	30.58	.16 0837	3
58	.83 9956	30.30	8958	.13	.84 0998	30.45	.15 9002	2
59	.84 1774	30.18	8950	.15	2825	30.32	7175	1
60	8.84 3585		9.99 8941		8.84 4644		11.15 5356	0
′	Cosine.	D.1″.	Sine.	D.1″.	Cotang.	D.1″.	Tang.	′

93°   86°

Table II. (Continued)

4°   175°

′	Sine.	D.1″.	Cosine.	D.1″.	Tang.	D.1″.	Cotang.	′
0	8.84 3585	30.03	9.99 8941	.15	8.84 4644	30.18	11.15 5356	60
1	5387	29.93	8932	.15	6455	30.08	3545	59
2	7183	29.80	8923	.15	.84 8260	29.95	.15 1740	58
3	.84 8971	29.67	8914	.15	.85 0057	29.82	.14 9943	57
4	.85 0751	29.57	8905	.15	1846	29.70	8154	56
5	8.85 2525	29.43	9.99 8896	.15	8.85 3628	29.58	11.14 6372	55
6	4291	29.30	8887	.15	5403	29.47	4597	54
7	6049	29.20	8878	.15	7171	29.35	2829	53
8	7801	29.08	8869	.15	.85 8932	29.23	.14 1068	52
9	.85 9546	28.95	8860	.15	.86 0686	29.12	.13 9314	51
10	8.86 1283	28.85	9.99 8851	.17	8.86 2433	29.00	11.13 7567	50
11	3014	28.73	8841	.15	4173	28.88	5827	49
12	4738	28.62	8832	.15	5906	28.77	4094	48
13	6455	28.50	8823	.17	7632	28.65	2368	47
14	8165	28.38	8813	.15	.86 9351	28.55	.13 0649	46
15	8.86 9868	28.28	9.99 8804	.15	8.87 1064	28.43	11.12 8936	45
16	.87 1565	28.17	8795	.17	2770	28.32	7230	44
17	3255	28.05	8785	.15	4469	28.22	5531	43
18	4938	27.95	8776	.17	6162	28.12	3838	42
19	6615	27.83	8766	.15	7849	28.00	2151	41
20	8.87 8285	27.73	9.99 8757	.17	8.87 9529	27.88	11.12 0471	40
21	.87 9949	27.63	8747	.15	.88 1202	27.78	.11 8798	39
22	.88 1607	27.52	8738	.17	2869	27.68	7131	38
23	3258	27.42	8728	.17	4530	27.58	5470	37
24	4903	27.32	8718	.17	6185	27.47	3815	36
25	8.88 6542	27.20	9.99 8708	.15	8.88 7833	27.38	11.11 2167	35
26	8174	27.12	8699	.17	.88 9476	27.27	.11 0524	34
27	.88 9801	27.00	8689	.17	.89 1112	27.17	.10 8888	33
28	.89 1421	26.90	8679	.17	2742	27.07	7258	32
29	3035	26.80	8669	.17	4366	26.97	5634	31
30	8.89 4643	26.72	9.99 8659	.17	8.89 5984	26.87	11.10 4016	30
31	6246	26.60	8649	.17	7596	26.78	2404	29
32	7842	26.50	8639	.17	.89 9203	26.67	.10 0797	28
33	.89 9432	26.42	8629	.17	.90 0803	26.58	.09 9197	27
34	.90 1017	26.32	8619	.17	2398	26.48	7602	26
35	8.90 2596	26.22	9.99 8609	.17	8.90 3987	26.38	11.09 6013	25
36	4169	26.12	8599	.17	5570	26.28	4430	24
37	5736	26.02	8589	.18	7147	26.20	2853	23
38	7297	25.93	8578	.17	.90 8719	26.10	.09 1281	22
39	.90 8853	25.85	8568	.17	.91 0285	26.02	.08 9715	21
40	8.91 0404	25.75	9.99 8558	.17	8.91 1846	25.92	11.08 8154	20
41	1949	25.65	8548	.18	3401	25.83	6599	19
42	3488	25.57	8537	.17	4951	25.73	5049	18
43	5022	25.47	8527	.18	6495	25.63	3505	17
44	6550	25.38	8516	.17	8034	25.57	1966	16
45	8.91 8073	25.30	9.99 8506	.18	8.91 9568	25.47	11.08 0432	15
46	.91 9591	25.20	8495	.17	.92 1096	25.38	.07 8904	14
47	.92 1103	25.12	8485	.18	2619	25.28	7381	13
48	2610	25.03	8474	.17	4136	25.22	5864	12
49	4112	24.95	8464	.18	5649	25.12	4351	11
50	8.92 5609	24.85	9.99 8453	.18	8.92 7156	25.03	11.07 2844	10
51	7100	24.78	8442	.18	.92 8658	24.95	.07 1342	9
52	.92 8587	24.68	8431	.17	.93 0155	24.87	.06 9845	8
53	.93 0068	24.60	8421	.18	1647	24.78	8353	7
54	1544	24.52	8410	.18	3134	24.70	6866	6
55	8.93 3015	24.43	9.99 8399	.18	8.93 4616	24.62	11.06 5384	5
56	4481	24.35	8388	.18	6093	24.53	3907	4
57	5942	24.27	8377	.18	7565	24.45	2435	3
58	7398	24.20	8366	.18	.93 9032	24.37	.06 0968	2
59	.93 8850	24.10	8355	.18	.94 0494	24.30	.05 9506	1
60	8.94 0296		9.99 8344		8.94 1952		11.05 8048	0
′	Cosine.	D.1″.	Sine.	D.1″.	Cotang.	D.1″.	Tang.	′

94°   85°

Table II. (Continued)

5°   174°

′	Sine.	D.1″.	Cosine.	D.1″.	Tang.	D.1″.	Cotang.	′
0	8.94 0296	24.03	9.99 8344	.18	8.94 1952	24.20	11.05 8048	60
1	1738	23.93	8333	.18	3404	24.13	6596	59
2	3174	23.87	8322	.18	4852	24.05	5148	58
3	4606	23.80	8311	.18	6295	23.98	3705	57
4	6034	23.70	8300	.18	7734	23.90	2266	56
5	8.94 7456	23.63	9.99 8289	.20	8.94 9168	23.82	11.05 0832	55
6	.94 8874	23.55	8277	.18	.95 0597	23.73	.04 9403	54
7	.95 0287	23.48	8266	.18	2021	23.67	7979	53
8	1696	23.40	8255	.20	3441	23.58	6559	52
9	3100	23.32	8243	.18	4856	23.52	5144	51
10	8.95 4499	23.25	9.99 8232	.20	8.95 6267	23.45	11.04 3733	50
11	5894	23.17	8220	.18	7674	23.35	2326	49
12	7284	23.10	8209	.20	.95 9075	23.30	.04 0925	48
13	.95 8670	23.03	8197	.18	.96 0473	23.22	.03 9527	47
14	.96 0052	22.95	8186	.20	1866	23.15	8134	46
15	8.96 1429	22.87	9.99 8174	.18	8.96 3255	23.07	11.03 6745	45
16	2801	22.82	8163	.20	4639	23.00	5361	44
17	4170	22.73	8151	.20	6019	22.92	3981	43
18	5534	22.65	8139	.18	7394	22.87	2606	42
19	6893	22.60	8128	.20	.96 8766	22.78	.03 1234	41
20	8.96 8249	22.52	9.99 8116	.20	8.97 0133	22.72	11.02 9867	40
21	.96 9600	22.45	8104	.20	1496	22.65	8504	39
22	.97 0947	22.37	8092	.20	2855	22.57	7145	38
23	2289	22.32	8080	.20	4209	22.52	5791	37
24	3628	22.23	8068	.20	5560	22.43	4440	36
25	8.97 4962	22.18	9.99 8056	.20	8.97 6906	22.37	11.02 3094	35
26	6293	22.10	8044	.20	8248	22.30	1752	34
27	7619	22.03	8032	.20	.97 9586	22.25	.02 0414	33
28	.97 8941	21.97	8020	.20	.98 0921	22.17	.01 9079	32
29	.98 0259	21.90	8008	.20	2251	22.10	7749	31
30	8.98 1573	21.83	9.99 7996	.20	8.98 3577	22.03	11.01 6423	30
31	2883	21.77	7984	.20	4899	21.97	5101	29
32	4189	21.72	7972	.22	6217	21.92	3783	28
33	5491	21.63	7959	.20	7532	21.83	2468	27
34	6789	21.57	7947	.20	.98 8842	21.78	.01 1158	26
35	8.98 8083	21.52	9.99 7935	.22	8.99 0149	21.70	11.00 9851	25
36	.98 9374	21.43	7922	.20	1451	21.65	8549	24
37	.99 0660	21.38	7910	.22	2750	21.58	7250	23
38	1943	21.32	7897	.20	4045	21.53	5955	22
39	3222	21.25	7885	.22	5337	21.45	4663	21
40	8.99 4497	21.18	9.99 7872	.20	8.99 6624	21.40	11.00 3376	20
41	5768	21.13	7860	.22	7908	21.33	2092	19
42	7036	21.05	7847	.20	8.99 9188	21.28	11.00 0812	18
43	8299	21.02	7835	.22	9.00 0465	21.22	10.99 9535	17
44	8.99 9560	20.93	7822	.22	1738	21.15	8262	16
45	9.00 0816	20.88	9.99 7809	.20	9.00 3007	21.08	10.99 6993	15
46	2069	20.82	7797	.22	4272	21.03	5728	14
47	3318	20.75	7784	.22	5534	20.97	4466	13
48	4563	20.70	7771	.22	6792	20.92	3208	12
49	5805	20.65	7758	.22	8047	20.85	1953	11
50	9.00 7044	20.57	9.99 7745	.22	9.00 9298	20.80	10.99 0702	10
51	8278	20.53	7732	.22	.01 0546	20.73	.98 9454	9
52	.00 9510	20.45	7719	.22	1790	20.68	8210	8
53	.01 0737	20.42	7706	.22	3031	20.62	6969	7
54	1962	20.33	7693	.22	4268	20.57	5732	6
55	9.01 3182	20.30	9.99 7680	.22	9.01 5502	20.50	10.98 4498	5
56	4400	20.22	7667	.22	6732	20.45	3268	4
57	5613	20.18	7654	.22	7959	20.40	2041	3
58	6824	20.12	7641	.22	.01 9183	20.33	.98 0817	2
59	8031	20.07	7628	.23	.02 0403	20.28	.97 9597	1
60	9.01 9235		9.99 7614		9.02 1620		10.97 8380	0
′	Cosine.	D.1″.	Sine.	D.1″.	Cotang.	D.1″.	Tang.	′

95°   84°

Table II. (Continued)

6°   173°

′	Sine.	D.1″.	Cosine.	D.1″.	Tang.	D.1″.	Cotang.	′
0	9.01 9235	20.00	9.99 7614	.22	9.02 1620	20.23	10.97 8380	60
1	.02 0435	19.95	7601	.22	2834	20.17	7166	59
2	1632	19.88	7588	.23	4044	20.12	5956	58
3	2825	19.85	7574	.22	5251	20.07	4749	57
4	4016	19.78	7561	.23	6455	20.00	3545	56
5	9.02 5203	19.72	9.99 7547	.22	9.02 7655	19.95	10.97 2345	55
6	6386	19.68	7534	.23	.02 8852	19.90	.97 1148	54
7	7567	19.62	7520	.22	.03 0046	19.85	.96 9954	53
8	8744	19.57	7507	.23	1237	19.80	8763	52
9	.02 9918	19.52	7493	.22	2425	19.73	7575	51
10	9.03 1089	19.47	9.99 7480	.23	9.03 3609	19.70	10.96 6391	50
11	2257	19.40	7466	.23	4791	19.63	5209	49
12	3421	19.35	7452	.22	5969	19.58	4031	48
13	4582	19.32	7439	.23	7144	19.53	2856	47
14	5741	19.25	7425	.23	8316	19.48	1684	46
15	9.03 6896	19.20	9.99 7411	.23	9.03 9485	19.43	10.96 0515	45
16	8048	19.15	7397	.23	.04 0651	19.37	.95 9349	44
17	.03 9197	19.08	7383	.23	1813	19.33	8187	43
18	.04 0342	19.05	7369	.23	2973	19.28	7027	42
19	1485	19.00	7355	.23	4130	19.23	5870	41
20	9.04 2625	18.95	9.99 7341	.23	9.04 5284	19.17	10.95 4716	40
21	3762	18.88	7327	.23	6434	19.13	3566	39
22	4895	18.85	7313	.23	7582	19.08	2418	38
23	6026	18.80	7299	.23	8727	19.03	1273	37
24	7154	18.75	7285	.23	.04 9869	18.98	.95 0131	36
25	9.04 8279	18.68	9.99 7271	.23	9.05 1008	18.93	10.94 8992	35
26	.04 9400	18.65	7257	.25	2144	18.88	7856	34
27	.05 0519	18.60	7242	.23	3277	18.83	6723	33
28	1635	18.57	7228	.23	4407	18.80	5593	32
29	2749	18.50	7214	.25	5535	18.73	4465	31
30	9.05 3859	18.45	9.99 7199	.23	9.05 6659	18.70	10.94 3341	30
31	4966	18.42	7185	.25	7781	18.65	2219	29
32	6071	18.35	7170	.23	.05 8900	18.60	.94 1100	28
33	7172	18.32	7156	.25	.06 0016	18.57	.93 9984	27
34	8271	18.27	7141	.23	1130	18.50	8870	26
35	9.05 9367	18.22	9.99 7127	.25	9.06 2240	18.47	10.93 7760	25
36	.06 0460	18.18	7112	.23	3348	18.42	6652	24
37	1551	18.13	7098	.25	4453	18.38	5547	23
38	2639	18.08	7083	.25	5556	18.32	4444	22
39	3724	18.03	7068	.25	6655	18.28	3345	21
40	9.06 4806	17.98	9.99 7053	.23	9.06 7752	18.25	10.93 2248	20
41	5885	17.95	7039	.25	8846	18.20	1154	19
42	6962	17.90	7024	.25	.06 9938	18.15	.93 0062	18
43	8036	17.85	7009	.25	.07 1027	18.10	.92 8973	17
44	.06 9107	17.82	6994	.25	2113	18.07	7887	16
45	9.07 0176	17.77	9.99 6979	.25	9.07 3197	18.02	10.92 6803	15
46	1242	17.73	6964	.25	4278	17.97	5722	14
47	2306	17.67	6949	.25	5356	17.93	4644	13
48	3366	17.63	6934	.25	6432	17.88	3568	12
49	4424	17.60	6919	.25	7505	17.85	2495	11
50	9.07 5480	17.55	9.99 6904	.25	9.07 8576	17.80	10.92 1424	10
51	6533	17.50	6889	.25	.07 9644	17.77	.92 0356	9
52	7583	17.47	6874	.27	.08 0710	17.72	.91 9290	8
53	8631	17.42	6858	.25	1773	17.67	8227	7
54	.07 9676	17.38	6843	.27	2833	17.63	7167	6
55	9.08 0719	17.33	9.99 6828	.27	9.08 3891	17.60	10.91 6109	5
56	1759	17.30	6812	.25	4947	17.55	5053	4
57	2797	17.25	6797	.25	6000	17.50	4000	3
58	3832	17.20	6782	.27	7050	17.47	2950	2
59	4864	17.17	6766	.25	8098	17.43	1902	1
60	9.08 5894		9.99 6751		9.08 9144		10.91 0856	0
′	Cosine.	D.1″.	Sine.	D.1″.	Cotang.	D.1″.	Tang.	′

96°   83°

Table II. (Continued)

7°   172°

′	Sine.	D.1″.	Cosine.	D.1″.	Tang.	D.1″.	Cotang.	′
0	9.08 5894	17.13	9.99 6751	.27	9.08 9144	17.38	10.91 0856	60
1	6922	17.08	6735	.25	.09 0187	17.35	.90 9813	59
2	7947	17.05	6720	.27	1228	17.30	8772	58
3	8970	17.00	6704	.27	2266	17.27	7734	57
4	.08 9990	16.97	6688	.25	3302	17.23	6698	56
5	9.09 1008	16.93	9.99 6673	.27	9.09 4336	17.18	10.90 5664	55
6	2024	16.88	6657	.27	5367	17.13	4633	54
7	3037	16.83	6641	.27	6395	17.12	3605	53
8	4047	16.82	6625	.25	7422	17.07	2578	52
9	5056	16.77	6610	.27	8446	17.03	1554	51
10	9.09 6062	16.72	9.99 6594	.27	9.09 9468	16.98	10.90 0532	50
11	7065	16.68	6578	.27	.10 0487	16.95	.89 9513	49
12	8066	16.65	6562	.27	1504	16.92	8496	48
13	.09 9065	16.62	6546	.27	2519	16.88	7481	47
14	.10 0062	16.57	6530	.27	3533	16.83	6468	46
15	9.10 1056	16.53	9.99 6514	.27	9.10 4542	16.80	10.89 5458	45
16	2048	16.48	6498	.27	5550	16.77	4450	44
17	3037	16.47	6482	.28	6556	16.72	3444	43
18	4025	16.42	6465	.27	7559	16.68	2441	42
19	5010	16.37	6449	.27	8560	16.65	1440	41
20	9.10 5992	16.35	9.99 6433	.27	9.10 9559	16.62	10.89 0441	40
21	6973	16.30	6417	.28	.11 0556	16.58	.88 9444	39
22	7951	16.27	6400	.28	1551	16.53	8449	38
23	8927	16.23	6384	.27	2543	16.50	7457	37
24	.10 9901	16.20	6368	.28	3533	16.47	6467	36
25	9.11 0873	16.15	9.99 6351	.27	9.11 4521	16.43	10.88 5479	35
26	1842	16.12	6335	.28	5507	16.40	4493	34
27	2809	16.08	6318	.27	6491	16.35	3509	33
28	3774	16.05	6302	.28	7472	16.33	2528	32
29	4737	16.02	6285	.27	8452	16.28	1548	31
30	9.11 5698	15.97	9.99 6269	.28	9.11 9429	16.25	10.88 0571	30
31	6656	15.95	6252	.28	.12 0404	16.22	.87 9596	29
32	7613	15.90	6235	.27	1377	16.18	8623	28
33	8567	15.87	6219	.28	2348	16.15	7652	27
34	.11 9519	15.83	6202	.28	3317	16.12	6683	26
35	9.12 0469	15.80	9.99 6185	.28	9.12 4284	16.08	10.87 5716	25
36	1417	15.75	6168	.28	5249	16.03	4751	24
37	2362	15.73	6151	.28	6211	16.02	3789	23
38	3306	15.70	6134	.28	7172	15.97	2828	22
39	4248	15.65	6117	.28	8130	15.95	1870	21
40	9.12 5187	15.63	9.99 6100	.28	9.12 9087	15.90	10.87 0913	20
41	6125	15.58	6083	.28	.13 0041	15.88	.86 9959	19
42	7060	15.55	6066	.28	0994	15.83	9006	18
43	7993	15.53	6049	.28	1944	15.82	8056	17
44	8925	15.48	6032	.28	2893	15.77	7107	16
45	9.12 9854	15.45	9.99 6015	.28	9.13 3839	15.75	10.86 6161	15
46	.13 0781	15.42	5998	.30	4784	15.70	5216	14
47	1706	15.40	5980	.28	5726	15.68	4274	13
48	2630	15.35	5963	.28	6667	15.63	3333	12
49	3551	15.32	5946	.30	7605	15.62	2395	11
50	9.13 4470	15.28	9.99 5928	.28	9.13 8542	15.57	10.86 1458	10
51	5387	15.27	5911	.28	.13 9476	15.55	.86 0524	9
52	6303	15.22	5894	.30	.14 0409	15.52	.85 9591	8
53	7216	15.20	5876	.28	1340	15.48	8660	7
54	8128	15.15	5859	.30	2269	15.45	7731	6
55	9.13 9037	15.12	9.99 5841	.30	9.14 3196	15.42	10.85 6804	5
56	.13 9944	15.10	5823	.28	4121	15.38	5879	4
57	.14 0850	15.07	5806	.30	5044	15.37	4956	3
58	1754	15.02	5788	.28	5966	15.32	4034	2
59	2655	15.00	5771	.30	6885	15.30	3115	1
60	9.14 3555		9.99 5753		9.14 7803		10.85 2197	0
′	Cosine.	D.1″.	Sine.	D.1″.	Cotang.	D.1″.	Tang.	′

97°   82°

Table II. (Continued)

8°   171°

′	Sine.	D.1″.	Cosine.	D.1″.	Tang.	D.1″.	Cotang.	′
0	9.14 3555		9.99 5753		9.14 7803		10.85 2197	60
1	4453	14.97	5735	.30	8718	15.25	1282	59
2	5349	14.93	5717	.30	.14 9632	15.23	.85 0368	58
3	6243	14.90	5699	.30	.15 0544	15.20	.84 9456	57
4	7136	14.88	5681	.30	1454	15.17	8546	56
		14.83		.28		15.15		
5	9.14 8026	14.82	9.99 5664	.30	9.15 2363	15.10	10.84 7637	55
6	8915	14.78	5646	.30	3269	15.08	6731	54
7	.14 9802	14.73	5628	.30	4174	15.05	5826	53
8	.15 0686	14.72	5610	.32	5077	15.02	4923	52
9	1569	14.70	5591	.30	5978	14.98	4022	51
10	9.15 2451	14.65	9.99 5573	.30	9.15 6877	14.97	10.84 3123	50
11	3330	14.63	5555	.30	7775	14.93	2225	49
12	4208	14.58	5537	.30	8671	14.90	1329	48
13	5083	14.57	5519	.30	.15 9565	14.87	.84 0435	47
14	5957	14.55	5501	.32	.16 0457	14.83	.83 9543	46
15	9.15 6830	14.50	9.99 5482	.30	9.16 1347	14.82	10.83 8653	45
16	7700	14.48	5464	.30	2236	14.78	7764	44
17	8569	14.43	5446	.32	3123	14.75	6877	43
18	.15 9435	14.43	5427	.30	4008	14.73	5992	42
19	.16 0301	14.38	5409	.32	4892	14.70	5108	41
20	9.16 1164	14.35	9.99 5390	.30	9.16 5774	14.67	10.83 4226	40
21	2025	14.33	5372	.32	6654	14.63	3346	39
22	2885	14.30	5353	.32	7532	14.62	2468	38
23	3743	14.28	5334	.30	8409	14.58	1591	37
24	4600	14.23	5316	.32	.16 9284	14.55	.83 0716	36
25	9.16 5454	14.22	9.99 5297	.32	9.17 0157	14.53	10.82 9843	35
26	6307	14.20	5278	.30	1029	14.50	8971	34
27	7159	14.15	5260	.32	1899	14.47	8101	33
28	8008	14.13	5241	.32	2767	14.45	7233	32
29	8856	14.10	5222	.32	3634	14.42	6366	31
30	9.16 9702	14.08	9.99 5203	.32	9.17 4499	14.38	10.82 5501	30
31	.17 0547	14.03	5184	.32	5362	14.37	4638	29
32	1389	14.02	5165	.32	6224	14.33	3776	28
33	2230	14.00	5146	.32	7084	14.30	2916	27
34	3070	13.97	5127	.32	7942	14.28	2058	26
35	9.17 3908	13.93	9.99 5108	.32	9.17 8799	14.27	10.82 1201	25
36	4744	13.90	5089	.32	.17 9655	14.22	.82 0345	24
37	5578	13.88	5070	.32	.18 0508	14.20	.81 9492	23
38	6411	13.85	5051	.32	1360	14.18	8640	22
39	7242	13.83	5032	.32	2211	14.13	7789	21
40	9.17 8072	13.80	9.99 5013	.33	9.18 3059	14.13	10.81 6941	20
41	8900	13.77	4993	.32	3907	14.08	6093	19
42	.17 9726	13.75	4974	.32	4752	14.08	5248	18
43	.18 0551	13.72	4955	.33	5597	14.03	4403	17
44	1374	13.70	4935	.32	6439	14.02	3561	16
45	9.18 2196	13.67	9.99 4916	.33	9.18 7280	14.00	10.81 2720	15
46	3016	13.63	4896	.32	8120	13.97	1880	14
47	3834	13.62	4877	.33	8958	13.93	1042	13
48	4651	13.58	4857	.32	.18 9794	13.92	.81 0206	12
49	5466	13.57	4838	.33	.19 0629	13.88	.80 9371	11
50	9.18 6280	13.53	9.99 4818	.33	9.19 1462	13.87	10.80 8538	10
51	7092	13.52	4798	.32	2294	13.83	7706	9
52	7903	13.48	4779	.33	3124	13.82	6876	8
53	8712	13.45	4759	.33	3953	13.78	6047	7
54	.18 9519	13.43	4739	.32	4780	13.77	5220	6
55	9.19 0325	13.38	9.99 4720	.33	9.19 5606	13.73	10.80 4394	5
56	1130	13.35	4700	.33	6430	13.72	3570	4
57	1933	13.33	4680	.33	7253	13.68	2747	3
58	2734	13.30	4660	.33	8074	13.67	1926	2
59	3534		4640	.33	8894	13.65	1106	1
60	9.19 4332		9.99 4620	.33	9.19 9713		10.80 0287	0

′	Cosine.	D.1″.	Sine.	D.1″.	Cotang.	D.1″.	Tang.	′

98°   81°

Table II. (Continued)

9°   170°

′	Sine.	D.1″.	Cosine.	D.1″.	Tang.	D.1″.	Cotang.	′
0	9.19 4332	13.28	9.99 4620	.33	9.19 9713	13.60	10.80 0287	60
1	5129	13.27	4600	.33	.20 0529	13.60	.79 9471	59
2	5925	13.23	4580	.33	1345	13.57	8655	58
3	6719	13.20	4560	.33	2159	13.53	7841	57
4	7511	13.18	4540	.35	2971	13.52	7029	56
5	9.19 8302	13.15	9.99 4519	.33	9.20 3782	13.50	10.79 6218	55
6	9091	13.13	4499	.33	4592	13.47	5408	54
7	.19 9879	13.12	4479	.33	5400	13.45	4600	53
8	.20 0666	13.08	4459	.35	6207	13.43	3793	52
9	1451	13.05	4438	.33	7013	13.40	2987	51
10	9.20 2234	13.05	9.99 4418	.33	9.20 7817	13.37	10.79 2183	50
11	3017	13.00	4398	.35	8619	13.35	1381	49
12	3797	13.00	4377	.33	.20 9420	13.33	.79 0580	48
13	4577	12.95	4357	.35	.21 0220	13.30	.78 9780	47
14	5354	12.95	4336	.33	1018	13.28	8982	46
15	9.20 6131	12.92	9.99 4316	.35	9.21 1815	13.27	10.78 8185	45
16	6906	12.88	4295	.35	2611	13.23	7389	44
17	7679	12.88	4274	.33	3405	13.22	6595	43
18	8452	12.83	4254	.35	4198	13.18	5802	42
19	9222	12.83	4233	.35	4989	13.18	5011	41
20	9.20 9992	12.80	9.99 4212	.35	9.21 5780	13.13	10.78 4220	40
21	.21 0760	12.77	4191	.33	6568	13.13	3432	39
22	1526	12.75	4171	.35	7356	13.10	2644	38
23	2291	12.73	4150	.35	8142	13.07	1858	37
24	3055	12.72	4129	.35	8926	13.07	1074	36
25	9.21 3818	12.68	9.99 4108	.35	9.21 9710	13.03	10.78 0290	35
26	4579	12.65	4087	.35	.22 0492	13.00	.77 9508	34
27	5338	12.65	4066	.35	1272	13.00	8728	33
28	6097	12.62	4045	.35	2052	12.97	7948	32
29	6854	12.58	4024	.35	2830	12.95	7170	31
30	9.21 7609	12.57	9.99 4003	.35	9.22 3607	12.92	10.77 6393	30
31	8363	12.55	3982	.37	4382	12.90	5618	29
32	9116	12.53	3960	.35	5156	12.88	4844	28
33	.21 9868	12.50	3939	.35	5929	12.85	4071	27
34	.22 0618	12.48	3918	.35	6700	12.85	3300	26
35	9.22 1367	12.47	9.99 3897	.37	9.22 7471	12.80	10.77 2529	25
36	2115	12.43	3875	.35	8239	12.80	1761	24
37	2861	12.42	3854	.37	9007	12.77	0993	23
38	3606	12.38	3832	.35	.22 9773	12.77	.77 0227	22
39	4349	12.38	3811	.37	.23 0539	12.72	.76 9461	21
40	9.22 5092	12.35	9.99 3789	.35	9.23 1302	12.72	10.76 8698	20
41	5833	12.33	3768	.37	2065	12.68	7935	19
42	6573	12.30	3746	.35	2826	12.67	7174	18
43	7311	12.28	3725	.37	3586	12.65	6414	17
44	8048	12.27	3703	.37	4345	12.63	5655	16
45	9.22 8784	12.23	9.99 3681	.35	9.23 5103	12.60	10.76 4897	15
46	.22 9518	12.23	3660	.37	5859	12.58	4141	14
47	.23 0252	12.20	3638	.37	6614	12.57	3386	13
48	0984	12.18	3616	.37	7368	12.53	2632	12
49	1715	12.15	3594	.37	8120	12.53	1880	11
50	9.23 2444	12.13	9.99 3572	.37	9.23 8872	12.50	10.76 1128	10
51	3172	12.12	3550	.37	.23 9622	12.48	.76 0378	9
52	3899	12.10	3528	.37	.24 0371	12.45	.75 9629	8
53	4625	12.07	3506	.37	1118	12.45	8882	7
54	5349	12.07	3484	.37	1865	12.42	8135	6
55	9.23 6073	12.03	9.99 3462	.37	9.24 2610	12.40	10.75 7390	5
56	6795	12.00	3440	.37	3354	12.38	6646	4
57	7515	12.00	3418	.37	4097	12.37	5903	3
58	8235	11.97	3396	.37	4839	12.33	5161	2
59	8953	11.95	3374	.38	5579	12.33	4421	1
60	9.23 9670		9.99 3351		9.24 6319		10.75 3681	0
′	Cosine.	D.1″.	Sine.	D.1″.	Cotang.	D.1″.	Tang.	′

99°   80°

Table II. (Continued)

10° 169°

′	Sine.	D.1″.	Cosine.	D.1″.	Tang.	D.1″.	Cotang.	′
0	9.23 9670	11.93	9.99 3351	.37	9.24 6319	12.30	10.75 3681	60
1	.24 0386	11.92	3329	.37	7057	12.28	2943	59
2	1101	11.88	3307	.38	7794	12.27	2206	58
3	1814	11.87	3284	.37	8530	12.23	1470	57
4	2526	11.85	3262	.37	9264	12.23	0736	56
5	9.24 3237	11.83	9.99 3240	.38	9.24 9998	12.20	10.75 0002	55
6	3947	11.82	3217	.37	.25 0730	12.18	.74 9270	54
7	4656	11.78	3195	.38	1461	12.17	8539	53
8	5363	11.77	3172	.38	2191	12.15	7809	52
9	6069	11.77	3149	.37	2920	12.13	7080	51
10	9.24 6775	11.72	9.99 3127	.38	9.25 3648	12.10	10.74 6352	50
11	7478	11.72	3104	.38	4374	12.10	5626	49
12	8181	11.70	3081	.37	5100	12.07	4900	48
13	8883	11.67	3059	.38	5824	12.05	4176	47
14	.24 9583	11.65	3036	.38	6547	12.03	3453	46
15	9.25 0282	11.63	9.99 3013	.38	9.25 7269	12.02	10.74 2731	45
16	0980	11.62	2990	.38	7990	12.00	2010	44
17	1677	11.60	2967	.38	8710	11.98	1290	43
18	2373	11.57	2944	.38	.25 9429	11.95	.74 0571	42
19	3067	11.57	2921	.38	.26 0146	11.95	.73 9854	41
20	9.25 3761	11.53	9.99 2898	.38	9.26 0863	11.92	10.73 9137	40
21	4453	11.52	2875	.38	1578	11.90	8422	39
22	5144	11.50	2852	.38	2292	11.88	7708	38
23	5834	11.48	2829	.38	3005	11.87	6995	37
24	6523	11.47	2806	.38	3717	11.85	6283	36
25	9.25 7211	11.45	9.99 2783	.40	9.26 4428	11.83	10.73 5572	35
26	7898	11.42	2759	.38	5138	11.82	4862	34
27	8583	11.42	2736	.38	5847	11.80	4153	33
28	9268	11.38	2713	.38	6555	11.77	3445	32
29	.25 9951	11.37	2690	.40	7261	11.77	2739	31
30	9.26 0633	11.35	9.99 2666	.38	9.26 7967	11.73	10.73 2033	30
31	1314	11.33	2643	.40	8671	11.73	1329	29
32	1994	11.32	2619	.38	.26 9375	11.70	.73 0625	28
33	2673	11.30	2596	.38	.27 0077	11.70	.72 9923	27
34	3351	11.27	2572	.38	0779	11.67	9221	26
35	9.26 4027	11.27	9.99 2549	.40	9.27 1479	11.65	10.72 8521	25
36	4703	11.23	2525	.40	2178	11.63	7822	24
37	5377	11.23	2501	.40	2876	11.62	7124	23
38	6051	11.20	2478	.38	3573	11.60	6427	22
39	6723	11.20	2454	.40	4269	11.58	5731	21
40	9.26 7395	11.17	9.99 2430	.40	9.27 4964	11.57	10.72 5036	20
41	8065	11.15	2406	.40	5658	11.55	4342	19
42	8734	11.13	2382	.40	6351	11.53	3649	18
43	.26 9402	11.12	2359	.38	7043	11.52	2957	17
44	.27 0069	11.10	2335	.40	7734	11.50	2266	16
45	9.27 0735	11.08	9.99 2311	.40	9.27 8424	11.48	10.72 1576	15
46	1400	11.07	2287	.40	9113	11.47	0887	14
47	2064	11.03	2263	.40	.27 9801	11.45	.72 0199	13
48	2726	11.03	2239	.42	.28 0488	11.43	.71 9512	12
49	3388	11.02	2214	.40	1174	11.40	8826	11
50	9.27 4049	10.98	9.99 2190	.40	9.28 1858	11.40	10.71 8142	10
51	4708	10.98	2166	.40	2542	11.38	7458	9
52	5367	10.97	2142	.40	3225	11.37	6775	8
53	6025	10.93	2118	.40	3907	11.35	6093	7
54	6681	10.93	2093	.42	4588	11.33	5412	6
55	9.27 7337	10.90	9.99 2069	.42	9.28 5268	11.32	10.71 4732	5
56	7991	10.90	2044	.40	5947	11.28	4053	4
57	8645	10.87	2020	.40	6624	11.28	3376	3
58	9297	10.85	1996	.40	7301	11.28	2699	2
59	.27 9948	10.85	1971	.42	7977	11.27	2023	1
60	9.28 0599		9.99 1947	.40	9.28 8652	11.25	10.71 1348	0
′	Cosine.	D.1″.	Sine.	D.1″.	Cotang.	D.1″.	Tang.	′

100° 79°

Table II. (Continued)

11°   168°

′	Sine.	D.1″.	Cosine.	D.1″.	Tang.	D.1″.	Cotang.	′
0	9.28 0599	10.82	9.99 1947	.42	9.28 8652	11.23	10.71 1348	60
1	1248	10.82	1922	.42	9326	11.22	0674	59
2	1897	10.78	1897	.40	.28 9999	11.20	.71 0001	58
3	2544	10.77	1873	.42	.29 0671	11.18	.70 9329	57
4	3190	10.77	1848	.42	1342	11.18	8658	56
5	9.28 3836	10.73	9.99 1823	.40	9.29 2013	11.15	10.70 7987	55
6	4480	10.73	1799	.42	2682	11.13	7318	54
7	5124	10.70	1774	.42	3350	11.12	6650	53
8	5766	10.70	1749	.42	4017	11.12	5983	52
9	6408	10.67	1724	.42	4684	11.08	5316	51
10	9.28 7048	10.67	9.99 1699	.42	9.29 5349	11.07	10.70 4651	50
11	7688	10.63	1674	.42	6013	11.07	3987	49
12	8326	10.63	1649	.42	6677	11.03	3323	48
13	8964	10.60	1624	.42	7339	11.03	2661	47
14	.28 9600	10.60	1599	.42	8001	11.02	1999	46
15	9.29 0236	10.57	9.99 1574	.42	9.29 8662	11.00	10.70 1338	45
16	0870	10.57	1549	.42	9322	10.97	0678	44
17	1504	10.55	1524	.43	.29 9980	10.97	.70 0020	43
18	2137	10.52	1498	.42	.30 0638	10.95	.69 9362	42
19	2768	10.52	1473	.42	1295	10.93	8705	41
20	9.29 3399	10.50	9.99 1448	.43	9.30 1951	10.93	10.69 8049	40
21	4029	10.48	1422	.42	2607	10.90	7393	39
22	4658	10.47	1397	.42	3261	10.88	6739	38
23	5286	10.45	1372	.43	3914	10.88	6086	37
24	5913	10.43	1346	.42	4567	10.85	5433	36
25	9.29 6539	10.42	9.99 1321	.43	9.30 5218	10.85	10.69 4782	35
26	7164	10.40	1295	.42	5869	10.83	4131	34
27	7788	10.40	1270	.43	6519	10.82	3481	33
28	8412	10.37	1244	.43	7168	10.80	2832	32
29	9034	10.35	1218	.42	7816	10.78	2184	31
30	9.29 9655	10.35	9.99 1193	.43	9.30 8463	10.77	10.69 1537	30
31	.30 0276	10.32	1167	.43	9109	10.75	0891	29
32	0895	10.32	1141	.43	.30 9754	10.75	.69 0246	28
33	1514	10.30	1115	.43	.31 0399	10.72	.68 9601	27
34	2132	10.27	1090	.43	1042	10.72	8958	26
35	9.30 2748	10.27	9.99 1064	.43	9.31 1685	10.70	10.68 8315	25
36	3364	10.25	1038	.43	2327	10.68	7673	24
37	3979	10.23	1012	.43	2968	10.67	7032	23
38	4593	10.23	0986	.43	3608	10.65	6392	22
39	5207	10.20	0960	.43	4247	10.63	5753	21
40	9.30 5819	10.18	9.99 0934	.43	9.31 4885	10.63	10.68 5115	20
41	6430	10.18	0908	.43	5523	10.60	4477	19
42	7041	10.15	0882	.43	6159	10.60	3841	18
43	7650	10.15	0855	.45	6795	10.58	3205	17
44	8259	10.13	0829	.43	7430	10.57	2570	16
45	9.30 8867	10.12	9.99 0803	.43	9.31 8064	10.55	10.68 1936	15
46	.30 9474	10.10	0777	.43	8697	10.55	1303	14
47	.31 0080	10.08	0750	.43	9330	10.52	0670	13
48	0685	10.07	0724	.45	.31 9961	10.52	.68 0039	12
49	1289	10.07	0697	.43	.32 0592	10.50	.67 9408	11
50	9.31 1893	10.03	9.99 0671	.43	9.32 1222	10.48	10.67 8778	10
51	2495	10.03	0645	.45	1851	10.47	8149	9
52	3097	10.02	0618	.45	2479	10.45	7521	8
53	3698	9.98	0591	.43	3106	10.45	6894	7
54	4297	10.00	0565	.45	3733	10.42	6267	6
55	9.31 4897	9.97	9.99 0538	.45	9.32 4358	10.42	10.67 5642	5
56	5495	9.95	0511	.43	4983	10.40	5017	4
57	6092	9.95	0485	.45	5607	10.40	4393	3
58	6689	9.92	0458	.45	6231	10.37	3769	2
59	7284	9.92	0431	.45	6853	10.37	3147	1
60	9.31 7879		9.99 0404		9.32 7475		10.67 2525	0
′	Cosine.	D.1″.	Sine.	D.1″.	Cotang.	D.1″.	Tang.	′

101°   78°

Table II. (Continued)

12°   167°

′	Sine.	D.1″.	Cosine.	D.1″.	Tang.	D.1″.	Cotang.	′
0	9.31 7879	9.90	9.99 0404	.43	9.32 7475	10.33	10.67 2525	60
1	8473	9.88	0378	.45	8095	10.33	1905	59
2	9066	9.87	0351	.45	8715	10.32	1285	58
3	.31 9658	9.85	0324	.45	9334	10.32	0666	57
4	.32 0249	9.85	0297	.45	.32 9953	10.28	.67 0047	56
5	9.32 0840	9.83	9.99 0270	.45	9.33 0570	10.28	10.66 9430	55
6	1430	9.82	0243	.47	1187	10.27	8813	54
7	2019	9.80	0215	.45	1803	10.25	8197	53
8	2607	9.78	0188	.45	2418	10.25	7582	52
9	3194	9.77	0161	.45	3033	10.22	6967	51
10	9.32 3780	9.77	9.99 0134	.45	9.33 3646	10.22	10.66 6354	50
11	4366	9.73	0107	.47	4259	10.20	5741	49
12	4950	9.73	0079	.45	4871	10.18	5129	48
13	5534	9.72	0052	.45	5482	10.18	4518	47
14	6117	9.72	.99 0025	.47	6093	10.15	3907	46
15	9.32 6700	9.68	9.98 9997	.45	9.33 6702	10.15	10.66 3298	45
16	7281	9.68	9970	.47	7311	10.13	2689	44
17	7862	9.67	9942	.45	7919	10.13	2081	43
18	8442	9.65	9915	.47	8527	10.10	1473	42
19	9021	9.63	9887	.45	9133	10.10	0867	41
20	9.32 9599	9.62	9.98 9860	.47	9.33 9739	10.08	10.66 0261	40
21	.33 0176	9.62	9832	.47	.34 0344	10.07	.65 9656	39
22	0753	9.60	9804	.45	0948	10.07	9052	38
23	1329	9.57	9777	.47	1552	10.05	8448	37
24	1903	9.58	9749	.47	2155	10.03	7845	36
25	9.33 2476	9.55	9.98 9721	.47	9.34 2757	10.02	10.65 7243	35
26	3051	9.55	9693	.47	3358	10.00	6642	34
27	3624	9.52	9665	.47	3958	10.00	6042	33
28	4195	9.53	9637	.45	4558	9.98	5442	32
29	4767	9.50	9610	.47	5157	9.97	4843	31
30	9.33 5337	9.48	9.98 9582	.48	9.34 5755	9.97	10.65 4245	30
31	5906	9.48	9553	.47	6353	9.93	3647	29
32	6475	9.47	9525	.47	6949	9.93	3051	28
33	7043	9.45	9497	.47	7545	9.93	2455	27
34	7610	9.43	9469	.47	8141	9.90	1859	26
35	9.33 8176	9.43	9.98 9441	.47	9.34 8735	9.90	10.65 1265	25
36	8742	9.42	9413	.47	9329	9.88	0671	24
37	9307	9.40	9385	.48	.34 9922	9.87	.65 0078	23
38	.33 9871	9.38	9356	.47	.35 0514	9.87	.64 9486	22
39	.34 0434	9.37	9328	.47	1106	9.85	8894	21
40	9.34 0996	9.37	9.98 9300	.48	9.35 1697	9.83	10.64 8303	20
41	1558	9.35	9271	.47	2287	9.82	7713	19
42	2119	9.33	9243	.48	2876	9.82	7124	18
43	2679	9.33	9214	.47	3465	9.80	6535	17
44	3239	9.30	9186	.48	4053	9.78	5947	16
45	9.34 3797	9.30	9.98 9157	.48	9.35 4640	9.78	10.64 5360	15
46	4355	9.28	9128	.47	5227	9.77	4773	14
47	4912	9.28	9100	.48	5813	9.75	4187	13
48	5469	9.25	9071	.48	6398	9.73	3602	12
49	6024	9.25	9042	.47	6982	9.73	3018	11
50	9.34 6579	9.25	9.98 9014	.48	9.35 7566	9.72	10.64 2434	10
51	7134	9.22	8985	.48	8149	9.70	1851	9
52	7687	9.22	8956	.48	8731	9.70	1269	8
53	8240	9.20	8927	.48	9313	9.67	0687	7
54	8792	9.18	8898	.48	.35 9893	9.68	.64 0107	6
55	9.34 9343	9.17	9.98 8869	.48	9.36 0474	9.65	10.63 9526	5
56	.34 9893	9.17	8840	.48	1053	9.65	8947	4
57	.35 0443	9.15	8811	.48	1632	9.63	8368	3
58	0992	9.13	8782	.48	2210	9.62	7790	2
59	1540	9.13	8753	.48	2787	9.62	7213	1
60	9.35 2088		9.98 8724		9.36 3364		10.63 6636	0
′	Cosine.	D.1″.	Sine.	D.1″.	Cotang.	D.1″.	Tang.	′

102°   77°

Table II. (Continued)

13° 166°

′	Sine.	D.1″.	Cosine.	D.1″.	Tang.	D.1″.	Cotang.	′
0	9.35 2088	9.12	9.98 8724	.48	9.36 3364	9.60	10.63 6636	50
1	2635	9.10	8695	.48	3940	9.58	6060	59
2	3181	9.08	8666	.48	4515	9.58	5485	58
3	3726	9.08	8636	.50	5090	9.57	4910	57
4	4271	9.07	8607	.48	5664	9.55	4336	56
5	9.35 4815	9.05	9.98 8578	.50	9.36 6237	9.55	10.63 3763	55
6	5358	9.05	8548	.48	6810	9.53	3190	54
7	5901	9.03	8519	.50	7382	9.52	2618	53
8	6443	9.02	8489	.48	7953	9.52	2047	52
9	6984	9.00	8460	.50	8524	9.50	1476	51
10	9.35 7524	9.00	9.98 8430	.48	9.36 9094	9.48	10.63 0906	50
11	8064	8.98	8401	.50	.36 9663	9.48	.63 0337	49
12	8603	8.97	8371	.48	.37 0232	9.45	.62 9768	48
13	9141	8.95	8342	.50	0799	9.47	9201	47
14	.35 9678	8.95	8312	.50	1367	9.43	8633	46
15	9.36 0215	8.95	9.98 8282	.50	9.37 1933	9.43	10.62 8067	45
16	0752	8.92	8252	.48	2499	9.42	7501	44
17	1287	8.92	8223	.50	3064	9.42	6936	43
18	1822	8.90	8193	.50	3629	9.40	6371	42
19	2356	8.88	8163	.50	4193	9.38	5807	41
20	9.36 2889	8.88	9.98 8133	.50	9.37 4756	9.38	10.62 5244	40
21	3422	8.87	8103	.50	5319	9.37	4681	39
22	3954	8.85	8073	.50	5881	9.35	4119	38
23	4485	8.85	8043	.50	6442	9.35	3558	37
24	5016	8.83	8013	.50	7003	9.33	2997	36
25	9.36 5546	8.82	9.98 7983	.50	9.37 7563	9.32	10.62 2437	35
26	6075	8.82	7953	.52	8122	9.32	1878	34
27	6604	8.78	7922	.50	8681	9.30	1319	33
28	7131	8.80	7892	.50	9239	9.30	0761	32
29	7659	8.77	7862	.50	.37 9797	9.28	.62 0203	31
30	9.36 8185	8.77	9.98 7832	.52	9.38 0354	9.27	10.61 9646	30
31	8711	8.75	7801	.50	0910	9.27	9090	29
32	9236	8.75	7771	.52	1466	9.23	8534	28
33	.36 9761	8.72	7740	.52	2020	9.25	7980	27
34	.37 0285	8.72	7710	.52	2575	9.23	7425	26
35	9.37 0808	8.70	9.98 7679	.50	9.38 3129	9.22	10.61 6871	25
36	1330	8.70	7649	.52	3682	9.20	6318	24
37	1852	8.68	7618	.50	4234	9.20	5766	23
38	2373	8.68	7588	.52	4786	9.18	5214	22
39	2894	8.67	7557	.52	5337	9.18	4663	21
40	9.37 3414	8.65	9.98 7526	.50	9.38 5888	9.17	10.61 4112	20
41	3933	8.65	7496	.52	6438	9.15	3562	19
42	4452	8.63	7465	.52	6987	9.15	3013	18
43	4970	8.62	7434	.52	7536	9.13	2464	17
44	5487	8.60	7403	.52	8084	9.12	1916	16
45	9.37 6003	8.60	9.98 7372	.52	9.38 8631	9.12	10.61 1369	15
46	6519	8.60	7341	.52	9178	9.10	0822	14
47	7035	8.57	7310	.52	.38 9724	9.10	.61 0276	13
48	7549	8.57	7279	.52	.39 0270	9.08	.60 9730	12
49	8063	8.57	7248	.52	0815	9.08	9185	11
50	9.37 8577	8.53	9.98 7217	.52	9.39 1360	9.05	10.60 8640	10
51	9089	8.53	7186	.52	1903	9.07	8097	9
52	.37 9601	8.53	7155	.52	2447	9.03	7553	8
53	.38 0113	8.52	7124	.52	2989	9.03	7011	7
54	0624	8.50	7092	.53	3531	9.03	6469	6
55	9.38 1134	8.48	9.98 7061	.52	9.39 4073	9.02	10.60 5927	5
56	1643	8.48	7030	.53	4614	9.00	5386	4
57	2152	8.48	6998	.52	5154	9.00	4846	3
58	2661	8.45	6967	.52	5694	8.98	4306	2
59	3168	8.45	6936	.53	6233	8.97	3767	1
60	9.38 3675		9.98 6904		9.39 6771		10.60 3229	0
′	Cosine.	D.1″.	Sine.	D.1″.	Cotang.	D.1″.	Tang.	′

103° 76°

Table II. (Continued)

14°   165°

′	Sine.	D.1″.	Cosine.	D.1″.	Tang.	D.1″.	Cotang.	′
0	9.38 3675	8.45	9.98 6904	.52	9.39 6771	8.97	10.60 3229	60
1	4182	8.42	6873	.53	7309	8.95	2691	59
2	4687	8.42	6841	.53	7846	8.95	2154	58
3	5192	8.42	6809	.52	8383	8.93	1617	57
4	5697	8.40	6778	.53	8919	8.93	1081	56
5	9.38 6201	8.38	9.98 6746	.53	9.39 9455	8.92	10.60 0545	55
6	6704	8.38	6714	.52	.39 9990	8.90	.60 0010	54
7	7207	8.37	6683	.53	.40 0524	8.90	.59 9476	53
8	7709	8.35	6651	.53	1058	8.88	8942	52
9	8210	8.35	6619	.53	1591	8.88	8409	51
10	9.38 8711	8.33	9.98 6587	.53	9.40 2124	8.87	10.59 7876	50
11	9211	8.33	6555	.53	2656	8.85	7344	49
12	.38 9711	8.32	6523	.53	3187	8.85	6813	48
13	.39 0210	8.30	6491	.53	3718	8.85	6282	47
14	0708	8.30	6459	.53	4249	8.82	5751	46
15	9.39 1206	8.28	9.98 6427	.53	9.40 4778	8.83	10.59 5222	45
16	1703	8.27	6395	.53	5308	8.80	4692	44
17	2199	8.27	6363	.53	5836	8.80	4164	43
18	2695	8.27	6331	.53	6364	8.80	3636	42
19	3191	8.23	6299	.55	6892	8.78	3108	41
20	9.39 3685	8.23	9.98 6266	.53	9.40 7419	8.77	10.59 2581	40
21	4179	8.23	6234	.53	7945	8.77	2055	39
22	4673	8.22	6202	.55	8471	8.75	1529	38
23	5166	8.20	6169	.53	8996	8.75	1004	37
24	5658	8.20	6137	.55	.40 9521	8.73	.59 0479	36
25	9.39 6150	8.18	9.98 6104	.53	9.41 0045	8.73	10.58 9955	35
26	6641	8.18	6072	.55	0569	8.72	9431	34
27	7132	8.15	6039	.55	1092	8.72	8908	33
28	7621	8.17	6007	.55	1615	8.70	8385	32
29	8111	8.15	5974	.53	2137	8.68	7863	31
30	9.39 8600	8.13	9.98 5942	.55	9.41 2658	8.68	10.58 7342	30
31	9088	8.12	5909	.55	3179	8.67	6821	29
32	.39 9575	8.12	5876	.55	3699	8.67	6301	28
33	.40 0062	8.12	5843	.53	4219	8.65	5781	27
34	0549	8.10	5811	.55	4738	8.65	5262	26
35	9.40 1035	8.08	9.98 5778	.55	9.41 5257	8.63	10.58 4743	25
36	1520	8.08	5745	.55	5775	8.63	4225	24
37	2005	8.07	5712	.55	6293	8.62	3707	23
38	2489	8.05	5679	.55	6810	8.60	3190	22
39	2972	8.05	5646	.55	7326	8.60	2674	21
40	9.40 3455	8.05	9.98 5613	.55	9.41 7842	8.60	10.58 2158	20
41	3938	8.03	5580	.55	8358	8.58	1642	19
42	4420	8.02	5547	.55	8873	8.57	1127	18
43	4901	8.02	5514	.57	9387	8.57	0613	17
44	5382	8.00	5480	.55	.41 9901	8.57	.58 0099	16
45	9.40 5862	7.98	9.98 5447	.55	9.42 0415	8.55	10.57 9585	15
46	6341	7.98	5414	.55	0927	8.55	9073	14
47	6820	7.98	5381	.57	1440	8.53	8560	13
48	7299	7.97	5347	.55	1952	8.52	8048	12
49	7777	7.95	5314	.57	2463	8.52	7537	11
50	9.40 8254	7.95	9.98 5280	.55	9.42 2974	8.50	10.57 7026	10
51	8731	7.93	5247	.57	3484	8.48	6516	9
52	9207	7.93	5213	.55	3993	8.50	6007	8
53	.40 9682	7.92	5180	.57	4503	8.47	5197	7
54	.41 0157	7.92	5146	.55	5011	8.47	4989	6
55	9.41 0632	7.90	9.98 5113	.57	9.42 5519	8.47	10.57 4481	5
56	1106	7.88	5079	.57	6027	8.45	3973	4
57	1579	7.88	5045	.57	6534	8.45	3466	3
58	2052	7.87	5011	.55	7041	8.43	2959	2
59	2524	7.87	4978	.57	7547	8.42	2453	1
60	9.41 2996		9.98 4944		9.42 8052		10.57 1948	0
′	Cosine.	D.1″.	Sine.	D.1″.	Cotang.	D.1″.	Tang.	′

104°   75°

Table II. (Continued)

15°   164°

′	Sine.	D.1″.	Cosine.	D.1″.	Tang.	D.1″.	Cotang.	′
0	9.41 2996	7.85	9.98 4944	.57	9.42 8052	8.43	10.57 1948	60
1	3467	7.85	4910	.57	8558	8.40	1442	59
2	3938	7.83	4876	.57	9062	8.40	0938	58
3	4408	7.83	4842	.57	.42 9566	8.40	.57 0434	57
4	4878	7.82	4808	.57	.43 0070	8.38	.56 9930	56
5	9.41 5347	7.80	9.98 4774	.57	9.43 0573	8.37	10.56 9427	55
6	5815	7.80	4740	.57	1075	8.37	8925	54
7	6283	7.80	4706	.57	1577	8.37	8423	53
8	6751	7.77	4672	.57	2079	8.35	7921	52
9	7217	7.78	4638	.58	2580	8.33	7420	51
10	9.41 7684	7.77	9.98 4603	.57	9.43 3080	8.33	10.56 6920	50
11	8150	7.75	4569	.57	3580	8.33	6420	49
12	8615	7.73	4535	.58	4080	8.32	5920	48
13	9079	7.75	4500	.57	4579	8.32	5421	47
14	.41 9544	7.72	4466	.57	5078	8.30	4922	46
15	9.42 0007	7.72	9.98 4432	.58	9.43 5576	8.28	10.56 4424	45
16	0470	7.72	4397	.57	6073	8.28	3927	44
17	0933	7.70	4363	.58	6570	8.28	3430	43
18	1395	7.70	4328	.57	7067	8.27	2933	42
19	1857	7.68	4294	.58	7563	8.27	2437	41
20	9.42 2318	7.67	9.98 4259	.58	9.43 8059	8.25	10.56 1941	40
21	2778	7.67	4224	.57	8554	8.23	1446	39
22	3238	7.65	4190	.58	9048	8.25	0952	38
23	3697	7.65	4155	.58	.43 9543	8.22	.56 0457	37
24	4156	7.65	4120	.58	.44 0036	8.22	.55 9964	36
25	9.42 4615	7.63	9.98 4085	.58	9.44 0529	8.22	10.55 9471	35
26	5073	7.62	4050	.58	1022	8.20	8978	34
27	5530	7.62	4015	.58	1514	8.20	8486	33
28	5987	7.60	3981	.57	2006	8.18	7994	32
29	6443	7.60	3946	.58	2497	8.18	7503	31
30	9.42 6899	7.58	9.98 3911	.58	9.44 2988	8.18	10.55 7012	30
31	7354	7.58	3875	.60	3479	8.15	6521	29
32	7809	7.57	3840	.58	3968	8.17	6032	28
33	8263	7.57	3805	.58	4458	8.15	5542	27
34	8717	7.55	3770	.58	4947	8.13	5053	26
35	9.42 9170	7.55	9.98 3735	.58	9.44 5435	8.13	10.55 4565	25
36	.42 9623	7.53	3700	.60	5923	8.13	4077	24
37	.43 0075	7.53	3664	.58	6411	8.12	3589	23
38	0527	7.52	3629	.58	6898	8.10	3102	22
39	0978	7.52	3594	.60	7384	8.10	2616	21
40	9.43 1429	7.50	9.98 3558	.58	9.44 7870	8.10	10.55 2130	20
41	1879	7.50	3523	.60	8356	8.08	1644	19
42	2329	7.48	3487	.58	8841	8.08	1159	18
43	2778	7.47	3452	.60	9326	8.07	0674	17
44	3226	7.48	3416	.58	.44 9810	8.07	.55 0190	16
45	9.43 3675	7.45	9.98 3381	.60	9.45 0294	8.05	10.54 9706	15
46	4122	7.45	3345	.60	0777	8.05	9223	14
47	4569	7.45	3309	.60	1260	8.05	8740	13
48	5016	7.43	3273	.58	1743	8.03	8257	12
49	5462	7.43	3238	.60	2225	8.02	7775	11
50	9.43 5908	7.42	9.98 3202	.60	9.45 2706	8.02	10.54 7294	10
51	6353	7.42	3166	.60	3187	8.02	6813	9
52	6798	7.40	3130	.60	3668	8.00	6332	8
53	7242	7.40	3094	.60	4148	8.00	5852	7
54	7686	7.38	3058	.60	4628	7.98	5372	6
55	9.43 8129	7.38	9.98 3022	.60	9.45 5107	7.98	10.54 4893	5
56	8572	7.37	2986	.60	5586	7.97	4414	4
57	9014	7.37	2950	.60	6064	7.97	3936	3
58	9456	7.35	2914	.60	6542	7.95	3458	2
59	.43 9897	7.35	2878	.60	7019	7.95	2981	1
60	9.44 0338		9.98 2842		9.45 7496		10.54 2404	0

′	Cosine.	D.1″.	Sine.	D.1″.	Cotang.	D.1″.	Tang.	′

105°   74°

Table II. (Continued)

16°   163°

′	Sine.	D.1″.	Cosine.	D.1″.	Tang.	D.1″.	Cotang.	′
0	9.44 0338	7.33	9.98 2842	.62	9.45 7496	7.95	10.54 2504	60
1	0778	7.33	2805	.60	7973	7.93	2027	59
2	1218	7.33	2769	.60	8449	7.93	1551	58
3	1658	7.30	2733	.62	8925	7.92	1075	57
4	2096	7.32	2696	.60	9400	7.92	0600	56
5	9.44 2535	7.30	9.98 2660	.60	9.45 9875	7.90	10.54 0125	55
6	2973	7.28	2624	.62	.46 0349	7.90	.53 9651	54
7	3410	7.28	2587	.60	0823	7.90	9177	53
8	3847	7.28	2551	.62	1297	7.88	8703	52
9	4284	7.27	2514	.62	1770	7.87	8230	51
10	9.44 4720	7.25	9.98 2477	.60	9.46 2242	7.88	10.53 7758	50
11	5155	7.25	2441	.62	2715	7.85	7285	49
12	5590	7.25	2404	.62	3186	7.87	6814	48
13	6025	7.23	2367	.62	3658	7.83	6342	47
14	6459	7.23	2331	.60	4128	7.85	5872	46
15	9.44 6893	7.22	9.98 2294	.62	9.46 4599	7.83	10.53 5401	45
16	7326	7.22	2257	.62	5069	7.83	4931	44
17	7759	7.20	2220	.62	5539	7.82	4461	43
18	8191	7.20	2183	.62	6008	7.82	3992	42
19	8623	7.18	2146	.62	6477	7.80	3523	41
20	9.44 9054	7.18	9.98 2109	.62	9.46 6945	7.80	10.53 3055	40
21	9485	7.17	2072	.62	7413	7.78	2587	39
22	.44 9915	7.17	2035	.62	7880	7.78	2120	38
23	.45 0345	7.17	1998	.62	8347	7.78	1653	37
24	0775	7.15	1961	.62	8814	7.77	1186	36
25	9.45 1204	7.13	9.98 1924	.63	9.46 9280	7.77	10.53 0720	35
26	1632	7.13	1886	.62	.46 9746	7.75	.53 0254	34
27	2060	7.13	1849	.62	.47 0211	7.75	.52 9789	33
28	2488	7.12	1812	.63	0676	7.75	9324	32
29	2915	7.12	1774	.62	1141	7.73	8859	31
30	9.45 3342	7.10	9.98 1737	.62	9.47 1605	7.73	10.52 8395	30
31	3768	7.10	1700	.63	2069	7.72	7931	29
32	4194	7.08	1662	.62	2532	7.72	7468	28
33	4619	7.08	1625	.63	2995	7.70	7005	27
34	5044	7.08	1587	.63	3457	7.70	6543	26
35	9.45 5469	7.07	9.98 1549	.62	9.47 3919	7.70	10.52 6081	25
36	5893	7.05	1512	.63	4381	7.68	5619	24
37	6316	7.05	1474	.63	4842	7.68	5158	23
38	6739	7.05	1436	.63	5303	7.67	4697	22
39	7162	7.05	1399	.62	5763	7.67	4237	21
40	9.45 7584	7.03	9.98 1361	.63	9.47 6223	7.67	10.52 3777	20
41	8006	7.03	1323	.63	6683	7.65	3317	19
42	8427	7.02	1285	.63	7142	7.65	2858	18
43	8848	7.02	1247	.63	7601	7.63	2399	17
44	9268	7.00	1209	.63	8059	7.63	1941	16
45	9.45 9688	7.00	9.98 1171	.63	9.47 8517	7.63	10.52 1483	15
46	.46 0108	6.98	1133	.63	8975	7.63	1025	14
47	0527	6.98	1095	.63	9432	7.62	0568	13
48	0946	6.97	1057	.63	.47 9889	7.62	.52 0111	12
49	1364	6.97	1019	.63	.48 0345	7.60	.51 9655	11
50	9.46 1782	6.95	9.98 0981	.65	9.48 0801	7.60	10.51 9199	10
51	2199	6.95	0942	.63	1257	7.60	8743	9
52	2616	6.93	0904	.63	1712	7.58	8288	8
53	3032	6.93	0866	.65	2167	7.58	7833	7
54	3448	6.93	0827	.63	2621	7.57	7379	6
55	9.46 3864	6.92	9.98 0789	.65	9.48 3075	7.57	10.51 6925	5
56	4279	6.92	0750	.65	3529	7.57	6471	4
57	4694	6.90	0712	.63	3982	7.55	6018	3
58	5108	6.90	0673	.65	4435	7.55	5565	2
59	5522	6.90	0635	.63	4887	7.53	5113	1
60	9.46 5935	6.88	9.98 0596	.65	9.48 5339	7.53	10.51 4661	0
′	Cosine.	D.1″.	Sine.	D.1″.	Cotang.	D.1″.	Tang.	′

106°   73°

Table II. (Continued)

17°   162°

′	Sine.	D.1″.	Cosine.	D.1″.	Tang.	D.1″.	Cotang.	′
0	9.46 5935	6.88	9.98 0596	.63	9.48 5339	7.53	10.51 4661	60
1	6348	6.88	0558	.65	5791	7.52	4209	59
2	6761	6.87	0519	.65	6242	7.52	3758	58
3	7173	6.87	0480	.63	6693	7.50	3307	57
4	7585	6.85	0442	.65	7143	7.50	2857	56
5	9.46 7996	6.85	9.98 0403	.65	9.48 7593	7.50	10.51 2407	55
6	8407	6.83	0364	.65	8043	7.48	1957	54
7	8817	6.83	0325	.65	8492	7.48	1508	53
8	9227	6.83	0286	.65	8941	7.48	1059	52
9	.46 9637	6.82	0247	.65	9390	7.47	0610	51
10	9.47 0046	6.82	9.98 0208	.65	9.48 9838	7.47	10.51 0162	50
11	0455	6.80	0169	.65	.49 0286	7.45	.50 9714	49
12	0863	6.80	0130	.65	0733	7.45	9267	48
13	1271	6.80	0091	.65	1180	7.45	8820	47
14	1679	6.78	0052	.67	1627	7.43	8373	46
15	9.47 2086	6.77	9.98 0012	.65	9.49 2073	7.43	10.50 7927	45
16	2492	6.77	.97 9973	.65	2519	7.43	7481	44
17	2898	6.77	9934	.65	2965	7.42	7035	43
18	3304	6.77	9895	.67	3410	7.40	6590	42
19	3710	6.75	9855	.65	3854	7.42	6146	41
20	9.47 4115	6.73	9.97 9816	.67	9.49 4299	7.40	10.50 5701	40
21	4519	6.73	9776	.65	4743	7.38	5257	39
22	4923	6.73	9737	.67	5186	7.40	4814	38
23	5327	6.72	9697	.65	5630	7.38	4370	37
24	5730	6.72	9658	.67	6073	7.37	3927	36
25	9.47 6133	6.72	9.97 9618	.65	9.49 6515	7.37	10.50 3485	35
26	6536	6.70	9579	.67	6957	7.37	3043	34
27	6938	6.70	9539	.67	7399	7.37	2601	33
28	7340	6.68	9499	.67	7841	7.35	2159	32
29	7741	6.68	9459	.65	8282	7.33	1718	31
30	9.47 8142	6.67	9.97 9420	.67	9.49 8722	7.35	10.50 1278	30
31	8542	6.67	9380	.67	9163	7.33	0837	29
32	8942	6.67	9340	.67	.49 9603	7.32	.50 0397	28
33	9342	6.65	9300	.67	.50 0042	7.32	.49 9958	27
34	.47 9741	6.65	9260	.67	0481	7.32	9519	26
35	9.48 0140	6.65	9.97 9220	.67	9.50 0920	7.32	10.49 9080	25
36	0539	6.63	9180	.67	1359	7.30	8641	24
37	0937	6.62	9140	.67	1797	7.30	8203	23
38	1334	6.62	9100	.68	2235	7.28	7765	22
39	1731	6.62	9059	.67	2672	7.28	7328	21
40	9.48 2128	6.62	9.97 9019	.67	9.50 3109	7.28	10.49 6891	20
41	2525	6.60	8979	.67	3546	7.27	6454	19
42	2921	6.58	8939	.68	3982	7.27	6018	18
43	3316	6.60	8898	.67	4418	7.25	5582	17
44	3712	6.58	8858	.68	4854		5146	16
45	9.48 4107	6.57	9.97 8817	.67	9.50 5289	7.25	10.49 4711	15
46	4501	6.57	8777	.67	5724	7.25	4276	14
47	4895	6.57	8737	.68	6159	7.23	3841	13
48	5289	6.55	8696	.68	6593	7.23	3407	12
49	5682	6.55	8655	.67	7027		2973	11
50	9.48 6075	6.53	9.97 8615	.68	9.50 7460	7.22	10.49 2540	10
51	6467	6.55	8574	.68	7893	7.22	2107	9
52	6860	6.52	8533	.67	8326	7.22	1674	8
53	7251	6.53	8493	.68	8759	7.20	1241	7
54	7643	6.52	8452	.68	9191	7.18	0809	6
55	9.48 8034	6.50	9.97 8411	.68	9.50 9622	7.20	10.49 0378	5
56	8424	6.50	8370	.68	.51 0054	7.18	.48 9946	4
57	8814	6.50	8329	.68	0485	7.18	9515	3
58	9204	6.48	8288	.68	0916	7.17	9084	2
59	9593	6.48	8247	.68	1346	7.17	8654	1
60	9.48 9982		9.97 8206		9.51 1776		10.48 8224	0
′	Cosine.	D.1″.	Sine.	D.1″.	Cotang.	D.1″.	Tang.	′

107°   72°

Table II. (Continued)

18°   161°

′	Sine.	D.1″.	Cosine.	D.1″.	Tang.	D.1″.	Cotang.	′
0	9.48 9982	6.48	9.97 8206	.68	9.51 1776	7.17	10.48 8224	60
1	.49 0371	6.47	8165	.68	2206	7.15	7794	59
2	0759	6.47	8124	.68	2635	7.15	7365	58
3	1147	6.47	8083	.68	3064	7.15	6936	57
4	1535	6.45	8042	.68	3493	7.13	6507	56
5	9.49 1922	6.43	9.97 8001	.70	9.51 3921	7.13	10.48 6079	55
6	2308	6.45	7959	.68	4349	7.13	5651	54
7	2695	6.43	7918	.68	4777	7.12	5223	53
8	3081	6.42	7877	.70	5204	7.12	4796	52
9	3466	6.42	7835	.68	5631	7.10	4369	51
10	9.49 3851	6.42	9.97 7794	.70	9.51 6057	7.12	10.48 3934	50
11	4236	6.42	7752	.68	6484	7.10	3516	49
12	4621	6.40	7711	.70	6910	7.08	3090	48
13	5005	6.38	7669	.68	7335	7.10	2665	47
14	5388	6.40	7628	.70	7761	7.08	2239	46
15	9.49 5772	6.37	9.97 7586	.70	9.51 8186	7.07	10.48 1814	45
16	6154	6.38	7544	.68	8610	7.07	1390	44
17	6537	6.37	7503	.70	9034	7.07	0966	43
18	6919	6.37	7461	.70	9458	7.07	0542	42
19	7301	6.35	7419	.70	.51 9882	7.05	.48 0118	41
20	9.49 7682	6.35	9.97 7377	.70	9.52 0305	7.05	10.47 9695	40
21	8064	6.33	7335	.70	0728	7.05	9272	39
22	8444	6.35	7293	.70	1151	7.03	8849	38
23	8825	6.32	7251	.70	1573	7.03	8427	37
24	9204	6.33	7209	.70	1995	7.03	8005	36
25	9.49 9584	6.32	9.97 7167	.70	9.52 2417	7.02	10.47 7583	35
26	.49 9963	6.32	7125	.70	2833	7.02	7162	34
27	.50 0342	6.32	7083	.70	3259	7.02	6741	33
28	0721	6.30	7041	.70	3680	7.00	6320	32
29	1099	6.28	6999	.70	4100	7.00	5900	31
30	9.50 1476	6.30	9.97 6957	.72	9.52 4520	7.00	10.47 5480	30
31	1854	6.28	6914	.70	4940	6.98	5060	29
32	2231	6.27	6872	.70	5359	6.98	4641	28
33	2607	6.28	6830	.72	5778	6.98	4222	27
34	2984	6.27	6787	.70	6197	6.97	3803	26
35	9.50 3360	6.25	9.97 6745	.72	9.52 6615	6.97	10.47 3385	25
36	3735	6.25	6702	.70	7033	6.97	2967	24
37	4110	6.25	6660	.72	7451	6.95	2549	23
38	4485	6.25	6617	.72	7868	6.95	2132	22
39	4860	6.23	6574	.70	8285	6.95	1715	21
40	9.50 5234	6.23	9.97 6532	.72	9.52 8702	6.95	10.47 1298	20
41	5608	6.22	6489	.72	9119	6.93	0881	19
42	5981	6.22	6446	.70.	9535	6.93	0465	18
43	6354	6.22	6104	.72	.52 9951	6.92	.47 0049	17
44	6727	6.20	6361	.72	.53 0366	6.92	.46 9634	16
45	9.50 7099	6.20	9.97 6318	.72	9.53 0781	6.92	10.46 9219	15
46	7471	6.20	6275	.72	1196	6.92	8804	14
47	7843	6.18	6232	.72	1611	6.90	8389	13
48	8214	6.18	6189	.72	2025	6.90	7975	12
49	8585	6.18	6146	.72	2439	6.90	7561	11
50	9.50 8956	6.17	9.97 6103	.72	9.53 2853	6.88	10.46 7147	10
51	9326	6.17	6060	.72	3266	6.88	6734	9
52	.50 9696	6.15	6017	.72	3679	6.88	6321	8
53	.51 0065	6.15	5974	.73	4092	6.87	5908	7
54	0434	6.15	5930	.72	4504	6.87	5496	6
55	9.51 0803	6.15	9.97 5887	.72	9.53 4916	6.87	10.46 5084	5
56	1172	6.13	5844	.73	5328	6.85	4672	4
57	1540	6.12	5800	.72	5739	6.85	4261	3
58	1907	6.13	5757	.72	6150	6.85	3850	2
59	2275	6.12	5714	.73	6561	6.85	3439	1
60	9.51 2642		9.97 5670		9.53 6972		10.46 3028	0

| ′ | Cosine. | D.1″. | Sine. | D.1″. | Cotang. | D.1″. | Tang. | ′ |

108°   71°

Table II. (Continued)

19°   160°

′	Sine.	D.1″.	Cosine.	D.1″.	Tang.	D.1″.	Cotang.	′
0	9.51 2642	6.12	9.97 5670	.72	9.53 6972	6.83	10.46 3028	60
1	3009	6.10	5627	.73	7382	6.83	2618	59
2	3375	6.10	5583	.73	7792	6.83	2208	58
3	3741	6.10	5539	.72	8202	6.82	1798	57
4	4107	6.08	5496	.73	8611	6.82	1389	56
5	9.51 4472	6.08	9.97 5452	.73	9.53 9020	6.82	10.46 0980	55
6	4837	6.08	5408	.72	9429	6.80	0571	54
7	5202	6.07	5365	.73	.53 9837	6.80	.46 0163	53
8	5566	6.07	5321	.73	.54 0245	6.80	.45 9755	52
9	5930	6.07	5277	.73	0653	6.80	9347	51
10	9.51 6294	6.05	9.97 5233	.73	9.54 1061	6.78	10 45 8939	50
11	6657	6.05	5189	.73	1468	6.78	8532	49
12	7020	6.03	5145	.73	1875	6.77	8125	48
13	7382	6.05	5101	.73	2281	6.78	7719	47
14	7745	6.03	5057	.73	2688	6.77	7312	46
15	9.51 8107	6.02	9.97 5013	.73	9.54 3094	6.75	10.45 6906	45
16	8468	6.02	4969	.73	3499	6.77	6501	44
17	8829	6.02	4925	.75	3905	6.75	6095	43
18	9190	6.02	4880	.73	4310	6.75	5690	42
19	9551	6.00	4836	.73	4715	6.73	5285	41
20	9.51 9911	6.00	9.97 4792	.73	9.54 5119	6.75	10.45 4881	40
21	.52 0271	6.00	4748	.75	5524	6.73	4476	39
22	0631	5.98	4703	.73	5928	6.72	4072	38
23	0990	5.98	4659	.75	6331	6.73	3669	37
24	1349	5.97	4614	.73	6735	6.72	3265	36
25	9.52 1707	5.98	9.97 4570	.75	9.54 7138	6.70	10.45 2862	35
26	2066	5.97	4525	.73	7540	6.72	2460	34
27	2424	5.95	4481	.75	7943	6.70	2057	33
28	2781	5.95	4436	.75	8345	6.70	1655	32
29	3138	5.95	4391	.73	8747	6.70	1253	31
30	9.52 3495	5.95	9.97 4347	.75	9.54 9149	6.68	10.45 0851	30
31	3852	5.93	4302	.75	9550	6.68	0450	29
32	4208	5.93	4257	.75	.54 9951	6.68	.45 0049	28
33	4564	5.93	4212	.75	.55 0352	6.67	.44 9648	27
34	4920	5.92	4167	.75	0752	6.68	9248	26
35	9.52 5275	5.92	9.97 4122	.75	9.55 1153	6.65	10.44 8847	25
36	5630	5.90	4077	.75	1552	6.67	8448	24
37	5984	5.92	4032	.75	1952	6.65	8048	23
38	6339	5.90	3987	.75	2351	6.65	7649	22
39	6693	5.88	3942	.75	2750	6.65	7250	21
40	9.52 7046	5.90	9.97 3897	.75	9.55 3149	6.65	10.44 6851	20
41	7400	5.88	3852	.75	3548	6.63	6452	19
42	7753	5.87	3807	.77	3946	6.63	6054	18
43	8105	5.88	3761	.75	4344	6.62	5656	17
44	8458	5.87	3716	.75	4741	6.63	5259	16
45	9.52 8810	5.85	9.97 3671	.77	9.55 5139	6.62	10.44 4861	15
46	9161	5.87	3625	.75	5536	6.62	4464	14
47	9513	5.85	3580	.75	5933	6.60	4067	13
48	.52 9864	5.85	3535	.77	6329	6.60	3671	12
49	.53 0215	5.83	3489	.75	6725	6.60	3275	11
50	9.53 0565	5.83	9.97 3444	.77	9.55 7121	6.60	10.44 2879	10
51	0915	5.83	3398	.77	7517	6.60	2483	9
52	1265	5.82	3352	.75	7913	6.58	2087	8
53	1614	5.82	3307	.77	8308	6.58	1692	7
54	1963	5.82	3261	.77	8703	6.57	1297	6
55	9.53 2312	5.82	9.97 3215	.77	9.55 9097	6.57	10.44 0903	5
56	2661	5.80	3169	.75	9491	6.57	0509	4
57	3009	5.80	3124	.77	.55 9885	6.57	.44 0115	3
58	3357	5.78	3078	.77	.56 0279	6.57	.43 9721	2
59	3704	5.80	3032	.77	0673	6.55	9327	1
60	9.53 4052		9.97 2986		9.56 1066		10.43 8934	0
′	Cosine.	D.1″.	Sine.	D.1″.	Cotang.	D.1″.	Tang.	′

109°   70°

Table II. (Continued)

20°   159°

′	Sine.	D.1″.	Cosine.	D.1″.	Tang.	D.1″.	Cotang.	′
0	9.53 4052	5.78	9.97 2986	.77	9.56 1066	6.55	10.43 8934	60
1	4399	5.77	2940	.77	1459	6.53	8541	59
2	4745	5.78	2894	.77	1851	6.55	8149	58
3	5092	5.77	2848	.77	2244	6.53	7756	57
4	5438	5.75	2802	.78	2636	6.53	7364	56
5	9.53 5783	5.77	9.97 2755	.77	9.56 3028	6.52	10.43 6972	55
6	6129	5.75	2709	.77	3419	6.53	6581	54
7	6474	5.73	2663	.77	3811	6.52	6189	53
8	6818	5.75	2617	.78	4202	6.52	5798	52
9	7163	5.73	2570	.77	4593	6.50	5407	51
10	9.53 7507	5.73	9.97 2524	.77	9.56 4983	6.50	10.43 5017	50
11	7851	5.72	2478	.78	5373	6.50	4627	49
12	8194	5.73	2431	.77	5763	6.50	4237	48
13	8538	5.70	2385	.78	6153	6.48	3847	47
14	8880	5.72	2338	.78	6542	6.50	3458	46
15	9.53 9223	5.70	9.97 2291	.77	9.56 6932	6.47	10.43 3068	45
16	9565	5.70	2245	.78	7320	6.48	2680	44
17	.53 9907	5.68	2198	.78	7709	6.48	2291	43
18	.54 0249	5.68	2151	.77	8098	6.47	1902	42
19	0590	5.68	2105	.78	8486	6.45	1514	41
20	9.54 0931	5.68	9.97 2058	.78	9.56 8873	6.47	10.43 1127	40
21	1272	5.68	2011	.78	9261	6.45	0739	39
22	1613	5.67	1964	.78	.56 9648	6.45	.43 0352	38
23	1953	5.67	1917	.78	.57 0035	6.45	.42 9965	37
24	2293	5.65	1870	.78	0422	6.45	9578	36
25	9.54 2632	5.65	9.97 1823	.78	9.57 0809	6.43	10.42 9191	35
26	2971	5.65	1776	.78	1195	6.43	8805	34
27	3310	5.65	1729	.78	1581	6.43	8419	33
28	3649	5.63	1682	.78	1967	6.42	8033	32
29	3987	5.63	1635	.78	2352	6.43	7648	31
30	9.54 4325	5.63	9.97 1588	.80	9.57 2738	6.42	10.42 7262	30
31	4663	5.62	1540	.78	3123	6.40	6877	29
32	5000	5.63	1493	.78	3507	6.42	6493	28
33	5338	5.60	1446	.80	3892	6.40	6108	27
34	5674	5.62	1398	.78	4276	6.40	5724	26
35	9.54 6011	5.60	9.97 1351	.80	9.57 4660	6.40	10.42 5340	25
36	6347	5.60	1303	.78	5044	6.38	4956	24
37	6683	5.60	1256	.80	5427	6.38	4573	23
38	7019	5.58	1208	.78	5810	6.38	4190	22
39	7354	5.58	1161	.80	6193	6.38	3807	21
40	9.54 7689	5.58	9.97 1113	.78	9.57 6576	6.38	10.42 3424	20
41	8024	5.58	1066	.80	6959	6.37	3041	19
42	8359	5.57	1018	.80	7341	6.37	2659	18
43	8693	5.57	0970	.80	7723	6.35	2277	17
44	9027	5.55	0922	.80	8104	6.37	1896	16
45	9.54 9360	5.55	9.97 0374	.78	9.57 8486	6.35	10.42 1514	15
46	.54 9693	5.55	0827	.80	8867	6.35	1133	14
47	.55 0026	5.55	0779	.80	9248	6.35	0752	13
48	0359	5.55	0731	.80	.57 9629	6.33	.42 0371	12
49	0692	5.53	0683	.80	.58 0009	6.33	.41 9991	11
50	9.55 1024	5.53	9.97 0635	.82	9.58 0389	6.33	10.41 9611	10
51	1356	5.52	0586	.80	0769	6.33	9231	9
52	1687	5.52	0538	.80	1149	6.32	8851	8
53	2018	5.52	0490	.80	1528	6.32	8472	7
54	2349	5.52	0442	.80	1907	6.32	8093	6
55	9.55 2680	5.50	9.97 0394	.82	9.58 2286	6.32	10.41 7714	5
56	3010	5.52	0345	.80	2665	6.32	7335	4
57	3341	5.48	0297	.80	3044	6.30	6956	3
58	3670	5.50	0249	.82	3422	6.30	6578	2
59	4000	5.48	0200	.80	3800	6.28	6200	1
60	9.55 4329		9.97 0152		9.58 4177		10.41 5823	0
′	Cosine.	D.1″.	Sine.	D.1″.	Cotang.	D.1″.	Tang.	′

110°   69°

Table II. (Continued)

21°   158°

′	Sine.	D.1″.	Cosine.	D.1″.	Tang.	D.1″.	Cotang.	′
0	9.55 4329	5.48	9.97 0152	.82	9.58 4177	6.30	10.41 5823	60
1	4658	5.48	0103	.80	4555	6.28	5445	59
2	4987	5.47	0055	.82	4932	6.28	5068	58
3	5315	5.47	.97 0006	.82	5309	6.28	4691	57
4	5643	5.47	.96 9957	.80	5686	6.27	4314	56
5	9.55 5971	5.47	9.96 9909	.82	9.58 6062	6.28	10.41 3938	55
6	6299	5.45	9860	.82	6439	6.27	3561	54
7	6626	5.45	9811	.82	6815	6.25	3185	53
8	6953	5.45	9762	.80	7190	6.27	2810	52
9	7280	5.43	9714	.82	7566	6.25	2434	51
10	9.55 7606	5.43	9.96 9665	.82	9.58 7941	6.25	10.41 2059	50
11	7932	5.43	9616	.82	8316	6.25	1684	49
12	8258	5.42	9567	.82	8691	6.25	1309	48
13	8583	5.43	9518	.82	9066	6.23	0934	47
14	8909	5.42	9469	.82	9440	6.23	0560	46
15	9.55 9234	5.40	9.96 9420	.83	9.58 9814	6.23	10.41 0186	45
16	9558	5.42	9370	.82	.59 0188	6.23	.40 9812	44
17	.55 9883	5.40	9321	.82	0562	6.22	9438	43
18	.56 0207	5.40	9272	.82	0935	6.22	9065	42
19	0531	5.40	9223	.83	1308	6.22	8692	41
20	9.56 0855	5.38	9.96 9173	.82	9.59 1681	6.22	10.40 8319	40
21	1178	5.38	9124	.82	2054	6.20	7946	39
22	1501	5.38	9075	.83	2426	6.22	7574	38
23	1824	5.37	9025	.82	2799	6.20	7201	37
24	2146	5.37	8976	.83	3171	6.18	6829	36
25	9.56 2468	5.37	9.96 8926	.82	9.59 3542	6.20	10.40 6458	35
26	2790	5.37	8877	.83	3914	6.18	6086	34
27	3112	5.35	8827	.83	4285	6.18	5715	33
28	3433	5.37	8777	.82	4656	6.18	5344	32
29	3755	5.33	8728	.83	5027	6.18	4973	31
30	9.56 4075	5.35	9.96 8678	.83	9.59 5398	6.17	10.40 4602	30
31	4396	5.33	8628	.83	5768	6.17	4232	29
32	4716	5.33	8578	.83	6138	6.17	3862	28
33	5036	5.33	8528	.82	6508	6.17	3492	27
34	5356	5.33	8479	.83	6878	6.15	3122	26
35	9.56 5676	5.32	9.96 8429	.83	9.59 7247	6.15	10.40 2753	25
36	5995	5.32	8379	.83	7616	6.15	2384	24
37	6314	5.32	8329	.85	7985	6.15	2015	23
38	6632	5.32	8278	.83	8354	6.13	1646	22
39	6951	5.30	8228	.83	8722	6.15	1278	21
40	9.56 7269	5.30	9.96 8178	.83	9.59 9091	6.13	10.40 0909	20
41	7587	5.28	8128	.83	9459	6.13	0541	19
42	7904	5.30	8078	.85	.59 9827	6.12	.40 0173	18
43	8222	5.28	8027	.83	.60 0194	6.13	.39 9806	17
44	8539	5.28	7977	.83	0562	6.12	9438	16
45	9.56 8856	5.27	9.96 7927	.85	9.60 0929	6.12	10.39 9071	15
46	9172	5.27	7876	.83	1296	6.12	8704	14
47	9488	5.27	7826	.85	1663	6.10	8337	13
48	.56 9804	5.27	7775	.83	2029	6.10	7971	12
49	.57 0120	5.25	7725	.85	2395	6.10	7605	11
50	9.57 0435	5.27	9.96 7674	.83	9.60 2761	6.10	10.39 7239	10
51	0751	5.25	7624	.85	3127	6.10	6873	9
52	1066	5.23	7573	.85	3493	6.08	6507	8
53	1380	5.25	7522	.85	3858	6.08	6142	7
54	1695	5.23	7471	.83	4223	6.08	5777	6
55	9.57 2009	5.23	9.96 7421	.85	9.60 4588	6.08	10.39 5412	5
56	2323	5.22	7370	.85	4953	6.07	5047	4
57	2636	5.23	7319	.85	5317	6.08	4683	3
58	2950	5.22	7268	.85	5682	6.07	4318	2
59	3263	5.20	7217	.85	6046	6.07	3954	1
60	9.57 3575		9.96 7166	.85	9.60 6410		10.39 3590	0
′	Cosine.	D.1″.	Sine.	D.1″.	Cotang.	D.1″.	Tang.	′

111°   68°

Table II. (Continued)

22°   157°

′	Sine.	D.1″.	Cosine.	D.1″.	Tang.	D.1″.	Cotang.	′
0	9.57 3575	5.22	9.96 7166	.85	9.60 6410	6.05	10.39 3590	60
1	3888	5.20	7115	.85	6773	6.07	3227	59
2	4200	5.20	7064	.85	7137	6.05	2863	58
3	4512	5.20	7013	.87	7500	6.05	2500	57
4	4824	5.20	6961	.85	7863	6.03	2137	56
5	9.57 5136	5.18	9.96 6910	.85	9.60 8225	6.05	10.39 1775	55
6	5447	5.18	6859	.85	8588	6.03	1412	54
7	5758	5.18	6808	.87	8950	6.03	1050	53
8	6069	5.17	6756	.85	9312	6.03	0688	52
9	6379	5.17	6705	.87	.60 9674	6.03	.39 0326	51
10	9.57 6689	5.17	9.96 6653	.85	9.61 0036	6.02	10.38 9964	50
11	6999	5.17	6602	.87	0397	6.03	9603	49
12	7309	5.15	6550	.85	0759	6.02	9241	48
13	7618	5.15	6499	.87	1120	6.00	8880	47
14	7927	5.15	6447	.87	1480	6.02	8520	46
15	9.57 8236	5.15	9.96 6395	.85	9.61 1841	6.00	10.38 8159	45
16	8545	5.13	6344	.87	2201	6.00	7799	44
17	8853	5.15	6292	.87	2561	6.00	7439	43
18	9162	5.13	6240	.87	2921	6.00	7079	42
19	9470	5.12	6188	.87	3281	6.00	6719	41
20	9.57 9777	5.13	9.96 6136	.85	9.61 3641	5.98	10.38 6359	40
21	.58 0085	5.12	6085	.87	4000	5.98	6000	39
22	0392	5.12	6033	.87	4359	5.98	5641	38
23	0699	5.10	5981	.87	4718	5.98	5282	37
24	1005	5.12	5929	.88	5077	5.97	4923	36
25	9.58 1312	5.10	9.96 5876	.87	9.61 5435	5.97	10.38 4565	35
26	1618	5.10	5824	.87	5793	5.97	4207	34
27	1924	5.08	5772	.87	6151	5.97	3849	33
28	2229	5.10	5720	.87	6509	5.97	3491	32
29	2535	5.08	5668	.88	6867	5.95	3133	31
30	9.58 2840	5.08	9.96 5615	.87	9.61 7224	5.97	10.38 2776	30
31	3145	5.07	5563	.87	7582	5.95	2418	29
32	3449	5.08	5511	.88	7939	5.93	2061	28
33	3754	5.07	5458	.87	8295	5.95	1705	27
34	4058	5.05	5406	.88	8652	5.93	1348	26
35	9.58 4361	5.07	9.96 5353	.87	9.61 9008	5.93	10.38 0992	25
36	4665	5.05	5301	.88	9364	5.93	′ 0636	24
37	4968	5.07	5248	.88	.61 9720	5.93	.38 0280	23
38	5272	5.03	5195	.87	.62 0076	5.93	.37 9924	22
39	5574	5.05	5143	.88	0432	5.92	9568	21
40	9.58 5877	5.03	9.96 5090	.88	9.62 0787	5.92	10.37 9213	20
41	6179	5.05	5037	.88	1142	5.92	8858	19
42	6482	5.02	4984	.88	1497	5.92	8503	18
43	6783	5.03	4931	.87	1852	5.92	8148	17
44	7085	5.02	4879	.88	2207	5.90	7793	16
45	9.58 7386	5.03	9.96 4826	.88	9.62 2561	5.90	10.37 7439	15
46	7688	5.02	4773	.88	2915	5.90	7085	14
47	7989	5.02	4720	.90	3269	5.90	6731	13
48	8289	5.02	4666	.88	3623	5.88	6377	12
49	8590	5.00	4613	.88	3976	5.90	6024	11
50	9.58 8890	5.00	9.96 4560	.88	9.62 4330	5.88	10.37 5670	10
51	9190	4.98	4507	.88	4683	5.88	5317	9
52	9489	5.00	4454	.90	5036	5.87	4964	8
53	.58 9789	4.98	4400	.88	5388	5.88	4612	7
54	.59 0088	4.98	4347	.88	5741	5.87	4259	6
55	9.59 0387	4.98	9.96 4294	.90	9.62 6093	5.87	10.37 3907	5
56	0686	4.97	4240	.88	6445	5.87	3555	4
57	0984	4.97	4187	.90	6797	5.87	3203	3
58	1282	4.97	4133	.88	7149	5.87	2851	2
59	1580	4.97	4080	.90	7501	5.85	2499	1
60	9.59 1878		9.96 4026		9.62 7852		10.37 2148	0
′	Cosine.	D.1″.	Sine.	D.1″.	Cotang.	D.1″.	Tang.	′

112°   67°

Table II. (Continued)

23°   156°

′	Sine.	D.1″.	Cosine.	D.1″.	Tang.	D.1″.	Cotang.	′
0	9.59 1878	4.97	9.96 4026	.90	9.62 7825	5.85	10.37 2148	60
1	2176	4.95	3972	.88	8203	5.85	1797	59
2	2473	4.95	3919	.90	8554	5.85	1446	58
3	2770	4.95	3865	.90	8905	5.83	1095	57
4	3067	4.93	3811	.90	9255	5.85	0745	56
5	9.59 3363	4.93	9.96 3757	.88	9.62 9606	5.83	10.37 0394	55
6	3659	4.93	3704	.90	.62 9956	5.83	.37 0044	54
7	3955	4.93	3650	.90	.63 0306	5.83	.36 9694	53
8	4251	4.93	3596	.90	0656	5.82	9344	52
9	4547	4.92	3542	.90	1005	5.83	8995	51
10	9.59 4842	4.92	9.96 3488	.90	9.63 1355	5.82	10.36 8645	50
11	5137	4.92	3434	.92	1704	5.82	8296	49
12	5432	4.92	3379	.90	2053	5.82	7947	48
13	5727	4.90	3325	.90	2402	5.80	7598	47
14	6021	4.90	3271	.90	2750	5.82	7250	46
15	9.59 6315	4.90	9.96 3217	.90	9.63 3099	5.80	10.36 6901	45
16	6609	4.90	3163	.92	3447	5.80	6553	44
17	6903	4.88	3108	.90	3795	5.80	6205	43
18	7196	4.90	3054	.92	4143	5.78	5857	42
19	7490	4.88	2999	.90	4490	5.80	5510	41
20	9.59 7783	4.87	9.96 2945	.92	9.63 4838	5.78	10.36 5162	40
21	8075	4.88	2890	.90	5185	5.78	4815	39
22	8368	4.87	2836	.92	5532	5.78	4468	38
23	8660	4.87	2781	.90	5879	5.78	4121	37
24	8952	4.87	2727	.92	6226	5.77	3774	36
25	9.59 9244	4.87	9.96 2672	.92	9.63 6572	5.78	10.36 3428	35
26	9536	4.85	2617	.92	6919	5.77	3081	34
27	.59 9827	4.85	2562	.90	7265	5.77	2735	33
28	.60 0118	4.85	2508	.92	7611	5.75	2389	32
29	0409	4.85	2453	.92	7956	5.77	2044	31
30	9.60 0700	4.83	0.96 2398	.92	9.63 8302	5.75	10.36 1698	30
31	0990	4.83	2343	.92	8647	5.75	1353	29
32	1280	4.83	2288	.92	8992	5.75	1008	28
33	1570	4.83	2233	.92	9337	5.75	0663	27
34	1860	4.83	2178	.92	.63 9682	5.75	.36 0318	26
35	9.60 2150	4.82	9.96 2123	.93	9.64 0027	5.73	10.35 9973	25
36	2439	4.82	2067	.92	0371	5.75	9629	24
37	2728	4.82	2012	.92	0716	5.73	9284	23
38	3017	4.80	1957	.92	1060	5.73	8940	22
39	3305	4.82	1902	.93	1404	5.72	8596	21
40	9.60 3594	4.80	0.96 1846	.92	9.64 1747	5.73	10.35 8253	20
41	3882	4.80	1791	.93	2091	5.72	7909	19
42	4170	4.78	1735	.92	2434	5.72	7566	18
43	4457	4.80	1680	.93	2777	5.72	7223	17
44	4745	4.78	1624	.92	3120	5.72	6880	16
45	9.60 5032	4.78	9.96 1569	.93	9.64 3463	5.72	10.35 6537	15
46	5319	4.78	1513	.92	3806	5.70	6194	14
47	5606	4.77	1458	.93	4148	5.70	5852	13
48	5892	4.78	1402	.93	4490	5.70	5510	12
49	6179	4.77	1346	.93	4832	5.70	5168	11
50	9.60 6465	4.77	9.96 1290	.92	9.64 5174	5.70	10.35 4826	10
51	6751	4.75	1235	.93	5516	5.68	4484	9
52	7036	4.77	1179	.93	5857	5.70	4143	8
53	7322	4.75	1123	.93	6199	5.68	3801	7
54	7607	4.75	1067	.93	6540	5.68	3460	6
55	9.60 7892	4.75	9.96 1011	.93	9.64 6881	5.68	10.35 3119	5
56	8177	4.73	0955	.93	7222	5.67	2778	4
57	8461	4.73	0899	.93	7562	5.68	2438	3
58	8745	4.73	0843	.95	7903	5.67	2097	2
59	9029	4.73	0786	.93	8243	5.67	1757	1
60	9.60 9313		9.96 0730		9.64 8583		10.35 1417	0
′	Cosine.	D.1″.	Sine.	D.1″.	Cotang.	D.1″.	Tang.	′

113°   66°

Table II. (Continued)

24°   155°

′	Sine.	D.1″.	Cosine.	D.1″.	Tang.	D.1″.	Cotang.	′
0	9.60 9313	4.73	9.96 0730	.93	9.64 8583	5.67	10.35 1417	60
1	9597	4.72	0674	.93	8923	5.67	1077	59
2	.60 9880	4.73	0618	.95	9263	5.65	0737	58
3	.61 0164	4.72	0561	.93	9602	5.67	0398	57
4	0447	4.70	0505	.95	.64 9942	5.65	.35 0058	56
5	9.61 0729	4.72	9.96 0448	.93	9.65 0281	5.65	10.34 9719	55
6	1012	4.70	0392	.95	0620	5.65	9380	54
7	1294	4.70	0335	.93	0959	5.63	9041	53
8	1576	4.70	0279	.95	1297	5.65	8703	52
9	1858	4.70	0222	.95	1636	5.63	8364	51
10	9.61 2140	4.68	9.96 0165	.93	9.65 1974	5.63	10.34 8026	50
11	2421	4.68	0109	.95	2312	5.63	7688	49
12	2702	4.68	.96 0052	.95	2650	5.63	7350	48
13	2983	4.68	.95 9995	.95	2988	5.63	7012	47
14	3264	4.68	9938	.93	3326	5.62	6674	46
15	9.61 3545	4.67	9.95 9882	.95	9.65 3663	5.62	10.34 6337	45
16	3825	4.67	9825	.95	4000	5.62	6000	44
17	4105	4.67	9768	.95	4337	5.62	5663	43
18	4385	4.67	9711	.95	4674	5.62	5326	42
19	4665	4.63	9654	.97	5011	5.62	4989	41
20	9.61 4944	4.65	9.95 9596	.95	9.65 5348	5.60	10.34 4652	40
21	5223	4.65	9539	.95	5684	5.60	4316	39
22	5502	4.65	9482	.95	6020	5.60	3980	38
23	5781	4.65	9425	.95	6356	5.60	3644	37
24	6060	4.63	9368	.97	6692	5.60	3308	36
25	9.61 6338	4.63	9.95 9310	.95	9.65 7028	5.60	10.34 2972	35
26	6616	4.63	9253	.97	7364	5.58	2636	34
27	6894	4.63	9195	.95	7699	5.58	2301	33
28	7172	4.63	9138	.97	8034	5.58	1966	32
29	7450	4.62	9080	.95	8369	5.58	1631	31
30	9.61 7727	4.62	9.95 9023	.97	9.65 8704	5.58	10.34 1296	30
31	8004	4.62	8965	.95	9039	5.57	0961	29
32	8281	4.62	8908	.97	9373	5.58	0627	28
33	8558	4.60	8850	.97	.65 9708	5.57	.34 0292	27
34	8834	4.60	8792	.97	.66 0042	5.57	.33 9958	26
35	9.61 9110	4.60	9.95 8734	.95	9.66 0376	5.57	10.33 9624	25
36	9386	4.60	8677	.97	0710	5.55	9290	24
37	9662	4.60	8619	.97	1043	5.57	8957	23
38	.61 9938	4.58	8561	.97	1377	5.55	8623	22
39	.62 0213	4.58	8503	.97	1710	5.55	8290	21
40	9.62 0488	4.58	9.95 8445	.97	9.66 2043	5.55	10.33 7957	20
41	0763	4.58	8387	.97	2376	5.55	7624	19
42	1038	4.58	8329	.97	2709	5.55	7291	18
43	1313	4.57	8271	.97	3042	5.55	6958	17
44	1587	4.57	8213	.98	3375	5.53	6625	16
45	9.62 1861	4.57	9.95 8154	.97	9.66 3707	5.53	10.33 6293	15
46	2135	4.57	8096	.97	4039	5.53	5961	14
47	2409	4.55	8038	.98	4371	5.53	5629	13
48	2682	4.57	7979	.97	4703	5.53	5297	12
49	2956	4.55	7921	.97	5035	5.52	4965	11
50	9.62 3229	4.55	9.95 7863	.98	9.66 5366	5.53	10.33 4634	10
51	3502	4.53	7804	.97	5698	5.52	4302	9
52	3774	4.55	7746	.98	6029	5.52	3971	8
53	4047	4.53	7687	.98	6360	5.52	3640	7
54	4319	4.53	7628	.97	6691	5.50	3309	6
55	9.62 4591	4.53	9.95 7570	.98	9.66 7021	5.52	10.33 2979	5
56	4863	4.53	7511	.98	7352	5.50	2648	4
57	5135	4.52	7452	.98	7682	5.52	2318	3
58	5406	4.52	7393	.97	8013	5.50	1987	2
59	5677	4.52	7335	.98	8343	5.50	1657	1
60	9.62 5948		9.95 7276		9.66 8673		10.33 1327	0
′	Cosine.	D.1″.	Sine.	D.1″.	Cotang.	D.1″.	Tang.	′

114°   65°

Table II. (Continued)

25°   **154°**

′	Sine.	D.1″.	Cosine.	D.1″.	Tang.	D.1″.	Cotang.	′
0	9.62 5948	4.52	9.95 7276	.98	9.66 8673	5.48	10.33 1327	60
1	6219	4.52	7217	.98	9002	5.50	0998	59
2	6490	4.50	7158	.98	9332	5.48	0668	58
3	6760	4.50	7099	.98	9661	5.50	0339	57
4	7030	4.50	7040	.98	.66 9991	5.48	.33 0009	56
5	9.62 7300	4.50	9.95 6981	1.00	9.67 0320	5.48	10.32 9680	55
6	7570	4.50	6921	.98	0649	5.47	9351	54
7	7840	4.48	6862	.98	0977	5.48	9023	53
8	8109	4.48	6803	.98	1306	5.48	8694	52
9	8378	4.48	6744	1.00	1635	5.47	8365	51
10	9.62 8647	4.48	9.95 6684	.98	9.67 1963	5.47	10.32 8037	50
11	8916	4.48	6625	.98	2291	5.47	7709	49
12	9185	4.47	6566	1.00	2619	5.47	7381	48
13	9453	4.47	6506	.98	2947	5.45	7053	47
14	9721	4.47	6447	1.00	3274	5.47	6726	46
15	9.62 9989	4.47	9.95 6387	1.00	9.67 3602	5.45	10.32 6398	45
16	.63 0257	4.45	6327	.98	3929	5.47	6071	44
17	0524	4.47	6268	1.00	4257	5.45	5743	43
18	0792	4.45	6208	1.00	4584	5.45	5416	42
19	1059	4.45	6148	.98	4911	5.43	5089	41
20	9.63 1326	4.45	9.95 6089	1.00	9.67 5237	5.45	10.32 4763	40
21	1593	4.43	6029	1.00	5564	5.43	4436	39
22	1859	4.43	5969	1.00	5890	5.45	4110	38
23	2125	4.45	5909	1.00	6217	5.43	3783	37
24	2392	4.43	5849	1.00	6543	5.43	3457	36
25	9.63 2658	4.42	9.95 5789	1.00	9.67 6869	5.42	10.32 3131	35
26	2923	4.43	5729	1.00	7194	5.43	2806	34
27	3189	4.42	5669	1.00	7520	5.43	2480	33
28	3454	4.42	5609	.98	7846	5.42	2154	32
29	3719	4.42	5548	1.00	8171	5.42	1829	31
30	9.63 3984	4.42	9.95 5483	1.00	9.67 8496	5.42	10.32 1504	30
31	4249	4.42	5428	1.00	8821	5.42	1179	29
32	4514	4.40	5368	1.02	9146	5.42	0854	28
33	4778	4.40	5307	1.00	9471	5.40	0529	27
34	5042	4.40	5247	1.02	.67 9795	5.42	.32 0205	26
35	9.63 5306	4.40	9.95 5186	1.00	9.68 0120	5.40	10.31 9880	25
36	5570	4.40	5126	1.02	0444	5.40	9556	24
37	5834	4.38	5065	1.00	0763	5.40	9232	23
38	6097	4.38	5005	1.02	1092	5.40	8908	22
39	6360	4.38	4944	1.02	1416	5.40	8584	21
40	9.63 6623	4.38	9.95 4883	1.00	9.68 1740	5.38	10.31 8260	20
41	6886	4.37	4823	1.02	2063	5.40	7937	19
42	7148	4.38	4762	1.02	2387	5.38	7613	18
43	7411	4.37	4701	1.02	2710	5.38	7290	17
44	7673	4.37	4640	1.02	3033	5.38	6967	16
45	9.63 7935	4.37	9.95 4579	1.02	9.68 3356	5.38	10.31 6644	15
46	8197	4.35	4518	1.02	3679	5.37	6321	14
47	8458	4.37	4457	1.02	4001	5.38	5999	13
48	8720	4.35	4396	1.02	4324	5.37	5676	12
49	8981	4.35	4335	1.02	4646	5.37	5354	11
50	9.63 9242	4.35	9.95 4274	1.02	9.68 4968	5.37	10.31 5032	10
51	9503	4.35	4213	1.02	5290	5.37	4710	9
52	.63 9764	4.33	4152	1.03	5612	5.37	4388	8
53	.64 0024	4.33	4090	1.02	5934	5.35	4066	7
54	0284	4.33	4029	1.02	6255	5.37	3745	6
55	9.64 0544	4.33	9.95 3963	1.03	9.68 6577	5.35	10.31 3423	5
56	0804	4.33	3906	1.02	6898	5.35	3102	4
57	1064	4.33	3845	1.03	7219	5.35	2781	3
58	1324	4.32	3783	1.03	7540	5.35	2460	2
59	1583	4.32	3722	1.03	7861	5.35	2139	1
60	9.64 1842		9.95 3660		9.68 8182		10.31 1818	0
′	Cosine.	D.1″.	Sine.	D.1″.	Cotang.	D.1″.	Tang.	′

115°   **64°**

Table II. (Continued)

26°   153°

′	Sine.	D.1″.	Cosine.	D.1″.	Tang.	D.1″.	Cotang.	′
0	9.64 1842	4.32	9.95 3660	1.02	9.68 8182	5.33	10.31 1818	60
1	2101	4.32	3599	1.03	8502	5.32	1498	59
2	2360	4.30	3537	1.03	8823	5.33	1177	58
3	2618	4.32	3475	1.03	9143	5.33	0857	57
4	2877	4.30	3413	1.02	9463	5.33	0537	56
5	9.64 3135	4.30	9.95 3352	1.03	9.68 9783	5.33	10.31 0217	55
6	3393	4.28	3290	1.03	.69 0103	5.33	.30 9897	54
7	3650	4.30	3228	1.03	0423	5.33	9577	53
8	3908	4.28	3166	1.03	0742	5.33	9258	52
9	4165	4.30	3104	1.03	1062	5.32	8938	51
10	9.64 4423	4.28	9.95 3042	1.03	9.69 1381	5.32	10.30 8619	50
11	4680	4.27	2980	1.03	1700	5.32	8300	49
12	4936	4.28	2918	1.05	2019	5.32	7981	48
13	5193	4.28	2855	1.03	2338	5.30	7662	47
14	5450	4.27	2793	1.03	2656	5.32	7344	46
15	9.64 5706	4.27	9.95 2731	1.03	9.69 2975	5.30	10.30 7025	45
16	5962	4.27	2669	1.05	3293	5.32	6707	44
17	6218	4.27	2606	1.03	3612	5.30	6388	43
18	6474	4.25	2544	1.05	3930	5.30	6070	42
19	6729	4.25	2481	1.03	4228	5.30	5752	41
20	9.64 6984	4.27	9.95 2419	1.05	9.69 4566	5.28	10.30 5434	40
21	7240	4.23	2356	1.03	4883	5.30	5117	39
22	7494	4.25	2294	1.05	5201	5.28	4799	38
23	7749	4.25	2231	1.05	5518	5.30	4482	37
24	8004	4.23	2168	1.03	5836	5.28	4164	36
25	9.64 8258	4.23	9.95 2106	1.05	9.69 6153	5.28	10.30 3847	35
26	8512	4.23	2043	1.05	6470	5.28	3530	34
27	8766	4.23	1980	1.05	6787	5.27	3213	33
28	9020	4.23	1917	1.05	7103	5.28	2897	32
29	9274	4.22	1854	1.05	7420	5.27	2580	31
30	9.64 9527	4.23	9.95 1791	1.05	9.69 7736	5.28	10.30 2264	30
31	.64 9781	4.22	1728	1.05	8053	5.27	1947	29
32	.65 0034	4.22	1665	1.05	8369	5.27	1631	28
33	0287	4.20	1602	1.05	8685	5.27	1315	27
34	0539	4.22	1539	1.05	9001	5.25	0999	26
35	9.65 0792	4.20	9.95 1476	1.07	9.69 9316	5.27	10.30 0684	25
36	1044	4.22	1412	1.05	9632	5.25	0368	24
37	1297	4.20	1349	1.05	.69 9947	5.27	.30 0053	23
38	1549	4.18	1286	1.07	.70 0263	5.25	.29 9737	22
39	1800	4.20	1222	1.05	0578	5.25	9422	21
40	9.65 2052	4.20	9.95 1159	1.05	9.70 0893	5.25	10.29 9107	20
41	2304	4.18	1096	1.07	1208	5.25	8792	19
42	2555	4.18	1032	1.07	1523	5.23	8477	18
43	2806	4.18	0968	1.05	1837	5.25	8163	17
44	3057	4.18	0905	1.07	2152	5.23	7848	16
45	9.65 3308	4.17	9.95 0841	1.05	9.70 2466	5.25	10.29 7534	15
46	3558	4.17	0778	1.07	2781	5.23	7219	14
47	3808	4.18	0714	1.07	3095	5.23	6905	13
48	4059	4.17	0650	1.07	3409	5.22	6591	12
49	4309	4.15	0586	1.07	3722	5.23	6278	11
50	9.65 4558	4.17	9.95 0522	1.07	9.70 4036	5.23	10.29 5964	10
51	4808	4.17	0458	1.07	4350	5.22	5650	9
52	5058	4.15	0394	1.07	4663	5.22	5337	8
53	5307	4.15	0330	1.07	4976	5.23	5024	7
54	5556	4.15	0266	1.07	5290	5.22	4710	6
55	9.65 5805	4.15	9.95 0202	1.07	9.70 5603	5.22	10.29 4397	5
56	6054	4.13	0138	1.07	5916	5.20	4084	4
57	6302	4.15	0074	1.07	6228	5.22	3772	3
58	6551	4.13	.95 0010	1.08	6541	5.22	3459	2
59	6799	4.13	.94 9945	1.07	6854	5.20	3146	1
60	9.65 7047		9.94 9881		9.70 7166		10.29 2834	0

′	Cosine.	D.1″.	Sine.	D.1″.	Cotang.	D.1″.	Tang.	′

116°   63°

Table II. (Continued)

27°   152°

′	Sine.	D.1″.	Cosine.	D.1″.	Tang.	D.1″.	Cotang.	′
0	9.65 7047	4.13	9.94 9881	1.08	9.70 7166	5.20	10.29 2834	60
1	7295	4.12	9816	1.07	7478	5.20	2522	59
2	7542	4.13	9752	1.07	7790	5.20	2210	58
3	7790	4.12	9688	1.08	8102	5.20	1898	57
4	8037	4.12	9623	1.08	8414	5.20	1586	56
5	9.65 8284	4.12	9.94 9558	1.07	9.70 8726	5.18	10.29 1274	55
6	8531	4.12	9494	1.08	9037	5.20	0963	54
7	8778	4.12	9429	1.08	9349	5.18	0651	53
8	9025	4.10	9364	1.07	9660	5.18	0340	52
9	9271	4.10	9300	1.08	9971	5.18	.29 0029	51
10	9.65 9517	4.10	9.94 9235	1.08	9.71 0282	5.18	10.28 9718	50
11	.65 9763	4.10	9170	1.08	0593	5.18	9407	49
12	.66 0009	4.10	9105	1.08	0904	5.18	9096	48
13	0255	4.10	9040	1.08	1215	5.17	8785	47
14	0501	4.08	8975	1.08	1525	5.18	8475	46
15	9.66 0746	4.08	9.94 8910	1.08	9.71 1836	5.17	10.28 8164	45
16	0991	4.08	8845	1.08	2146	5.17	7854	44
17	1236	4.08	8780	1.08	2456	5.17	7544	43
18	1481	4.08	8715	1.08	2766	5.17	7234	42
19	1726	4.07	8650	1.10	3076	5.17	6924	41
20	9.66 1970	4.07	9.94 8584	1.08	9.71 3386	5.17	10.28 6614	40
21	2214	4.08	8519	1.08	3696	5.15	6304	39
22	2459	4.07	8454	1.10	4005	5.15	5995	38
23	2703	4.05	8388	1.08	4314	5.17	5686	37
24	2946	4.07	8323	1.10	4624	5.15	5376	36
25	9.66 3190	4.05	9.94 8257	1.08	9.71 4933	5.15	10.28 5067	35
26	3433	4.07	8192	1.10	5242	5.15	4758	34
27	3677	4.05	8126	1.10	5551	5.15	4449	33
28	3920	4.05	8060	1.08	5860	5.13	4140	32
29	4163	4.05	7995	1.10	6168	5.15	3832	31
30	9.66 4406	4.03	9.94 7929	1.10	9.71 6477	5.13	10.28 3523	30
31	4648	4.05	7863	1.10	6785	5.13	3215	29
32	4891	4.03	7797	1.10	7093	5.13	2907	28
33	5133	4.03	7731	1.10	7401	5.13	2599	27
34	5375	4.03	7665	1.08	7709	5.13	2291	26
35	9.66 5617	4.03	9.94 7600	1.12	9.71 8017	5.13	10.28 1983	25
36	5859	4.02	7533	1.10	8325	5.13	1675	24
37	6100	4.03	7467	1.10	8633	5.13	1367	23
38	6342	4.02	7401	1.10	8940	5.13	1060	22
39	6583	4.02	7335	1.10	9248	5.12	0752	21
40	9.66 6824	4.02	9.94 7269	1.10	9.71 9555	5.12	10.28 0445	20
41	7065	4.00	7203	1.12	.71 9862	5.12	.28 0138	19
42	7305	4.02	7136	1.10	.72 0169	5.12	.27 9831	18
43	7546	4.00	7070	1.10	0476	5.12	9524	17
44	7786	4.02	7004	1.12	0783	5.10	9217	16
45	9.66 8027	4.00	9.94 6937	1.10	9.72 1089	5.12	10.27 8911	15
46	8267	3.98	6871	1.12	1396	5.10	8604	14
47	8506	4.00	6804	1.10	1702	5.12	8298	13
48	8746	4.00	6738	1.12	2009	5.10	7991	12
49	8986	4.02	6671	1.12	2315	5.10	7685	11
50	9.66 9225	3.98	9.94 6604	1.10	9.72 2621	5.10	10.27 7379	10
51	9464	3.98	6538	1.12	2927	5.08	7073	9
52	9703	3.98	6471	1.12	3232	5.10	6768	8
53	.66 9942	3.98	6404	1.12	3538	5.10	6462	7
54	.67 0181	3.97	6337	1.12	3844	5.08	6156	6
55	9.67 0419	3.98	9.94 6270	1.12	9.72 4149	5.08	10.27 5851	5
56	0658	3.97	6203	1.12	4454	5.10	5546	4
57	0896	3.97	6136	1.12	4760	5.08	5240	3
58	1134	3.97	6069	1.12	5065	5.08	4935	2
59	1372	3.95	6002	1.12	5370	5.07	4630	1
60	9.67 1609		9.94 5935		9.72 5674		10.27 4326	0
′	Cosine.	D.1″.	Sine.	D.1″.	Cotang.	D.1″.	Tang.	′

117°   62°

Table II. (Continued)

28°   151°

′	Sine.	D.1″.	Cosine.	D.1″.	Tang.	D.1″.	Cotang.	′
0	9.67 1609	3.97	9.94 5935	1.12	9.72 5674	5.08	10.27 4326	60
1	1847	3.95	5868	1.13	5979	5.08	4021	59
2	2084	3.95	5800	1.12	6284	5.07	3716	58
3	2321	3.95	5733	1.12	6588	5.07	3412	57
4	2558	3.95	5666	1.13	6892	5.05	3108	56
5	9.67 2795	3.95	9.94 5598	1.12	9.72 7197	5.07	10.27 2803	55
6	3032	3.93	5531	1.12	7501	5.07	2499	54
7	3268	3.95	5464	1.13	7805	5.07	2195	53
8	3505	3.93	5396	1.13	8109	5.05	1891	52
9	3741	3.93	5328	1.12	8412	5.07	1588	51
10	9.67 3977	3.93	9.94 5261	1.13	9.72 8716	5.07	10.27 1284	50
11	4213	3.92	5193	1.13	9020	5.05	0980	49
12	4448	3.93	5125	1.12	9323	5.05	0677	48
13	4684	3.92	5058	1.13	9626	5.05	0374	47
14	4919	3.93	4990	1.13	.72 9929	5.07	.27 0071	46
15	9.67 5155	3.92	9.94 4922	1.13	9.73 0233	5.03	10.26 9767	45
16	5390	3.90	4854	1.13	0535	5.05	9465	44
17	5624	3.92	4786	1.13	0838	5.05	9162	43
18	5859	3.92	4718	1.13	1141	5.05	8859	42
19	6094	3.90	4650	1.13	1444	5.03	8556	41
20	9.67 6328	3.90	9.94 4582	1.13	9.73 1746	5.03	10.26 8254	40
21	6562	3.90	4514	1.13	2048	5.05	7952	39
22	6796	3.90	4446	1.15	2351	5.03	7649	38
23	7030	3.90	4377	1.13	2653	5.03	7347	37
24	7264	3.90	4309	1.13	2955	5.03	7045	36
25	9.67 7498	3.88	9.94 4241	1.15	9.73 3257	5.02	10.26 6743	35
26	7731	3.88	4172	1.13	3558	5.03	6442	34
27	7964	3.88	4104	1.13	3860	5.03	6140	33
28	8197	3.88	4036	1.15	4162	5.02	5838	32
29	8430	3.88	3967	1.13	4463	5.02	5537	31
30	9.67 8663	3.87	9.94 3899	1.15	9.73 4764	5.03	10.26 5236	30
31	8895	3.88	3830	1.15	5066	5.02	4934	29
32	9128	3.87	3761	1.13	5367	5.02	4633	28
33	9360	3.87	3693	1.15	5668	5.02	4332	27
34	9592	3.87	3624	1.15	5969	5.00	4031	26
35	9.67 9824	3.87	9.94 3555	1.15	9.73 6269	5.02	10.26 3731	25
36	.68 0056	3.87	3486	1.15	6570	5.00	3430	24
37	0288	3.85	3417	1.15	6870	5.02	3130	23
38	0519	3.85	3348	1.15	7171	5.00	2829	22
39	0750	3.87	3279	1.15	7471	5.00	2529	21
40	9.68 0982	3.85	9.94 3210	1.15	9.73 7771	5.00	10.26 2229	20
41	1213	3.83	3141	1.15	8071	5.00	1929	19
42	1443	3.85	3072	1.15	8371	5.00	1629	18
43	1674	3.85	3003	1.15	8671	5.00	1329	17
44	1905	3.83	2934	1.17	8971	5.00	1029	16
45	9.68 2135	3.83	9.94 2864	1.15	9.73 9271	4.98	10.26 0729	15
46	2365	3.83	2795	1.15	9570	5.00	0430	14
47	2595	3.83	2726	1.17	.73 9870	4.98	.26 0130	13
48	2825	3.83	2656	1.15	.74 0169	4.98	.25 9831	12
49	3055	3.82	2587	1.17	0468	4.98	9532	11
50	9.68 3284	3.83	9.94 2517	1.15	9.74 0767	4.98	10.25 9233	10
51	3514	3.82	2448	1.17	1066	4.98	8934	9
52	3743	3.82	2378	1.17	1365	4.98	8635	8
53	3972	3.82	2308	1.15	1664	4.97	8336	7
54	4201	3.82	2239	1.17	1962	4.98	8038	6
55	9.68 4430	3.80	9.94 2169	1.17	9.74 2261	4.97	10.25 7739	5
56	4658	3.82	2099	1.17	2559	4.98	7441	4
57	4887	3.80	2029	1.17	2858	4.97	7142	3
58	5115	3.80	1959	1.17	3156	4.97	6844	2
59	5343	3.80	1889	1.17	3454	4.97	6546	1
60	9.68 5571		9.94 1819		9.74 3752		10.25 6248	0
′	Cosine.	D.1″.	Sine.	D.1″.	Cotang.	D.1″.	Tang.	′

118°   61°

Table II. (Continued)

29°   150°

′	Sine.	D.1″.	Cosine.	D.1″.	Tang.	D.1″.	Cotang.	′
0	9.68 5571	3.80	9.94 1819	1.17	9.74 3752	4.97	10.25 6248	60
1	5799	3.80	1749	1.17	4050	4.97	5950	59
2	6027	3.78	1679	1.17	4348	4.95	5652	58
3	6254	3.80	1609	1.17	4645	4.97	5355	57
4	6482	3.78	1539	1.17	4943	4.95	5057	56
5	9.68 6709	3.78	9.94 1469	1.18	9.74 5240	4.97	10.25 4760	55
6	6936	3.78	1398	1.17	5538	4.95	4462	54
7	7163	3.77	1328	1.17	5835	4.95	4165	53
8	7389	3.78	1258	1.18	6132	4.95	3868	52
9	7616	3.78	1187	1.17	6429	4.95	3571	51
10	9.68 7843	3.77	9.94 1117	1.18	9.74 6726	4.95	10.25 3274	50
11	8069	3.77	1046	1.18	7023	4.93	2977	49
12	8295	3.77	0975	1.17	7319	4.95	2681	48
13	8521	3.77	0905	1.18	7616	4.95	2384	47
14	8747	3.75	0834	1.18	7913	4.93	2087	46
15	9.68 8972	3.77	9.94 0763	1.17	9.74 8209	4.93	10.25 1791	45
16	9198	3.75	0693	1.18	8505	4.93	1495	44
17	9423	3.75	0622	1.18	8801	4.93	1199	43
18	9648	3.75	0551	1.18	9097	4.93	0903	42
19	.68 9873	3.75	0480	1.18	9393	4.93	0607	41
20	9.69 0098	3.75	9.94 0409	1.18	9.74 9689	4.93	10.25 0311	40
21	0323	3.75	0338	1.18	.74 9985	4.93	.25 0015	39
22	0548	3.73	0267	1.18	.75 0281	4.92	.24 9719	38
23	0772	3.73	0196	1.18	0576	4.93	9424	37
24	0996	3.73	0125	1.18	0872	4.92	9128	36
25	9.69 1220	3.73	9.94 0054	1.20	9.75 1167	4.92	10.24 8833	35
26	1444	3.73	.93 9982	1.18	1462	4.92	8538	34
27	1668	3.73	9911	1.18	1757	4.92	8243	33
28	1892	3.72	9840	1.20	2052	4.92	7948	32
29	2115	3.73	9768	1.18	2347	4.92	7653	31
30	9.69 2339	3.72	9.93 9697	1.20	9.75 2642	4.92	10.24 7358	30
31	2562	3.72	9625	1.18	2937	4.90	7063	29
32	2785	3.72	9554	1.20	3231	4.92	6769	28
33	3008	3.72	9482	1.20	3526	4.90	6474	27
34	3231	3.70	9410	1.18	3820	4.92	6180	26
35	9.69 3453	3.72	9.93 9339	1.20	9.75 4115	4.90	10.24 5885	25
36	3676	3.70	9267	1.20	4409	4.90	5591	24
37	3898	3.70	9195	1.20	4703	4.90	5297	23
38	4120	3.70	9123	1.18	4997	4.90	5003	22
39	4342	3.70	9052	1.20	5291	4.90	4709	21
40	9.69 4564	3.70	9.93 8980	1.20	9.75 5585	4.88	10.24 4415	20
41	4786	3.68	8908	1.20	5878	4.90	4122	19
42	5007	3.70	8836	1.22	6172	4.88	3828	18
43	5229	3.68	8763	1.20	6465	4.90	3535	17
44	5450	3.68	8691	1.20	6759	4.88	3241	16
45	9.69 5671	3.68	9.93 8619	1.20	9.75 7052	4.88	10.24 2948	15
46	5892	3.68	8547	1.20	7345	4.88	2655	14
47	6113	3.68	8475	1.22	7638	4.88	2362	13
48	6334	3.67	8402	1.20	7931	4.88	2069	12
49	6554	3.68	8330	1.20	8224	4.88	1776	11
50	9.69 6775	3.67	9.93 8258	1.22	9.75 8517	4.88	10.24 1483	10
51	6995	3.67	8185	1.20	8810	4.87	1190	9
52	7215	3.67	8113	1.22	9102	4.88	0898	8
53	7435	3.65	8040	1.22	9395	4.87	0605	7
54	7654	3.67	7967	1.20	9687	4.87	0313	6
55	9.69 7874	3.65	9.93 7895	1.22	9.75 9979	4.88	10.24 0021	5
56	8094	3.65	7822	1.22	.76 0272	4.87	.23 9728	4
57	8313	3.65	7749	1.22	0564	4.87	9436	3
58	8532	3.65	7676	1.20	0856	4.87	9144	2
59	8751	3.65	7604	1.22	1148	4.85	8852	1
60	9.69 8970		9.93 7531		9.76 1439		10.23 8561	0
′	Cosine.	D.1″.	Sine.	D.1″.	Cotang.	D.1″.	Tang.	′

119°   60°

Table II. (Continued)

30°   149°

′	Sine.	D.1″.	Cosine.	D.1″.	Tang.	D.1″.	Cotang.	′
0	9.69 8970	3.65	9.93 7531	1.22	9.76 1439	4.87	10.23 8561	60
1	9189	3.63	7458	1.22	1731	4.87	8269	59
2	9407	3.65	7385	1.22	2023	4.85	7977	58
3	9626	3.63	7312	1.23	2314	4.87	7686	57
4	.69 9844	3.63	7238	1.22	2606	4.85	7394	56
5	9.70 0062	3.63	9.93 7165	1.22	9.76 2897	4.85	10.23 7103	55
6	0280	3.63	7092	1.22	3188	4.85	6812	54
7	0498	3.63	7019	1.22	3479	4.85	6521	53
8	0716	3.62	6946	1.23	3770	4.85	6230	52
9	0933	3.63	6872	1.22	4061	4.85	5939	51
10	9.70 1151	3.62	9.93 6799	1.23	9.76 4352	4.85	10.23 5648	50
11	1368	3.62	6725	1.22	4643	4.83	5357	49
12	1585	3.62	6652	1.23	4933	4.85	5067	48
13	1802	3.62	6578	1.22	5224	4.83	4776	47
14	2019	3.62	6505	1.22	5514	4.85	4486	46
15	9.70 2236	3.60	9.93 6431	1.23	9.76 5805	4.83	10.23 4195	45
16	2452	3.62	6357	1.22	6095	4.83	3905	44
17	2669	3.60	6284	1.23	6385	4.83	3615	43
18	2885	3.60	6210	1.23	6675	4.83	3325	42
19	3101	3.60	6136	1.23	6965	4.83	3035	41
20	9.70 3317	3.60	9.93 6062	1.23	9.76 7255	4.83	10.23 2745	40
21	3533	3.60	5988	1.23	7545	4.82	2455	39
22	3749	3.58	5914	1.23	7834	4.83	2166	38
23	3964	3.58	5840	1.23	8124	4.83	1876	37
24	4179	3.60	5766	1.23	8414	4.82	1586	36
25	9.70 4395	3.58	9.93 5692	1.23	9.76 8703	4.82	10.23 1297	35
26	4610	3.58	5618	1.25	8992	4.82	1008	34
27	4825	3.58	5543	1.23	9281	4.83	0719	33
28	5040	3.57	5469	1.23	9571	4.82	0429	32
29	5254	3.58	5395	1.25	.76 9860	4.80	.23 0140	31
30	9.70 5469	3.57	9.93 5320	1.23	9.77 0148	4.82	10.22 9852	30
31	5683	3.58	5246	1.25	0437	4.82	9563	29
32	5898	3.57	5171	1.25	0726	4.82	9274	28
33	6112	3.57	5097	1.25	1015	4.80	8985	27
34	6326	3.55	5022	1.23	1303	4.82	8697	26
35	9.70 6539	3.57	9.93 4948	1.25	9.77 1592	4.80	10.22 8408	25
36	6753	3.57	4873	1.25	1880	4.80	8120	24
37	6967	3.55	4798	1.25	2168	4.82	7832	23
38	7180	3.55	4723	1.23	2457	4.80	7543	22
39	7393	3.55	4649	1.25	2745	4.80	7255	21
40	9.70 7606	3.55	9.93 4574	1.25	9.77 3033	4.80	10.22 6967	20
41	7819	3.55	4499	1.25	3321	4.78	6679	19
42	8032	3.55	4424	1.25	3608	4.80	6392	18
43	8245	3.55	4349	1.25	3896	4.80	6104	17
44	8458	3.53	4274	1.25	4184	4.78	5816	16
45	9.70 8670	3.53	9.93 4199	1.27	9.77 4471	4.80	10.22 5529	15
46	8882	3.53	4123	1.25	4759	4.78	5241	14
47	9094	3.53	4048	1.25	5046	4.78	4954	13
48	9306	3.53	3973	1.25	5333	4.80	4667	12
49	9518	3.53	3898	1.27	5621	4.78	4379	11
50	9.70 9730	3.52	9.93 3822	1.25	9.77 5908	4.78	10.22 4092	10
51	.70 9941	3.53	3747	1.27	6195	4.78	3805	9
52	.71 0153	3.52	3671	1.25	6482	4.77	3518	8
53	0364	3.52	3596	1.27	6768	4.78	3232	7
54	0575	3.52	3520	1.25	7055	4.78	2945	6
55	9.71 0786	3.52	9.93 3445	1.27	9.77 7342	4.77	10.22 2658	5
56	0997	3.52	3369	1.27	7628	4.78	2372	4
57	1208	3.52	3293	1.27	7915	4.77	2085	3
58	1419	3.50	3217	1.27	8201	4.78	1799	2
59	1629	3.50	3141	1.25	8488	4.77	1512	1
60	9.71 1839		9.93 3066		9.77 8774		10.22 1226	0
′	Cosine.	D.1″.	Sine.	D.1″.	Cotang.	D.1″.	Tang.	′

120°   59°

Table II. (Continued)

31°   148°

′	Sine.	D.1″.	Cosine.	D.1″.	Tang.	D.1″.	Cotang.	′
0	9.71 1839	3.52	9.93 3066	1.27	9.77 8774	4.77	10.22 1226	60
1	2050	3.50	2990	1.27	9060	4.77	0940	59
2	2260	3.48	2914	1.27	9346	4.77	0654	58
3	2469	3.50	2838	1.27	9632	4.77	0368	57
4	2679	3.50	2762	1.28	.77 9918	4.75	.22 0082	56
5	9.71 2889	3.48	9.93 2685	1.27	9.78 0203	4.77	10.21 9797	55
6	3098	3.50	2609	1.27	0489	4.77	9511	54
7	3298	3.48	2533	1.27	0775	4.75	9225	53
8	3517	3.48	2457	1.28	1060	4.77	8940	52
9	3726	3.48	2380	1.27	1346	4.75	8654	51
10	9.71 3935	3.48	9.93 2304	1.27	9.78 1631	4.75	10.21 8369	50
11	4144	3.47	2228	1.28	1916	4.75	8084	49
12	4352	3.48	2151	1.27	2201	4.75	7799	48
13	4561	3.47	2075	1.28	2486	4.75	7514	47
14	4769	3.48	1998	1.28	2771	4.75	7229	46
15	9.71 4978	3.47	9.93 1921	1.27	9.78 3056	4.75	10.21 6944	45
16	5186	3.47	1845	1.28	3341	4.75	6659	44
17	5394	3.47	1768	1.28	3626	4.73	6374	43
18	5602	3.45	1691	1.28	3910	4.75	6090	42
19	5809	3.47	1614	1.28	4195	4.73	5805	41
20	9.71 6017	3.45	9.93 1537	1.28	9.78 4479	4.75	10.21 5521	40
21	6224	3.47	1460	1.28	4764	4.73	5236	39
22	6432	3.45	1383	1.28	5048	4.73	4952	38
23	6639	3.45	1306	1.28	5332	4.73	4668	37
24	6846	3.45	1229	1.28	5616	4.73	4384	36
25	9.71 7053	3.43	9.93 1152	1.28	9.78 5900	4.73	10.21 4100	35
26	7259	3.45	1075	1.28	6184	4.73	3816	34
27	7466	3.45	0998	1.28	6468	4.73	3532	33
28	7673	3.43	0921	1.30	6752	4.73	3248	32
29	7879	3.43	0843	1.28	7036	4.72	2964	31
30	9.71 8085	3.43	9.93 0766	1.30	9.78 7319	4.73	10.21 2681	30
31	8291	3.43	0688	1.28	7603	4.72	2397	29
32	8497	3.43	0611	1.30	7886	4.73	2114	28
33	8703	3.43	0533	1.28	8170	4.72	1830	27
34	8909	3.42	0456	1.30	8453	4.72	1547	26
35	9.71 9114	3.43	9.93 0378	1.30	9.78 8736	4.72	10.21 1264	25
36	9320	3.42	0300	1.28	9019	4.72	0981	24
37	9525	3.42	0223	1.30	9302	4.72	0698	23
38	9730	3.42	0145	1.30	9585	4.72	0415	22
39	.71 9935	3.42	.93 0067	1.30	.78 9868	4.72	.21 0132	21
40	9.72 0140	3.42	9.92 9989	1.30	9.79 0151	4.72	10.20 9849	20
41	0345	3.42	9911	1.30	0434	4.70	9566	19
42	0549	3.40	9833	1.30	0716	4.72	9284	18
43	0754	3.40	9755	1.30	0999	4.70	9001	17
44	0958	3.40	9677	1.30	1281	4.70	8719	16
45	9.72 1162	3.40	9.92 9599	1.30	9.79 1563	4.72	10.20 8437	15
46	1366	3.40	9521	1.32	1846	4.70	8154	14
47	1570	3.40	9442	1.30	2128	4.70	7872	13
48	1774	3.40	9364	1.30	2410	4.70	7590	12
49	1978	3.38	9286	1.32	2692	4.70	7308	11
50	9.72 2181	3.40	9.92 9207	1.30	9.79 2974	4.70	10.20 7026	10
51	2385	3.38	9129	1.32	3256	4.70	6744	9
52	2588	3.38	9050	1.32	3538	4.68	6462	8
53	2791	3.38	8972	1.32	3819	4.70	6181	7
54	2994	3.38	8893	1.30	4101	4.70	5899	6
55	9.72 3197	3.38	9.92 8815	1.32	9.79 4383	4.68	10.20 5617	5
56	3400	3.38	8736	1.32	4664	4.70	5336	4
57	3603	3.37	8657	1.32	4946	4.68	5054	3
58	3805	3.37	8578	1.32	5227	4.68	4773	2
59	4007	3.38	8499	1.32	5508	4.68	4492	1
60	9.72 4210		9.92 8420		9.79 5789		10.20 4211	0
′	Cosine.	D.1″.	Sine.	D.1″.	Cotang.	D.1″.	Tang.	′

121°   58°

Table II. (Continued)

32°   **147°**

′	Sine.	D.1″.	Cosine.	D.1″.	Tang.	D.1″.	Cotang.	′
0	9.72 4210	3.37	9.92 8420	1.30	9.79 5789	4.68	10.20 4211	60
1	4412	3.37	8342	1.32	6070	4.68	3930	59
2	4614	3.37	8263	1.33	6351	4.68	3649	58
3	4816	3.35	8183	1.32	6632	4.68	3368	57
4	5017	3.37	8104	1.32	6913	4.68	3087	56
5	9.72 5219	3.35	9.92 8025	1.32	9.79 7194	4.67	10.20 2806	55
6	5420	3.35	7946	1.32	7474	4.68	2526	54
7	5622	3.35	7867	1.33	7755	4.68	2245	53
8	5823	3.35	7787	1.32	8036	4.67	1964	52
9	6024	3.35	7708	1.32	8316	4.67	1684	51
10	9.72 6225	3.35	9.92 7629	1.33	9.79 8596	4.68	10.20 1404	50
11	6426	3.33	7549	1.32	8877	4.67	1123	49
12	6626	3.35	7470	1.33	9157	4.67	0843	48
13	6827	3.33	7390	1.33	9437	4.67	0563	47
14	7027	3.35	7310	1.32	9717	4.67	0283	46
15	9.72 7223	3.33	9.92 7231	1.33	9.79 9997	4.67	10.20 0003	45
16	7428	3.33	7151	1.33	.80 0277	4.67	.19 9723	44
17	7628	3.33	7071	1.33	0557	4.65	9443	43
18	7828	3.32	6991	1.33	0836	4.67	9164	42
19	8027	3.33	6911	1.33	1116	4.67	8884	41
20	9.72 8227	3.33	9.92 6831	1.33	9.80 1396	4.65	10.19 8604	40
21	8427	3.32	6751	1.33	1675	4.67	8325	39
22	8626	3.32	6671	1.33	1955	4.65	8045	38
23	8825	3.32	6591	1.33	2234	4.65	7766	37
24	9024	3.32	6511	1.33	2513	4.65	7487	36
25	9.72 9223	3.32	9.92 6431	1.33	9.80 2792	4.67	10.19 7208	35
26	9422	3.32	6351	1.35	3072	4.65	6928	34
27	9621	3.32	6270	1.33	3351	4.65	6649	33
28	.72 9820	3.30	6190	1.33	3630	4.65	6370	32
29	.73 0018	3.32	6110	1.35	3909	4.63	6091	31
30	9.73 0217	3.30	9.92 6029	1.33	9.80 4187	4.65	10.19 5813	30
31	0415	3.30	5949	1.35	4466	4.65	5534	29
32	0613	3.30	5868	1.33	4745	4.63	5255	28
33	0811	3.30	5788	1.35	5023	4.65	4977	27
34	1009	3.28	5707	1.35	5302	4.63	4698	26
35	9.73 1206	3.30	9.92 5626	1.35	9.80 5580	4.65	10.19 4420	25
36	1404	3.30	5545	1.33	5859	4.63	4141	24
37	1602	3.28	5465	1.35	6137	4.63	3863	23
38	1799	3.28	5384	1.35	6415	4.63	3585	22
39	1996	3.28	5303	1.35	6693	4.63	3307	21
40	9.73 2193	3.28	9.92 5222	1.35	9.80 6971	4.63	10.19 3029	20
41	2390	3.28	5141	1.35	7249	4.63	2751	19
42	2587	3.28	5060	1.35	7527	4.63	2473	18
43	2784	3.27	4979	1.37	7805	4.63	2195	17
44	2980	3.28	4897	1.35	8083	4.63	1917	16
45	9.73 3177	3.27	9.92 4816	1.35	9.80 8361	4.62	10.19 1639	15
46	3373	3.27	4735	1.35	8638	4.63	1362	14
47	3569	3.27	4654	1.37	8916	4.62	1084	13
48	3765	3.27	4572	1.35	9193	4.63	0807	12
49	3961	3.27	4491	1.37	9471	4.62	0529	11
50	9.73 4157	3.27	9.92 4409	1.35	9.80 9748	4.62	10.19 0252	10
51	4353	3.27	4328	1.37	.81 0025	4.62	.18 9975	9
52	4549	3.25	4246	1.37	0302	4.63	9698	8
53	4744	3.25	4164	1.35	0580	4.62	9420	7
54	4939	3.27	4083	1.37	0857	4.62	9143	6
55	9.73 5135	3.25	9.92 4001	1.37	9.81 1134	4.60	10.18 8866	5
56	5330	3.25	3919	1.37	1410	4.62	8590	4
57	5525	3.23	3837	1.37	1687	4.62	8313	3
58	5719	3.25	3755	1.37	1964	4.62	8036	2
59	5914	3.25	3673	1.37	2241	4.60	7759	1
60	9.73 6109		9.92 3591		9.81 2517		10.18 7483	0
′	Cosine.	D.1″.	Sine.	D.1″.	Cotang.	D.1″.	Tang.	′

122°   **57°**

Table II. (Continued)

33°   146°

′	Sine.	D.1″.	Cosine.	D.1″.	Tang.	D.1″.	Cotang.	′
0	9.73 6109	3.23	9.92 3591	1.37	9.81 2517	4.62	10.18 7483	60
1	6303	3.25	3509	1.37	2794	4.60	7206	59
2	6498	3.23	3427	1.37	3070	4.62	6930	58
3	6692	3.23	3345	1.37	3347	4.60	6653	57
4	6886	3.23	3263	1.37	3623	4.60	6377	56
5	9.73 7080	3.23	9.92 3181	1.38	9.81 3899	4.6.	10.18 6101	55
6	7274	3.22	3098	1.37	4176	4.60	5824	54
7	7467	3.23	3016	1.38	4452	4.60	5548	53
8	7661	3.23	2933	1.37	4728	4.60	5272	52
9	7855	3.22	2851	1.38	5004	4.60	4996	51
10	9.73 8048	3.22	9.92 2768	1.37	9.81 5280	4.58	10.18 4720	50
11	8241	3.22	2686	1.38	5555	4.60	4445	49
12	8434	3.22	2603	1.38	5831	4.60	4169	48
13	8627	3.22	2520	1.37	6107	4.58	3893	47
14	8820	3.22	2438	1.38	6382	4.60	3618	46
15	9.73 9013	3.22	9.92 2355	1.38	9.81 6658	4.58	10.18 3342	45
16	9206	3.20	2272	1.38	6933	4.60	3067	44
17	9398	3.20	2189	1.38	7209	4.58	2791	43
18	9590	3.22	2106	1.38	7484	4.58	2516	42
19	9783	3.20	2023	1.38	7759	4.60	2241	41
20	9.73 9975	3.20	9.92 1940	1.38	9.81 8035	4.58	10.18 1965	40
21	.74 0167	3.20	1857	1.38	8310	4.58	1690	39
22	0359	3.18	1774	1.38	8585	4.58	1415	38
23	0550	3.20	1691	1.40	8860	4.58	1140	37
24	0742	3.20	1607	1.38	9135	4.58	0865	36
25	9.74 0934	3.18	9.92 1524	1.38	9.81 9410	4.57	10.18 0590	35
26	1125	3.18	1441	1.40	9684	4.58	0316	34
27	1316	3.20	1357	1.38	.81 9959	4.58	.18 0041	33
28	1508	3.18	1274	1.40	.82 0234	4.57	.17 9766	32
29	1699	3.17	1190	1.38	0508	4.58	9492	31
30	9.74 1889	3.18	9.92 1107	1.40	9.82 0783	4.57	10.17 9217	30
31	2080	3.18	1023	1.40	1057	4.58	8943	29
32	2271	3.18	0939	1.38	1332	4.57	8668	28
33	2462	3.17	0856	1.40	1606	4.57	8394	27
34	2652	3.17	0772	1.40	1880	4.57	8120	26
35	9.74 2842	3.18	9.92 0688	1.40	9.82 2154	4.58	10.17 7846	25
36	3033	3.17	0604	1.40	2429	4.57	7571	24
37	3223	3.17	0520	1.40	2703	4.57	7297	23
38	3413	3.15	0436	1.40	2977	4.57	7023	22
39	3602	3.17	0352	1.40	3251	4.55	6749	21
40	9.74 3792	3.17	9.92 0268	1.40	9.82 3524	4.57	10.17 6476	20
41	3982	3.15	0184	1.42	3798	4.57	6202	19
42	4171	3.17	0099	1.40	4072	4.55	5928	18
43	4361	3.15	.92 0015	1.40	4345	4.57	5655	17
44	4550	3.15	.91 9931	1.42	4619	4.57	5381	16
45	9.74 4739	3.15	9.91 9846	1.40	9.82 4893	4.55	10.17 5107	15
46	4928	3.15	9762	1.42	5166	4.55	4834	14
47	5117	3.15	9677	1.40	5439	4.57	4561	13
48	5306	3.13	9593	1.42	5713	4.55	4287	12
49	5494	3.15	9508	1.40	5986	4.55	4014	11
50	9.74 5683	3.13	9.91 9424	1.42	9.82 6254	4.55	10.17 3741	10
51	5871	3.15	9339	1.42	6532	4.55	3468	9
52	6060	3.13	9254	1.42	6805	4.55	3195	8
53	6248	3.13	9169	1.40	7078	4.55	2922	7
54	6436	3.13	9085	1.42	7351	4.55	2649	6
55	9.74 6624	3.13	9.91 9000	1.42	9.82 7624	4.55	10.17 2376	5
56	6812	3.12	8915	1.42	7897	4.55	2103	4
57	6999	3.13	8830	1.42	8170	4.53	1830	3
58	7187	3.12	8745	1.43	8442	4.55	1558	2
59	7374	3.13	8659	1.42	8715	4.53	1285	1
60	9.74 7562		9.91 8574		9.82 8987		10.17 1013	0
′	Cosine.	D.1″.	Sine.	D.1″.	Cotang.	D.1″.	Tang.	′

123°   56°

Table II. (Continued)

34°   145°

′	Sine.	D.1″	Cosine.	D.1″	Tang.	D.1″	Cotang.	′
0	9.74 7562	3.12	9.91 8574	1.42	9.82 8987	4.55	10.17 1013	60
1	7749	3.12	8489	1.42	9260	4.53	0740	59
2	7936	3.12	8404	1.43	9532	4.55	0468	58
3	8123	3.12	8318	1.42	.82 9805	4.53	.17 0195	57
4	8310	3.12	8233	1.43	.83 0077	4.53	.16 9923	56
5	9.74 8497	3.10	9.91 8147	1.42	9.83 0349	4.53	10.16 9651	55
6	8683	3.12	8062	1.43	0621	4.53	9379	54
7	8870	3.10	7976	1.42	0893	4.53	9107	53
8	9056	3.12	7891	1.43	1165	4.53	8835	52
9	9243	3.10	7805	1.43	1437	4.53	8563	51
10	9.74 9429	3.10	9.91 7719	1.42	9.83 1709	4.53	10.16 8291	50
11	9615	3.10	7634	1.43	1981	4.53	8019	49
12	9801	3.10	7548	1.43	2253	4.53	7747	48
13	.74 9987	3.08	7462	1.43	2525	4.52	7475	47
14	.75 0172	3.10	7376	1.43	2796	4.53	7204	46
15	9.75 0358	3.08	9.91 7290	1.43	9.83 3068	4.52	10.16 6932	45
16	0543	3.10	7204	1.43	3339	4.53	6661	44
17	0729	3.08	7118	1.43	3611	4.52	6389	43
18	0914	3.08	7032	1.43	3882	4.53	6118	42
19	1099	3.08	6946	1.45	4154	4.52	5846	41
20	9.75 1284	3.08	9.91 6859	1.43	9.83 4425	4.52	10.16 5575	40
21	1469	3.08	6773	1.43	4696	4.52	5304	39
22	1654	3.08	6687	1.45	4967	4.52	5033	38
23	1839	3.07	6600	1.43	5238	4.52	4762	37
24	2023	3.08	6514	1.45	5509	4.52	4491	36
25	9.75 2208	3.07	9.91 6427	1.43	9.83 5780	4.52	10.16 4220	35
26	2392	3.07	6341	1.45	6051	4.52	3949	34
27	2576	3.07	6254	1.45	6322	4.52	3678	33
28	2760	3.07	6167	1.43	6593	4.52	3407	32
29	2944	3.07	6081	1.45	6864	4.50	3136	31
30	9.75 3128	3.07	9.91 5994	1.45	9.83 7134	4.52	10.16 2866	30
31	3512	3.05	5907	1.45	7405	4.50	2595	29
32	3495	3.07	5820	1.45	7675	4.52	2325	28
33	3679	3.07	5733	1.45	7946	4.50	2054	27
34	3862	3.07	5646	1.45	8216	4.50	1784	26
35	9.75 4046	3.05	9.91 5559	1.45	9.83 8487	4.50	10.16 1513	25
36	4229	3.05	5472	1.45	8757	4.50	1243	24
37	4412	3.05	5385	1.47	9027	4.50	0973	23
38	4595	3.05	5297	1.45	9297	4.52	0703	22
39	4778	3.03	5210	1.45	9568	4.50	0432	21
40	9.75 4960	3.05	9.91 5123	1.47	9.83 9838	4.50	10.16 0162	20
41	5143	3.05	5035	1.45	.84 0108	4.50	.15 9892	19
42	5326	3.03	4948	1.47	0378	4.50	9622	18
43	5508	3.03	4860	1.45	0648	4.48	9352	17
44	5690	3.03	4773	1.47	0917	4.50	9083	16
45	9.75 5872	3.03	9.91 4685	1.45	9.84 1187	4.50	10.15 8813	15
46	6054	3.03	4598	1.47	1457	4.50	8543	14
47	6236	3.03	4510	1.47	1727	4.48	8273	13
48	6418	3.03	4422	1.47	1996	4.50	8004	12
49	6600	3.02	4334	1.47	2266	4.48	7734	11
50	9.75 6782	3.02	9.91 4246	1.47	9.84 2535	4.50	10.15 7465	10
51	6963	3.02	4158	1.47	2805	4.48	7195	9
52	7144	3.03	4070	1.47	3074	4.48	6926	8
53	7326	3.02	3982	1.47	3343	4.48	6657	7
54	7507	3.02	3894	1.47	3612	4.50	6388	6
55	9.75 7688	3.02	9.91 3806	1.47	9.84 3882	4.48	10.15 6118	5
56	7869	3.02	3718	1.47	4151	4.48	5849	4
57	8050	3.00	3630	1.48	4420	4.48	5580	3
58	8230	3.02	3541	1.47	4689	4.48	5311	2
59	8411	3.00	3453	1.47	4958	4.48	5042	1
60	9.75 8591		9.91 3365		9.84 5227		10.15 4773	0

′	Cosine.	D.1″.	Sine.	D.1″.	Cotang.	D.1″.	Tang.	′

124°   55°

Table II. (Continued)

35°   144°

′	Sine.	D.1″.	Cosine.	D.1″.	Tang.	D.1″.	Cotang.	′
0	9.75 8591	3.02	9.91 3365	1.48	9.84 5227	4.48	10 15 4773	60
1	8772	3.00	3276	1.48	5496	4.47	4504	59
2	8952	3.00	3187	1.47	5764	4.48	4236	58
3	9132	3.00	3099	1.48	6033	4.48	3967	57
4	9312	3.00	3010	1.47	6302	4.47	3698	56
5	9.75 9492	3.00	9.91 2922	1.48	9.84 6570	4.48	10.15 3430	55
6	9672	3.00	2833	1.48	6839	4.48	3161	54
7	.75 9852	2.98	2744	1.48	7108	4.47	2892	53
8	.76 0031	3.00	2655	1.48	7376	4.47	2624	52
9	0211	2.98	2566	1.48	7644	4.48	2356	51
10	9.76 0390	2.98	9.91 2477	1.48	9.84 7913	4.47	10.15 2087	50
11	0569	2.98	2388	1.48	8181	4.47	1819	49
12	0748	2.98	2299	1.48	8449	4.47	1551	48
13	0927	2.98	2210	1.48	8717	4.48	1283	47
14	1106	2.98	2121	1.50	8986	4.47	1014	46
15	9.76 1285	2.98	9.91 2031	1.48	9.84 9254	4.47	10.15 0746	45
16	1464	2.97	1942	1.48	9522	4.47	0478	44
17	1642	2.98	1853	1.50	.84 9790	4.45	.15 0210	43
18	1821	2.97	1763	1.48	.85 0057	4.47	.14 9943	42
19	1999	2.97	1674	1.50	0325	4.47	9675	41
20	9.76 2177	2.98	9.91 1584	1.48	9.85 0593	4.47	10.14 9407	40
21	2356	2.97	1495	1.50	0861	4.47	9139	39
22	2534	2.97	1405	1.50	1129	4.45	8871	38
23	2712	2.95	1315	1.48	1396	4.47	8604	37
24	2889	2.97	1226	1.50	1664	4.45	8336	36
25	9.76 3067	2.97	9.91 1136	1.50	9.85 1931	4.47	10.14 8069	35
26	3245	2.95	1046	1.50	2199	4.45	7801	34
27	3422	2.97	0956	1.50	2466	4.45	7534	33
28	3600	2.95	0866	1.50	2733	4.47	7267	32
29	3777	2.95	0776	1.50	3001	4.45	6999	31
30	9.76 3954	2.95	9.91 0686	1.50	9.85 3268	4.45	10.14 6732	30
31	4131	2.95	0596	1.50	3535	4.45	6465	29
32	4308	2.95	0506	1.52	3802	4.45	6198	28
33	4485	2.95	0415	1.50	4069	4.45	5931	27
34	4662	2.93	0325	1.50	4336	4.45	5664	26
35	9.76 4838	2.95	9.91 0235	1.52	9.85 4603	4.45	10.14 5397	25
36	5015	2.93	0144	1.50	4870	4.45	5130	24
37	5191	2.93	.91 0054	1.52	5137	4.45	4863	23
38	5367	2.95	.90 9963	1.50	5404	4.45	4596	22
39	5544	2.93	9873	1.52	5671	4.45	4329	21
40	9.76 5720	2.93	9.90 9782	1.52	9.85 5938	4.43	10.14 4062	20
41	5896	2.93	9691	1.50	6204	4.45	3796	19
42	6072	2.92	9601	1.52	6471	4.43	3529	18
43	6247	2.93	9510	1.52	6737	4.45	3263	17
44	6423	2.92	9419	1.52	7004	4.43	2996	16
45	9.76 6598	2.93	9.90 9328	1.52	9.85 7270	4.45	10.14 2730	15
46	6774	2.92	9237	1.52	7537	4.43	2463	14
47	6949	2.92	9146	1.52	7803	4.43	2197	13
48	7124	2.93	9055	1.52	8069	4.45	1931	12
49	7300	2.92	8964	1.52	8336	4.43	1664	11
50	9.76 7475	2.90	9.90 8873	1.53	9.85 8602	4.43	10.14 1398	10
51	7649	2.92	8781	1.52	8868	4.43	1132	9
52	7824	2.92	8690	1.52	9134	4.43	0866	8
53	7999	2.90	8599	1.53	9400	4.43	0600	7
54	8173	2.92	8507	1.52	9666	4.43	0334	6
55	9.76 8348	2.90	9.90 8416	1.53	9.85 9932	4.43	10.14 0068	5
56	8522	2.92	8324	1.52	.86 0198	4.43	.13 9802	4
57	8697	2.90	8233	1.53	0464	4.43	9536	3
58	8871	2.90	8141	1.53	0730	4.42	9270	2
59	9045	2.90	8049	1.52	0995	4.43	9005	1
60	9.76 9219		9.90 7958		9.86 1261		10.13 8739	0
′	Cosine.	D.1″.	Sine.	D.1″.	Cotang.	D.1″.	Tang.	′

125°   54°

Table II. (Continued)

36°   143°

′	Sine.	D.1″.	Cosine.	D.1″.	Tang.	D.1″.	Cotang.	′
0	9.76 9219	2.90	9.90 7958	1.53	9.86 1261	4.43	10.13 8739	60
1	9393	2.88	7866	1.53	1527	4.42	8473	59
2	9566	2.90	7774	1.53	1792	4.43	8208	58
3	9740	2.88	7682	1.53	2058	4.42	7942	57
4	.76 9913	2.88	7590	1.53	2323	4.43	7677	56
5	9.77 0087	2.88	9.90 7498	1.53	9.86 2589	4.42	10.13 7411	55
6	0260	2.88	7406	1.53	2854	4.42	7146	54
7	0433	2.88	7314	1.53	3119	4.43	6881	53
8	0606	2.88	7222	1.55	3385	4.42	6615	52
9	0779	2.88	7129	1.53	3650	4.42	6350	51
10	9.77 0952	2.88	9.90 7037	1.53	9.86 3915	4.42	10.13 6085	50
11	1125	2.88	6945	1.55	4180	4.42	5820	49
12	1298	2.87	6852	1.53	4445	4.42	5555	48
13	1470	2.88	6760	1.55	4710	4.42	5290	47
14	1643	2.87	6667	1.53	4975	4.42	5025	46
15	9.77 1815	2.87	9.90 6575	1.55	9.86 5240	4.42	10.13 4760	45
16	1987	2.87	6482	1.55	5505	4.42	4495	44
17	2159	2.87	6389	1.55	5770	4.42	4230	43
18	2331	2.87	6296	1.53	6035	4.42	3965	42
19	2503	2.87	6204	1.55	6300	4.40	3700	41
20	9.77 2675	2.87	9.90 6111	1.55	9.86 6564	4.42	10.13 3436	40
21	2847	2.85	6018	1.55	6829	4.42	3171	39
22	3018	2.87	5925	1.55	7094	4.40	2906	38
23	3190	2.85	5832	1.55	7358	4.42	2642	37
24	3361	2.87	5739	1.57	7623	4.40	2377	36
25	9.77 3533	2.85	9.90 5645	1.55	9.86 7887	4.42	10.13 2113	35
26	3704	2.85	5552	1.55	8152	4.40	1848	34
27	3875	2.85	5459	1.55	8416	4.40	1584	33
28	4046	2.85	5366	1.57	8680	4.42	1320	32
29	4217	2.85	5272	1.55	8945	4.40	1055	31
30	9.77 4388	2.83	9.90 5179	1.57	9.86 9209	4.40	10.13 0791	30
31	4558	2.85	5085	1.55	9473	4.40	0527	29
32	4729	2.83	4992	1.57	.86 9737	4.40	.13 0263	28
33	4899	2.85	4898	1.57	.87 0001	4.40	.12 9999	27
34	5070	2.83	4804	1.55	0265	4.40	9735	26
35	9.77 5240	2.83	9.90 4711	1.57	9.87 0529	4.40	10.12 9471	25
36	5410	2.83	4617	1.57	0793	4.40	9207	24
37	5580	2.83	4523	1.57	1057	4.40	8943	23
38	5750	2.83	4429	1.57	1321	4.40	8679	22
39	5920	2.83	4335	1.57	1585	4.40	8415	21
40	9.77 6090	2.82	9.90 4241	1.57	9.87 1849	4.38	10.12 8151	20
41	6259	2.83	4147	1.57	2112	4.40	7888	19
42	6429	2.82	4053	1.57	2376	4.40	7624	18
43	6598	2.83	3959	1.58	2640	4.38	7360	17
44	6768	2.82	3864	1.57	2903	4.40	7097	16
45	9.77 6937	2.82	9.90 3770	1.57	9.87 3167	4.38	10.12 6833	15
46	7106	2.82	3676	1.58	3430	4.40	6570	14
47	7275	2.82	3581	1.57	3694	4.38	6306	13
48	7444	2.82	3487	1.58	3957	4.38	6043	12
49	7613	2.80	3392	1.57	4220	4.40	5780	11
50	9.77 7781	2.82	9.90 3298	1.58	9.87 4484	4.38	10.12 5516	10
51	7950	2.82	3203	1.58	4747	4.38	5253	9
52	8119	2.80	3108	1.57	5010	4.38	4990	8
53	8287	2.80	3014	1.58	5273	4.40	4727	7
54	8455	2.82	2919	1.58	5537	4.38	4463	6
55	9.77 8624	2.80	9.90 2824	1.58	9.87 5800	4.38	10.12 4200	5
56	8792	2.80	2729	1.58	6063	4.38	3937	4
57	8960	2.80	2634	1.58	6326	4.38	3674	3
58	9128	2.78	2539	1.58	6589	4.38	3411	2
59	9295	2.80	2444	1.58	6852	4.37	3148	1
60	9.77 9463		9.90 2349		9.87 7114		10.12 2886	0
′	Cosine.	D.1″.	Sine.	D.1″.	Cotang.	D.1″.	Tang.	′

126°   53°

Table II. (Continued)

37°   142°

′	Sine.	D.1″.	Cosine.	D.1″.	Tang.	D.1″.	Cotang.	′
0	9.77 9463	2.80	9.90 2349	1.60	9.87 7114	4.38	10.12 2886	60
1	9631	2.78	2253	1.58	7377	4.38	2623	59
2	9798	2.80	2158	1.58	7640	4.38	2360	58
3	.77 9966	2.78	2063	1.60	7903	4.37	2097	57
4	.78 0133	2.78	1967	1.58	8165	4.38	1835	56
5	9.78 0300	2.78	9.90 1872	1.60	9.87 8428	4.38	10.12 1572	55
6	0467	2.78	1776	1.58	8691	4.37	1309	54
7	0634	2.78	1681	1.60	8953	4.38	1047	53
8	0801	2.78	1585	1.58	9216	4.37	0784	52
9	0968	2.77	1490	1.60	9478	4.38	0522	51
10	9.78 1134	2.78	9.90 1394	1.60	9.87 9741	4.37	10.12 0259	50
11	1301	2.78	1298	1.60	.88 0003	4.37	.11 9997	49
12	1468	2.77	1202	1.60	0265	4.38	9735	48
13	1634	2.77	1106	1.60	0528	4.37	9472	47
14	1800	2.77	1010	1.60	0790	4.37	9210	46
15	9.78 1966	2.77	9.90 0914	1.60	9.88 1052	4.37	10.11 8948	45
16	2132	2.77	0818	1.60	1314	4.38	8686	44
17	2298	2.77	0722	1 60	1577	4.37	8423	43
18	2464	2.77	0626	1.32	1839	4.37	8161	42
19	2630	2.77	0529	1.60	2101	4.37	7899	41
20	9.78 2796	2.75	9.90 0433	1.60	9.88 2363	4.37	10.11 7637	40
21	2961	2.77	0337	1.62	2625	4.37	7375	39
22	3127	2.75	0240	1.60	2887	4.35	7113	38
23	3292	2.77	0144	1.62	3148	4.37	6852	37
24	3458	2.75	.90 0047	1.60	3410	4.37	6590	36
25	9.78 3623	2.75	9.89 9951	1.62	9.88 3672	4.37	10.11 6328	35
26	3788	2.75	9854	1.62	3934	4.37	6066	34
27	3953	2.75	9757	1.62	4196	4.35	5804	33
28	4118	2.73	9660	1.60	4457	4.37	5543	32
29	4282	2.75	9564	1.62	4719	4.35	5281	31
30	9.78 4447	2.75	9.89 9467	1.62	9.88 4980	4.37	10.11 5020	30
31	4612	2.73	9370	1.62	5242	4.37	4758	29
32	4776	2.75	9273	1.62	5504	4.35	4496	28
33	4941	2.73	9176	1.63	5765	4.37	4235	27
34	5105	2.73	9078	1.62	6026	4.37	3974	26
35	9.78 5269	2.73	9.89 8981	1.62	9.88 6288	4.35	10.11 3712	25
36	5433	2.73	8884	1.62	6549	4.37	3451	24
37	5597	2.73	8787	1.63	6811	4.35	3189	23
38	5761	2.73	8689	1.62	7072	4.35	2928	22
39	5925	2.73	8592	1.63	7333	4.35	2667	21
40	9.78 6089	2.72	9.89 8494	1.62	9.88 7594	4.35	10.11 2406	20
41	6252	2.73	8397	1.63	7855	4.35	2145	19
42	6416	2.72	8299	1.62	8116	4.37	1884	18
43	6579	2.72	8202	1.63	8378	4.35	1622	17
44	6742	2.73	8104	1.63	8639	4.35	1361	16
45	9.78 6906	2.72	9.89 8006	1.63	9.88 8900	4.35	10.11 1100	15
46	7069	2.72	7908	1.63	9161	4.35	0839	14
47	7232	2.72	7810	1.63	9421	4.35	0579	13
48	7395	2.70	7712	1.63	9682	4.33	0318	12
49	7557	2.72	7614	1.63	.88 9943	4.35	.11 0057	11
50	9.78 7720	2.72	9.89 7516	1.63	9.89 0204	4.35	10.10 9796	10
51	7883	2.70	7418	1.63	0465	4.35	9535	9
52	8045	2.72	7320	1.63	0725	4.35	9275	8
53	8208	2.70	7222	1.65	0986	4.35	9014	7
54	8370	2.70	7123	1.63	1247	4.33	8753	6
55	9.78 8532	2.70	9.89 7025	1.65	9.89 1507	4.35	10.10 8493	5
56	8694	2.70	6926	1.63	1768	4.35	8232	4
57	8856	2.70	6828	1.65	2028	4.35	7972	3
58	9018	2.70	6729	1.63	2289	4.33	7711	2
59	9180	2.70	6631	1.65	2549	4.35	7451	1
60	9.78 9342		9.89 6532		9.89 2810		10.10 7190	0
′	Cosine.	D.1″.	Sine.	D.1″.	Cotang.	D.1″.	Tang.	′

127°   52°

Table II. (Continued)

38°   141°

′	Sine.	D.1″.	Cosine.	D.1″.	Tang.	D.1″.	Cotang.	′
0	9.78 9342	2.70	9.89 6532	1.65	9.89 2810	4.33	10.10 7190	60
1	9504	2.68	6433	1.63	3070	4.35	6930	59
2	9665	2.70	6335	1.65	3331	4.33	6669	58
3	9827	2.68	6236	1.65	3591	4.33	6409	57
4	.78 9988	2.68	6137	1.65	3851	4.33	6149	56
5	9.79 0149	2.68	9.89 6038	1.65	9.89 4111	4.35	10.10 5889	55
6	0310	2.68	5939	1.65	4372	4.33	5628	54
7	0471	2.68	5840	1.65	4632	4.33	5368	53
8	0632	2.68	5741	1.67	4892	4.33	5108	52
9	0793	2.68	5641	1.65	5152	4.33	4848	51
10	9.79 0954	2.68	9.89 5542	1.65	9.89 5412	4.33	10.10 4588	50
11	1115	2.67	5443	1.67	5672	4.33	4328	49
12	1275	2.68	5343	1.65	5932	4.33	4068	48
13	1436	2.67	5244	1.65	6192	4.33	3808	47
14	1596	2.68	5145	1.67	6452	4.33	3548	46
15	9.79 1757	2.67	9.89 5045	1.67	9.89 6712	4.32	10.10 3288	45
16	1917	2.67	4945	1.65	6971	4.33	3029	44
17	2077	2.67	4846	1.67	7231	4.33	2769	43
18	2237	2.67	4746	1.67	7491	4.33	2509	42
19	2397	2.67	4646	1.67	7751	4.32	2249	41
20	9.79 2557	2.65	9.89 4546	1.67	9.89 8010	4.33	10.10 1990	40
21	2716	2.67	4446	1.67	8270	4.33	1730	39
22	2876	2.65	4346	1.67	8530	4.32	1470	38
23	3035	2.67	4246	1.67	8789	4.32	1211	37
24	3195	2.65	4146	1.67	9049	4.33	0951	36
25	9.79 3354	2.67	9.89 4046	1.67	9.89 9308	4.33	10.10 0692	35
26	3514	2.65	3946	1.67	9568	4.32	0432	34
27	3673	2.65	3846	1.68	.89 9827	4.33	.10 0173	33
28	3832	2.65	3745	1.67	.90 0087	4.32	.09 9913	32
29	3991	2.65	3645	1.68	0346	4.32	9654	31
30	9.79 4150	2.63	9.89 3544	1.67	9.90 0605	4.32	10.09 9395	30
31	4308	2.65	3444	1.68	0864	4.33	9136	29
32	4467	2.65	3343	1.67	1124	4.32	8876	28
33	4626	2.63	3243	1.68	1383	4.32	8617	27
34	4784	2.63	3142	1.68	1642	4.32	8358	26
35	9.79 4942	2.65	9.89 3041	1.68	9.90 1901	4.32	10.09 8099	25
36	5101	2.63	2940	1.68	2160	4.33	7840	24
37	5259	2.63	2839	1.67	2420	4.32	7580	23
38	5417	2.63	2739	1.68	2679	4.32	7321	22
39	5575	2.63	2638	1.70	2938	4.32	7062	21
40	9.79 5733	2.63	9.89 2536	1.68	9.90 3197	4.32	10.09 6803	20
41	5891	2.63	2435	1.68	3456	4.30	6544	19
42	6049	2.62	2334	1.68	3714	4.32	6286	18
43	6206	2.63	2233	1.68	3973	4.32	6027	17
44	6364	2.62	2132	1.70	4232	4.32	5768	16
45	9.79 6521	2.63	9.89 2030	1.68	9.90 4491	4.32	10.09 5509	15
46	6679	2.62	1929	1.70	4750	4.30	5250	14
47	6836	2.62	1827	1.68	5008	4.32	4992	13
48	6993	2.62	1726	1.70	5267	4.32	4733	12
49	7150	2.62	1624	1.68	5526	4.32	4474	11
50	9.79 7307	2.62	9.89 1523	1.70	9.90 5785	4.30	10.09 4215	10
51	7464	2.62	1421	1.70	6043	4.32	3957	9
52	7621	2.60	1319	1.70	6302	4.32	3698	8
53	7777	2.62	1217	1.70	6560	4.30	3440	7
54	7934	2.62	1115	1.70	6819	4.30	3181	6
55	9.79 8091	2.60	9.89 1013	1.70	9.90 7077	4.32	10.09 2923	5
56	8247	2.60	0911	1.70	7336	4.30	2664	4
57	8403	2.62	0809	1.70	7594	4.32	2406	3
58	8560	2.60	0707	1.70	7853	4.32	2147	2
59	8716	2.60	0605	1.70	8111	4.30	1889	1
60	9.79 8872		9.89 0503		9.90 8369		10.09 1631	0
′	Cosine.	D.1″.	Sine.	D.1″.	Cotang.	D.1″.	Tang.	′

128°   51°

Table II. (Continued)

39°   **140°**

′	Sine.	D.1″.	Cosine.	D.1″.	Tang.	D.1″.	Cotang.	′
0	9.79 8872	2.60	9.89 0503	1.72	9.90 8369	4.32	10.09 1631	60
1	9028	2.60	0400	1.70	8628	4.30	1372	59
2	9184	2.58	0298	1.72	8886	4.30	1114	58
3	9339	2.60	0195	1.70	9144	4.30	0856	57
4	9495	2.60	.89 0093	1.72	9402	4.30	0598	56
5	9.79 9651	2.58	9.88 9990	1.70	9.90 9660	4.30	10.09 0340	55
6	9806	2.60	9888	1.72	.90 9918	4.32	.09 0082	54
7	.79 9962	2.58	9785	1.72	.91 0177	4.30	.08 9823	53
8	.80 0117	2.58	9682	1.72	0435	4.30	9565	52
9	0272	2.58	9579	1.70	0693	4.30	9307	51
10	9.80 0427	2.58	9.88 9477	1.72	9.91 0951	4.30	10.08 9049	50
11	0582	2.58	9374	1.72	1209	4.30	8791	49
12	0737	2.58	9271	1.72	1467	4.30	8533	48
13	0892	2.58	9168	1.73	1725	4.30	8275	47
14	1047	2.57	9064	1.72	1982	4.30	8018	46
15	9.80 1201	2.58	9.88 8961	1.72	9.91 2240	4.30	10.08 7760	45
16	1356	2.58	8858	1.72	2498	4.30	7502	44
17	1511	2.57	8755	1.73	2756	4.30	7244	43
18	1665	2.57	8651	1.72	3014	4.30	6986	42
19	1819	2.57	8548	1.73	3271	4.30	6729	41
20	9.80 1973	2.58	9.88 8444	1.72	9.91 3529	4.30	10.08 6471	40
21	2128	2.57	8341	1.73	3787	4.28	6213	39
22	2282	2.57	8237	1.72	4044	4.30	5956	38
23	2436	2.55	8134	1.73	4302	4.30	5698	37
24	2589	2.57	8030	1.73	4560	4.28	5440	36
25	9.80 2743	2.57	9.88 7926	1.73	9.91 4817	4.30	10.08 5183	35
26	2897	2.55	7822	1.73	5075	4.28	4925	34
27	3050	2.57	7718	1.73	5332	4.30	4668	33
28	3204	2.55	7614	1.73	5590	4.28	4410	32
29	3357	2.57	7510	1.73	5847	4.28	4153	31
30	9.80 3511	2.55	9.88 7406	1.73	9.91 6104	4.30	10.08 3896	30
31	3664	2.55	7302	1.73	6362	4.28	3638	29
32	3817	2.55	7198	1.75	6619	4.30	3381	28
33	3970	2.55	7093	1.73	6877	4.28	3123	27
34	4123	2.55	6989	1.73	7134	4.28	2866	26
35	9.80 4276	2.55	9.88 6885	1.75	9.91 7391	4.28	10.08 2609	25
36	4428	2.55	6780	1.73	7648	4.30	2352	24
37	4581	2.55	6676	1.75	7906	4.28	2094	23
38	4734	2.53	6571	1.75	8163	4.28	1837	22
39	4886	2.55	6466	1.73	8420	4.28	1580	21
40	9.80 5039	2.53	9.88 6362	1.75	9.91 8677	4.28	10.08 1323	20
41	5191	2.53	6257	1.75	8934	4.28	1066	19
42	5343	2.53	6152	1.75	9191	4.28	0809	18
43	5495	2.53	6047	1.75	9448	4.28	0552	17
44	5647	2.53	5942	1.75	9705	4.28	0295	16
45	9.80 5799	2.53	9.88 5837	1.75	9.91 9962	4.28	10.08 0038	15
46	5951	2.53	5732	1.75	.92 0219	4.28	.07 9781	14
47	6103	2.52	5627	1.75	0476	4.28	9524	13
48	6254	2.53	5522	1.77	0733	4.28	9267	12
49	6406	2.52	5416	1.75	0990	4.28	9010	11
50	9.80 6557	2.53	9.88 5311	1.77	9.92 1247	4.27	10.07 8753	10
51	6709	2.52	5205	1.75	1503	4.28	8497	9
52	6860	2.52	5100	1.77	1760	4.28	8240	8
53	7011	2.53	4994	1.75	2017	4.28	7983	7
54	7163	2.52	4889	1.77	2274	4.27	7726	6
55	9.80 7314	2.52	9.88 4783	1.77	9.92 2530	4.28	10.07 7470	5
56	7465	2.50	4677	1.75	2787	4.28	7213	4
57	7615	2.52	4572	1.77	3044	4.27	6956	3
58	7766	2.52	4466	1.77	3300	4.28	6700	2
59	7917	2.50	4360	1.77	3557	4.28	6443	1
60	9.80 8067		9.88 4254		9.92 3814		10.07 6186	0
′	Cosine.	D.1″.	Sine.	D.1″.	Cotang.	D.1″.	Tang.	′

129°   **50°**

Table II. (Continued)

40°   139°

′	Sine.	D.1″.	Cosine.	D.1″.	Tang.	D.1″.	Cotang.	′
0	9.80 8067	2.52	9.88 4254	1.77	9.92 3814	4.27	10.07 6186	60
1	8218	2.50	4148	1.77	4070	4.28	5930	59
2	8368	2.52	4042	1.77	4327	4.27	5673	58
3	8519	2.50	3936	1.78	4583	4.28	5417	57
4	8669	2.50	3829	1.77	4840	4.27	5160	56
5	9.80 8819	2.50	9.88 3723	1.77	9.92 5096	4.27	10.07 4904	55
6	8969	2.50	3617	1.78	5352	4.28	4648	54
7	9119	2.50	3510	1.77	5609	4.27	4391	53
8	9269	2.50	3404	1.78	5865	4.28	4135	52
9	9419	2.50	3297	1.77	6122	4.27	3878	51
10	9.80 9569	2.48	9.88 3191	1.78	9.92 6378	4.27	10.07 3622	50
11	9718	2.50	3084	1.78	6634	4.27	3366	49
12	.80 9868	2.48	2977	1.77	6890	4.28	3110	48
13	.81 0017	2.50	2871	1.78	7147	4.27	2853	47
14	0167	2.48	2764	1.78	7403	4.27	2597	46
15	9.81 0316	2.48	9.88 2657	1.78	9.92 7659	4.27	10.07 2341	45
16	0465	2.48	2550	1.78	7915	4.27	2085	44
17	0614	2.48	2443	1.78	8171	4.27	1829	43
18	0763	2.48	2336	1.78	8427	4.28	1573	42
19	0912	2.48	2229	1.80	8684	4.27	1316	41
20	9.81 1061	2.48	9.88 2121	1.78	9.92 8940	4.27	10.07 1060	40
21	1210	2.47	2014	1.78	9196	4.27	0804	39
22	1358	2.48	1907	1.80	9452	4.27	0548	38
23	1507	2.47	1799	1.78	9708	4.27	0292	37
24	1655	2.48	1692	1.80	.92 9964	4.27	.07 0036	36
25	9.81 1804	2.47	9.88 1584	1.78	9.93 0220	4.25	10.06 9780	35
26	1952	2.47	1477	1.80	0475	4.27	9525	34
27	2100	2.47	1369	1.80	0731	4.27	9269	33
28	2248	2.47	1261	1.80	0987	4.27	9013	32
29	2396	2.47	1153	1.78	1243	4.27	8757	31
30	9.81 2544	2.47	9.88 1046	1.80	9.93 1499	4.27	10.06 8501	30
31	2692	2.47	0938	1.80	1755	4.25	8245	29
32	2840	2.47	0830	1.80	2010	4.27	7990	28
33	2988	2.45	0722	1.82	2266	4.27	7734	27
34	3135	2.47	0613	1.80	2522	4.27	7478	26
35	9.81 3283	2.45	9.88 0505	1.80	9.93 2778	4.25	10.06 7222	25
36	3430	2.47	0397	1.80	3033	4.27	6967	24
37	3578	2.45	0289	1.82	3289	4.27	6711	23
38	3725	2.45	0180	1.80	3545	4.25	6455	22
39	3872	2.45	.88 0072	1.82	3800	4.27	6200	21
40	9.81 4019	2.45	9.87 9963	1.80	9.93 4056	4.25	10.06 5944	20
41	4166	2.45	9855	1.82	4311	4.27	5689	19
42	4313	2.45	9746	1.82	4567	4.25	5433	18
43	4460	2.45	9637	1.80	4822	4.27	5178	17
44	4607	2.43	9529	1.82	5078	4.25	4922	16
45	9.81 4753	2.45	9.87 9420	1.82	9.93 5333	4.27	10.06 4667	15
46	4900	2.43	9311	1.82	5589	4.25	4411	14
47	5046	2.45	9202	1.82	5844	4.27	4156	13
48	5193	2.43	9093	1.82	6100	4.25	3900	12
49	5339	2.43	8984	1.82	6355	4.27	3645	11
50	9.81 5485	2.45	9.87 8875	1.82	9.93 6611	4.25	10.06 3389	10
51	5632	2.43	8766	1.83	6866	4.25	3134	9
52	5778	2.43	8656	1.82	7121	4.27	2879	8
53	5924	2.42	8547	1.82	7377	4.25	2623	7
54	6069	2.43	8438	1.83	7632	4.25	2368	6
55	9.81 6215	2.43	9.87 8328	1.82	9.93 7887	4.25	10.06 2113	5
56	6361	2.43	8219	1.83	8142	4.27	1858	4
57	6507	2.42	8109	1.83	8398	4.25	1602	3
58	6652	2.43	7999	1.82	8653	4.25	1347	2
59	6798	2.42	7890	1.83	8908	4.25	1092	1
60	9.81 6943		9.87 7780		9.93 9163		10.06 0837	0

′	Cosine.	D.1″.	Sine.	D.1″.	Cotang.	D.1″.	Tang.	′

130°   49°

Table II. (Continued)

41° 138°

′	Sine.	D.1″.	Cosine.	D.1″.	Tang.	D.1″.	Cotang.	′
0	9.81 6943	2.42	9.87 7780	1.83	9.93 9163	4.25	10.06 0837	60
1	7088	2.42	7670	1.83	9418	4.25	0582	59
2	7233	2.43	7560	1.83	9673	4.25	0327	58
3	7379	2.42	7450	1.83	.93 9928	4.25	.06 0072	57
4	7524	2.40	7340	1.83	.94 0183	4.27	.05 9817	56
5	9.81 7668	2.42	9.87 7230	1.83	9.94 0439	4.25	10.05 9561	55
6	7813	2.42	7120	1.83	0694	4.25	9306	54
7	7958	2.42	7010	1.85	0949	4.25	9051	53
8	8103	2.40	6899	1.83	1204	4.25	8796	52
9	8247	2.42	6789	1.85	1459	4.23	8541	51
10	9.81 8392	2.40	9.87 6678	1.83	9.94 1713	4.25	10.05 8287	50
11	8536	2.42	6568	1.85	1968	4.25	8032	49
12	8681	2.40	6457	1.83	2223	4.25	7777	48
13	8825	2.40	6347	1.85	2478	4.25	7522	47
14	8969	2.40	6236	1.85	2733	4.25	7267	46
15	9.81 9113	2.40	9.87 6125	1.85	9.94 2988	4.25	10.05 7012	45
16	9257	2.40	6014	1.83	3243	4.25	6757	44
17	9401	2.40	5904	1.85	3498	4.23	6502	43
18	9545	2.40	5793	1.85	3752	4.25	6248	42
19	9689	2.38	5682	1.85	4007	4.25	5993	41
20	9.81 9832	2.40	9.87 5571	1.87	9.94 4262	4.25	10.05 5738	40
21	.81 9976	2.40	5459	1.85	4517	4.23	5483	39
22	.82 0120	2.38	5348	1.85	4771	4.25	5229	38
23	0263	2.38	5237	1.85	5026	4.25	4974	37
24	0406	2.40	5126	1.87	5281	4.23	4719	36
25	9.82 0550	2.38	9.87 5014	1.85	9.94 5535	4.25	10.05 4465	35
26	0693	2.38	4903	1.87	5790	4.25	4210	34
27	0836	2.38	4791	1.85	6045	4.25	3955	33
28	0979	2.38	4680	1.87	6299	4.25	3701	32
29	1122	2.38	4568	1.87	6554	4.23	3446	31
30	9.82 1265	2.37	9.87 4456	1.87	9.94 6808	4.25	10.05 3192	30
31	1407	2.38	4344	1.87	7063	4.25	2937	29
32	1550	2.38	4232	1.85	7318	4.25	2682	28
33	1693	2.37	4121	1.87	7572	4.25	2428	27
34	1835	2.37	4009	1.88	7827	4.23	2173	26
35	9.82 1977	2.38	9.87 3896	1.87	9.94 8081	4.23	10.05 1919	25
36	2120	2.37	3784	1.87	8335	4.25	1665	24
37	2262	2.37	3672	1.87	8590	4.25	1410	23
38	2404	2.37	3560	1.87	8844	4.25	1156	22
39	2546	2.37	3448	1.88	9099	4.23	0901	21
40	9.82 2688	2.37	9.87 3335	1.87	9.94 9353	4.25	10.05 0647	20
41	2830	2.27	3223	1.88	9608	4.25	0392	19
42	2972	2.27	3110	1.87	.94 9862	4.25	.05 0138	18
43	3114	2.25	2998	1.88	.95 0116	4.25	.04 9884	17
44	3255	2.57	2885	1.88	0371	4.23	9629	16
45	9.82 3397	2.37	9.87 2772	1.88	9.95 0625	4.23	10.04 9375	15
46	3539	2.35	2659	1.87	0879	4.25	9121	14
47	3680	2.35	2547	1.88	1133	4.25	8867	13
48	3821	2.37	2434	1.88	1388	4.23	8612	12
49	3963	2.35	2321	1.88	1642	4.23	8358	11
50	9.82 4104	2.35	9.87 2208	1.88	9.95 1896	4.23	10.04 8104	10
51	4245	2.35	2095	1.90	2150	4.25	7850	9
52	4386	2.35	1981	1.88	2405	4.23	7595	8
53	4527	2.35	1868	1.88	2659	4.23	7341	7
54	4668	2.33	1755	1.90	2913	4.23	7087	6
55	9.82 4808	2.35	9.87 1641	1.88	9.95 3167	4.23	10.04 6833	5
56	4949	2.35	1528	1.90	3421	4.23	6579	4
57	5090	2.33	1414	1.88	3675	4.23	6325	3
58	5230	2.35	1301	1.90	3929	4.23	6071	2
59	5371	2.33	1187	1.90	4183	4.23	5817	1
60	9.82 5511		9.87 1073		9.95 4437		10.04 5563	0
′	Cosine.	D.1″.	Sine.	D.1″.	Cotang.	D.1″.	Tang.	′

131° 48°

Table II. (Continued)

42°   **137°**

′	Sine.	D.1″.	Cosine.	D.1″.	Tang.	D.1″.	Cotang.	′
0	9.82 5511	2.33	9.87 1073	1.88	9.95 4437	4.23	10.04 5563	60
1	5651	2.33	0960	1.90	4691	4.25	5309	59
2	5791	2.33	0846	1.90	4946	4.23	5054	58
3	5931	2.33	0732	1.90	5200	4.23	4800	57
4	6071	2.33	0618	1.90	5454	4.23	4546	56
5	9.82 6211	2.33	9.87 0504	1.90	9.95 5708	4.22	10.04 4292	55
6	6351	2.33	0390	1.90	5961	4.23	4039	54
7	6491	2.33	0276	1.92	6215	4.23	3785	53
8	6631	2.32	0161	1.90	6469	4.23	3531	52
9	6770	2.33	.87 0047	1.90	6723	4.23	3277	51
10	9.82 6910	2.32	9.86 9933	1.92	9.95 6977	4.23	10.04 3023	50
11	7049	2.33	9818	1.90	7231	4.23	2769	49
12	7189	2.32	9704	1.92	7485	4.23	2515	48
13	7328	2.32	9589	1.92	7739	4.23	2261	47
14	7467	2.32	9474	1.90	7993	4.23	2007	46
15	9.82 7606	2.32	9.86 9360	1.92	9.95 8247	4.22	10.04 1753	45
16	7745	2.32	9245	1.92	8500	4.23	1500	44
17	7884	2.32	9130	1.92	8754	4.23	1246	43
18	8023	2.32	9015	1.92	9008	4.23	0992	42
19	8162	2.32	8900	1.92	9262	4.23	0738	41
20	9.82 8301	2.30	9.86 8785	1.92	9.95 9516	4.22	10.04 0484	40
21	8439	2.32	8670	1.92	.95 9769	4.23	.04 0231	39
22	8578	2.30	8555	1.92	.96 0023	4.23	.03 9977	38
23	8716	2.32	8440	1.93	0277	4.22	9723	37
24	8855	2.30	8324	1.92	0530	4.23	9470	36
25	9.82 8993	2.30	9.86 8209	1.93	9.96 0784	4.23	10.03 9216	35
26	9131	2.30	8093	1.92	1038	4.23	8962	34
27	9269	2.30	7978	1.93	1292	4.22	8708	33
28	9407	2.30	7862	1.92	1545	4.23	8455	32
29	9545	2.30	7747	1.93	1799	4.22	8201	31
30	9.82 9683	2.30	9.86 7631	1.93	9.96 2052	4.23	10.03 7948	30
31	9621	2.30	7515	1.93	2306	4.22	7694	29
32	.82 9959	2.30	7399	1.93	2560	4.22	7440	28
33	.83 0097	2.28	7283	1.93	2813	4.23	7187	27
34	0234	2.30	7167	1.93	3067	4.22	6933	26
35	9.83 0372	2.28	9.86 7051	1.93	9.96 3320	4.23	10.03 6680	25
36	0509	2.28	6935	1.93	3574	4.22	6426	24
37	0646	2.30	6819	1.93	3828	4.22	6172	23
38	0784	2.28	6703	1.95	4081	4.23	5919	22
39	0921	2.28	6586	1.93	4335	4.22	5665	21
40	9.83 1058	2.28	9.86 6470	1.95	9.96 4588	4.23	10.03 5412	20
41	1195	2.28	6353	1.93	4842	4.22	5158	19
42	1332	2.28	6237	1.95	5095	4.23	4905	18
43	1469	2.28	6120	1.93	5349	4.22	4651	17
44	1606	2.27	6004	1.95	5602	4.22	4398	16
45	9.83 1742	2.28	9.86 5887	1.95	9.96 5855	4.23	10.03 4145	15
46	1879	2.27	5770	1.95	6109	4.22	3891	14
47	2015	2.28	5653	1.95	6362	4.23	3638	13
48	2152	2.27	5536	1.95	6616	4.22	3384	12
49	2288	2.28	5419	1.95	6869	4.23	3131	11
50	9.83 2425	2.27	9.86 5302	1.95	9.96 7123	4.22	10.03 2877	10
51	2561	2.27	5185	1.95	7376	4.23	2624	9
52	2697	2.27	5068	1.97	7629	4.23	2371	8
53	2833	2.27	4950	1.95	7883	4.22	2117	7
54	2969	2.27	4833	1.95	8136	4.22	1864	6
55	9.83 3105	2.27	9.86 4716	1.97	9.96 8389	4.23	10.03 1611	5
56	3241	2.27	4598	1.95	8643	4.22	1357	4
57	3377	2.25	4481	1.97	8896	4.22	1104	3
58	3512	2.27	4363	1.97	9149	4.23	0851	2
59	3648	2.25	4245	1.97	9403	4.22	0597	1
60	9.83 3783		9.86 4127		9.96 9656		10.03 0344	0
′	Cosine.	D.1″.	Sine.	D.1″.	Cotang.	D.1″.	Tang.	′

132°   **47°**

Table II. (Continued)

43°   136°

′	Sine.	D.1″.	Cosine.	D.1″.	Tang.	D.1″.	Cotang.	′
0	9.83 3783	2.27	9.86 4127	1.95	9.96 9656	4.22	10.03 0344	60
1	3919	2.25	4010	1.97	.96 9909	4.22	.03 0091	59
2	4054	2.25	3892	1.97	.97 0162	4.23	.02 9838	58
3	4189	2.27	3774	1.97	0416	4.22	9584	57
4	4325	2.25	3656	1.97	0669	4.22	9331	56
5	9.83 4460	2.25	9.86 3538	1.98	9.97 0922	4.22	10.02 9078	55
6	4595	2.25	3419	1.97	1175	4.23	8825	54
7	4730	2.25	3301	1.97	1429	4.22	8571	53
8	4865	2.23	3183	1.98	1682	4.22	8318	52
9	4999	2.25	3064	1.97	1935	4.22	8065	51
10	9.83 5134	2.25	9.86 2946	1.98	9.97 2188	4.22	10.02 7812	50
11	5269	2.23	2827	1.97	2441	4.23	7559	49
12	5403	2.23	2709	1.98	2695	4.22	7305	48
13	5538	2.23	2590	1.98	2948	4.22	7052	47
14	5672	2.25	2471	1.97	3201	4.22	6799	46
15	9.83 5807	2.23	9.86 2353	1.98	9.97 3454	4.22	10.02 6546	45
16	5941	2.23	2234	1.98	3707	4.22	6293	44
17	6075	2.23	2115	1.98	3960	4.22	6040	43
18	6209	2.23	1996	1.98	4213	4.23	5787	42
19	6343	2.23	1877	1.98	4466	4.23	5534	41
20	9.83 6477	2.23	9.86 1758	2.00	9.97 4720	4.22	10.02 5280	40
21	6611	2.23	1638	1.98	4973	4.22	5027	39
22	6745	2.22	1519	1.98	5226	4.22	4774	38
23	6878	2.23	1400	2.00	5479	4.22	4521	37
24	7012	2.23	1280	1.98	5732	4.22	4268	36
25	9.83 7146	2.22	9.86 1161	2.00	9.97 5985	4.22	10.02 4015	35
26	7279	2.22	1041	1.98	6238	4.22	3762	34
27	7412	2.23	0922	2.00	6491	4.22	3509	33
28	7546	2.22	0802	2.00	6744	4.22	3256	32
29	7679	2.22	0682	2.00	6997	4.22	3003	31
30	9.83 7812	2.22	9.86 0562	2.00	9.97 7250	4.22	10.02 2750	30
31	7945	2.22	0442	2.00	7503	4.22	2497	29
32	8078	2.22	0322	2.00	7756	4.22	2244	28
33	8211	2.22	0202	2.00	8009	4.22	1991	27
34	8344	2.22	.86 0082	2.00	8262	4.22	1738	26
35	9.83 8477	2.22	9.85 9962	2.00	9.97 8515	4.22	10.02 1485	25
36	8610	2.20	9842	2.02	8768	4.22	1232	24
37	8742	2.22	9721	2.00	9021	4.22	0979	23
38	8875	2.20	9601	2.02	9274	4.22	0726	22
39	9007	2.22	9480	2.00	9527	4.22	0473	21
40	9.83 9140	2.20	9.85 9360	2.00	9.97 9780	4.22	10.02 0220	20
41	9272	2.20	9239	2.00	.98 0033	4.22	.01 9967	19
42	9404	2.20	9119	2.02	0286	4.20	9714	18
43	9536	2.20	8998	2.02	0538	4.22	9462	17
44	9668	2.20	8877	2.02	0791	4.22	9209	16
45	9.83 9800	2.20	9.85 8756	2.02	9.98 1044	4.22	10.01 8956	15
46	.83 9932	2.20	8635	2.02	1297	4.22	8703	14
47	.84 0064	2.20	8514	2.02	1550	4.22	8450	13
48	0196	2.20	8393	2.02	1803	4.22	8197	12
49	0328	2.18	8272	2.02	2056	4.22	7944	11
50	9.84 0459	2.20	9.85 8151	2.03	9.98 2309	4.22	10.01 7691	10
51	0591	2.18	8029	2.02	2562	4.20	7438	9
52	0722	2.20	7908	2.03	2814	4.22	7186	8
53	0854	2.18	7786	2.02	3067	4.22	6933	7
54	0985	2.18	7665	2.03	3320	4.22	6680	6
55	9.84 1116	2.18	9.85 7543	2.02	9.98 3573	4.22	10.01 6427	5
56	1247	2.18	7422	2.03	3826	4.22	6174	4
57	1378	2.18	7300	2.03	4079	4.22	5921	3
58	1509	2.18	7178	2.03	4332	4.20	5668	2
59	1640	2.18	7056	2.03	4584	4.22	5416	1
60	9.84 1771		9.85 6934		9.98 4837		10.01 5163	0
′	Cosine.	D.1″.	Sine.	D.1″.	Cotang.	D.1″.	Tang.	′

133°   46°

Table II. (Concluded)

44°   135°

'	Sine.	D.1".	Cosine.	D.1".	Tang.	D.1".	Cotang.	'
0	9.84 1771	2.18	9.85 6934	2.03	9.98 4837	4.22	10.01 5163	60
1	1902	2.18	6812	2.03	5090	4.22	4910	59
2	2033	2.17	6690	2.03	5343	4.22	4657	58
3	2163	2.18	6568	2.03	5596	4.20	4404	57
4	2294	2.17	6446	2.05	5848	4.22	4152	56
5	9.84 2424	2.18	9.85 6323	2.03	9.98 6101	4.22	10.01 3899	55
6	2555	2.17	6201	2.05	6354	4.22	3646	54
7	2685	2.17	6078	2.C3	6607	4.22	3393	53
8	2815	2.18	5956	2.C5	6860	4.20	3140	52
9	2946	2.17	5833	2.03	7112	4.22	2888	51
10	9.84 3076	2.17	9.85 5711	2.05	9.98 7365	4.22	10.01 2635	50
11	3206	2.17	5588	2.05	7618	4.22	2382	49
12	3336	2.17	5465	2.05	7871	4.20	2129	48
13	3466	2.15	5342	2.05	8123	4.22	1877	47
14	3595	2.17	5219	2.05	8376	4.22	1624	46
15	9.84 3725	2.17	9.85 5096	2.05	9.98 8629	4.22	10.01 1371	45
16	3855	2.15	4973	2.05	8882	4.20	1118	44
17	3984	2.17	4850	2.05	9134	4.22	0866	43
18	4114	2.15	4727	2.07	9387	4.22	0613	42
19	4243	2.15	4603	2.05	9640	4.22	0360	41
20	9.84 4372	2.17	9.85 4480	2.07	9.98 9893	4.20	10.01 0107	40
21	4502	2.15	4356	2.05	.99 0145	4.22	.00 9855	39
22	4631	2.15	4233	2.07	0398	4.22	9602	38
23	4760	2.15	4109	2.05	0651	4.20	9349	37
24	4889	2.15	3986	2.07	0903	4.22	9097	36
25	9.84 5018	2.15	9.85 3862	2.07	9.99 1156	4.22	10.00 8844	35
26	5147	2.15	3738	2.07	1409	4.22	8591	34
27	5276	2.15	3614	2.07	1662	4.20	8338	33
28	5405	2.13	3490	2.07	1914	4.22	8086	32
29	5533	2.15	3366	2.07	2167	4.22	7833	31
30	9.84 5662	2.13	9.85 3242	2.07	9.99 2420	4.20	10.00 7580	30
31	5790	2.15	3118	2.07	2672	4.22	7328	29
32	5919	2.13	2994	2.08	2925	4.22	7075	28
33	6047	2.13	2869	2.07	3178	4.22	6822	27
34	6175	2.15	2745	2.08	3431	4.22	6569	26
35	9.84 6304	2.13	9.85 2620	2.07	9.99 3683	4.22	10.00 6317	25
36	6432	2.13	2496	2.08	3936	4.22	6064	24
37	6560	2.13	2371	2.07	4189	4.20	5811	23
38	6688	2.13	2247	2.08	4441	4.22	5559	22
39	6816	2.13	2122	2.08	4694	4.22	5306	21
40	9.84 6944	2.12	9.85 1997	2.08	9.99 4947	4.20	10.00 5053	20
41	7071	2.13	1872	2.08	5199	4.22	4801	19
42	7199	2.13	1747	2.08	5452	4.20	4548	18
43	7327	2.12	1622	2.08	5705	4.22	4295	17
44	7454	2.13	1497	2.08	5957	4.22	4043	16
45	9.84 7582	2.12	9.85 1372	2.10	9.99 6210	4.22	10.00 3790	15
46	7709	2.12	1246	2.08	6463	4.20	3537	14
47	7836	2.13	1121	2.C8	6715	4.22	3285	13
48	7964	2.12	0996	2.10	6968	4.22	3032	12
49	8091	2.12	0870	2.08	7221	4.20	2779	11
50	9.84 8218	2.12	9.85 0745	2.10	9.99 7473	4.22	10.00 2527	10
51	8345	2.12	0619	2.10	7726	4.22	2274	9
52	8472	2.12	0493	2.08	7979	4.20	2021	8
53	8599	2.12	0368	2.10	8231	4.22	1769	7
54	8726	2.10	0242	2.10	8484	4.22	1516	6
55	9.84 8852	2.12	9.85 0116	2.10	9.99 8737	4.20	10.00 1263	5
56	8979	2.12	84 9990	2.10	8989	4.22	1011	4
57	9106	2.10	9864	2.10	9242	4.22	0758	3
58	9232	2.12	9738	2.12	9495	4.20	0505	2
59	9359	2.10	9611	2.10	9.99 9747	4.22	0253	1
60	9.84 9485		9.84 9485		10.00 0000		10.00 0000	0
'	Cosine.	D.1".	Sine.	D.1".	Cotang.	D.1".	Tang.	'

134°   45°

Table III. Natural Sines and Cosines

'	0° SINE	0° COSINE	'	'	0° SINE	0° COSINE	'	'	0° SINE	0° COSINE	'
0	.00000	1	60	21	.00611	.99998	39	41	.01193	.99993	19
1	.00029	1	59	22	.00640	.99998	38	42	.01222	.99993	18
2	.00058	1	58	23	.00669	.99998	37	43	.01251	.99992	17
3	.00087	1	57	24	.00698	.99998	36	44	.01280	.99992	16
4	.00116	1	56	25	.00727	.99997	35	45	.01309	.99991	15
5	.00145	1	55	26	.00756	.99997	34	46	.01338	.99991	14
6	.00175	1	54	27	.00785	.99997	33	47	.01367	.99991	13
7	.00204	1	53	28	.00814	.99997	32	48	.01396	.99990	12
8	.00233	1	52	29	.00844	.99996	31	49	.01425	.99990	11
9	.00262	1	51	30	.00873	.99996	30	50	.01454	.99989	10
10	.00291	1	50	31	.00902	.99996	29	51	.01483	.99989	9
11	.00320	.99999	49	32	.00931	.99996	28	52	.01513	.99989	8
12	.00349	.99999	48	33	.00960	.99995	27	53	.01542	.99988	7
13	.00378	.99999	47	34	.00989	.99995	26	54	.01571	.99988	6
14	.00407	.99999	46	35	.01018	.99995	25	55	.01600	.99987	5
15	.00436	.99999	45	36	.01047	.99995	24	56	.01629	.99987	4
16	.00465	.99999	44	37	.01076	.99994	23	57	.01658	.99986	3
17	.00495	.99999	43	38	.01105	.99994	22	58	.01687	.99986	2
18	.00524	.99999	42	39	.01134	.99994	21	59	.01716	.99985	1
19	.00553	.99998	41	40	.01164	.99993	20	60	.01745	.99985	0
20	.00582	.99998	40								
'	COSINE	SINE	'	'	COSINE	SINE	'	'	COSINE	SINE	'
	89°				89°				89°		

Table III. (Continued)

′	1° Sine	1° Cosine	2° Sine	2° Cosine	3° Sine	3° Cosine	4° Sine	4° Cosine	′
0	.01745	.99985	.03490	.99939	.05234	.99863	.06976	.99756	60
1	.01774	.99984	.03519	.99938	.05263	.99861	.07005	.99754	59
2	.01803	.99984	.03548	.99937	.05292	.99860	.07034	.99752	58
3	.01832	.99983	.03577	.99936	.05321	.99858	.07063	.99750	57
4	.01862	.99983	.03606	.99935	.05350	.99857	.07092	.99748	56
5	.01891	.99982	.03635	.99934	.05379	.99855	.07121	.99746	55
6	.01920	.99982	.03664	.99933	.05408	.99854	.07150	.99744	54
7	.01949	.99981	.03693	.99932	.05437	.99852	.07179	.99742	53
8	.01978	.99980	.03723	.99931	.05466	.99851	.07208	.99740	52
9	.02007	.99980	.03752	.99930	.05495	.99849	.07237	.99738	51
10	.02036	.99979	.03781	.99929	.05524	.99847	.07266	.99736	50
11	.02065	.99979	.03810	.99927	.05553	.99846	.07295	.99734	49
12	.02094	.99978	.03839	.99926	.05582	.99844	.07324	.99731	48
13	.02123	.99977	.03868	.99925	.05611	.99842	.07353	.99729	47
14	.02152	.99977	.03897	.99924	.05640	.99841	.07382	.99727	46
15	.02181	.99976	.03926	.99923	.05669	.99839	.07411	.99725	45
16	.02211	.99976	.03955	.99922	.05698	.99838	.07440	.99723	44
17	.02240	.99975	.03984	.99921	.05727	.99836	.07469	.99721	43
18	.02269	.99974	.04013	.99919	.05756	.99834	.07498	.99719	42
19	.02298	.99974	.04042	.99918	.05785	.99833	.07527	.99716	41
20	.02327	.99973	.04071	.99917	.05814	.99831	.07556	.99714	40
21	.02356	.99972	.04100	.99916	.05844	.99829	.07585	.99712	39
22	.02385	.99972	.04129	.99915	.05873	.99827	.07614	.99710	38
23	.02414	.99971	.04159	.99913	.05902	.99826	.07643	.99708	37
24	.02443	.99970	.04188	.99912	.05931	.99824	.07672	.99705	36
25	.02472	.99969	.04217	.99911	.05960	.99822	.07701	.99703	35
26	.02501	.99969	.04246	.99910	.05989	.99821	.07730	.99701	34
27	.02530	.99968	.04275	.99909	.06018	.99819	.07759	.99699	33
28	.02560	.99967	.04304	.99907	.06047	.99817	.07788	.99696	32
29	.02589	.99966	.04333	.99906	.06076	.99815	.07817	.99694	31
30	.02618	.99966	.04362	.99905	.06105	.99813	.07846	.99692	30
31	.02647	.99965	.04391	.99904	.06134	.99812	.07875	.99689	29
32	.02676	.99964	.04420	.99902	.06163	.99810	.07904	.99687	28
33	.02705	.99963	.04449	.99901	.06192	.99808	.07933	.99685	27
34	.02734	.99963	.04478	.99900	.06221	.99806	.07962	.99683	26
35	.02763	.99962	.04507	.99898	.06250	.99804	.07991	.99680	25
36	.02792	.99961	.04536	.99897	.06279	.99803	.08020	.99678	24
37	.02821	.99960	.04565	.99896	.06308	.99801	.08049	.99676	23
38	.02850	.99959	.04594	.99894	.06337	.99799	.08078	.99673	22
39	.02879	.99959	.04623	.99893	.06366	.99797	.08107	.99671	21
40	.02908	.99958	.04653	.99892	.06395	.99795	.08136	.99668	20
41	.02938	.99957	.04682	.99890	.06424	.99793	.08165	.99666	19
42	.02967	.99956	.04711	.99889	.06453	.99792	.08194	.99664	18
43	.02996	.99955	.04740	.99888	.06482	.99790	.08223	.99661	17
44	.03025	.99954	.04769	.99886	.06511	.99788	.08252	.99659	16
45	.03054	.99953	.04798	.99885	.06540	.99786	.08281	.99657	15
46	.03083	.99952	.04827	.99883	.06569	.99784	.08310	.99654	14
47	.03112	.99952	.04856	.99882	.06598	.99782	.08339	.99652	13
48	.03141	.99951	.04885	.99881	.06627	.99780	.08368	.99649	12
49	.03170	.99950	.04914	.99879	.06656	.99778	.08397	.99647	11
50	.03199	.99949	.04943	.99878	.06685	.99776	.08426	.99644	10
51	.03228	.99948	.04972	.99876	.06714	.99774	.08455	.99642	9
52	.03257	.99947	.05001	.99875	.06743	.99772	.08484	.99639	8
53	.03286	.99946	.05030	.99873	.06773	.99770	.08513	.99637	7
54	.03316	.99945	.05059	.99872	.06802	.99768	.08542	.99635	6
55	.03345	.99944	.05088	.99870	.06831	.99766	.08571	.99632	5
56	.03374	.99943	.05117	.99869	.06860	.99764	.08600	.99630	4
57	.03403	.99942	.05146	.99867	.06889	.99762	.08629	.99627	3
58	.03432	.99941	.05175	.99866	.06918	.99760	.08658	.99625	2
59	.03461	.99940	.05205	.99864	.06947	.99758	.08687	.99622	1
60	.03490	.99939	.05234	.99863	.06976	.99756	.08716	.99619	0
′	Cosine	Sine 88°	Cosine	Sine 87°	Cosine	Sine 86°	Cosine	Sine 85°	′

Table III. (Continued)

′	5° Sine	5° Cosine	6° Sine	6° Cosine	7° Sine	7° Cosine	8° Sine	8° Cosine	′
0	.08716	.99619	.10453	.99452	.12187	.99255	.13917	.99027	60
1	.08745	.99617	.10482	.99449	.12216	.99251	.13946	.99023	59
2	.08774	.99614	.10511	.99446	.12245	.99248	.13975	.99019	58
3	.08803	.99612	.10540	.99443	.12274	.99244	.14004	.99015	57
4	.08831	.99609	.10569	.99440	.12302	.99240	.14033	.99011	56
5	.08860	.99607	.10597	.99437	.12331	.99237	.14061	.99006	55
6	.08889	.99604	.10626	.99434	.12360	.99233	.14090	.99002	54
7	.08918	.99602	.10655	.99431	.12389	.99230	.14119	.98998	53
8	.08947	.99599	.10684	.99428	.12418	.99226	.14148	.98994	52
9	.08976	.99596	.10713	.99424	.12447	.99222	.14177	.98990	51
10	.09005	.99594	.10742	.99421	.12476	.99219	.14205	.98986	50
11	.09034	.99591	.10771	.99418	.12504	.99215	.14234	.98982	49
12	.09063	.99588	.10800	.99415	.12533	.99211	.14263	.98978	48
13	.09092	.99586	.10829	.99412	.12562	.99208	.14292	.98973	47
14	.09121	.99583	.10858	.99409	.12591	.99204	.14320	.98969	46
15	.09150	.99580	.10887	.99406	.12620	.99200	.14349	.98965	45
16	.09179	.99578	.10916	.99402	.12649	.99197	.14378	.98961	44
17	.09208	.99575	.10945	.99399	.12678	.99193	.14407	.98957	43
18	.09237	.99572	.10973	.99396	.12706	.99189	.14436	.98953	42
19	.09266	.99570	.11002	.99393	.12735	.99186	.14464	.98948	41
20	.09295	.99567	.11031	.99390	.12764	.99182	.14493	.98944	40
21	.09324	.99564	.11060	.99386	.12793	.99178	.14522	.98940	39
22	.09353	.99562	.11089	.99383	.12822	.99175	.14551	.98936	38
23	.09382	.99559	.11118	.99380	.12851	.99171	.14580	.98931	37
24	.09411	.99556	.11147	.99377	.12880	.99167	.14608	.98927	36
25	.09440	.99553	.11176	.99374	.12908	.99163	.14637	.98923	35
26	.09469	.99551	.11205	.99370	.12937	.99160	.14666	.98919	34
27	.09498	.99548	.11234	.99367	.12966	.99156	.14695	.98914	33
28	.09527	.99545	.11263	.99364	.12995	.99152	.14723	.98910	32
29	.09556	.99542	.11291	.99360	.13024	.99148	.14752	.98906	31
30	.09585	.99540	.11320	.99357	.13053	.99144	.14781	.98902	30
31	.09614	.99537	.11349	.99354	.13081	.99141	.14810	.98897	29
32	.09642	.99534	.11378	.99351	.13110	.99137	.14838	.98893	28
33	.09671	.99531	.11407	.99347	.13139	.99133	.14867	.98889	27
34	.09700	.99528	.11436	.99344	.13168	.99129	.14896	.98884	26
35	.09729	.99526	.11465	.99341	.13197	.99125	.14925	.98880	25
36	.09758	.99523	.11494	.99337	.13226	.99122	.14954	.98876	24
37	.09787	.99520	.11523	.99334	.13254	.99118	.14982	.98871	23
38	.09816	.99517	.11552	.99331	.13283	.99114	.15011	.98867	22
39	.09845	.99514	.11580	.99327	.13312	.99110	.15040	.98863	21
40	.09874	.99511	.11609	.99324	.13341	.99106	.15069	.98858	20
41	.09903	.99508	.11638	.99320	.13370	.99102	.15097	.98854	19
42	.09932	.99506	.11667	.99317	.13399	.99098	.15126	.98849	18
43	.09961	.99503	.11696	.99314	.13427	.99094	.15155	.98845	17
44	.09990	.99500	.11725	.99310	.13456	.99091	.15184	.98841	16
45	.10019	.99497	.11754	.99307	.13485	.99087	.15212	.98836	15
46	.10048	.99494	.11783	.99303	.13514	.99083	.15241	.98832	14
47	.10077	.99491	.11812	.99300	.13543	.99079	.15270	.98827	13
48	.10106	.99488	.11840	.99297	.13572	.99075	.15299	.98823	12
49	.10135	.99485	.11869	.99293	.13600	.99071	.15327	.98818	11
50	.10164	.99482	.11898	.99290	.13629	.99067	.15356	.98814	10
51	.10192	.99479	.11927	.99286	.13658	.99063	.15385	.98809	9
52	.10221	.99476	.11956	.99283	.13687	.99059	.15414	.98805	8
53	.10250	.99473	.11985	.99279	.13716	.99055	.15442	.98800	7
54	.10279	.99470	.12014	.99276	.13744	.99051	.15471	.98796	6
55	.10308	.99467	.12043	.99272	.13773	.99047	.15500	.98791	5
56	.10337	.99464	.12071	.99269	.13802	.99043	.15529	.98787	4
57	.10366	.99461	.12100	.99265	.13831	.99039	.15557	.98782	3
58	.10395	.99458	.12129	.99262	.13860	.99035	.15586	.98778	2
59	.10424	.99455	.12158	.99258	.13889	.99031	.15615	.98773	1
60	.10453	.99452	.12187	.99255	.13917	.99027	.15643	.98769	0
′	Cosine	Sine	Cosine	Sine	Cosine	Sine	Cosine	Sine	′
	84°		83°		82°		81°		

Table III. (Continued)

′	9° SINE	9° COSINE	10° SINE	10° COSINE	11° SINE	11° COSINE	12° SINE	12° COSINE	′
0	.15643	.98769	.17365	.98481	.19081	.98163	.20791	.97815	60
1	.15672	.98764	.17393	.98476	.19109	.98157	.20820	.97809	59
2	.15701	.98760	.17422	.98471	.19138	.98152	.20848	.97803	58
3	.15730	.98755	.17451	.98466	.19167	.98146	.20877	.97797	57
4	.15758	.98751	.17479	.98461	.19195	.98140	.20905	.97791	56
5	.15787	.98746	.17508	.98455	.19224	.98135	.20933	.97784	55
6	.15816	.98741	.17537	.98450	.19252	.98129	.20962	.97778	54
7	.15845	.98737	.17565	.98445	.19281	.98124	.20990	.97772	53
8	.15873	.98732	.17594	.98440	.19309	.98118	.21019	.97766	52
9	.15902	.98728	.17623	.98435	.19338	.98112	.21047	.97760	51
10	.15931	.98723	.17651	.98430	.19366	.98107	.21076	.97754	50
11	.15959	.98718	.17680	.98425	.19395	.98101	.21104	.97748	49
12	.15988	.98714	.17708	.98420	.19423	.98096	.21132	.97742	48
13	.16017	.98709	.17737	.98414	.19452	.98090	.21161	.97735	47
14	.16046	.98704	.17766	.98409	.19481	.98084	.21189	.97729	46
15	.16074	.98700	.17794	.98404	.19509	.98079	.21218	.97723	45
16	.16103	.98695	.17823	.98399	.19538	.98073	.21246	.97717	44
17	.16132	.98690	.17852	.98394	.19566	.98067	.21275	.97711	43
18	.16160	.98686	.17880	.98389	.19595	.98061	.21303	.97705	42
19	.16189	.98681	.17909	.98383	.19623	.98056	.21331	.97698	41
20	.16218	.98676	.17937	.98378	.19652	.98050	.21360	.97692	40
21	.16246	.98671	.17966	.98373	.19680	.98044	.21388	.97686	39
22	.16275	.98667	.17995	.98368	.19709	.98039	.21417	.97680	38
23	.16304	.98662	.18023	.98362	.19737	.98033	.21445	.97673	37
24	.16333	.98657	.18052	.98357	.19766	.98027	.21474	.97667	36
25	.16361	.98652	.18081	.98352	.19794	.98021	.21502	.97661	35
26	.16390	.98648	.18109	.98347	.19823	.98016	.21530	.97655	34
27	.16419	.98643	.18138	.98341	.19851	.98010	.21559	.97648	33
28	.16447	.98638	.18166	.98336	.19880	.98004	.21587	.97642	32
29	.16476	.98633	.18195	.98331	.19908	.97998	.21616	.97636	31
30	.16505	.98629	.18224	.98325	.19937	.97992	.21644	.97630	30
31	.16533	.98624	.18252	.98320	.19965	.97987	.21672	.97623	29
32	.16562	.98619	.18281	.98315	.19994	.97981	.21701	.97617	28
33	.16591	.98614	.18309	.98310	.20022	.97975	.21729	.97611	27
34	.16620	.98609	.18338	.98304	.20051	.97969	.21758	.97604	26
35	.16648	.98604	.18367	.98299	.20079	.97963	.21786	.97598	25
36	.16677	.98600	.18395	.98294	.20108	.97958	.21814	.97592	24
37	.16706	.98595	.18424	.98288	.20136	.97952	.21843	.97585	23
38	.16734	.98590	.18452	.98283	.20165	.97946	.21871	.97579	22
39	.16763	.98585	.18481	.98277	.20193	.97940	.21899	.97573	21
40	.16792	.98580	.18509	.98272	.20222	.97934	.21928	.97566	20
41	.16820	.98575	.18538	.98267	.20250	.97928	.21956	.97560	19
42	.16849	.98570	.18567	.98261	.20279	.97922	.21985	.97553	18
43	.16878	.98565	.18595	.98256	.20307	.97916	.22013	.97547	17
44	.16906	.98561	.18624	.98250	.20336	.97910	.22041	.97541	16
45	.16935	.98556	.18652	.98245	.20364	.97905	.22070	.97534	15
46	.16964	.98551	.18681	.98240	.20393	.97899	.22098	.97528	14
47	.16992	.98546	.18710	.98234	.20421	.97893	.22126	.97521	13
48	.17021	.98541	.18738	.98229	.20450	.97887	.22155	.97515	12
49	.17050	.98536	.18767	.98223	.20478	.97881	.22183	.97508	11
50	.17078	.98531	.18795	.98218	.20507	.97875	.22212	.97502	10
51	.17107	.98526	.18824	.98212	.20535	.97869	.22240	.97496	9
52	.17136	.98521	.18852	.98207	.20563	.97863	.22268	.97489	8
53	.17164	.98516	.18881	.98201	.20592	.97857	.22297	.97483	7
54	.17193	.98511	.18910	.98196	.20620	.97851	.22325	.97476	6
55	.17222	.98506	.18938	.98190	.20649	.97845	.22353	.97470	5
56	.17250	.98501	.18967	.98185	.20677	.97839	.22382	.97463	4
57	.17279	.98496	.18995	.98179	.20706	.97833	.22410	.97457	3
58	.17308	.98491	.19024	.98174	.20734	.97827	.22438	.97450	2
59	.17336	.98486	.19052	.98168	.20763	.97821	.22467	.97444	1
60	.17365	.98481	.19081	.98163	.20791	.97815	.22495	.97437	0
′	COSINE	SINE 80°	COSINE	SINE 79°	COSINE	SINE 78°	COSINE	SINE 77°	′

Table III. (Continued)

′	13° SINE	13° COSINE	14° SINE	14° COSINE	15° SINE	15° COSINE	16° SINE	16° COSINE	′
0	.22495	.97437	.24192	.97030	.25882	.96593	.27564	.96126	60
1	.22523	.97430	.24220	.97023	.25910	.96585	.27592	.96118	59
2	.22552	.97424	.24249	.97015	.25938	.96578	.27620	.96110	58
3	.22580	.97417	.24277	.97008	.25966	.96570	.27648	.96102	57
4	.22608	.97411	.24305	.97001	.25994	.96562	.27676	.96094	56
5	.22637	.97404	.24333	.96994	.26022	.96555	.27704	.96086	55
6	.22665	.97398	.24362	.96987	.26050	.96547	.27731	.96078	54
7	.22693	.97391	.24390	.96980	.26079	.96540	.27759	.96070	53
8	.22722	.97384	.24418	.96973	.26107	.96532	.27787	.96062	52
9	.22750	.97378	.24446	.96966	.26135	.96524	.27815	.96054	51
10	.22778	.97371	.24474	.96959	.26163	.96517	.27843	.96046	50
11	.22807	.97365	.24503	.96952	.26191	.96509	.27871	.96037	49
12	.22835	.97358	.24531	.96945	.26219	.96502	.27899	.96029	48
13	.22863	.97351	.24559	.96937	.26247	.96494	.27927	.96021	47
14	.22892	.97345	.24587	.96930	.26275	.96486	.27955	.96013	46
15	.22920	.97338	.24615	.96923	.26303	.96479	.27983	.96005	45
16	.22948	.97331	.24644	.96916	.26331	.96471	.28011	.95997	44
17	.22977	.97325	.24672	.96909	.26359	.96463	.28039	.95989	43
18	.23005	.97318	.24700	.96902	.26387	.96456	.28067	.95981	42
19	.23033	.97311	.24728	.96894	.26415	.96448	.28095	.95972	41
20	.23062	.97304	.24756	.96887	.26443	.96440	.28123	.95964	40
21	.23090	.97298	.24784	.96880	.26471	.96433	.28150	.95956	39
22	.23118	.97291	.24813	.96873	.26500	.96425	.28178	.95948	38
23	.23146	.97284	.24841	.96866	.26528	.96417	.28206	.95940	37
24	.23175	.97278	.24869	.96858	.26556	.96410	.28234	.95931	36
25	.23203	.97271	.24897	.96851	.26584	.96402	.28262	.95923	35
26	.23231	.97264	.24925	.96844	.26612	.96394	.28290	.95915	34
27	.23260	.97257	.24954	.96837	.26640	.96386	.28318	.95907	33
28	.23288	.97251	.24982	.96829	.26668	.96379	.28346	.95898	32
29	.23316	.97244	.25010	.96822	.26696	.96371	.28374	.95890	31
30	.23345	.97237	.25038	.96815	.26724	.96363	.28402	.95882	30
31	.23373	.97230	.25066	.96807	.26752	.96355	.28429	.95874	29
32	.23401	.97223	.25094	.96800	.26780	.96347	.28457	.95865	28
33	.23429	.97217	.25122	.96793	.26808	.96340	.28485	.95857	27
34	.23458	.97210	.25151	.96786	.26836	.96332	.28513	.95849	26
35	.23486	.97203	.25179	.96778	.26864	.96324	.28541	.95841	25
36	.23514	.97196	.25207	.96771	.26892	.96316	.28569	.95832	24
37	.23542	.97189	.25235	.96764	.26920	.96308	.28597	.95824	23
38	.23571	.97182	.25263	.96756	.26948	.96301	.28625	.95816	22
39	.23599	.97176	.25291	.96749	.26976	.96293	.28652	.95807	21
40	.23627	.97169	.25320	.96742	.27004	.96285	.28680	.95799	20
41	.23656	.97162	.25348	.96734	.27032	.96277	.28708	.95791	19
42	.23684	.97155	.25376	.96727	.27060	.96269	.28736	.95782	18
43	.23712	.97148	.25404	.96719	.27088	.96261	.28764	.95774	17
44	.23740	.97141	.25432	.96712	.27116	.96253	.28792	.95766	16
45	.23769	.97134	.25460	.96705	.27144	.96246	.28820	.95757	15
46	.23797	.97127	.25488	.96697	.27172	.96238	.28847	.95749	14
47	.23825	.97120	.25516	.96690	.27200	.96230	.28875	.95740	13
48	.23853	.97113	.25545	.96682	.27228	.96222	.28903	.95732	12
49	.23882	.97106	.25573	.96675	.27256	.96214	.28931	.95724	11
50	.23910	.97100	.25601	.96667	.27284	.96206	.28959	.95715	10
51	.23938	.97093	.25629	.96660	.27312	.96198	.28987	.95707	9
52	.23966	.97086	.25657	.96653	.27340	.96190	.29015	.95698	8
53	.23995	.97079	.25685	.96645	.27368	.96182	.29042	.95690	7
54	.24023	.97072	.25713	.96638	.27396	.96174	.29070	.95681	6
55	.24051	.97065	.25741	.96630	.27424	.96166	.29098	.95673	5
56	.24079	.97058	.25769	.96623	.27452	.96158	.29126	.95664	4
57	.24108	.97051	.25798	.96615	.27480	.96150	.29154	.95656	3
58	.24136	.97044	.25826	.96608	.27508	.96142	.29182	.95647	2
59	.24164	.97037	.25854	.96600	.27536	.96134	.29209	.95639	1
60	.24192	.97030	.25882	.96593	.27564	.96126	.29237	.95630	0
′	COSINE	SINE 76°	COSINE	SINE 75°	COSINE	SINE 74°	COSINE	SINE 73°	′

Table III. (Continued)

′	17° SINE	17° COSINE	18° SINE	18° COSINE	19° SINE	19° COSINE	20° SINE	20° COSINE	′
0	.29237	.95630	.30902	.95106	.32557	.94552	.34202	.93969	60
1	.29265	.95622	.30929	.95097	.32584	.94542	.34229	.93959	59
2	.29293	.95613	.30957	.95088	.32612	.94533	.34257	.93949	58
3	.29321	.95605	.30985	.95079	.32639	.94523	.34284	.93939	57
4	.29348	.95596	.31012	.95070	.32667	.94514	.34311	.93929	56
5	.29376	.95588	.31040	.95061	.32694	.94504	.34339	.93919	55
6	.29404	.95579	.31068	.95052	.32722	.94495	.34366	.93909	54
7	.29432	.95571	.31095	.95043	.32749	.94485	.34393	.93899	53
8	.29460	.95562	.31123	.95033	.32777	.94476	.34421	.93889	52
9	.29487	.95554	.31151	.95024	.32804	.94466	.34448	.93879	51
10	.29515	.95545	.31178	.95015	.32832	.94457	.34475	.93869	50
11	.29543	.95536	.31206	.95006	.32859	.94447	.34503	.93859	49
12	.29571	.95528	.31233	.94997	.32887	.94438	.34530	.93849	48
13	.29599	.95519	.31261	.94988	.32914	.94428	.34557	.93839	47
14	.29626	.95511	.31289	.94979	.32942	.94418	.34584	.93829	46
15	.29654	.95502	.31316	.94970	.32969	.94409	.34612	.93819	45
16	.29682	.95493	.31344	.94961	.32997	.94399	.34639	.93809	44
17	.29710	.95485	.31372	.94952	.33024	.94390	.34666	.93799	43
18	.29737	.95476	.31399	.94943	.33051	.94380	.34694	.93789	42
19	.29765	.95467	.31427	.94933	.33079	.94370	.34721	.93779	41
20	.29793	.95459	.31454	.94924	.33106	.94361	.34748	.93769	40
21	.29821	.95450	.31482	.94915	.33134	.94351	.34775	.93759	39
22	.29849	.95441	.31510	.94906	.33161	.94342	.34803	.93748	38
23	.29876	.95433	.31537	.94897	.33189	.94332	.34830	.93738	37
24	.29904	.95424	.31565	.94888	.33216	.94322	.34857	.93728	36
25	.29932	.95415	.31593	.94878	.33244	.94313	.34884	.93718	35
26	.29960	.95407	.31620	.94869	.33271	.94303	.34912	.93708	34
27	.29987	.95398	.31648	.94860	.33298	.94293	.34939	.93698	33
28	.30015	.95389	.31675	.94851	.33326	.94284	.34966	.93688	32
29	.30043	.95380	.31703	.94842	.33353	.94274	.34993	.93677	31
30	.30071	.95372	.31730	.94832	.33381	.94264	.35021	.93667	30
31	.30098	.95363	.31758	.94823	.33408	.94254	.35048	.93657	29
32	.30126	.95354	.31786	.94814	.33436	.94245	.35075	.93647	28
33	.30154	.95345	.31813	.94805	.33463	.94235	.35102	.93637	27
34	.30182	.95337	.31841	.94795	.33490	.94225	.35130	.93626	26
35	.30209	.95328	.31868	.94786	.33518	.94215	.35157	.93616	25
36	.30237	.95319	.31896	.94777	.33545	.94206	.35184	.93606	24
37	.30265	.95310	.31923	.94768	.33573	.94196	.35211	.93596	23
38	.30292	.95301	.31951	.94758	.33600	.94186	.35239	.93585	22
39	.30320	.95293	.31979	.94749	.33627	.94176	.35266	.93575	21
40	.30348	.95284	.32006	.94740	.33655	.94167	.35293	.93565	20
41	.30376	.95275	.32034	.94730	.33682	.94157	.35320	.93555	19
42	.30403	.95266	.32061	.94721	.33710	.94147	.35347	.93544	18
43	.30431	.95257	.32089	.94712	.33737	.94137	.35375	.93534	17
44	.30459	.95248	.32116	.94702	.33764	.94127	.35402	.93524	16
45	.30486	.95240	.32144	.94693	.33792	.94118	.35429	.93514	15
46	.30514	.95231	.32171	.94684	.33819	.94108	.35456	.93503	14
47	.30542	.95222	.32199	.94674	.33846	.94098	.35484	.93493	13
48	.30570	.95213	.32227	.94665	.33874	.94088	.35511	.93483	12
49	.30597	.95204	.32254	.94656	.33901	.94078	.35538	.93472	11
50	.30625	.95195	.32282	.94646	.33929	.94068	.35565	.93462	10
51	.30653	.95186	.32309	.94637	.33956	.94058	.35592	.93452	9
52	.30680	.95177	.32337	.94627	.33983	.94049	.35619	.93441	8
53	.30708	.95168	.32364	.94618	.34011	.94039	.35647	.93431	7
54	.30736	.95159	.32392	.94609	.34038	.94029	.35674	.93420	6
55	.30763	.95150	.32419	.94599	.34065	.94019	.35701	.93410	5
56	.30791	.95142	.32447	.94590	.34093	.94009	.35728	.93400	4
57	.30819	.95133	.32474	.94580	.34120	.93999	.35755	.93389	3
58	.30846	.95124	.32502	.94571	.34147	.93989	.35782	.93379	2
59	.30874	.95115	.32529	.94561	.34175	.93979	.35810	.93368	1
60	.30902	.95106	.32557	.94552	.34202	.93969	.35837	.93358	0
′	COSINE	SINE 72°	COSINE	SINE 71°	COSINE	SINE 70°	COSINE	SINE 69°	

Table III. (Continued)

′	21° SINE	21° COSINE	22° SINE	22° COSINE	23° SINE	23° COSINE	24° SINE	24° COSINE	′
0	.35837	.93358	.37461	.92718	.39073	.92050	.40674	.91355	60
1	.35864	.93348	.37488	.92707	.39100	.92039	.40700	.91343	59
2	.35891	.93337	.37515	.92697	.39127	.92028	.40727	.91331	58
3	.35918	.93327	.37542	.92686	.39153	.92016	.40753	.91319	57
4	.35945	.93316	.37569	.92675	.39180	.92005	.40780	.91307	56
5	.35973	.93306	.37595	.92664	.39207	.91994	.40806	.91295	55
6	.36000	.93295	.37622	.92653	.39234	.91982	.40833	.91283	54
7	.36027	.93285	.37649	.92642	.39260	.91971	.40860	.91272	53
8	.36054	.93274	.37676	.92631	.39287	.91959	.40886	.91260	52
9	.36081	.93264	.37703	.92620	.39314	.91948	.40913	.91248	51
10	.36108	.93253	.37730	.92609	.39341	.91936	.40939	.91236	50
11	.36135	.93243	.37757	.92598	.39367	.91925	.40966	.91224	49
12	.36162	.93232	.37784	.92587	.39394	.91914	.40992	.91212	48
13	.36190	.93222	.37811	.92576	.39421	.91902	.41019	.91200	47
14	.36217	.93211	.37838	.92565	.39448	.91891	.41045	.91188	46
15	.36244	.93201	.37865	.92554	.39474	.91879	.41072	.91176	45
16	.36271	.93190	.37892	.92543	.39501	.91868	.41098	.91164	44
17	.36298	.93180	.37919	.92532	.39528	.91856	.41125	.91152	43
18	.36325	.93169	.37946	.92521	.39555	.91845	.41151	.91140	42
19	.36352	.93159	.37973	.92510	.39581	.91833	.41178	.91128	41
20	.36379	.93148	.37999	.92499	.39608	.91822	.41204	.91116	40
21	.36406	.93137	.38026	.92488	.39635	.91810	.41231	.91104	39
22	.36434	.93127	.38053	.92477	.39661	.91799	.41257	.91092	38
23	.36461	.93116	.38080	.92466	.39688	.91787	.41284	.91080	37
24	.36488	.93106	.38107	.92455	.39715	.91775	.41310	.91068	36
25	.36515	.93095	.38134	.92444	.39741	.91764	.41337	.91056	35
26	.36542	.93084	.38161	.92432	.39768	.91752	.41363	.91044	34
27	.36569	.93074	.38188	.92421	.39795	.91741	.41390	.91032	33
28	.36596	.93063	.38215	.92410	.39822	.91729	.41416	.91020	32
29	.36623	.93052	.38241	.92399	.39848	.91718	.41443	.91008	31
30	.36650	.93042	.38268	.92388	.39875	.91706	.41469	.90996	30
31	.36677	.93031	.38295	.92377	.39902	.91694	.41496	.90984	29
32	.36704	.93020	.38322	.92366	.39928	.91683	.41522	.90972	28
33	.36731	.93010	.38349	.92355	.39955	.91671	.41549	.90960	27
34	.36758	.92999	.38376	.92343	.39982	.91660	.41575	.90948	26
35	.36785	.92988	.38403	.92332	.40008	.91648	.41602	.90936	25
36	.36812	.92978	.38430	.92321	.40035	.91636	.41628	.90924	24
37	.36839	.92967	.38456	.92310	.40062	.91625	.41655	.90911	23
38	.36867	.92956	.38483	.92299	.40088	.91613	.41681	.90899	22
39	.36894	.92945	.38510	.92287	.40115	.91601	.41707	.90887	21
40	.36921	.92935	.38537	.92276	.40141	.91590	.41734	.90875	20
41	.36948	.92924	.38564	.92265	.40168	.91578	.41760	.90863	19
42	.36975	.92913	.38591	.92254	.40195	.91566	.41787	.90851	18
43	.37002	.92902	.38617	.92243	.40221	.91555	.41813	.90839	17
44	.37029	.92892	.38644	.92231	.40248	.91543	.41840	.90826	16
45	.37056	.92881	.38671	.92220	.40275	.91531	.41866	.90814	15
46	.37083	.92870	.38698	.92209	.40301	.91519	.41892	.90802	14
47	.37110	.92859	.38725	.92198	.40328	.91508	.41919	.90790	13
48	.37137	.92849	.38752	.92186	.40355	.91496	.41945	.90778	12
49	.37164	.92838	.38778	.92175	.40381	.91484	.41972	.90766	11
50	.37191	.92827	.38805	.92164	.40408	.91472	.41998	.90753	10
51	.37218	.92816	.38832	.92152	.40434	.91461	.42024	.90741	9
52	.37245	.92805	.38859	.92141	.40461	.91449	.42051	.90729	8
53	.37272	.92794	.38886	.92130	.40488	.91437	.42077	.90717	7
54	.37299	.92784	.38912	.92119	.40514	.91425	.42104	.90704	6
55	.37326	.92773	.38939	.92107	.40541	.91414	.42130	.90692	5
56	.37353	.92762	.38966	.92096	.40567	.91402	.42156	.90680	4
57	.37380	.92751	.38993	.92085	.40594	.91390	.42183	.90668	3
58	.37407	.92740	.39020	.92073	.40621	.91378	.42209	.90655	2
59	.37434	.92729	.39046	.92062	.40647	.91366	.42235	.90643	1
60	.37461	.92718	.39073	.92050	.40674	.91355	.42262	.90631	0
′	COSINE	SINE 68°	COSINE	SINE 67°	COSINE	SINE 66°	COSINE	SINE 65°	′

Table III. (Continued)

′	25° SINE	25° COSINE	26° SINE	26° COSINE	27° SINE	27° COSINE	28° SINE	28° COSINE	′
0	.42262	.90631	.43837	.89879	.45399	.89101	.46947	.88295	60
1	.42288	.90618	.43863	.89867	.45425	.89087	.46973	.88281	59
2	.42315	.90606	.43889	.89854	.45451	.89074	.46999	.88267	58
3	.42341	.90594	.43916	.89841	.45477	.89061	.47024	.88254	57
4	.42367	.90582	.43942	.89828	.45503	.89048	.47050	.88240	56
5	.42394	.90569	.43968	.89816	.45529	.89035	.47076	.88226	55
6	.42420	.90557	.43994	.89803	.45554	.89021	.47101	.88213	54
7	.42446	.90545	.44020	.89790	.45580	.89008	.47127	.88199	53
8	.42473	.90532	.44046	.89777	.45606	.88995	.47153	.88185	52
9	.42499	.90520	.44072	.89764	.45632	.88981	.47178	.88172	51
10	.42525	.90507	.44098	.89752	.45658	.88968	.47204	.88158	50
11	.42552	.90495	.44124	.89739	.45684	.88955	.47229	.88144	49
12	.42578	.90483	.44151	.89726	.45710	.88942	.47255	.88130	48
13	.42604	.90470	.44177	.89713	.45736	.88928	.47281	.88117	47
14	.42631	.90458	.44203	.89700	.45762	.88915	.47306	.88103	46
15	.42657	.90446	.44229	.89687	.45787	.88902	.47332	.88089	45
16	.42683	.90433	.44255	.89674	.45813	.88888	.47358	.88075	44
17	.42709	.90421	.44281	.89662	.45839	.88875	.47383	.88062	43
18	.42736	.90408	.44307	.89649	.45865	.88862	.47409	.88048	42
19	.42762	.90396	.44333	.89636	.45891	.88848	.47434	.88034	41
20	.42788	.90383	.44359	.89623	.45917	.88835	.47460	.88020	40
21	.42815	.90371	.44385	.89610	.45942	.88822	.47486	.88006	39
22	.42841	.90358	.44411	.89597	.45968	.88808	.47511	.87993	38
23	.42867	.90346	.44437	.89584	.45994	.88795	.47537	.87979	37
24	.42894	.90334	.44464	.89571	.46020	.88782	.47562	.87965	36
25	.42920	.90321	.44490	.89558	.46046	.88768	.47588	.87951	35
26	.42946	.90309	.44516	.89545	.46072	.88755	.47614	.87937	34
27	.42972	.90296	.44542	.89532	.46097	.88741	.47639	.87923	33
28	.42999	.90284	.44568	.89519	.46123	.88728	.47665	.87909	32
29	.43025	.90271	.44594	.89506	.46149	.88715	.47690	.87896	31
30	.43051	.90259	.44620	.89493	.46175	.88701	.47716	.87882	30
31	.43077	.90246	.44646	.89480	.46201	.88688	.47741	.87868	29
32	.43104	.90233	.44672	.89467	.46226	.88674	.47767	.87854	28
33	.43130	.90221	.44698	.89454	.46252	.88661	.47793	.87840	27
34	.43156	.90208	.44724	.89441	.46278	.88647	.47818	.87826	26
35	.43182	.90196	.44750	.89428	.46304	.88634	.47844	.87812	25
36	.43209	.90183	.44776	.89415	.46330	.88620	.47869	.87798	24
37	.43235	.90171	.44802	.89402	.46355	.88607	.47895	.87784	23
38	.43261	.90158	.44828	.89389	.46381	.88593	.47920	.87770	22
39	.43287	.90146	.44854	.89376	.46407	.88580	.47946	.87756	21
40	.43313	.90133	.44880	.89363	.46433	.88566	.47971	.87743	20
41	.43340	.90120	.44906	.89350	.46458	.88553	.47997	.87729	19
42	.43366	.90108	.44932	.89337	.46484	.88539	.48022	.87715	18
43	.43392	.90095	.44958	.89324	.46510	.88526	.48048	.87701	17
44	.43418	.90082	.44984	.89311	.46536	.88512	.48073	.87687	16
45	.43445	.90070	.45010	.89298	.46561	.88499	.48099	.87673	15
46	.43471	.90057	.45036	.89285	.46587	.88485	.48124	.87659	14
47	.43497	.90045	.45062	.89272	.46613	.88472	.48150	.87645	13
48	.43523	.90032	.45088	.89259	.46639	.88458	.48175	.87631	12
49	.43549	.90019	.45114	.89245	.46664	.88445	.48201	.87617	11
50	.43575	.90007	.45140	.89232	.46690	.88431	.48226	.87603	10
51	.43602	.89994	.45166	.89219	.46716	.88417	.48252	.87589	9
52	.43628	.89981	.45192	.89206	.46742	.88404	.48277	.87575	8
53	.43654	.89968	.45218	.89193	.46767	.88390	.48303	.87561	7
54	.43680	.89956	.45243	.89180	.46793	.88377	.48328	.87546	6
55	.43706	.89943	.45269	.89167	.46819	.88363	.48354	.87532	5
56	.43733	.89930	.45295	.89153	.46844	.88349	.48379	.87518	4
57	.43759	.89918	.45321	.89140	.46870	.88336	.48405	.87504	3
58	.43785	.89905	.45347	.89127	.46896	.88322	.48430	.87490	2
59	.43811	.89892	.45373	.89114	.46921	.88308	.48456	.87476	1
60	.43837	.89879	.45399	.89101	.46947	.88295	.48481	.87462	0
′	COSINE	SINE	COSINE	SINE	COSINE	SINE	COSINE	SINE	′
	64°		63°		62°		61°		

Table III. (Continued)

′	29° SINE	29° COSINE	30° SINE	30° COSINE	31° SINE	31° COSINE	32° SINE	32° COSINE	′
0	.48481	.87462	.50000	.86603	.51504	.85717	.52992	.84805	60
1	.48506	.87448	.50025	.86588	.51529	.85702	.53017	.84789	59
2	.48532	.87434	.50050	.86573	.51554	.85687	.53041	.84774	58
3	.48557	.87420	.50076	.86559	.51579	.85672	.53066	.84759	57
4	.48583	.87406	.50101	.86544	.51604	.85657	.53091	.84743	56
5	.48608	.87391	.50126	.86530	.51628	.85642	.53115	.84728	55
6	.48634	.87377	.50151	.86515	.51653	.85627	.53140	.84712	54
7	.48659	.87363	.50176	.86501	.51678	.85612	.53164	.84697	53
8	.48684	.87349	.50201	.86486	.51703	.85597	.53189	.84681	52
9	.48710	.87335	.50227	.86471	.51728	.85582	.53214	.84666	51
10	.48735	.87321	.50252	.86457	.51753	.85567	.53238	.84650	50
11	.48761	.87306	.50277	.86442	.51778	.85551	.53263	.84635	49
12	.48786	.87292	.50302	.86427	.51803	.85536	.53288	.84619	48
13	.48811	.87278	.50327	.86413	.51828	.85521	.53312	.84604	47
14	.48837	.87264	.50352	.86398	.51852	.85506	.53337	.84588	46
15	.48862	.87250	.50377	.86384	.51877	.85491	.53361	.84573	45
16	.48888	.87235	.50403	.86369	.51902	.85476	.53386	.84557	44
17	.48913	.87221	.50428	.86354	.51927	.85461	.53411	.84542	43
18	.48938	.87207	.50453	.86340	.51952	.85446	.53435	.84526	42
19	.48964	.87193	.50478	.86325	.51977	.85431	.53460	.84511	41
20	.48989	.87178	.50503	.86310	.52002	.85416	.53484	.84495	40
21	.49014	.87164	.50528	.86295	.52026	.85401	.53509	.84480	39
22	.49040	.87150	.50553	.86281	.52051	.85385	.53534	.84464	38
23	.49065	.87136	.50578	.86266	.52076	.85370	.53558	.84448	37
24	.49090	.87121	.50603	.86251	.52101	.85355	.53583	.84433	36
25	.49116	.87107	.50628	.86237	.52126	.85340	.53607	.84417	35
26	.49141	.87093	.50654	.86222	.52151	.85325	.53632	.84402	34
27	.49166	.87079	.50679	.86207	.52175	.85310	.53656	.84386	33
28	.49192	.87064	.50704	.86192	.52200	.85294	.53681	.84370	32
29	.49217	.87050	.50729	.86178	.52225	.85279	.53705	.84355	31
30	.49242	.87036	.50754	.86163	.52250	.85264	.53730	.84339	30
31	.49268	.87021	.50779	.86148	.52275	.85249	.53754	.84324	29
32	.49293	.87007	.50804	.86133	.52299	.85234	.53779	.84308	28
33	.49318	.86993	.50829	.86119	.52324	.85218	.53804	.84292	27
34	.49344	.86978	.50854	.86104	.52349	.85203	.53828	.84277	26
35	.49369	.86964	.50879	.86089	.52374	.85188	.53853	.84261	25
36	.49394	.86949	.50904	.86074	.52399	.85173	.53877	.84245	24
37	.49419	.86935	.50929	.86059	.52423	.85157	.53902	.84230	23
38	.49445	.86921	.50954	.86045	.52448	.85142	.53926	.84214	22
39	.49470	.86906	.50979	.86030	.52473	.85127	.53951	.84198	21
40	.49495	.86892	.51004	.86015	.52498	.85112	.53975	.84182	20
41	.49521	.86878	.51029	.86000	.52522	.85096	.54000	.84167	19
42	.49546	.86863	.51054	.85985	.52547	.85081	.54024	.84151	18
43	.49571	.86849	.51079	.85970	.52572	.85066	.54049	.84135	17
44	.49596	.86834	.51104	.85956	.52597	.85051	.54073	.84120	16
45	.49622	.86820	.51129	.85941	.52621	.85035	.54097	.84104	15
46	.49647	.86805	.51154	.85926	.52646	.85020	.54122	.84088	14
47	.49672	.86791	.51179	.85911	.52671	.85005	.54146	.84072	13
48	.49697	.86777	.51204	.85896	.52696	.84989	.54171	.84057	12
49	.49723	.86762	.51229	.85881	.52720	.84974	.54195	.84041	11
50	.49748	.86748	.51254	.85866	.52745	.84959	.54220	.84025	10
51	.49773	.86733	.51279	.85851	.52770	.84943	.54244	.84009	9
52	.49798	.86719	.51304	.85836	.52794	.84928	.54269	.83994	8
53	.49824	.86704	.51329	.85821	.52819	.84913	.54293	.83978	7
54	.49849	.86690	.51354	.85806	.52844	.84897	.54317	.83962	6
55	.49874	.86675	.51379	.85792	.52869	.84882	.54342	.83946	5
56	.49899	.86661	.51404	.85777	.52893	.84866	.54366	.83930	4
57	.49924	.86646	.51429	.85762	.52918	.84851	.54391	.83915	3
58	.49950	.86632	.51454	.85747	.52943	.84836	.54415	.83899	2
59	.49975	.86617	.51479	.85732	.52967	.84820	.54440	.83883	1
60	.50000	.86603	.51504	.85717	.52992	.84805	.54464	.83867	0
′	COSINE	SINE 60°	COSINE	SINE 59°	COSINE	SINE 58°	COSINE	SINE 57°	′

Table III. (Continued)

′	33° SINE	33° COSINE	34° SINE	34° COSINE	35° SINE	35° COSINE	36° SINE	36° COSINE	′
0	.54464	.83867	.55919	.82904	.57358	.81915	.58779	.80902	60
1	.54488	.83851	.55943	.82887	.57381	.81899	.58802	.80885	59
2	.54513	.83835	.55968	.82871	.57405	.81882	.58826	.80867	58
3	.54537	.83819	.55992	.82855	.57429	.81865	.58849	.80850	57
4	.54561	.83804	.56016	.82839	.57453	.81848	.58873	.80833	56
5	.54586	.83788	.56040	.82822	.57477	.81832	.58896	.80816	55
6	.54610	.83772	.56064	.82806	.57501	.81815	.58920	.80799	54
7	.54635	.83756	.56088	.82790	.57524	.81798	.58943	.80782	53
8	.54659	.83740	.56112	.82773	.57548	.81782	.58967	.80765	52
9	.54683	.83724	.56136	.82757	.57572	.81765	.58990	.80748	51
10	.54708	.83708	.56160	.82741	.57596	.81748	.59014	.80730	50
11	.54732	.83692	.56184	.82724	.57619	.81731	.59037	.80713	49
12	.54756	.83676	.56208	.82708	.57643	.81714	.59061	.80696	48
13	.54781	.83660	.56232	.82692	.57667	.81698	.59084	.80679	47
14	.54805	.83645	.56256	.82675	.57691	.81681	.59108	.80662	46
15	.54829	.83629	.56280	.82659	.57715	.81664	.59131	.80644	45
16	.54854	.83613	.56305	.82643	.57738	.81647	.59154	.80627	44
17	.54878	.83597	.56329	.82626	.57762	.81631	.59178	.80610	43
18	.54902	.83581	.56353	.82610	.57786	.81614	.59201	.80593	42
19	.54927	.83565	.56377	.82593	.57810	.81597	.59225	.80576	41
20	.54951	.83549	.56401	.82577	.57833	.81580	.59248	.80558	40
21	.54975	.83533	.56425	.82561	.57857	.81563	.59272	.80541	39
22	.54999	.83517	.56449	.82544	.57881	.81546	.59295	.80524	38
23	.55024	.83501	.56473	.82528	.57904	.81530	.59318	.80507	37
24	.55048	.83485	.56497	.82511	.57928	.81513	.59342	.80489	36
25	.55072	.83469	.56521	.82495	.57952	.81496	.59365	.80472	35
26	.55097	.83453	.56545	.82478	.57976	.81479	.59389	.80455	34
27	.55121	.83437	.56569	.82462	.57999	.81462	.59412	.80438	33
28	.55145	.83421	.56593	.82446	.58023	.81445	.59436	.80420	32
29	.55169	.83405	.56617	.82429	.58047	.81428	.59459	.80403	31
30	.55194	.83389	.56641	.82413	.58070	.81412	.59482	.80386	30
31	.55218	.83373	.56665	.82396	.58094	.81395	.59506	.80368	29
32	.55242	.83356	.56689	.82380	.58118	.81378	.59529	.80351	28
33	.55266	.83340	.56713	.82363	.58141	.81361	.59552	.80334	27
34	.55291	.83324	.56736	.82347	.58165	.81344	.59576	.80316	26
35	.55315	.83308	.56760	.82330	.58189	.81327	.59599	.80299	25
36	.55339	.83292	.56784	.82314	.58212	.81310	.59622	.80282	24
37	.55363	.83276	.56808	.82297	.58236	.81293	.59646	.80264	23
38	.55388	.83260	.56832	.82281	.58260	.81276	.59669	.80247	22
39	.55412	.83244	.56856	.82264	.58283	.81259	.59693	.80230	21
40	.55436	.83228	.56880	.82248	.58307	.81242	.59716	.80212	20
41	.55460	.83212	.56904	.82231	.58330	.81225	.59739	.80195	19
42	.55484	.83195	.56928	.82214	.58354	.81208	.59763	.80178	18
43	.55509	.83179	.56952	.82198	.58378	.81191	.59786	.80160	17
44	.55533	.83163	.56976	.82181	.58401	.81174	.59809	.80143	16
45	.55557	.83147	.57000	.82165	.58425	.81157	.59832	.80125	15
46	.55581	.83131	.57024	.82148	.58449	.81140	.59856	.80108	14
47	.55605	.83115	.57047	.82132	.58472	.81123	.59879	.80091	13
48	.55630	.83098	.57071	.82115	.58496	.81106	.59902	.80073	12
49	.55654	.83082	.57095	.82098	.58519	.81089	.59926	.80056	11
50	.55678	.83066	.57119	.82082	.58543	.81072	.59949	.80038	10
51	.55702	.83050	.57143	.82065	.58567	.81055	.59972	.80021	9
52	.55726	.83034	.57167	.82048	.58590	.81038	.59995	.80003	8
53	.55750	.83017	.57191	.82032	.58614	.81021	.60019	.79986	7
54	.55775	.83001	.57215	.82015	.58637	.81004	.60042	.79968	6
55	.55799	.82985	.57238	.81999	.58661	.80987	.60065	.79951	5
56	.55823	.82969	.57262	.81982	.58684	.80970	.60089	.79934	4
57	.55847	.82953	.57286	.81965	.58708	.80953	.60112	.79916	3
58	.55871	.82936	.57310	.81949	.58731	.80936	.60135	.79899	2
59	.55895	.82920	.57334	.81932	.58755	.80919	.60158	.79881	1
60	.55919	.82904	.57358	.81915	.58779	.80902	.60182	.79864	0
′	COSINE	SINE 56°	COSINE	SINE 55°	COSINE	SINE 54°	COSINE	SINE 53°	′

Table III. (Continued)

′	37° Sine	Cosine	38° Sine	Cosine	39° Sine	Cosine	40° Sine	Cosine	′
0	.60182	.79864	.61566	.78801	.62932	.77715	.64279	.76604	60
1	.60205	.79846	.61589	.78783	.62955	.77696	.64301	.76586	59
2	.60228	.79829	.61612	.78765	.62977	.77678	.64323	.76567	58
3	.60251	.79811	.61635	.78747	.63000	.77660	.64346	.76548	57
4	.60274	.79793	.61658	.78729	.63022	.77641	.64368	.76530	56
5	.60298	.79776	.61681	.78711	.63045	.77623	.64390	.76511	55
6	.60321	.79758	.61704	.78694	.63068	.77605	.64412	.76492	54
7	.60344	.79741	.61726	.78676	.63090	.77586	.64435	.76473	53
8	.60367	.79723	.61749	.78658	.63113	.77568	.64457	.76455	52
9	.60390	.79706	.61772	.78640	.63135	.77550	.64479	.76436	51
10	.60414	.79688	.61795	.78622	.63158	.77531	.64501	.76417	50
11	.60437	.79671	.61818	.78604	.63180	.77513	.64524	.76398	49
12	.60460	.79653	.61841	.78586	.63203	.77494	.64546	.76380	48
13	.60483	.79635	.61864	.78568	.63225	.77476	.64568	.76361	47
14	.60506	.79618	.61887	.78550	.63248	.77458	.64590	.76342	46
15	.60529	.79600	.61909	.78532	.63271	.77439	.64612	.76323	45
16	.60553	.79583	.61932	.78514	.63293	.77421	.64635	.76304	44
17	.60576	.79565	.61955	.78496	.63316	.77402	.64657	.76286	43
18	.60599	.79547	.61978	.78478	.63338	.77384	.64679	.76267	42
19	.60622	.79530	.62001	.78460	.63361	.77366	.64701	.76248	41
20	.60645	.79512	.62024	.78442	.63383	.77347	.64723	.76229	40
21	.60668	.79494	.62046	.78424	.63406	.77329	.64746	.76210	39
22	.60691	.79477	.62069	.78405	.63428	.77310	.64768	.76192	38
23	.60714	.79459	.62092	.78387	.63451	.77292	.64790	.76173	37
24	.60738	.79441	.62115	.78369	.63473	.77273	.64812	.76154	36
25	.60761	.79424	.62138	.78351	.63496	.77255	.64834	.76135	35
26	.60784	.79406	.62160	.78333	.63518	.77236	.64856	.76116	34
27	.60807	.79388	.62183	.78315	.63540	.77218	.64878	.76097	33
28	.60830	.79371	.62206	.78297	.63563	.77199	.64901	.76078	32
29	.60853	.79353	.62229	.78279	.63585	.77181	.64923	.76059	31
30	.60876	.79335	.62251	.78261	.63608	.77162	.64945	.76041	30
31	.60899	.79318	.62274	.78243	.63630	.77144	.64967	.76022	29
32	.60922	.79300	.62297	.78225	.63653	.77125	.64989	.76003	28
33	.60945	.79282	.62320	.78206	.63675	.77107	.65011	.75984	27
34	.60968	.79264	.62342	.78188	.63698	.77088	.65033	.75965	26
35	.60991	.79247	.62365	.78170	.63720	.77070	.65055	.75946	25
36	.61015	.79229	.62388	.78152	.63742	.77051	.65077	.75927	24
37	.61038	.79211	.62411	.78134	.63765	.77033	.65100	.75908	23
38	.61061	.79193	.62433	.78116	.63787	.77014	.65122	.75889	22
39	.61084	.79176	.62456	.78098	.63810	.76996	.65144	.75870	21
40	.61107	.79158	.62479	.78079	.63832	.76977	.65166	.75851	20
41	.61130	.79140	.62502	.78061	.63854	.76959	.65188	.75832	19
42	.61153	.79122	.62524	.78043	.63877	.76940	.65210	.75813	18
43	.61176	.79105	.62547	.78025	.63899	.76921	.65232	.75794	17
44	.61199	.79087	.62570	.78007	.63922	.76903	.65254	.75775	16
45	.61222	.79069	.62592	.77988	.63944	.76884	.65276	.75756	15
46	.61245	.79051	.62615	.77970	.63966	.76866	.65298	.75738	14
47	.61268	.79033	.62638	.77952	.63989	.76847	.65320	.75719	13
48	.61291	.79016	.62660	.77934	.64011	.76828	.65342	.75700	12
49	.61314	.78998	.62683	.77916	.64033	.76810	.65364	.75680	11
50	.61337	.78980	.62706	.77897	.64056	.76791	.65386	.75661	10
51	.61360	.78962	.62728	.77879	.64078	.76772	.65408	.75642	9
52	.61383	.78944	.62751	.77861	.64100	.76754	.65430	.75623	8
53	.61406	.78926	.62774	.77843	.64123	.76735	.65452	.75604	7
54	.61429	.78908	.62796	.77824	.64145	.76717	.65474	.75585	6
55	.61451	.78891	.62819	.77806	.64167	.76698	.65496	.75566	5
56	.61474	.78873	.62842	.77788	.64190	.76679	.65518	.75547	4
57	.61497	.78855	.62864	.77769	.64212	.76661	.65540	.75528	3
58	.61520	.78837	.62887	.77751	.64234	.76642	.65562	.75509	2
59	.61543	.78819	.62909	.77733	.64256	.76623	.65584	.75490	1
60	.61566	.78801	.62932	.77715	.64279	.76604	.65606	.75471	0
′	Cosine	Sine 52°	Cosine	Sine 51°	Cosine	Sine 50°	Cosine	Sine 49°	′

Table III. (Concluded)

′	41° SINE	41° COSINE	42° SINE	42° COSINE	43° SINE	43° COSINE	44° SINE	44° COSINE	′
0	.65606	.75471	.66913	.74314	.68200	.73135	.69466	.71934	60
1	.65628	.75452	.66935	.74295	.68221	.73116	.69487	.71914	59
2	.65650	.75433	.66956	.74276	.68242	.73096	.69508	.71894	58
3	.65672	.75414	.66978	.74256	.68264	.73076	.69529	.71873	57
4	.65694	.75395	.66999	.74237	.68285	.73056	.69549	.71853	56
5	.65716	.75375	.67021	.74217	.68306	.73036	.69570	.71833	55
6	.65738	.75356	.67043	.74198	.68327	.73016	.69591	.71813	54
7	.65759	.75337	.67064	.74178	.68349	.72996	.69612	.71792	53
8	.65781	.75318	.67086	.71159	.68370	.72976	.69633	.71772	52
9	.65803	.75299	.67107	.74139	.68391	.72957	.69654	.71752	51
10	.65825	.75280	.67129	.74120	.68412	.72937	.69675	.71732	50
11	.65847	.75261	.67151	.74100	.68434	.72917	.69696	.71711	49
12	.65869	.75241	.67172	.74080	.68455	.72897	.69717	.71691	48
13	.65891	.75222	.67194	.74061	.68476	.72877	.69737	.71671	47
14	.65913	.75203	.67215	.74041	.68497	.72857	.69758	.71650	46
15	.65935	.75184	.67237	.74022	.68518	.72837	.69779	.71630	45
16	.65956	.75165	.67258	.74002	.68539	.72817	.69800	.71610	44
17	.65978	.75146	.67280	.73983	.68561	.72797	.69821	.71590	43
18	.66000	.75126	.67301	.73963	.68582	.72777	.69842	.71569	42
19	.66022	.75107	.67323	.73944	.68603	.72757	.69862	.71549	41
20	.66044	.75088	.67344	.73924	.68624	.72737	.69883	.71529	40
21	.66066	.75069	.67366	.73904	.68645	.72717	.69904	.71508	39
22	.66088	.75050	.67387	.73885	.68666	.72697	.69925	.71488	38
23	.66109	.75030	.67409	.73865	.68688	.72677	.69946	.71468	37
24	.66131	.75011	.67430	.73846	.68709	.72657	.69966	.71447	36
25	.66153	.74992	.67452	.73826	.68730	.72637	.69987	.71427	35
26	.66175	.74973	.67473	.73806	.68751	.72617	.70008	.71407	34
27	.66197	.74953	.67495	.73787	.68772	.72597	.70029	.71386	33
28	.66218	.74934	.67516	.73767	.68793	.72577	.70049	.71366	32
29	.66240	.74915	.67538	.73747	.68814	.72557	.70070	.71345	31
80	.66262	.74896	.67559	.73728	.68835	.72537	.70091	.71325	30
31	.66284	.74876	.67580	.73708	.68857	.72517	.70112	.71305	29
32	.66306	.74857	.67602	.73688	.68878	.72497	.70132	.71284	28
33	.66327	.74838	.67623	.73669	.68899	.72477	.70153	.71264	27
34	.66349	.74818	.67645	.73649	.68920	.72457	.70174	.71243	26
35	.66371	.74799	.67666	.73629	.68941	.72437	.70195	.71223	25
36	.66393	.74780	.67688	.73610	.68962	.72417	.70215	.71203	24
37	.66414	.74760	.67709	.73590	.68983	.72397	.70236	.71182	23
38	.66436	.74741	.67730	.73570	.69004	.72377	.70257	.71162	22
39	.66458	.74722	.67752	.73551	.69025	.72357	.70277	.71141	21
40	.66480	.74703	.67773	.73531	.69046	.72337	.70298	.71121	20
41	.66501	.74683	.67795	.73511	.69067	.72317	.70319	.71100	19
42	.66523	.74664	.67816	.73491	.69088	.72297	.70339	.71080	18
43	.66545	.74644	.67837	.73472	.69109	.72277	.70360	.71059	17
44	.66566	.74625	.67859	.73452	.69130	.72257	.70381	.71039	16
45	.66588	.74606	.67880	.73432	.69151	.72236	.70401	.71019	15
46	.66610	.74586	.67901	.73413	.69172	.72216	.70422	.70998	14
47	.66632	.74567	.67923	.73393	.69193	.72196	.70443	.70978	13
48	.66653	.74548	.67944	.73373	.69214	.72176	.70463	.70957	12
49	.66675	.74528	.67965	.73353	.69235	.72156	.70484	.70937	11
50	.66697	.74509	.67987	.73333	.69256	.72136	.70505	.70916	10
51	.66718	.74489	.68008	.73314	.69277	.72116	.70525	.70896	9
52	.66740	.74470	.68029	.73294	.69298	.72095	.70546	.70875	8
53	.66762	.74451	.68051	.73274	.69319	.72075	.70567	.70855	7
54	.66783	.74431	.68072	.73254	.69340	.72055	.70587	.70834	6
55	.66805	.74412	.68093	.73234	.69361	.72035	.70608	.70813	5
56	.66827	.74392	.68115	.73215	.69382	.72015	.70628	.70793	4
57	.66848	.74373	.68136	.73195	.69403	.71995	.70649	.70772	3
58	.66870	.74353	.68157	.73175	.69424	.71974	.70670	.70752	2
59	.66891	.74334	.68179	.73155	.69445	.71954	.70690	.70731	1
60	.66913	.74314	.68200	.73135	.69466	.71934	.70711	.70711	0
′	COSINE	SINE	COSINE	SINE	COSINE	SINE	COSINE	SINE	′
	48°		47°		46°		45°		

Table IV. Natural Tangents and Cotangents

′	0° TAN.	0° CO-TAN.	1° TAN.	1° CO-TAN.	2° TAN.	2° CO-TAN.	3° TAN.	3° CO-TAN.	′
0	.00000	Infinite.	.01746	57.2900	.03492	28.6363	.05241	19.0811	60
1	.00029	3437.750	.01775	56.3506	.03521	28.3994	.05270	18.9755	59
2	.00058	1718.870	.01804	55.4415	.03550	28.1664	.05299	18.8711	58
3	.00087	1145.920	.01833	54.5613	.03579	27.9372	.05328	18.7678	57
4	.00116	859.436	.01862	53.7086	.03609	27.7117	.05357	18.6656	56
5	.00145	687.549	.01891	52.8821	.03638	27.4899	.05387	18.5645	55
6	.00175	572.957	.01920	52.0807	.03667	27.2715	.05416	18.4645	54
7	.00204	491.106	.01949	51.3032	.03696	27.0566	.05445	18.3655	53
8	.00233	429.718	.01978	50.5485	.03725	26.8450	.05474	18.2677	52
9	.00262	381.971	.02007	49.8157	.03754	26.6367	.05503	18.1708	51
10	.00291	343.774	.02036	49.1039	.03783	26.4316	.05533	18.0750	50
11	.00320	312.521	.02066	48.4121	.03812	26.2296	.05562	17.9802	49
12	.00349	286.478	.02095	47.7395	.03842	26.0307	.05591	17.8863	48
13	.00378	264.441	.02124	47.0853	.03871	25.8348	.05620	17.7934	47
14	.00407	245.552	.02153	46.4489	.03900	25.6418	.05649	17.7015	46
15	.00436	229.182	.02182	45.8294	.03929	25.4517	.05678	17.6106	45
16	.00465	214.858	.02211	45.2261	.03958	25.2644	.05708	17.5205	44
17	.00495	202.219	.02240	44.6386	.03987	25.0798	.05737	17.4314	43
18	.00524	190.984	.02269	44.0661	.04016	24.8978	.05766	17.3432	42
19	.00553	180.932	.02298	43.5081	.04046	24.7185	.05795	17.2558	41
20	.00582	171.885	.02328	42.9641	.04075	24.5418	.05824	17.1693	40
21	.00611	163.700	.02357	42.4335	.04104	24.3675	.05854	17.0837	39
22	.00640	156.259	.02386	41.9158	.04133	24.1957	.05883	16.9990	38
23	.00669	149.465	.02415	41.4106	.04162	24.0263	.05912	16.9150	37
24	.00698	143.237	.02444	40.9174	.04191	23.8593	.05941	16.8319	36
25	.00727	137.507	.02473	40.4358	.04220	23.6945	.05970	16.7496	35
26	.00756	132.219	.02502	39.9655	.04250	23.5321	.05999	16.6681	34
27	.00785	127.321	.02531	39.5059	.04279	23.3718	.06029	16.5874	33
28	.00814	122.774	.02560	39.0568	.04308	23.2137	.06058	16.5075	32
29	.00844	118.540	.02589	38.6177	.04337	23.0577	.06087	16.4283	31
30	.00873	114.589	.02619	38.1885	.04366	22.9038	.06116	16.3499	30
31	.00902	110.892	.02648	37.7686	.04395	22.7519	.06145	16.2722	29
32	.00931	107.426	.02677	37.3579	.04424	22.6020	.06175	16.1952	28
33	.00960	104.171	.02706	36.9560	.04454	22.4541	.06204	16.1190	27
34	.00989	101.107	.02735	36.5627	.04483	22.3081	.06233	16.0435	26
35	.01018	98.2179	.02764	36.1776	.04512	22.1640	.06262	15.9687	25
36	.01047	95.4895	.02793	35.8006	.04541	22.0217	.06291	15.8945	24
37	.01076	92.9085	.02822	35.4313	.04570	21.8813	.06321	15.8211	23
38	.01105	90.4633	.02851	35.0695	.04599	21.7426	.06350	15.7483	22
39	.01135	88.1436	.02881	34.7151	.04628	21.6056	.06379	15.6762	21
40	.01164	85.9398	.02910	34.3678	.04658	21.4704	.06408	15.6048	20
41	.01193	83.8435	.02939	34.0273	.04687	21.3369	.06437	15.5340	19
42	.01222	81.8470	.02968	33.6935	.04716	21.2049	.06467	15.4638	18
43	.01251	79.9434	.02997	33.3662	.04745	21.0747	.06496	15.3943	17
44	.01280	78.1263	.03026	33.0452	.04774	20.9460	.06525	15.3254	16
45	.01309	76.3900	.03055	32.7303	.04803	20.8188	.06554	15.2571	15
46	.01338	74.7292	.03084	32.4213	.04832	20.6932	.06584	15.1893	14
47	.01367	73.1390	.03114	32.1181	.04862	20.5691	.06613	15.1222	13
48	.01396	71.6151	.03143	31.8205	.04891	20.4465	.06642	15.0557	12
49	.01425	70.1533	.03172	31.5284	.04920	20.3253	.06671	14.9898	11
50	.01455	68.7501	.03201	31.2416	.04949	20.2056	.06700	14.9244	10
51	.01484	67.4019	.03230	30.9599	.04978	20.0872	.06730	14.8596	9
52	.01513	66.1055	.03259	30.6833	.05007	19.9702	.06759	14.7954	8
53	.01542	64.8580	.03288	30.4116	.05037	19.8546	.06788	14.7317	7
54	.01571	63.6567	.03317	30.1446	.05066	19.7403	.06817	14.6685	6
55	.01600	62.4992	.03346	29.8823	.05095	19.6273	.06847	14.6059	5
56	.01629	61.3829	.03376	29.6245	.05124	19.5156	.06876	14.5438	4
57	.01658	60.3058	.03405	29.3711	.05153	19.4051	.06905	14.4823	3
58	.01687	59.2659	.03434	29.1220	.05182	19.2959	.06934	14.4212	2
59	.01716	58.2612	.03463	28.8771	.05212	19.1879	.06963	14.3607	1
60	.01746	57.2900	.03492	28.6363	.05241	19.0811	.06993	14.3007	0
′	CO-TAN.	TAN.	CO-TAN.	TAN.	CO-TAN.	TAN.	CO-TAN.	TAN.	′
		89°		88°		87°		86°	

Table IV. (Continued)

′	4° TAN.	CO-TAN.	5° TAN.	CO-TAN.	6° TAN.	CO-TAN.	7° TAN.	CO-TAN.	′
0	.06993	14.3007	.08749	11.4301	.10510	9.51436	.12278	8.14435	60
1	.07022	14.2411	.08778	11.3919	.10540	9.48781	.12308	8.12481	59
2	.07051	14.1821	.08807	11.3540	.10569	9.46141	.12338	8.10536	58
3	.07080	14.1235	.08837	11.3163	.10599	9.43515	.12367	8.08600	57
4	.07110	14.0655	.08866	11.2789	.10628	9.40904	.12397	8.06674	56
5	.07139	14.0079	.08895	11.2417	.10657	9.38307	.12426	8.04756	55
6	.07168	13.9507	.08925	11.2048	.10687	9.35724	.12456	8.02848	54
7	.07197	13.8940	.08954	11.1681	.10716	9.33154	.12485	8.00948	53
8	.07227	13.8378	.08983	11.1316	.10746	9.30599	.12515	7.99058	52
9	.07256	13.7821	.09013	11.0954	.10775	9.28058	.12544	7.97176	51
10	.07285	13.7267	.09042	11.0594	.10805	9.25530	.12574	7.95302	50
11	.07314	13.6719	.09071	11.0237	.10834	9.23016	.12603	7.93438	49
12	.07344	13.6174	.09101	10.9882	.10863	9.20516	.12633	7.91582	48
13	.07373	13.5634	.09130	10.9529	.10893	9.18028	.12662	7.89734	47
14	.07402	13.5098	.09159	10.9178	.10922	9.15554	.12692	7.87895	46
15	.07431	13.4566	.09189	10.8829	.10952	9.13093	.12722	7.86064	45
16	.07461	13.4039	.09218	10.8483	.10981	9.10646	.12751	7.84242	44
17	.07490	13.3515	.09247	10.8139	.11011	9.08211	.12781	7.82428	43
18	.07519	13.2996	.09277	10.7797	.11040	9.05789	.12810	7.80622	42
19	.07548	13.2480	.09306	10.7457	.11070	9.03379	.12840	7.78825	41
20	.07578	13.1969	.09335	10.7119	.11099	9.00983	.12869	7.77035	40
21	.07607	13.1461	.09365	10.6783	.11128	8.98598	.12899	7.75254	39
22	.07636	13.0958	.09394	10.6450	.11158	8.96227	.12929	7.73480	38
23	.07665	13.0458	.09423	10.6118	.11187	8.93867	.12958	7.71715	37
24	.07695	12.9962	.09453	10.5789	.11217	8.91520	.12988	7.69957	36
25	.07724	12.9469	.09482	10.5462	.11246	8.89185	.13017	7.68208	35
26	.07753	12.8981	.09511	10.5136	.11276	8.86862	.13047	7.66466	34
27	.07782	12.8496	.09541	10.4813	.11305	8.84551	.13076	7.64732	33
28	.07812	12.8014	.09570	10.4491	.11335	8.82252	.13106	7.63005	32
29	.07841	12.7536	.09600	10.4172	.11364	8.79964	.13136	7.61287	31
30	.07870	12.7062	.09629	10.3854	.11394	8.77689	.13165	7.59575	30
31	.07899	12.6591	.09658	10.3538	.11423	8.75425	.13195	7.57872	29
32	.07929	12.6124	.09688	10.3224	.11452	8.73172	.13224	7.56176	28
33	.07958	12.5660	.09717	10.2913	.11482	8.70931	.13254	7.54487	27
34	.07987	12.5199	.09746	10.2602	.11511	8.68701	.13284	7.52806	26
35	.08017	12.4742	.09776	10.2294	.11541	8.66482	.13313	7.51132	25
36	.08046	12.4288	.09805	10.1988	.11570	8.64275	.13343	7.49465	24
37	.08075	12.3838	.09834	10.1683	.11600	8.62078	.13372	7.47806	23
38	.08104	12.3390	.09864	10.1381	.11629	8.59893	.13402	7.46154	22
39	.08134	12.2946	.09893	10.1080	.11659	8.57718	.13432	7.44509	21
40	.08163	12.2505	.09923	10.0780	.11688	8.55555	.13461	7.42871	20
41	.08192	12.2067	.09952	10.0483	.11718	8.53402	.13491	7.41240	19
42	.08221	12.1632	.09981	10.0187	.11747	8.51259	.13521	7.39616	18
43	.08251	12.1201	.10011	9.98931	.11777	8.49128	.13550	7.37999	17
44	.08280	12.0772	.10040	9.96007	.11806	8.47007	.13580	7.36389	16
45	.08309	12.0346	.10069	9.93101	.11836	8.44896	.13609	7.34786	15
46	.08339	11.9923	.10099	9.90211	.11865	8.42795	.13639	7.33190	14
47	.08368	11.9504	.10128	9.87338	.11895	8.40705	.13669	7.31600	13
48	.08397	11.9087	.10158	9.84482	.11924	8.38625	.13698	7.30018	12
49	.08427	11.8673	.10187	9.81641	.11954	8.36555	.13728	7.28442	11
50	.08456	11.8262	.10216	9.78817	.11983	8.34496	.13758	7.26873	10
51	.08485	11.7853	.10246	9.76009	.12013	8.32446	.13787	7.25310	9
52	.08514	11.7448	.10275	9.73217	.12042	8.30406	.13817	7.23754	8
53	.08544	11.7045	.10305	9.70441	.12072	8.28376	.13846	7.22204	7
54	.08573	11.6645	.10334	9.67680	.12101	8.26355	.13876	7.20661	6
55	.08602	11.6248	.10363	9.64935	.12131	8.24345	.13906	7.19125	5
56	.08632	11.5853	.10393	9.62205	.12160	8.22344	.13935	7.17594	4
57	.08661	11.5461	.10422	9.59490	.12190	8.20352	.13965	7.16071	3
58	.08690	11.5072	.10452	9.56791	.12219	8.18370	.13995	7.14553	2
59	.08720	11.4685	.10481	9.54106	.12249	8.16398	.14024	7.13042	1
60	.08749	11.4301	.10510	9.51436	.12278	8.14435	.14054	7.11537	0
′	CO-TAN.	TAN. 85°	CO-TAN.	TAN. 84°	CO-TAN.	TAN. 83°	CO-TAN.	TAN. 82°	′

Table IV. (Continued)

′	8° TAN.	8° CO-TAN.	9° TAN.	9° CO-TAN.	10° TAN.	10° CO-TAN.	11° TAN.	11° CO-TAN.	′
0	.14054	7.11537	.15838	6.31375	.17633	5.67128	.19438	5.14455	60
1	.14084	7.10038	.15868	6.30189	.17663	5.66165	.19468	5.13658	59
2	.14113	7.08546	.15898	6.29007	.17693	5.65205	.19498	5.12862	58
3	.14143	7.07059	.15928	6.27829	.17723	5.64248	.19529	5.12069	57
4	.14173	7.05579	.15958	6.26655	.17753	5.63295	.19559	5.11279	56
5	.14202	7.04105	.15988	6.25486	.17783	5.62344	.19589	5.10490	55
6	.14232	7.02637	.16017	6.24321	.17813	5.61397	.19619	5.09704	54
7	.14262	7.01174	.16047	6.23160	.17843	5.60452	.19649	5.08921	53
8	.14291	6.99718	.16077	6.22003	.17873	5.59511	.19680	5.08139	52
9	.14321	6.98268	.16107	6.20851	.17903	5.58573	.19710	5.07360	51
10	.14351	6.96823	.16137	6.19703	.17933	5.57638	.19740	5.06584	50
11	.14381	6.95385	.16167	6.18559	.17963	5.56706	.19770	5.05809	49
12	.14410	6.93952	.16196	6.17419	.17993	5.55777	.19801	5.05037	48
13	.14440	6.92525	.16226	6.16283	.18023	5.54851	.19831	5.04267	47
14	.14470	6.91104	.16256	6.15151	.18053	5.53927	.19861	5.03499	46
15	.14499	6.89688	.16286	6.14023	.18083	5.53007	.19891	5.02734	45
16	.14529	6.88278	.16316	6.12899	.18113	5.52090	.19921	5.01971	44
17	.14559	6.86874	.16346	6.11779	.18143	5.51176	.19952	5.01210	43
18	.14588	6.85475	.16376	6.10664	.18173	5.50264	.19982	5.00451	42
19	.14618	6.84082	.16405	6.09552	.18203	5.49356	.20012	4.99695	41
20	.14648	6.82694	.16435	6.08444	.18233	5.48451	.20042	4.98940	40
21	.14678	6.81312	.16465	6.07340	.18263	5.47548	.20073	4.98188	39
22	.14707	6.79936	.16495	6.06240	.18293	5.46648	.20103	4.97438	38
23	.14737	6.78564	.16525	6.05143	.18323	5.45751	.20133	4.96690	37
24	.14767	6.77199	.16555	6.04051	.18353	5.44857	.20164	4.95945	36
25	.14796	6.75838	.16585	6.02962	.18383	5.43966	.20194	4.95201	35
26	.14826	6.74483	.16615	6.01878	.18414	5.43077	.20224	4.94460	34
27	.14856	6.73133	.16645	6.00797	.18444	5.42192	.20254	4.93721	33
28	.14886	6.71789	.16674	5.99720	.18474	5.41309	.20285	4.92984	32
29	.14915	6.70450	.16704	5.98646	.18504	5.40429	.20315	4.92249	31
30	.14945	6.69116	.16734	5.97576	.18534	5.39552	.20345	4.91516	30
31	.14975	6.67787	.16764	5.96510	.18564	5.38677	.20376	4.90785	29
32	.15005	6.66463	.16794	5.95448	.18594	5.37805	.20406	4.90056	28
33	.15034	6.65144	.16824	5.94390	.18624	5.36936	.20436	4.89330	27
34	.15064	6.63831	.16854	5.93335	.18654	5.36070	.20466	4.88605	26
35	.15094	6.62523	.16884	5.92283	.18684	5.35206	.20497	4.87882	25
36	.15124	6.61219	.16914	5.91235	.18714	5.34345	.20527	4.87162	24
37	.15153	6.59921	.16944	5.90191	.18745	5.33487	.20557	4.86444	23
38	.15183	6.58627	.16974	5.89151	.18775	5.32631	.20588	4.85727	22
39	.15213	6.57339	.17004	5.88114	.18805	5.31778	.20618	4.85013	21
40	.15243	6.56055	.17033	5.87080	.18835	5.30928	.20648	4.84300	20
41	.15272	6.54777	.17063	5.86051	.18865	5.30080	.20679	4.83590	19
42	.15302	6.53503	.17093	5.85024	.18895	5.29235	.20709	4.82882	18
43	.15332	6.52234	.17123	5.84001	.18925	5.28393	.20739	4.82175	17
44	.15362	6.50970	.17153	5.82982	.18955	5.27553	.20770	4.81471	16
45	.15391	6.49710	.17183	5.81966	.18986	5.26715	.20800	4.80769	15
46	.15421	6.48456	.17213	5.80953	.19016	5.25880	.20830	4.80068	14
47	.15451	6.47206	.17243	5.79944	.19046	5.25048	.20861	4.79370	13
48	.15481	6.45961	.17273	5.78938	.19076	5.24218	.20891	4.78673	12
49	.15511	6.44720	.17303	5.77936	.19106	5.23391	.20921	4.77978	11
50	.15540	6.43484	.17333	5.76937	.19136	5.22566	.20952	4.77286	10
51	.15570	6.42253	.17363	5.75941	.19166	5.21744	.20982	4.76595	9
52	.15600	6.41026	.17393	5.74949	.19197	5.20925	.21013	4.75906	8
53	.15630	6.39804	.17423	5.73960	.19227	5.20107	.21043	4.75219	7
54	.15660	6.38587	.17453	5.72974	.19257	5.19293	.21073	4.74534	6
55	.15689	6.37374	.17483	5.71992	.19287	5.18480	.21104	4.73851	5
56	.15719	6.36165	.17513	5.71013	.19317	5.17671	.21134	4.73170	4
57	.15749	6.34961	.17543	5.70037	.19347	5.16863	.21164	4.72490	3
58	.15779	6.33761	.17573	5.69064	.19378	5.16058	.21195	4.71813	2
59	.15809	6.32566	.17603	5.68094	.19408	5.15256	.21225	4.71137	1
60	.15838	6.31375	.17633	5.67128	.19438	5.14455	.21256	4.70463	0
′	CO-TAN.	TAN. 81°	CO-TAN.	TAN. 80°	CO-TAN.	TAN. 79°	CO-TAN.	TAN. 78°	′

Table IV. (Continued)

′	12° Tan.	Co-tan.	13° Tan.	Co-tan.	14° Tan.	Co-tan.	15° Tan.	Co-tan.	′
0	.21256	4.70463	.23087	4.33148	.24933	4.01078	.26795	3.73205	60
1	.21286	4.69791	.23117	4.32573	.24964	4.00582	.26826	3.72771	59
2	.21316	4.69121	.23148	4.32001	.24995	4.00086	.26857	3.72338	58
3	.21347	4.68452	.23179	4.31430	.25026	3.99592	.26888	3.71907	57
4	.21377	4.67786	.23209	4.30860	.25056	3.99099	.26920	3.71476	56
5	.21408	4.67121	.23240	4.30291	.25087	3.98607	.26951	3.71046	55
6	.21438	4.66458	.23271	4.29724	.25118	3.98117	.26982	3.70616	54
7	.21469	4.65797	.23301	4.29159	.25149	3.97627	.27013	3.70188	53
8	.21499	4.65138	.23332	4.28595	.25180	3.97139	.27044	3.69761	52
9	.21529	4.64480	.23363	4.28032	.25211	3.96651	.27076	3.69335	51
10	.21560	4.63825	.23393	4.27471	.25242	3.96165	.27107	3.68909	50
11	.21590	4.63171	.23424	4.26911	.25273	3.95680	.27138	3.68485	49
12	.21621	4.62518	.23455	4.26352	.25304	3.95196	.27169	3.68061	48
13	.21651	4.61868	.23485	4.25795	.25335	3.94713	.27201	3.67638	47
14	.21682	4.61219	.23516	4.25239	.25366	3.94232	.27232	3.67217	46
15	.21712	4.60572	.23547	4.24685	.25397	3.93751	.27263	3.66796	45
16	.21743	4.59927	.23578	4.24132	.25428	3.93271	.27294	3.66376	44
17	.21773	4.59283	.23608	4.23580	.25459	3.92793	.27326	3.65957	43
18	.21804	4.58641	.23639	4.23030	.25490	3.92316	.27357	3.65538	42
19	.21834	4.58001	.23670	4.22481	.25521	3.91839	.27388	3.65121	41
20	.21864	4.57363	.23700	4.21933	.25552	3.91364	.27419	3.64705	40
21	.21895	4.56726	.23731	4.21387	.25583	3.90890	.27451	3.64289	39
22	.21925	4.56091	.23762	4.20842	.25614	3.90417	.27482	3.63874	38
23	.21956	4.55458	.23793	4.20298	.25645	3.89945	.27513	3.63461	37
24	.21986	4.54826	.23823	4.19756	.25676	3.89474	.27545	3.63048	36
25	.22017	4.54196	.23854	4.19215	.25707	3.89004	.27576	3.62636	35
26	.22047	4.53568	.23885	4.18675	.25738	3.88536	.27607	3.62224	34
27	.22078	4.52941	.23916	4.18137	.25769	3.88068	.27638	3.61814	33
28	.22108	4.52316	.23946	4.17600	.25800	3.87601	.27670	3.61405	32
29	.22139	4.51693	.23977	4.17064	.25831	3.87136	.27701	3.60996	31
30	.22169	4.51071	.24008	4.16530	.25862	3.86671	.27732	3.60588	30
31	.22200	4.50451	.24039	4.15997	.25893	3.86208	.27764	3.60181	29
32	.22231	4.49832	.24069	4.15465	.25924	3.85745	.27795	3.59775	28
33	.22261	4.49215	.24100	4.14934	.25955	3.85284	.27826	3.59370	27
34	.22292	4.48600	.24131	4.14405	.25986	3.84824	.27858	3.58966	26
35	.22322	4.47986	.24162	4.13877	.26017	3.84364	.27889	3.58562	25
36	.22353	4.47374	.24193	4.13350	.26048	3.83906	.27920	3.58160	24
37	.22383	4.46764	.24223	4.12825	.26079	3.83449	.27952	3.57758	23
38	.22414	4.46155	.24254	4.12301	.26110	3.82992	.27983	3.57357	22
39	.22444	4.45548	.24285	4.11778	.26141	3.82537	.28015	3.56957	21
40	.22475	4.44942	.24316	4.11256	.26172	3.82083	.28046	3.56557	20
41	.22505	4.44338	.24347	4.10736	.26203	3.81630	.28077	3.56159	19
42	.22536	4.43735	.24377	4.10216	.26235	3.81177	.28109	3.55761	18
43	.22567	4.43134	.24408	4.09699	.26266	3.80726	.28140	3.55364	17
44	.22597	4.42534	.24439	4.09182	.26297	3.80276	.28172	3.54968	16
45	.22628	4.41936	.24470	4.08666	.26328	3.79827	.28203	3.54573	15
46	.22658	4.41340	.24501	4.08152	.26359	3.79378	.28234	3.54179	14
47	.22689	4.40745	.24532	4.07639	.26390	3.78931	.28266	3.53785	13
48	.22719	4.40152	.24562	4.07127	.26421	3.78485	.28297	3.53393	12
49	.22750	4.39560	.24593	4.06616	.26452	3.78040	.28329	3.53001	11
50	.22781	4.38969	.24624	4.06107	.26483	3.77595	.28360	3.52609	10
51	.22811	4.38381	.24655	4.05599	.26515	3.77152	.28391	3.52219	9
52	.22842	4.37793	.24686	4.05092	.26546	3.76709	.28423	3.51829	8
53	.22872	4.37207	.24717	4.04586	.26577	3.76268	.28454	3.51441	7
54	.22903	4.36623	.24747	4.04081	.26608	3.75828	.28486	3.51053	6
55	.22934	4.36040	.24778	4.03578	.26639	3.75388	.28517	3.50666	5
56	.22964	4.35459	.24809	4.03075	.26670	3.74950	.28549	3.50279	4
57	.22995	4.34879	.24840	4.02574	.26701	3.74512	.28580	3.49894	3
58	.23026	4.34300	.24871	4.02074	.26733	3.74075	.28612	3.49509	2
59	.23056	4.33723	.24902	4.01576	.26764	3.73640	.28643	3.49125	1
60	.23087	4.33148	.24933	4.01078	.26795	3.73205	.28675	3.48741	0
′	Co-tan.	Tan. 77°	Co-tan.	Tan. 76°	Co-tan.	Tan. 75°	Co-tan.	Tan. 74°	′

Table IV. (Continued)

′	16° TAN.	CO-TAN.	17° TAN.	CO-TAN.	18° TAN.	CO-TAN.	19° TAN.	CO-TAN.	′
0	.28675	3.48741	.30573	3.27085	.32492	3.07768	.34433	2.90421	60
1	.28706	3.48359	.30605	3.26745	.32524	3.07464	.34465	2.90147	59
2	.28738	3.47977	.30637	3.26406	.32556	3.07160	.34498	2.89873	58
3	.28769	3.47596	.30669	3.26067	.32588	3.06857	.34530	2.89600	57
4	.28800	3.47216	.30700	3.25729	.32621	3.06554	.34563	2.89327	56
5	.28832	3.46837	.30732	3.25392	.32653	3.06252	.34596	2.89055	55
6	.28864	3.46458	.30764	3.25055	.32685	3.05950	.34628	2.88783	54
7	.28895	3.46080	.30796	3.24719	.32717	3.05649	.34661	2.88511	53
8	.28927	3.45703	.30828	3.24383	.32749	3.05349	.34693	2.88240	52
9	.28958	3.45327	.30860	3.24049	.32782	3.05049	.34726	2.87970	51
10	.28990	3.44951	.30891	3.23714	.32814	3.04749	.34758	2.87700	50
11	.29021	3.44576	.30923	3.23381	.32846	3.04450	.34791	2.87430	49
12	.29053	3.44202	.30955	3.23048	.32878	3.04152	.34824	2.87161	48
13	.29084	3.43829	.30987	3.22715	.32911	3.03854	.34856	2.86892	47
14	.29116	3.43456	.31019	3.22384	.32943	3.03556	.34889	2.86624	46
15	.29147	3.43084	.31051	3.22053	.32975	3.03260	.34922	2.86356	45
16	.29179	3.42713	.31083	3.21722	.33007	3.02963	.34954	2.86089	44
17	.29210	3.42343	.31115	3.21392	.33040	3.02667	.34987	2.85822	43
18	.29242	3.41973	.31147	3.21063	.33072	3.02372	.35019	2.85555	42
19	.29274	3.41604	.31178	3.20734	.33104	3.02077	.35052	2.85289	41
20	.29305	3.41236	.31210	3.20406	.33136	3.01783	.35085	2.85023	40
21	.29337	3.40869	.31242	3.20079	.33160	3.01489	.35117	2.84758	39
22	.29368	3.40502	.31274	3.19752	.33201	3.01196	.35150	2.84494	38
23	.29400	3.40136	.31306	3.19426	.33233	3.00903	.35183	2.84229	37
24	.29432	3.39771	.31338	3.19100	.33266	3.00611	.35216	2.83965	36
25	.29463	3.39406	.31370	3.18775	.33298	3.00319	.35248	2.83702	35
26	.29495	3.39042	.31402	3.18451	.33330	3.00028	.35281	2.83439	34
27	.29526	3.38679	.31434	3.18127	.33363	2.99738	.35314	2.83176	33
28	.29558	3.38317	.31466	3.17804	.33395	2.99447	.35346	2.82914	32
29	.29590	3.37955	.31498	3.17481	.33427	2.99158	.35379	2.82653	31
30	.29621	3.37594	.31530	3.17159	.33460	2.98868	.35412	2.82391	30
31	.29653	3.37234	.31562	3.16838	.33492	2.98580	.35445	2.82130	29
32	.29685	3.36875	.31594	3.16517	.33524	2.98292	.35477	2.81870	28
33	.29716	3.36516	.31626	3.16197	.33557	2.98004	.35510	2.81610	27
34	.29748	3.36158	.31658	3.15877	.33589	2.97717	.35543	2.81350	26
35	.29780	3.35800	.31690	3.15558	.33621	2.97430	.35576	2.81091	25
36	.29811	3.35443	.31722	3.15240	.33654	2.97144	.35608	2.80833	24
37	.29843	3.35087	.31754	3.14922	.33686	2.96858	.35641	2.80574	23
38	.29875	3.34732	.31786	3.14605	.33718	2.96573	.35674	2.80316	22
39	.29906	3.34377	.31818	3.14288	.33751	2.96288	.35707	2.80059	21
40	.29938	3.34023	.31850	3.13972	.33783	2.96004	.35740	2.79802	20
41	.29970	3.33670	.31882	3.13656	.33816	2.95721	.35772	2.79545	19
42	.30001	3.33317	.31914	3.13341	.33848	2.95437	.35805	2.79289	18
43	.30033	3.32965	.31946	3.13027	.33881	2.95155	.35838	2.79033	17
44	.30065	3.32614	.31978	3.12713	.33913	2.94872	.35871	2.78778	16
45	.30097	3.32264	.32010	3.12400	.33945	2.94590	.35904	2.78523	15
46	.30128	3.31914	.32042	3.12087	.33978	2.94309	.35937	2.78269	14
47	.30160	3.31565	.32074	3.11775	.34010	2.94028	.35969	2.78014	13
48	.30192	3.31216	.32106	3.11464	.34043	2.93748	.36002	2.77761	12
49	.30224	3.30868	.32139	3.11153	.34075	2.93468	.36035	2.77507	11
50	.30255	3.30521	.32171	3.10842	.34108	2.93189	.36068	2.77254	10
51	.30287	3.30174	.32203	3.10532	.34140	2.92910	.36101	2.77002	9
52	.30319	3.29829	.32235	3.10223	.34173	2.92632	.36134	2.76750	8
53	.30351	3.29483	.32267	3.09914	.34205	2.92354	.36167	2.76498	7
54	.30382	3.29139	.32299	3.09606	.34238	2.92076	.36199	2.76247	6
55	.30414	3.28795	.32331	3.09298	.34270	2.91799	.36232	2.75996	5
56	.30446	3.28452	.32363	3.08991	.34303	2.91523	.36265	2.75746	4
57	.30478	3.28109	.32396	3.08685	.34335	2.91246	.36298	2.75496	3
58	.30509	3.27767	.32428	3.08379	.34368	2.90971	.36331	2.75246	2
59	.30541	3.27426	.32460	3.08073	.34400	2.90696	.36364	2.74997	1
60	.30573	3.27085	.32492	3.07768	.34433	2.90421	.36397	2.74748	0
′	CO-TAN.	TAN. 73°	CO-TAN.	TAN. 72°	CO-TAN.	TAN. 71°	CO-TAN.	TAN. 70°	′

Table IV. (Continued)

′	20° TAN.	20° CO-TAN.	21° TAN.	21° CO-TAN.	22° TAN.	22° CO-TAN.	23° TAN.	23° CO-TAN.	′
0	.36397	2.74748	.38386	2.60509	.40403	2.47509	.42447	2.35585	60
1	.36430	2.74499	.38420	2.60283	.40436	2.47302	.42482	2.35395	59
2	.36463	2.74251	.38453	2.60057	.40470	2.47095	.42516	2.35205	58
3	.36496	2.74004	.38487	2.59831	.40504	2.46888	.42551	2.35015	57
4	.36529	2.73756	.38520	2.59606	.40538	2.46682	.42585	2.34825	56
5	.36562	2.73509	.38553	2.59381	.40572	2.46476	.42619	2.34636	55
6	.36595	2.73263	.38587	2.59156	.40606	2.46270	.42654	2.34447	54
7	.36628	2.73017	.38620	2.58932	.40640	2.46065	.42688	2.34258	53
8	.36661	2.72771	.38654	2.58708	.40674	2.45860	.42722	2.34069	52
9	.36694	2.72526	.38687	2.58484	.40707	2.45655	.42757	2.33881	51
10	.36727	2.72281	.38721	2.58261	.40741	2.45451	.42791	2.33693	50
11	.36760	2.72036	.38754	2.58038	.40775	2.45246	.42826	2.33505	49
12	.36793	2.71792	.38787	2.57815	.40809	2.45043	.42860	2.33317	48
13	.36826	2.71548	.38821	2.57593	.40843	2.44839	.42894	2.33130	47
14	.36859	2.71305	.38854	2.57371	.40877	2.44636	.42929	2.32943	46
15	.36892	2.71062	.38888	2.57150	.40911	2.44433	.42963	2.32756	45
16	.36925	2.70819	.38921	2.56928	.40945	2.44230	.42998	2.32570	44
17	.36958	2.70577	.38955	2.56707	.40979	2.44027	.43032	2.32383	43
18	.36991	2.70335	.38988	2.56487	.41013	2.43825	.43067	2.32197	42
19	.37024	2.70094	.39022	2.56266	.41047	2.43623	.43101	2.32012	41
20	.37057	2.69853	.39055	2.56046	.41081	2.43422	.43136	2.31826	40
21	.37090	2.69612	.39089	2.55827	.41115	2.43220	.43170	2.31641	39
22	.37124	2.69371	.39122	2.55608	.41149	2.43019	.43205	2.31456	38
23	.37157	2.69131	.39156	2.55389	.41183	2.42819	.43239	2.31271	37
24	.37190	2.68892	.39190	2.55170	.41217	2.42618	.43274	2.31086	36
25	.37223	2.68653	.39223	2.54952	.41251	2.42418	.43308	2.30902	35
26	.37256	2.68414	.39257	2.54734	.41285	2.42218	.43343	2.30718	34
27	.37289	2.68175	.39290	2.54516	.41319	2.42019	.43378	2.30534	33
28	.37322	2.67937	.39324	2.54299	.41353	2.41819	.43412	2.30351	32
29	.37355	2.67700	.39357	2.54082	.41387	2.41620	.43447	2.30167	31
30	.37388	2.67462	.39391	2.53865	.41421	2.41421	.43481	2.29984	30
31	.37422	2.67225	.39425	2.53648	.41455	2.41223	.43516	2.29801	29
32	.37455	2.66989	.39458	2.53432	.41490	2.41025	.43550	2.29619	28
33	.37488	2.66752	.39492	2.53217	.41524	2.40827	.43585	2.29437	27
34	.37521	2.66516	.39526	2.53001	.41558	2.40629	.43620	2.29254	26
35	.37554	2.66281	.39559	2.52786	.41592	2.40432	.43654	2.29073	25
36	.37588	2.66046	.39593	2.52571	.41626	2.40235	.43689	2.28891	24
37	.37621	2.65811	.39626	2.52357	.41660	2.40038	.43724	2.28710	23
38	.37654	2.65576	.39660	2.52142	.41694	2.39841	.43758	2.28528	22
39	.37687	2.65342	.39694	2.51929	.41728	2.39645	.43793	2.28348	21
40	.37720	2.65109	.39727	2.51715	.41763	2.39449	.43828	2.28167	20
41	.37754	2.64875	.39761	2.51502	.41797	2.39253	.43862	2.27987	19
42	.37787	2.64642	.39795	2.51289	.41831	2.39058	.43897	2.27806	18
43	.37820	2.64410	.39829	2.51076	.41865	2.38862	.43932	2.27626	17
44	.37853	2.64177	.39862	2.50864	.41899	2.38668	.43966	2.27447	16
45	.37887	2.63945	.39896	2.50652	.41933	2.38473	.44001	2.27267	15
46	.37920	2.63714	.39930	2.50440	.41968	2.38279	.44036	2.27088	14
47	.37953	2.63483	.39963	2.50229	.42002	2.38084	.44071	2.26909	13
48	.37986	2.63252	.39997	2.50018	.42036	2.37891	.44105	2.26730	12
49	.38020	2.63021	.40031	2.49807	.42070	2.37697	.44140	2.26552	11
50	.38053	2.62791	.40065	2.49597	.42105	2.37504	.44175	2.26374	10
51	.38086	2.62561	.40098	2.49386	.42139	2.37311	.44210	2.26196	9
52	.38120	2.62332	.40132	2.49177	.42173	2.37118	.44244	2.26018	8
53	.38153	2.62103	.40166	2.48967	.42207	2.36925	.44279	2.25840	7
54	.38186	2.61874	.40200	2.48758	.42242	2.36733	.44314	2.25663	6
55	.38220	2.61646	.40234	2.48549	.42276	2.36541	.44349	2.25486	5
56	.38253	2.61418	.40267	2.48340	.42310	2.36349	.44384	2.25309	4
57	.38286	2.61190	.40301	2.48132	.42345	2.36158	.44418	2.25132	3
58	.38320	2.60963	.40335	2.47924	.42379	2.35967	.44453	2.24956	2
59	.38353	2.60736	.40369	2.47716	.42413	2.35776	.44488	2.24780	1
60	.38386	2.60509	.40403	2.47509	.42447	2.35585	.44523	2.24604	0
′	CO-TAN.	TAN.	CO-TAN.	TAN.	CO-TAN.	TAN.	CO-TAN.	TAN.	′
	69°		68°		67°		66°		

Table IV. (Continued)

′	24° TAN.	CO-TAN.	25° TAN.	CO-TAN.	26° TAN.	CO-TAN.	27° TAN.	CO-TAN.	
0	.44523	2.24604	.46631	2.14451	.48773	2.05030	.50953	1.96261	60
1	.44558	2.24428	.46666	2.14288	.48809	2.04879	.50989	1.96120	59
2	.44593	2.24252	.46702	2.14125	.48845	2.04728	.51026	1.95979	58
3	.44627	2.24077	.46737	2.13963	.48881	2.04577	.51063	1.95838	57
4	.44662	2.23902	.46772	2.13801	.48917	2.04426	.51099	1.95698	56
5	.44697	2.23727	.46808	2.13639	.48953	2.04276	.51136	1.95557	55
6	.44732	2.23553	.46843	2.13477	.48989	2.04125	.51173	1.95417	54
7	.44767	2.23378	.46879	2.13316	.49026	2.03975	.51209	1.95277	53
8	.44802	2.23204	.46914	2.13154	.49062	2.03825	.51246	1.95137	52
9	.44837	2.23030	.46950	2.12993	.49098	2.03675	.51283	1.94997	51
10	.44872	2.22857	.46985	2.12832	.49134	2.03526	.51319	1.94858	50
11	.44907	2.22683	.47021	2.12671	.49170	2.03376	.51356	1.94718	49
12	.44942	2.22510	.47056	2.12511	.49206	2.03227	.51393	1.94579	48
13	.44977	2.22337	.47092	2.12350	.49242	2.03078	.51430	1.94440	47
14	.45012	2.22164	.47128	2.12190	.49278	2.02929	.51467	1.94301	46
15	.45047	2.21992	.47163	2.12030	.49315	2.02780	.51503	1.94162	45
16	.45082	2.21819	.47199	2.11871	.49351	2.02631	.51540	1.94023	44
17	.45117	2.21647	.47234	2.11711	.49387	2.02483	.51577	1.93885	43
18	.45152	2.21475	.47270	2.11552	.49423	2.02335	.51614	1.93746	42
19	.45187	2.21304	.47305	2.11392	.49459	2.02187	.51651	1.93608	41
20	.45222	2.21132	.47341	2.11233	.49495	2.02039	.51688	1.93470	40
21	.45257	2.20961	.47377	2.11075	.49532	2.01891	.51724	1.93332	39
22	.45292	2.20790	.47412	2.10916	.49568	2.01743	.51761	1.93195	38
23	.45327	2.20619	.47448	2.10758	.49604	2.01596	.51798	1.93057	37
24	.45362	2.20449	.47483	2.10600	.49640	2.01449	.51835	1.92920	36
25	.45397	2.20278	.47519	2.10442	.49677	2.01302	.51872	1.92782	35
26	.45432	2.20108	.47555	2.10284	.49713	2.01155	.51909	1.92645	34
27	.45467	2.19938	.47590	2.10126	.49749	2.01008	.51946	1.92508	33
28	.45502	2.19769	.47626	2.09969	.49786	2.00862	.51983	1.92371	32
29	.45537	2.19599	.47662	2.09811	.49822	2.00715	.52020	1.92235	31
30	.45573	2.19430	.47698	2.09654	.49858	2.00569	.52057	1.92098	30
31	.45608	2.19261	.47733	2.09498	.49894	2.00423	.52094	1.91962	29
32	.45643	2.19092	.47769	2.09341	.49931	2.00277	.52131	1.91826	28
33	.45678	2.18923	.47805	2.09184	.49967	2.00131	.52168	1.91690	27
34	.45713	2.18755	.47840	2.09028	.50004	1.99986	.52205	1.91554	26
35	.45748	2.18587	.47876	2.08872	.50040	1.99841	.52242	1.91418	25
36	.45784	2.18419	.47912	2.08716	.50076	1.99695	.52279	1.91282	24
37	.45819	2.18251	.47948	2.08560	.50113	1.99550	.52316	1.91147	23
38	.45854	2.18084	.47984	2.08405	.50149	1.99406	.52353	1.91012	22
39	.45889	2.17916	.48019	2.08250	.50185	1.99261	.52390	1.90876	21
40	.45924	2.17749	.48055	2.08094	.50222	1.99116	.52427	1.90741	20
41	.45960	2.17582	.48091	2.07939	.50258	1.98972	.52464	1.90607	19
42	.45995	2.17416	.48127	2.07785	.50295	1.98828	.52501	1.90472	18
43	.46030	2.17249	.48163	2.07630	.50331	1.98684	.52538	1.90337	17
44	.46065	2.17083	.48198	2.07476	.50368	1.98540	.52575	1.90203	16
45	.46101	2.16917	.48234	2.07321	.50404	1.98396	.52613	1.90069	15
46	.46136	2.16751	.48270	2.07167	.50441	1.98253	.52650	1.89935	14
47	.46171	2.16585	.48306	2.07014	.50477	1.98110	.52687	1.89801	13
48	.46206	2.16420	.48342	2.06860	.50514	1.97966	.52724	1.89667	12
49	.46242	2.16255	.48378	2.06706	.50550	1.97823	.52761	1.89533	11
50	.46277	2.16090	.48414	2.06553	.50587	1.97680	.52798	1.89400	10
51	.46312	2.15925	.48450	2.06400	.50623	1.97538	.52836	1.89266	9
52	.46348	2.15760	.48486	2.06247	.50660	1.97395	.52873	1.89133	8
53	.46383	2.15596	.48521	2.06094	.50696	1.97253	.52910	1.89000	7
54	.46418	2.15432	.48557	2.05942	.50733	1.97111	.52947	1.88867	6
55	.46454	2.15268	.48593	2.05790	.50769	1.96969	.52984	1.88734	5
56	.46489	2.15104	.48629	2.05637	.50806	1.96827	.53022	1.88602	4
57	.46525	2.14940	.48665	2.05485	.50843	1.96685	.53059	1.88469	3
58	.46560	2.14777	.48701	2.05333	.50879	1.96544	.53096	1.88337	2
59	.46595	2.14614	.48737	2.05182	.50916	1.96402	.53134	1.88205	1
60	.46631	2.14451	.48773	2.05030	.50953	1.96261	.53171	1.88073	0
′	CO-TAN.	TAN. 65°	CO-TAN.	TAN. 64°	CO-TAN.	TAN. 63°	CO-TAN.	TAN. 62°	′

Table IV. (Continued)

	28° Tan.	28° Co-tan.	29° Tan.	29° Co-tan.	30° Tan.	30° Co-tan.	31° Tan.	31° Co-tan.	
0	.53171	1.88073	.55431	1.80405	.57735	1.73205	.60086	1.66428	60
1	.53208	1.87941	.55469	1.80281	.57774	1.73089	.60126	1.66318	59
2	.53246	1.87809	.55507	1.80158	.57813	1.72973	.60165	1. 6209	58
3	.53283	1.87677	.55545	1.80034	.57851	1.72857	.60205	1.66099	57
4	.53320	1.87546	.55583	1.79911	.57890	1.72741	.60245	1.65990	56
5	.53358	1.87415	.55621	1.79788	.57929	1.72625	.60284	1.65881	55
6	.53395	1.87283	.55659	1.79665	.57968	1.72509	.60324	1.65772	54
7	.53432	1.87152	.55697	1.79542	.58007	1.72393	.60364	1.65668	53
8	.53470	1.87021	.55736	1.79419	.58046	1.72278	.60403	1.65534	52
9	.53507	1.86891	.55774	1.79296	.58085	1.72163	.60443	1.65445	51
10	.53545	1.86760	.55812	1.79174	.58124	1.72047	.60483	1.65337	50
11	.53582	1.86630	.55850	1.79051	.58162	1.71932	.60522	1.65228	49
12	.53620	1.86499	.55888	1.78929	.58201	1.71817	.60562	1.65120	48
13	.53657	1.86369	.55926	1.78807	.58240	1.71702	.60602	1.65011	47
14	.53694	1.86239	.55964	1.78685	.58279	1.71588	.60642	1.64903	46
15	.53732	1.86109	.56003	1.78563	.58318	1.71473	.60681	1.64795	45
16	.53769	1.85979	.56041	1.78441	.58357	1.71358	.60721	1.64687	44
17	.53807	1.85850	.56079	1.78319	.58396	1.71244	.60761	1.64579	43
18	.53844	1.85720	.56117	1.78198	.58435	1.71129	.60801	1.64471	42
19	.53882	1.85591	.56156	1.78077	.58474	1.71015	.60841	1.64363	41
20	.53920	1.85462	.56194	1.77955	.58513	1.70901	.60881	1.64256	40
21	.53957	1.85333	.56232	1.77834	.58552	1.70787	.60921	1.64148	39
22	.53995	1.85204	.56270	1.77713	.58591	1.70673	.60960	1.64041	38
23	.54032	1.85075	.56309	1.77592	.58631	1.70560	.61000	1.63934	37
24	.54070	1.84946	.56347	1.77471	.58670	1.70446	.61040	1.63826	36
25	.54107	1.84818	.56385	1.77351	.58709	1.70332	.61080	1.63719	35
26	.54145	1.84689	.56424	1.77230	.58748	1.70219	.61120	1.63612	34
27	.54183	1.84561	.56462	1.77110	.58787	1.70106	.61160	1.63505	33
28	.54220	1.84433	.56500	1.76990	.58826	1.69992	.61200	1.63398	32
29	.54258	1.84305	.56539	1.76869	.58865	1.69879	.61240	1.63292	31
30	.54296	1.84177	.56577	1.76749	.58904	1.69766	.61280	1.63185	30
31	.54333	1.84049	.56616	1.76630	.58944	1.69653	.61320	1.63079	29
32	.54371	1.83922	.56654	1.76510	.58983	1.69541	.61360	1.62972	28
33	.54409	1.83794	.56693	1.76390	.59022	1.69428	.61400	1.62866	27
34	.54446	1.83667	.56731	1.76271	.59061	1.69316	.61440	1.62760	26
35	.54484	1.83540	.56769	1.76151	.59101	1.69203	.61480	1.62654	25
36	.54522	1.83413	.56808	1'76032	.59140	1.69091	.61520	1.62548	24
37	.54560	1.83286	.56846	1.75913	.59179	1.68979	.61561	1.62442	23
38	.54597	1.83159	.56885	1.75794	.59218	1.68866	.61601	1.62336	22
39	.54635	1.83033	.56923	1.75675	.59258	1.68754	.61641	1.62230	21
40	.54673	1.82906	.56962	1.75556	.59297	1.68643	.61681	1.62125	20
41	.54711	1.82780	.57000	1.75437	.59336	1.68531	.61721	1.62019	19
42	.54748	1.82654	.57039	1.75319	.59376	1.68419	.61761	1.61914	18
43	.54786	1.82528	.57078	1.75200	.59415	1.68308	.61801	1.61808	17
44	.54824	1.82402	.57116	1.75082	.59454	1.68196	.61842	1.61703	16
45	.54862	1.82276	.57155	1.74964	.59494	1.68085	.61882	1.61598	15
46	.54900	1.82150	.57193	1.74846	.59533	1.67974	.61922	1.61493	14
47	.54938	1.82025	.57232	1.74728	.59573	1.67863	.61962	1.61388	13
48	.54975	1.81899	.57271	1.74610	.59612	1.67752	.62003	1.61283	12
49	.55013	1.81774	.57309	1.74492	.59651	1.67641	.62043	1.61179	11
50	.55051	1.81649	.57348	1.74375	.59691	1.67530	.62083	1.61074	10
51	.55089	1.81524	.57386	1.74257	.59730	1.67419	.62124	1.60970	9
52	.55127	1.81399	.57425	1.74140	.59770	1.67309	.62164	1.60865	8
53	.55165	1.81274	.57464	1.74022	.59809	1.67198	.62204	1.60761	7
54	.55203	1.81150	.57503	1.73905	.59849	1.67088	.62245	1.60657	6
55	.55241	1.81025	.57541	1.73788	.59888	1.66978	.62285	1.60553	5
56	.55279	1.80901	.57580	1.73671	.59928	1.66867	.62325	1.60449	4
57	.55317	1.80777	.57619	1.73555	.59967	1.66757	.62366	1.60345	3
58	.55355	1.80653	.57657	1.73438	.60007	1.66647	.62406	1.60241	2
59	.55393	1.80529	.57696	1.73321	.60046	1.66538	.62446	1.60137	1
60	.55431	1.80405	.57735	1.73205	.60086	1.66428	.62487	1.60033	0
	Co-tan.	Tan. 61°	Co-tan.	Tan. 60°	Co-tan.	Tan. 59°	Co-tan.	Tan. 58°	

Table IV. (Continued)

′	32° TAN.	CO-TAN.	33° TAN.	CO-TAN.	34° TAN.	CO-TAN.	35° TAN.	CO-TAN.	′
0	.62487	1.60033	.64941	1.53986	.67451	1.48256	.70021	1.42815	60
1	.62527	1.59930	.64982	1.53888	.67493	1.48163	.70064	1.42726	59
2	.62568	1.59826	.65023	1.53791	.67536	1.48070	.70107	1.42638	58
3	.62608	1.59723	.65065	1.53693	.67578	1.47977	.70151	1.42550	57
4	.62649	1.59620	.65106	1.53595	.67620	1.47885	.70194	1.42462	56
5	.62689	1.59517	.65148	1.53497	.67663	1.47792	.70238	1.42374	55
6	.62730	1.59414	.65189	1.53400	.67705	1.47699	.70281	1.42286	54
7	.62770	1.59311	.65231	1.53302	.67748	1.47607	.70325	1.42198	53
8	.62811	1.59208	.65272	1.53205	.67790	1.47514	.70368	1.42110	52
9	.62852	1.59105	.65314	1.53107	.67832	1.47422	.70412	1.42022	51
10	.62892	1.59002	.65355	1.53010	.67875	1.47330	.70455	1.41934	50
11	.62933	1.58900	.65397	1.52913	.67917	1.47238	.70499	1.41847	49
12	.62973	1.58797	.65438	1.52816	.67960	1.47146	.70542	1.41759	48
13	.63014	1.58695	.65480	1.52719	.68002	1.47053	.70586	1.41672	47
14	.63055	1.58593	.65521	1.52622	.68045	1.46962	.70629	1.41584	46
15	.63095	1.58490	.65563	1.52525	.68088	1.46870	.70673	1.41497	45
16	.63136	1.58388	.65604	1.52429	.68130	1.46778	.70717	1.41409	44
17	.63177	1.58286	.65646	1.52332	.68173	1.46686	.70760	1.41322	43
18	.63217	1.58184	.65688	1.52235	.68215	1.46595	.70804	1.41235	42
19	.63258	1.58083	.65729	1.52139	.68258	1.46503	.70848	1.41148	41
20	.63299	1.57981	.65771	1.52043	.68301	1.46411	.70891	1.41061	40
21	.63340	1.57870	.65813	1.51946	.68343	1.46320	.70935	1.40974	39
22	.63380	1.57778	.65854	1.51850	.68386	1.46229	.70979	1.40887	38
23	.63421	1.57676	.65896	1.51754	.68429	1.46137	.71023	1.40800	37
24	.63462	1.57575	.65938	1.51658	.68471	1.46046	.71066	1.40714	36
25	.63503	1.57474	.65980	1.51562	.68514	1.45955	.71110	1.40627	35
26	.63544	1.57372	.66021	1.51466	.68557	1.45864	.71154	1.40540	34
27	.63584	1.57271	.66063	1.51370	.68600	1.45773	.71198	1.40454	33
28	.63625	1.57170	.66105	1.51275	.68642	1.45682	.71242	1.40367	32
29	.63666	1.57069	.66147	1.51179	.68685	1.45592	.71285	1.40281	31
30	.63707	1.56969	.66189	1.51084	.68728	1.45501	.71329	1.40195	30
31	.63748	1.56868	.66230	1.50988	.68771	1.45410	.71373	1.40109	29
32	.63789	1.56767	.66272	1.50893	.68814	1.45320	.71417	1.40022	28
33	.63830	1.56667	.66314	1.50797	.68857	1.45229	.71461	1.39936	27
34	.63871	1.56566	.66356	1.50702	.68900	1.45139	.71505	1.39850	26
35	.63912	1.56466	.66398	1.50607	.68942	1.45049	.71549	1.39764	25
36	.63953	1.56366	.66440	1.50512	.68985	1.44958	.71593	1.39679	24
37	.63994	1.56265	.66482	1.50417	.69028	1.44868	.71637	1.39593	23
38	.64035	1.56165	.66524	1.50322	.69071	1.44778	.71681	1.39507	22
39	.64076	1.56065	.66566	1.50228	.69114	1.44688	.71725	1.39421	21
40	.64117	1.55966	.66608	1.50133	.69157	1.44598	.71769	1.39336	20
41	.64158	1.55866	.66650	1.50038	.69200	1.44508	.71813	1.39250	19
42	.64199	1.55766	.66692	1.49944	.69243	1.44418	.71857	1.39165	18
43	.64240	1.55666	.66734	1.49849	.69286	1.44329	.71901	1.39079	17
44	.64281	1.55567	.66776	1.49755	.69329	1.44239	.71946	1.38994	16
45	.64322	1.55467	.66818	1.49661	.69372	1.44149	.71990	1.38909	15
46	.64363	1.55368	.66860	1.49566	.69416	1.44060	.72034	1.38824	14
47	.64404	1.55269	.66902	1.49472	.69459	1.43970	.72078	1.38738	13
48	.64446	1.55170	.66944	1.49378	.69502	1.43881	.72122	1.38653	12
49	.64487	1.55071	.66986	1.49284	.69545	1.43792	.72166	1.38568	11
50	.64528	1.54972	.67028	1.49190	.69588	1.43703	.72211	1.38484	10
51	.64569	1.54873	.67071	1.49097	.69631	1.43614	.72255	1.38399	9
52	.64610	1.54774	.67113	1.49003	.69675	1.43525	.72299	1.38314	8
53	.64652	1.54675	.67155	1.48909	.69718	1.43436	.72344	1.38229	7
54	.64693	1.54576	.67197	1.48816	.69761	1.43347	.72388	1.38145	6
55	.64734	1.54478	.67239	1.48722	.69804	1.43258	.72432	1.38060	5
56	.64775	1.54379	.67282	1.48629	.69847	1.43169	.72477	1.37976	4
57	.64817	1.54281	.67324	1.48536	.69891	1.43080	.72521	1.37891	3
58	.64858	1.54183	.67366	1.48442	.69934	1.42992	.72565	1.37807	2
59	.64899	1.54085	.67409	1.48349	.69977	1.42903	.72610	1.37722	1
60	.64941	1.53986	.67451	1.48256	.70021	1.42815	.72654	1.37638	0
′	CO-TAN.	TAN.	CO-TAN	TAN.	CO-TAN.	TAN.	CO-TAN.	TAN.	′
	57°		56°		55°		54°		

Table IV. (Continued)

′	36° Tan.	36° Co-tan.	37° Tan.	37° Co-tan.	38° Tan.	38° Co-tan.	39° Tan.	39° Co-tan.	′
0	.72654	1.37638	.75355	1.32704	.78129	1.27994	.80978	1.23490	60
1	.72699	1.37554	.75401	1.32624	.78175	1.27917	.81027	1.23416	59
2	.72743	1.37470	.75447	1.32544	.78222	1.27841	.81075	1.23343	58
3	.72788	1.37386	.75492	1.32464	.78269	1.27764	.81123	1.23270	57
4	.72832	1.37302	.75538	1.32384	.78316	1.27688	.81171	1.23196	56
5	.72877	1.37218	.75584	1.32304	.78363	1.27611	.81220	1.23123	55
6	.72921	1.37134	.75629	1.32224	.78410	1.27535	.81268	1.23050	54
7	.72966	1.37050	.75675	1.32144	.78457	1.27458	.81316	1.22977	53
8	.73010	1.36967	.75721	1.32064	.78504	1.27382	.81364	1.22904	52
9	.73055	1.36883	.75767	1.31984	.78551	1.27306	.81413	1.22831	51
10	.73100	1.36800	.75812	1.31904	.78598	1.27230	.81461	1.22758	50
11	.73144	1.36716	.75858	1.31825	.78645	1.27153	.81510	1.22685	49
12	.73189	1.36633	.75904	1.31745	.78692	1.27077	.81558	1.22612	48
13	.73234	1.36549	.75950	1.31666	.78739	1.27001	.81606	1.22539	47
14	.73278	1.36466	.75996	1.31586	.78786	1.26925	.81655	1.22467	46
15	.73323	1.36383	.76042	1.31507	.78834	1.26849	.81703	1.22394	45
16	.73368	1.36300	.76088	1.31427	.78881	1.26774	.81752	1.22321	44
17	.73413	1.36217	.76134	1.31348	.78928	1.26698	.81800	1.22249	43
18	.73457	1.36133	.76180	1.31269	.78975	1.26622	.81849	1.22176	42
19	.73502	1.36051	.76226	1.31190	.79022	1.26546	.81898	1.22104	41
20	.73547	1.35968	.76272	1.31110	.79070	1.26471	.81946	1.22031	40
21	.73592	1.35885	.76318	1.31031	.79117	1.26395	.81995	1.21959	39
22	.73637	1.35802	.76364	1.30952	.79164	1.26319	.82044	1.21886	38
23	.73681	1.35719	.76410	1.30873	.79212	1.26244	.82092	1.21814	37
24	.73726	1.35637	.76456	1.30795	.79259	1.26169	.82141	1.21742	36
25	.73771	1.35554	.76502	1.30716	.79306	1.26093	.82190	1.21670	35
26	.73816	1.35472	.76548	1.30637	.79354	1.26018	.82238	1.21598	34
27	.73861	1.35389	.76594	1.30558	.79401	1.25943	.82287	1.21526	33
28	.73906	1.35307	.76640	1.30480	.79449	1.25867	.82336	1.21454	32
29	.73951	1.35224	.76686	1.30401	.79496	1.25792	.82385	1.21382	31
30	.73996	1.35142	.76733	1.30323	.79544	1.25717	.82434	1.21310	30
31	.74041	1.35060	.76779	1.30244	.79591	1.25642	.82483	1.21238	29
32	.74086	1.34978	.76825	1.30166	.79639	1.25567	.82531	1.21166	28
33	.74131	1.34896	.76871	1.30087	.79686	1.25492	.82580	1.21094	27
34	.74176	1.34814	.76918	1.30009	.79734	1.25417	.82629	1.21023	26
35	.74221	1.34732	.76964	1.29931	.79781	1.25343	.82678	1.20951	25
36	.74267	1.34650	.77010	1.29853	.79829	1.25268	.82727	1.20879	24
37	.74312	1.34568	.77057	1.29775	.79877	1.25193	.82776	1.20808	23
38	.74357	1.34487	.77103	1.29696	.79924	1.25118	.82825	1.20736	22
39	.74402	1.34405	.77149	1.29618	.79972	1.25044	.82874	1.20665	21
40	.74447	1.34323	.77196	1.29541	.80020	1.24969	.82923	1.20593	20
41	.74492	1.34242	.77242	1.29463	.80067	1.24895	.82972	1.20522	19
42	.74538	1.34160	.77289	1.29385	.80115	1.24820	.83022	1.20451	18
43	.74583	1.34079	.77335	1.29307	.80163	1.24746	.83071	1.20379	17
44	.74628	1.33998	.77382	1.29229	.80211	1.24672	.83120	1.20308	16
45	.74674	1.33916	.77428	1.29152	.80258	1.24597	.83169	1.20237	15
46	.74719	1.33835	.77475	1.29074	.80306	1.24523	.83218	1.20166	14
47	.74764	1.33754	.77521	1.28997	.80354	1.24449	.83268	1.20095	13
48	.74810	1.33673	.77568	1.28919	.80402	1.24375	.83317	1.20024	12
49	.74855	1.33592	.77615	1.28842	.80450	1.24301	.83366	1.19953	11
50	.74900	1.33511	.77661	1.28764	.80498	1.24227	.83415	1.19882	10
51	.74946	1.33430	.77708	1.28687	.80546	1.24153	.83465	1.19811	9
52	.74991	1.33349	.77754	1.28610	.80594	1.24079	.83514	1.19740	8
53	.75037	1.33268	.77801	1.28533	.80642	1.24005	.83564	1.19669	7
54	.75082	1.33187	.77848	1.28456	.80690	1.23931	.83613	1.19599	6
55	.75128	1.33107	.77895	1.28379	.80738	1.23858	.83662	1.19528	5
56	.75173	1.33026	.77941	1.28302	.80786	1.23784	.83712	1.19457	4
57	.75219	1.32946	.77988	1.28225	.80834	1.23710	.83761	1.19387	3
58	.75264	1.32865	.78035	1.28148	.80882	1.23637	.83811	1.19316	2
59	.75310	1.32785	.78082	1.28071	.80930	1.23563	.83860	1.19246	1
60	.75355	1.32704	.78129	1.27994	.80978	1.23490	.83910	1.19175	0
′	Co-tan.	Tan. 53°	Co-tan.	Tan. 52°	Co-tan.	Tan. 51°	Co-tan.	Tan. 50°	′

Table IV. (Continued)

′	40° TAN.	CO-TAN.	41° TAN.	CO-TAN.	42° TAN.	CO-TAN.	43° TAN.	CO-TAN.	′
0	.83910	1.19175	.86929	1.15037	.90040	1.11061	.93252	1.07237	60
1	.83960	1.19105	.86980	1.14969	.90093	1.10996	.93306	1.07174	59
2	.84009	1.19035	.87031	1.14902	.90146	1.10931	.93360	1.07112	58
3	.84059	1.18964	.87082	1.14834	.90199	1.10867	.93415	1.07049	57
4	.84108	1.18894	.87133	1.14767	.90251	1.10802	.93469	1.06987	56
5	.84158	1.18824	.87184	1.14699	.90304	1.10737	.93524	1.06925	55
6	.84208	1.18754	.87236	1.14632	.90357	1.10672	.93578	1.06862	54
7	.84258	1.18684	.87287	1.14565	.90410	1.10607	.93633	1.06800	53
8	.84307	1.18614	.87338	1.14498	.90463	1.10543	.93688	1.06738	52
9	.84357	1.18544	.87389	1.14430	.90516	1.10478	.93742	1.06676	51
10	.84407	1.18474	.87441	1.14363	.90569	1.10414	.93797	1.06613	50
11	.84457	1.18404	.87492	1.14296	.90621	1.10349	.93852	1.06551	49
12	.84507	1.18334	.87543	1.14229	.90674	1.10285	.93906	1.06489	48
13	.84556	1.18264	.87595	1.14162	.90727	1.10220	.93961	1.06427	47
14	.84606	1.18194	.87646	1.14095	.90781	1.10156	.94016	1.06365	46
15	.84656	1.18125	.87698	1.14028	.90834	1.10091	.94071	1.06303	45
16	.84706	1.18055	.87749	1.13961	.90887	1.10027	.94125	1.06241	44
17	.84756	1.17986	.87801	1.13894	.90940	1.09963	.94180	1.06179	43
18	.84806	1.17916	.87852	1.13828	.90993	1.09899	.94235	1.06117	42
19	.84856	1.17846	.87904	1.13761	.91046	1.09834	.94290	1.06056	41
20	.84906	1.17777	.87955	1.13694	.91099	1.09770	.94345	1.05994	40
21	.84956	1.17708	.88007	1.13627	.91153	1.09706	.94400	1.05932	39
22	.85006	1.17638	.88059	1.13561	.91206	1.09642	.94455	1.05870	38
23	.85057	1.17569	.88110	1.13494	.91259	1.09578	.94510	1.05809	37
24	.85107	1.17500	.88162	1.13428	.91313	1.09514	.94565	1.05747	36
25	.85157	1.17430	.88214	1.13361	.91366	1.09450	.94620	1.05685	35
26	.85207	1.17361	.88265	1.13295	.91419	1.09386	.94676	1.05624	34
27	.85257	1.17292	.88317	1.13228	.91473	1.09322	.94731	1.05562	33
28	.85307	1.17223	.88369	1.13162	.91526	1.09258	.94786	1.05501	32
29	.85358	1.17154	.88421	1.13096	.91580	1.09195	.94841	1.05439	31
30	.85408	1.17085	.88473	1.13029	.91633	1.09131	.94896	1.05378	30
31	.85458	1.17016	.88524	1.12963	.91687	1.09067	.94952	1.05317	29
32	.85509	1.16947	.88576	1.12897	.91740	1.09003	.95007	1.05255	28
33	.85559	1.16878	.88628	1.12831	.91794	1.08940	.95062	1.05194	27
34	.85609	1.16809	.88680	1.12765	.91847	1.08876	.95118	1.05133	26
35	.85660	1.16741	.88732	1.12699	.91901	1.08813	.95173	1.05072	25
36	.85710	1.16672	.88784	1.12633	.91955	1.08749	.95229	1.05010	24
37	.85761	1.16603	.88836	1.12567	.92008	1.08686	.95284	1.04949	23
38	.85811	1.16535	.88888	1.12501	.92062	1.08622	.95340	1.04888	22
39	.85862	1.16466	.88940	1.12435	.92116	1.08559	.95395	1.04827	21
40	.85912	1.16398	.88992	1.12369	.92170	1.08496	.95451	1.04766	20
41	.85963	1.16329	.89045	1.12303	.92224	1.08432	.95506	1.04705	19
42	.86014	1.16261	.89097	1.12238	.92277	1.08369	.95562	1.04644	18
43	.86064	1.16192	.89149	1.12172	.92331	1.08306	.95618	1.04583	17
44	.86115	1.16124	.89201	1.12106	.92385	1.08243	.95673	1.04522	16
45	.86166	1.16056	.89253	1.12041	.92439	1.08179	.95729	1.04461	15
46	.86216	1.15987	.89306	1.11975	.92493	1.08116	.95785	1.04401	14
47	.86267	1.15919	.89358	1.11909	.92547	1.08053	.95841	1.04340	13
48	.86318	1.15851	.89410	1.11844	.92601	1.07990	.95897	1.04279	12
49	.86368	1.15783	.89463	1.11778	.92655	1.07927	.95952	1.04218	11
50	.86419	1.15715	.89515	1.11713	.92709	1.07864	.96008	1.04158	10
51	.86470	1.15647	.89567	1.11648	.92763	1.07801	.96064	1.04097	9
52	.86521	1.15579	.89620	1.11582	.92817	1.07738	.96120	1.04036	8
53	.86572	1.15511	.89672	1.11517	.92872	1.07676	.96176	1.03976	7
54	.86623	1.15443	.89725	1.11452	.92926	1.07613	.96232	1.03915	6
55	.86674	1.15375	.89777	1.11387	.92980	1.07550	.96288	1.03855	5
56	.86725	1.15308	.89830	1.11321	.93034	1.07487	.96344	1.03794	4
57	.86776	1.15240	.89883	1.11256	.93088	1.07425	.96400	1.03734	3
58	.86827	1.15172	.89935	1.11191	.93143	1.07362	.96457	1.03674	2
59	.86878	1.15104	.89988	1.11126	.93197	1.07299	.96513	1.03613	1
60	.86929	1.15037	.90040	1.11061	.93252	1.07237	.96569	1.03553	0
′	CO-TAN.	TAN. 49°	CO-TAN.	TAN. 48°	CO-TAN.	TAN. 47°	CO-TAN.	TAN. 46°	′

Table IV. (Concluded)

'	44° TAN.	44° CO-TAN.	'	'	44° TAN.	44° CO-TAN.	'	'	44° TAN.	44° CO-TAN.	'
0	.96569	1.03553	60	21	.97756	1.02295	39	41	.98901	1.01112	19
1	.96625	1.03493	59	22	.97813	1.02236	38	42	.98958	1.01053	18
2	.96681	1.03433	58	23	.97870	1.02176	37	43	.99016	1.00994	17
3	.96738	1.03372	57	24	.97927	1.02117	36	44	.99073	1.00935	16
4	.96794	1.03312	56	25	.97984	1.02057	35	45	.99131	1.00876	15
5	.96850	1.03252	55	26	.98041	1.01998	34	46	.99189	1.00818	14
6	.96907	1.03192	54	27	.98098	1.01939	33	47	.99247	1.00759	13
7	.96963	1.03132	53	28	.98155	1.01879	32	48	.99304	1.00701	12
8	.97020	1.03072	52	29	.98213	1.01820	31	49	.99362	1.00642	11
9	.97076	1.03012	51	30	.98270	1.01761	30	50	.99420	1.00583	10
10	.97133	1.02952	50	31	.98327	1.01702	29	51	.99478	1.00525	9
11	.97189	1.02892	49	32	.98384	1.01642	28	52	.99536	1.00467	8
12	.97246	1.02832	48	33	.98441	1.01583	27	53	.99594	1.00408	7
13	.97302	1.02772	47	34	.98499	1.01524	26	54	.99652	1.00350	6
14	.97359	1.02713	46	35	.98556	1.01465	25	55	.99710	1.00291	5
15	.97416	1.02653	45	36	.98613	1.01406	24	56	.99768	1.00233	4
16	.97472	1.02593	44	37	.98671	1.01347	23	57	.99826	1.00175	3
17	.97529	1.02533	43	38	.98728	1.01288	22	58	.99884	1.00116	2
18	.97586	1.02474	42	39	.98786	1.01229	21	59	.99942	1.00058	1
19	.97643	1.02414	41	40	.98843	1.01170	20	60	1	1	0
20	.97700	1.02355	40								
'	CO-TAN.	TAN.	'	'	CO-TAN.	TAN.	'	'	CO-TAN.	TAN.	'
	45°				45°				45°		

Table V. Order of Accuracy Defined by Board of Surveys and Maps of the Federal Government, May 9, 1933

Type of survey	Type of error	Limits of error Order of accuracy			
		First	Second	Third	Fourth
Triangu-lation	Maximum angular closure per triangle	3″	5″	10″	No appreciable error in resulting map
	Average angular closure per triangle	1″	3″	5″	
	Error in length of base as computed through tri-angles after angles are adjusted	1/25,000	1/10,000	1/5000	
	Probable error* of base measurement	1/1,000,000	1/500,000	1/200,000	
Traverse	Position closure after angles are adjusted	1/25,000	1/10,000	1/5000	
Leveling	Error of closure, ft, divided by square root of distance leveled in miles	0.012	0.025	0.050	

* Probable error of the average of the individual measurements

$$= 0.6745 \sqrt{\frac{\Sigma v^2}{n(n-1)}}$$

where n = number of measurements

Σv^2 = sum of the squares of the quantities by which each measurement differs from the mean of the measurements

Table VI. Temperature Corrections* Per Foot for Steel Tapes

°F Sub-tract cor.	Cor.	°F Add cor.	°F Sub-tract cor.	Cor.	°F Add cor.	°F Sub-tract cor.	Cor.	°F Add cor.	°F Sub-tract cor.	Cor.
68	0.00000000	68	48	0.00012900	88	28	0.00025800	108	9	0.00038055
67	0.00000645	69	47	0.00013545	89	27	0.00026445	109	8	0.00038700
66	0.00001290	70	46	0.00014190	90	26	0.00027090	110	7	0.00039345
65	0.00001935	71	45	0.00014835	91	25	0.00027735	111	6	0.00039990
64	0.00002580	72	44	0.00015480	92	24	0.00028380	112	5	0.00040635
63	0.00003225	73	43	0.00016125	93	23	0.00029025	113	4	0.00041280
62	0.00003870	74	42	0.00016770	94	22	0.00029670	114	3	0.00041925
61	0.00004515	75	41	0.00017415	95	21	0.00030315	115	2	0.00042570
60	0.00005160	76	40	0.00018060	96	20	0.00030960	116	1	0.00043215
59	0.00005805	77	39	0.00018705	97	19	0.00031605	117	0	0.00043860
58	0.00006450	78	38	0.00019350	98	18	0.00032250	118	−1	0.00044505
57	0.00007095	79	37	0.00019995	99	17	0.00032895	119	−2	0.00045150
56	0.00007740	80	36	0.00020640	100	16	0.00033540	120	−3	0.00045795
55	0.00008385	81	35	0.00021285	101	15	0.00034185	121	−4	0.00046440
54	0.00009030	82	34	0.00021930	102	14	0.00034830	122	−5	0.00047085
53	0.00009675	83	33	0.00022575	103	13	0.00035475	123	−6	0.00047730
52	0.00010320	84	32	0.00023220	104	12	0.00036120	124	−7	0.00048375
51	0.00010965	85	31	0.00023865	105	11	0.00036765	125	−8	0.00049020
50	0.00011610	86	30	0.00024510	106	10	0.00037410	126	−9	0.00049665
49	0.00012255	87	29	0.00025155	107	9	0.00038055	127	−10	0.00050310

* Based on a coefficient of expansion of 0.00000645 per °F.

Table VII. Slope Corrections for 100 ft (Subtract); Given: Height Difference

h*	0.0	0.1	0.2	0.3	0.4	0.5	0.6	0.7	0.8	0.9
0	.000	.000	.000	.000	.001	.001	.002	.002	.003	.004
1	.005	.006	.007	.008	.010	.011	.013	.014	.016	.018
2	.020	.022	.024	.026	.029	.031	.034	.036	.039	.042
3	.045	.048	.051	.054	.058	.061	.065	.068	.072	.076
4	.080	.084	.088	.092	.097	.101	.106	.111	.115	.120
5	.125	.130	.135	.141	.146	.151	.157	.163	.168	.174
6	.180	.186	.192	.199	.205	.211	.218	.225	.231	.238
7	.245	.252	.260	.267	.274	.282	.289	.297	.305	.313
8	.321	.329	.337	.345	.353	.362	.370	.379	.388	.397
9	.406	.415	.424	.433	.443	.452	.462	.472	.481	.491
10	.501	.511	.522	.532	.542	.553	.563	.574	.585	.596
11	.607	.618	.629	.641	.652	.663	.675	.687	.699	.711
12	.723	.735	.747	.759	.772	.784	.797	.810	.823	.836
13	.849	.862	.875	.888	.902	.915	.929	.943	.957	.971
14	.985	.999								

* h = Difference in height in 100 ft on slope. *Example:* h = 6.5 ft, correction = 0.211, corrected distance = 99.789 ft.

Table VIII. Slope Corrections per 100 ft (Subtract);
Given: Angle of Slope

Slope	Cor.	Slope	Cor.	Slope	Cor.	Slope	Cor.	Slope	Cor.	Slope	Cor.
0°00′		3°18′		4°41′		5°46′		6°40′		7°27′	
	.00		.17		.34		.51		.68		.85
0 34		3 23		4 46		5 49		6 43		7 30	
	.01		.18		.35		.52		.69		.86
1 00		3 29		4 50		5 52		6 46		7 32	
	.02		.19		.36		.53		.70		.87
1 17		3 35		4 54		5 56		6 48		7 35	
	.03		.20		.37		.54		.71		.88
1 31		3 40		4 58		5 59		6 51		7 38	
	.04		.21		.38		.55		.72		.89
1 43		3 45		5 02		6 02		6 54		7 40	
	.05		.22		.39		.56		.73		.90
1 54		3 51		5 06		6 06		6 57		7 43	
	.06		.23		.40		.57		.74		.91
2 04		3 56		5 10		6 09		7 00		7 45	
	.07		.24		.41		.58		.75		.92
2 13		4 01		5 13		6 12		7 03		7 48	
	.08		.25		.42		.59		.76		.93
2 22		4 06		5 17		6 15		7 05		7 50	
	.09		.26		.43		.60		.77		.94
2 30		4 11		5 21		6 18		7 08		7 53	
	.10		.27		.44		.61		.78		.95
2 38		4 15		5 24		6 21		7 11		7 55	
	.11		.28		.45		.62		.79		.96
2 45		4 20		5 28		6 25		7 14		7 58	
	.12		.29		.46		.63		.80		.97
2 52		4 24		5 32		6 28		7 17		8 00	
	.13		.30		.47		.64		.81		.98
2 59		4 29		5 35		6 31		7 19		8 03	
	.14		.31		.48		.65		.82		.99
3 05		4 33		5 39		6 34		7 22		8 05	
	.15		.32		.49		.66		.83		1.00
3 11		4 37		5 42		6 37		7 25		8 08	
	.16		.33		.50		.67		.84		1.01
3 18		4 41		5 46		6 40		7 27		8 10	

Example: Slope measurement = 500 ft, slope = 5°12′, 5.00 × 0.41 = 2.05, corrected measurement = 497.95.

Table IX. Horizontal Corrections for Stadia Intercept of 1.00 ft

Zenith angle	Vert. angle	Hor. cor. for 1.00 ft	Zenith angle	Vert. angle	Hor. cor. for 1.00 ft	Zenith angle	Vert. angle	Hor. cor. for 1.00 ft
90°00′	0°00′		84°24′	5°36′		81°58′	8°02′	
		0.0 ft			1.0 ft			2.0 ft
88 43	1 17		84 07	5 53		81 46	8 14	
		0.1 ft			1.1 ft			2.1 ft
87 47	2 13		83 51	6 09		81 34	8 26	
		0.2 ft			1.2 ft			2.2 ft
87 08	2 52		83 35	6 25		81 22	8 38	
		0.3 ft			1.3 ft			2.3 ft
86 37	3 23		83 20	6 40		81 11	8 49	
		0.4 ft			1.4 ft			2.4 ft
86 09	3 51		83 05	6 55		81 00	9 00	
		0.5 ft			1.5 ft			2.5 ft
85 45	4 15		82 51	7 09		80 49	9 11	
		0.6 ft			1.6 ft			2.6 ft
85 23	4 37		82 37	7 23		80 38	9 22	
		0.7 ft			1.7 ft			2.7 ft
85 02	4 58		82 24	7 36		80 27	9 33	
		0.8 ft			1.8 ft			2.8 ft
84 43	5 17		82 11	7 49		80 17	9 43	
		0.9 ft			1.9 ft			2.9 ft
84 24	5 36		81 58	8 02		80 07	9 53	
								3.0 ft
						79 57	10 03	

For angles outside of range of table, use formulas:

$$\text{Hor. dist.} = 100S \cos^2 a \qquad \text{Vert. height} = 100S \sin a \cos a$$

where S = stadia intercept and a = vertical angle.

For zenith angles greater than 90°, subtract 90° to find vertical angle.

Results from Table IX are correct to the nearest foot at 1000 ft and to the nearest $\frac{1}{10}$ ft at 100 ft, etc.

With a slide rule, multiply the stadia intercept by the tabular value and subtract the product from the horizontal distance.

Example: Vertical angle, 4°22′, or zenith angle, 85°38′; stadia intercept, 3.58 ft.

$$\text{Corrected hor. dist.} = 358 - (3.58 \times 0.6) = 356 \text{ ft}$$

Table X gives the vertical heights for a stadia intercept of 1.00 ft. With a slide rule, multiply the stadia intercept by the tabular value.

Example: Vertical angle, 4°22′, or zenith angle, 85°38′; stadia intercept, 3.58 ft.

$$\text{Vertical height} = 3.58 \times 7.59 = 27.2 \text{ ft}$$

Table X. Vertical Heights for Stadia Intercept of 1.00 Ft

↙	90° 0°	91° 1°	92° 2°	93° 3°	94° 4°	95° 5°	96° 6°	97° 7°	98° 8°	99° 9°	
Min.											
0	0.00	1.74	3.49	5.23	6.96	8.68	10.40	12.10	13.78	15.45	60
1	0.03	1.77	3.52	5.26	6.99	8.71	10.42	12.12	13.81	15.48	59
2	0.06	1.80	3.55	5.28	7.02	8.74	10.45	12.15	13.84	15.51	58
3	0.09	1.83	3.57	5.31	7.05	8.77	10.48	12.18	13.87	15.53	57
4	0.12	1.86	3.60	5.34	7.07	8.80	10.51	12.21	13.89	15.56	56
5	0.14	1.89	3.63	5.37	7.10	8.83	10.54	12.24	13.92	15.59	55
6	0.17	1.92	3.66	5.40	7.13	8.85	10.57	12.27	13.95	15.62	54
7	0.20	1.95	3.69	5.43	7.16	8.88	10.59	12.29	13.98	15.64	53
8	0.23	1.98	3.72	5.46	7.19	8.91	10.62	12.32	14.01	15.67	52
9	0.26	2.01	3.75	5.49	7.22	8.94	10.65	12.35	14.03	15.70	51
10	0.29	2.04	3.78	5.52	7.25	8.97	10.68	12.38	14.06	15.73	50
11	0.32	2.06	3.81	5.54	7.28	9.00	10.71	12.41	14.09	15.75	49
12	0.35	2.09	3.84	5.57	7.30	9.03	10.74	12.43	14.12	15.78	48
13	0.38	2.12	3.87	5.60	7.33	9.05	10.77	12.46	14.15	15.81	47
14	0.41	2.15	3.89	5.63	7.36	9.08	10.79	12.49	14.17	15.84	46
15	0.44	2.18	3.92	5.66	7.39	9.11	10.82	12.52	14.20	15.87	45
16	0.47	2.21	3.95	5.69	7.42	9.14	10.85	12.55	14.23	15.89	44
17	0.49	2.24	3.98	5.72	7.45	9.17	10.88	12.58	14.26	15.92	43
18	0.52	2.27	4.01	5.75	7.48	9.20	10.91	12.60	14.28	15.95	42
19	0.55	2.30	4.04	5.78	7.51	9.23	10.94	12.63	14.31	15.98	41
20	0.58	2.33	4.07	5.80	7.53	9.25	10.96	12.66	14.34	16.00	40
21	0.61	2.36	4.10	5.83	7.56	9.28	10.99	12.69	14.37	16.03	39
22	0.64	2.38	4.13	5.86	7.59	9.31	11.02	12.72	14.40	16.06	38
23	0.67	2.41	4.16	5.89	7.62	9.34	11.05	12.74	14.42	16.09	37
24	0.70	2.44	4.18	5.92	7.65	9.37	11.08	12.77	14.45	16.11	36
25	0.73	2.47	4.21	5.95	7.68	9.40	11.11	12.80	14.48	16.14	35
26	0.76	2.50	4.24	5.98	7.71	9.43	11.13	12.83	14.51	16.17	34
27	0.78	2.53	4.27	6.01	7.74	9.46	11.16	12.86	14.54	16.20	33
28	0.81	2.56	4.30	6.04	7.76	9.48	11.19	12.88	14.56	16.22	32
29	0.84	2.59	4.33	6.06	7.79	9.51	11.22	12.91	14.59	16.25	31
30	0.87	2.62	4.36	6.09	7.82	9.54	11.25	12.94	14.62	16.28	30
31	0.90	2.65	4.39	6.12	7.85	9.57	11.28	12.97	14.65	16.31	29
32	0.93	2.67	4.42	6.15	7.88	9.60	11.30	13.00	14.67	16.33	28
33	0.96	2.70	4.44	6.18	7.91	9.63	11.33	13.03	14.70	16.36	27
34	0.99	2.73	4.47	6.21	7.94	9.65	11.36	13.05	14.73	16.39	26
35	1.02	2.76	4.50	6.24	7.97	9.68	11.39	13.08	14.76	16.42	25
36	1.05	2.79	4.53	6.27	7.99	9.71	11.42	13.11	14.79	16.44	24
37	1.08	2.82	4.56	6.30	8.02	9.74	11.44	13.14	14.81	16.47	23
38	1.11	2.85	4.59	6.32	8.05	9.77	11.47	13.17	14.84	16.50	22
39	1.14	2.88	4.62	6.35	8.08	9.80	11.50	13.19	14.87	16.53	21
40	1.16	2.91	4.65	6.38	8.11	9.83	11.53	13.22	14.90	16.55	20
41	1.19	2.94	4.68	6.41	8.14	9.85	11.56	13.25	14.92	16.58	19
42	1.22	2.97	4.71	6.44	8.17	9.88	11.59	13.28	14.95	16.61	18
43	1.25	2.99	4.73	6.47	8.20	9.91	11.62	13.31	14.98	16.64	17
44	1.28	3.02	4.76	6.50	8.22	9.94	11.64	13.33	15.01	16.66	16
45	1.31	3.05	4.79	6.53	8.25	9.97	11.67	13.36	15.03	16.69	15
46	1.34	3.08	4.82	6.56	8.28	10.00	11.70	13.39	15.06	16.72	14
47	1.37	3.11	4.85	6.58	8.31	10.03	11.73	13.42	15.09	16.74	13
48	1.40	3.14	4.88	6.61	8.34	10.05	11.76	13.45	15.12	16.77	12
49	1.42	3.17	4.91	6.64	8.37	10.08	11.79	13.47	15.15	16.80	11
50	1.45	3.20	4.94	6.67	8.40	10.11	11.81	13.50	15.17	16.83	10
51	1.48	3.23	4.97	6.70	8.42	10.14	11.84	13.53	15.20	16.85	9
52	1.51	3.26	4.99	6.73	8.45	10.17	11.87	13.56	15.23	16.88	8
53	1.54	3.28	5.02	6.76	8.48	10.20	11.90	13.59	15.26	16.91	7
54	1.57	3.31	5.05	6.79	8.51	10.22	11.93	13.61	15.28	16.94	6
55	1.60	3.34	5.08	6.82	8.54	10.25	11.96	13.64	15.31	16.96	5
56	1.63	3.37	5.11	6.84	8.57	10.28	11.98	13.67	15.34	16.99	4
57	1.66	3.40	5.14	6.87	8.60	10.31	12.01	13.70	15.37	17.02	3
58	1.69	3.43	5.17	6.90	8.63	10.34	12.04	13.73	15.40	17.05	2
59	1.71	3.46	5.20	6.93	8.65	10.37	12.07	13.75	15.42	17.07	1
60	1.74	3.49	5.23	6.96	8.68	10.40	12.10	13.78	15.45	17.10	0
											Min.
	89°	88°	87°	86°	85°	84°	83°	82°	81°	80°	↗

Table XI. Conversion from Zenith Angles to Vertical Angles for Stadia Slide Rule

90 to 99°, replace first 9 by a minus sign.
80 to 90°, use tables.

Degrees		Minutes					
Zenith angle	*Vertical angle*	0	60	10	50	20	40
		1	59	11	49	21	39
		2	58	12	48	22	38
80	+9	3	57	13	47	23	37
81	+8	4	56	14	46	24	36
82	+7						
83	+6	5	55	15	45	25	35
84	+5	6	54	16	44	26	34
85	+4	7	53	17	43	27	33
86	+3	8	52	18	42	28	32
87	+2	9	51	19	41	29	31
88	+1	10	50	20	40	30	30
89	+0						

Beside the degrees in the zenith angle are the degrees in the corresponding vertical angle.

Beside the minutes in the zenith angle to the left or right are the minutes in the vertical angle.

Examples:	Zenith angle	Vertical angle
	96°13′	−6°13′
	91 17	−1 17
	83 46	+6 14
	87 25	+2 35

Table XII. Length of Curve for Radius 1.00 ft
Given: △

△	Length	△	Length	△	Length
10°	0.17453	10′	0.00291	10″	0.00005
20°	0.34907	20′	0.00582	20″	0.00010
30°	0.52360	30′	0.00873	30″	0.00015
40°	0.69813	40′	0.01164	40″	0.00019
50°	0.87266	50′	0.01454	50″	0.00024
60°	1.04720	60′	0.01745	60″	0.00029
70°	1.22173	70′	0.02036	70″	0.00034
80°	1.39626	80′	0.02327	80″	0.00039
90°	1.57080	90′	0.02618	90″	0.00044
100°	1.74533	100′	0.02910	100″	0.00048

Example: Given: $\triangle = 35°27'47''$ $R = 600$

$$
\begin{aligned}
30° &= 0.52360 \\
5° &= 0.08727 \\
20' &= 0.00582 \\
7' &= 0.00204 \\
40'' &= 0.00019 \\
7'' &= 0.00003 \\
\text{Sum} &= \overline{0.61895}
\end{aligned}
$$

$$
\begin{aligned}
&\;\;0.61895 \\
&\underline{\times 600} \\
&371.37000
\end{aligned}
$$

$L = 371.37$ ft

Table XIII. Corrections to Subarcs

R	Lengths of Arcs in Feet								
	5	10	15	20	25	30	35	40	45	50
100	0.001	0.004	0.014	0.033	0.065	0.112	0.178	0.266	0.379	0.519
110	.000	.003	.012	.028	.054	.093	.147	.220	.313	.429
120	.000	.003	.010	.023	.045	.078	.124	.185̄	.263	.361
130	.000	.002	.008	.020	.038	.066	.106	.158	.224	.308
140	.000	.002	.007	.017	.033	.057	.091	.136	.193	.265
150	.000	.002	.006	.015̄	.029	.050	.079	.119	.168	.231
160	.000	.002	.005	.013	.025	.044	.070	.104	.148	.203
170	.000	.001	.005̄	.011	.023	.039	.062	.092	.131	.180
180	.000	.001	.004	.010	.020	.035̄	.055̄	.082	.117	.161
190	.000	.001	.004	.009	.018	.031	.049	.074	.105	.144
200	.000	.001	.004	.008	.016	.028	.045̄	.067	.095̄	.130
225	.000	.001	.003	.007	.013	.022	.035	.053	.075	.103
250	.000	.001	.002	.005	.010	.018	.029	.043	.061	.083
275	.000	.001	.002	.004	.009	.015̄	.024	.035	.050	.069
300	.000	.000	.002	.004	.007	.013	.020	.030	.042	.058
325	.000	.000	.001	.003	.006	.011	.017	.025	.036	.049
350	.000	.000	.001	.003	.005	.009	.015̄	.022	.031	.042
375	.000	.000	.001	.002	.005̄	.008	.013	.019	.027	.037
400	.000	.000	.001	.002	.004	.007	.011	.017	.024	.033
425	.000	.000	.001	.002	.004	.006	.010	.015̄	.021	.029
450	.000	.000	.001	.002	.003	.006	.009	.013	.018	.026
475	.000	.000	.001	.001	.003	.005̄	.008	.012	.017	.023
500	.000	.000	.001	.001	.003	.004	.007	.011	.015	.021
550	.000	.000	.000	.001	.002	.004	.006	.009	.012	.017
600	.000	.000	.000	.001	.002	.003	.005̄	.007	.011	.014
650	.000	.000	.000	.001	.002	.003	.004	.006	.009	.012
700	.000	.000	.000	.001	.001	.002	.004	.005	.008	.011
750	.000	.000	.000	.001	.001	.002	.003	.005̄	.007	.009
800	.000	.000	.000	.000	.001	.002	.003	.004	.006	.008
850	.000	.000	.000	.000	.001	.002	.002	.004	.005	.007
900	.000	.000	.000	.000	.001	.001	.002	.003	.005̄	.006
950	.000	.000	.000	.000	.000	.001	.002	.003	.004	.006
1000	.000	.000	.000	.000	.001	.001	.002	.003	.004	.005
1100	.000	.000	.000	.000	.001	.001	.001	.002	.003	.004
1200	.000	.000	.000	.000	.000	.001	.001	.002	.003	.004
1300	.000	.000	.000	.000	.000	.001	.001	.002	.002	.003
1400	.000	.000	.000	.000	.000	.001	.001	.001	.002	.003
1500	.000	.000	.000	.000	.000	.000	.001	.001	.002	.002
1600	.000	.000	.000	.000	.000	.000	.001	.001	.001	.002
1700	.000	.000	.000	.000	.000	.000	.001	.001	.001	.002
1800	.000	.000	.000	.000	.000	.000	.001	.001	.001	.002
1900	.000	.000	.000	.000	.000	.000	.000	.001	.001	.001
2000	.000	.000	.000	.000	.000	.000	.000	.001	.001	.001
2100	.000	.000	.000	.000	.000	.000	.000	.001	.001	.001
2200	.000	.000	.000	.000	.000	.000	.000	.001	.001	.001
2300	.000	.000	.000	.000	.000	.000	.000	.001	.001	.001
2400	.000	.000	.000	.000	.000	.000	.000	.000	.001	.001
2500	.000	.000	.000	.000	.000	.000	.000	.000	.001	.001
2600	.000	.000	.000	.000	.000	.000	.000	.000	.001	.001
2700	.000	.000	.000	.000	.000	.000	.000	.000	.001	.001
2800	.000	.000	.000	.000	.000	.000	.000	.000	.000	.001
2900	.000	.000	.000	.000	.000	.000	.000	.000	.000	.001
3000	.000	.000	.000	.000	.000	.000	.000	.000	.000	.001
3100	.000	.000	.000	.000	.000	.000	.000	.000	.000	.001
3200	.000	.000	.000	.000	.000	.000	.000	.000	.000	.001
3300	.000	.000	.000	.000	.000	.000	.000	.000	.000	.000
3400	.000	.000	.000	.000	.000	.000	.000	.000	.000	.000
3500	.000	.000	.000	.000	.000	.000	.000	.000	.000	.000
3600	.000	.000	.000	.000	.000	.000	.000	.000	.000	.000
3700	.000	.000	.000	.000	.000	.000	.000	.000	.000	.000
3800	.000	.000	.000	.000	.000	.000	.000	.000	.000	.000
3900	.000	.000	.000	.000	.000	.000	.000	.000	.000	.000
4000	.000	.000	.000	.000	.000	.000	.000	.000	.000	.000
5000	.000	.000	.000	.000	.000	.000	.000	.000	.000	.000

* From "Highway Curves, 4th Edition" by Howard C. Ives and Philip Kissam; used by permission

for Subchords; Given R (Subtract)*

R	Lengths of Arcs in Feet									
	55	60	65	70	75	80	85	90	95	100
100	0.691	0.896	1.138	1.420	1.745	2.116	2.536	3.007	3.532	4.115
110	.571	.741	.942	1.175	1.444	1.752	2.099	2.489	2.925	3.408
120	.480	.623	.792	.988	1.21$\bar{5}$	1.473	1.766	2.09$\bar{5}$	2.462	2.869
130	.409	.531	.67$\bar{5}$	.843	1.036	1.256	1.506	1.787	2.100	2.447
140	.353	.458	.582	.727	.894	1.084	1.300	1.542	1.812	2.112
150	.308	.399	.507	.633	.779	.94$\bar{5}$	1.133	1.344	1.580	1.842
160	.270	.351	.446	.557	.68$\bar{5}$	.831	.996	1.182	1.389	1.620
170	.240	.311	.395	.494	.607	.736	.883	1.047	1.231	1.436
180	.214	.277	.353	.440	.541	.657	.788	.93$\bar{5}$	1.099	1.281
190	.192	.249	.317	.395	.486	.590	.707	.839	.987	1.150
200	.173	.22$\bar{5}$	.286	.357	.439	.532	.638	.757	.891	1.038
225	.137	.178	.226	.282	.347	.421	.50$\bar{5}$	.599	.704	.821
250	.111	.144	.183	.228	.281	.341	.409	.485	.571	.665
275	.092	.119	.151	.189	.232	.282	.338	.401	.472	.550
300	.077	.100	.127	.159	.195	.23$\bar{5}$	.284	.337	.396	.462
325	.066	.085	.108	.135	.166	.202	.242	.287	.338	.394
350	.056	.073	.093	.117	.143	.174	.209	.248	.291	.340
375	.049	.064	.081	.102	.12$\bar{5}$	.152	.182	.216	.254	.294
400	.043	.056	.072	.089	.110	.133	.160	.190	.223	.260
425	.038	.050	.063	.079	.097	.118	.142	.168	.198	.231
450	.034	.044	.057	.071	.087	.10$\bar{5}$	.126	.150	.176	.206
475	.031	.040	.051	.063	.078	.09$\bar{5}$	.113	.13$\bar{5}$	.158	.18$\bar{5}$
500	.028	.036	.046	.057	.070	.085	.102	.121	.143	.167
550	.023	.030	.038	.047	.058	.071	.08$\bar{5}$	.100	.118	.138
600	.019	.025	.032	.040	.049	.059	.071	.08$\bar{5}$	.099	.116
650	.016	.021	.027	.034	.042	.050	.061	.072	.08$\bar{5}$	.099
700	.014	.018	.023	.029	.036	.044	.052	.062	.073	.085
750	.012	.016	.020	.025	.031	.038	.046	.054	.063	.074
800	.011	.014	.018	.022	.027	.033	.040	.048	.056	.065
850	.010	.012	.016	.020	.024	.030	.03$\bar{5}$	.042	.049	.057
900	.009	.011	.014	.018	.022	.026	.032	.037	.044	.052
950	.008	.010	.013	.016	.020	.024	.028	.034	.040	.046
1000	.007	.009	.011	.014	.018	.021	.026	.030	.036	.042
1100	.006	.007	.009	.012	.01$\bar{5}$	.018	.021	.025	.030	.03$\bar{5}$
1200	.00$\bar{5}$	.006	.008	.010	.012	.01$\bar{5}$	.018	.021	.02$\bar{5}$	.029
1300	.004	.005	.007	.008	.010	.013	.015	.018	.021	.02$\bar{5}$
1400	.004	.00$\bar{5}$	.006	.007	.009	.011	.013	.015	.018	.021
1500	.003	.004	.005	.006	.008	.010	.011	.013	.016	.018
1600	.003	.004	.004	.006	.007	.008	.010	.012	.014	.016
1700	.002	.003	.004	.00$\bar{5}$	.006	.007	.009	.011	.012	.014
1800	.002	.003	.004	.004	.005	.007	.008	.009	.011	.013
1900	.002	.002	.003	.004	.00$\bar{5}$	.006	.007	.008	.010	.012
2000	.002	.002	.003	.004	.004	.005	.006	.008	.009	.011
2100	.002	.002	.003	.003	.004	.005	.006	.007	.008	.009
2200	.001	.002	.002	.003	.004	.004	.005	.006	.007	.009
2300	.001	.002	.002	.003	.003	.004	.00$\bar{5}$	.006	.007	.008
2400	.001	.002	.002	.002	.003	.004	.004	.005	.006	.007
2500	.001	.001	.002	.002	.003	.003	.004	.00$\bar{5}$	.006	.007
2600	.001	.001	.002	.002	.003	.003	.004	.004	.005	.006
2700	.001	.001	.002	.002	.002	.003	.003	.004	.00$\bar{5}$	.006
2800	.001	.001	.001	.002	.002	.003	.003	.004	.00$\bar{5}$	.005
2900	.001	.001	.001	.002	.002	.003	.003	.004	.004	.00$\bar{5}$
3000	.001	.001	.001	.002	.002	.002	.003	.003	.004	.00$\bar{5}$
3100	.001	.001	.001	.001	.002	.002	.003	.003	.004	.004
3200	.001	.001	.001	.001	.002	.002	.003	.003	.003	.004
3300	.001	.001	.001	.001	.002	.002	.002	.003	.003	.004
3400	.001	.001	.001	.001	.002	.002	.002	.003	.003	.004
3500	.001	.001	.001	.001	.001	.002	.002	.003	.003	.003
3600	.001	.001	.001	.001	.001	.002	.002	.002	.003	.003
3700	.001	.001	.001	.001	.001	.002	.002	.002	.003	.003
3800	.000	.001	.001	.001	.001	.001	.002	.002	.002	.003
3900	.000	.001	.001	.001	.001	.001	.002	.002	.002	.003
4000	.000	.001	.001	.001	.001	.001	.002	.002	.002	.003
5000	.000	.000	.000	.001	.001	.001	.001	.001	.001	.002

of the publishers, John Wiley & Sons, Inc., New York.

Answers to Even-Numbered Problems

2-2.

	Bearing	Azimuth$_N$	Azimuth$_S$
a.	N30°40′E	30°40′	210°40′
b.	N59°50′W	300 10	120 10
c.	N 9°20′W	350 40	170 40
d.	N40°20′W	319 40	139 40
e.	N10°30′E	10 30	190 30
f.	S 0°30′E	179 30	359 30
g.	S89°40′E	90 20	270 20
h.	S70°00′W	250 00	70 00
i.	S20°40′W	200 40	20 40
j.	S29°30′E	150 30	330 30
k.	N19°50′W	340 10	160 10
l.	S49°30′E	130 30	310 30

2-4. a. S30°40′W   b. 300°10′   c. 170°40′
     d. S40°20′E   e. 190°30′   f. N0°30′W
     g. 270°20′   h. 70°00′   i. N20°40′E
     j. 330°30′   k. 340°10′   l. 310°30′

2-6. 100°40′   b. 49°30′   c. 89°50′   d. 94°50′

2-8. +8.47

2-10. +53.47

2-12. −8.74

2-14. a. S or A   b. A   c. A   d. S   e. B
     f. S   g. B   h. A   i. S or B   j. A

2-16. a. 1:3820; fourth   b. 1:28,571; first
     c. 1:6409; third   d. 1:13,146; second
     e. 1:17,543; second   f. 1:2500; fourth
     g. .012 first   h. .014 second
     i. .031 third   j. .013 second
     k. .034 third   l. .004 first

CHAPTER 3

3-2. Recorded distance $\qquad = \quad$ 982.750 ft
$\quad C_t = (9.83)(-0.004) \qquad = - \quad$ 0.039
$\quad C_f = (0.0983)(13 - 68)/15.5 = - \quad$ 0.349
$\quad C_h = (21)^2/800 \qquad = - \quad$ 0.551
$\quad C_h = (8)^2/1166 \qquad = - \quad$ 0.055
$\qquad\qquad$ Corrected distance $= \quad \overline{981.756}$
$\qquad\qquad\qquad$ Call it $\quad$ 981.76 ft

3-4. Recorded distance $\qquad$ 1062.140 ft
$\quad C_t = (10.62)(+0.016) \qquad = + \quad$ 0.170
$\quad C_f = (0.1062)(82 - 68)/15.5 = + \quad$ 0.096
$\quad C_h = (1.6)^2/400 \qquad = - \quad$ 0.006
$\quad C_h = (5.6)^2/1724 \qquad = - \quad$ 0.018
$\qquad\qquad$ Corrected distance $= \quad \overline{1062.382}$
$\qquad\qquad\qquad$ Call it $\quad$ 1062.38 ft

3-6. Recorded distance $\qquad$ 3259.580 ft
$\quad C_t = (32.60)(+0.008) \qquad = + \quad$ 0.261
$\quad C_f = (0.3260)(87 - 68)/15.5 = + \quad$ 0.400
$\quad C_h = -1000 \text{ vers } 1°09' \qquad = - \quad$ 0.200
$\quad C_h = -2260 \text{ vers } 1°03' \qquad = - \quad$ 0.384
$\qquad\qquad$ Corrected distance $= \quad \overline{3259.657}$
$\qquad\qquad\qquad$ Call it $\quad$ 3259.66 ft

CHAPTER 4

4-2. See Fig. Ans. Problem 4-2.

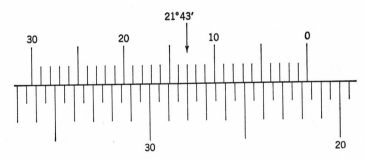

Fig. Ans. Problem 4-2

4-4. See Fig. Ans. Problem 4-4.

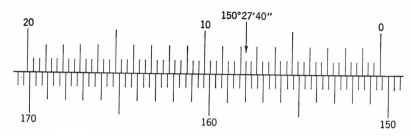

Fig. Ans. Problem 4-4

4-6. See Fig. Ans. Problem 4-6.

Fig. Ans. Problem 4-6

4-8. (*a*) Horizontal angles by negligible amount. (*b*) No others. (*c*) Yes. (*d*) Measure D. and R.

4.10. (*a*) Horizontal angles between points widely separated by vertical angles. (*b*) Vertical angles slightly. (*c*) No. (*d*) Impossible for vertical angles; measure D. and R. for horizontal angles.

4-12. (*a*) and (*b*) Either or both horizontal or vertical angles. (*c*) Yes. (*d*) Measure D. and R.

4-14. (*a*) and (*b*) No angles. (*c*) No. (*d*) Unnecessary.

4-16. (*a*) Vertical angles. (*b*) No other. (*c*) No. (*d*) Measuring D. and R. helps but does not eliminate.

4-18. (*a*) Vertical angles. (*b*) Horizontal angles when vertical angles to point differ widely. (*c*) No. (*d*) Level with upper clamp tight. Repeat any operation so that circle starts 180° from original position.

4-20. Secular, diurnal, annual, irregular.

4-22.

°	′	″	″	Av. ″	°	′	″	°	′	″	Cor. ″	°	′	″
0	0	00	30	15										
117	58	15	$\overline{45}$	00	117	57	45							
347	47	00	$\overline{45}$	$\overline{52.5}$	347	46	37.5							
					57	57	46.2	117	57	46.2	+2.5	117	57	48.7
0	0	00	15	7.5										
174	19	30	45	37.5	174	19	30							
325	57	15	$\overline{30}$	$\overline{52.5}$	325	56	45							
					54	19	27.5	174	19	27.5	+2.5	174	19	30.0
0	0	00	15	7.5										
67	43	00	$\overline{45}$	$\overline{52.5}$	67	42	45							
46	16	15	$\overline{45}$	00	46	15	52.5							
					7	42	38.8	67	42	38.8	+2.5	67	42	41.3
								358	118	112.5	+7.5	358	118	120.0

4-24.

°	′	A	B	Av.	°	′	″	°	′	″	Cor.	°	′	″
		″	″	″							″			
0	0	00	$\overline{45}$	$\overline{52.5}$										
83	54	15	45	30	83	54	37.5							
143	27	30	30	30	143	27	37.5							
					23	54	36.2	83	54	36.2	−0.6	83	54	35.6
0	0	00	30	15										
23	02	30	30	30	23	02	15							
138	14	15	45	30	138	14	15							
					23	02	22.5	23	02	22.5	−0.6	23	02	21.9
0	0	00	15	7.5										
18	09	30	60	45	18	09	37.5							
108	59	00	30	15	108	59	07.5							
					18	09	51.2	18	09	51.2	−0.6	18	09	50.6
0	0	00	$\overline{30}$	$\overline{45}$										
234	52	45	75	60	234	53	15							
329	19	15	$\overline{45}$	00	329	19	15							
					54	53	12.5	234	53	12.5	−0.6	234	53	11.9
								358	118	122.4	−2.4	358	118	120.0

4-26.

°	′	A	B	Av.	°	′	″	°	′	″	Cor.	°	′	″
		″	″	″							″			
0	00	00	20	10										
82	10	20	30	25	82	10	15							
133	02	00	$\overline{40}$	$\overline{50}$	133	01	40							
					22	10	16.7	82	10	16.7	− 4.7	82	10	12.0
0	00	00	$\overline{40}$	$\overline{50}$										
67	29	00	$\overline{50}$	$\overline{55}$	67	29	05							
44	54	40	30	35	44	54	45							
					7	29	07.5	67	29	07.5	− 4.7	67	29	02.8
0	00	00	$\overline{50}$	$\overline{55}$										
210	20	40	50	45	210	20	50							
182	05	00	$\overline{50}$	$\overline{55}$	182	05	00							
					30	20	50	210	20	50.0	− 4.8	210	20	45.2
								359	59	74.2	−14.2	359	59	60.0

4-28.   *Interior Angles*

A	B	C	D	E
88 30	40 15	40 15	50 15	89 00
−22 15	− 22 45	+51 45	+ 31 45	+ 32 15
66 15	− 17 30	92 00	− 82 00	121 15
	179 60		180	
	162 30		98 00	

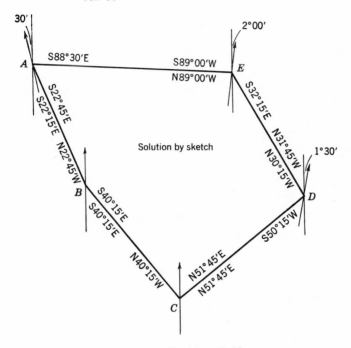

Fig. Ans. Problem 4-28

Computation by Interior Angles

A = 66 15	BC − S 40°15′E
B = 162 30	92 00
C = 92 00	CD N 51 45 E
D = 98 00	+ 98 00
E = 121 15	−149 45
Sum = 539 60	179 60
	DE − N 30 15 W
180	121 15
×3	S 91 00 W =
540	EA − N 89 00 W
	66 15
	AB − S 22 45 E
	162 30
	−139 45
	179 60

CHAPTER 5

5-2.

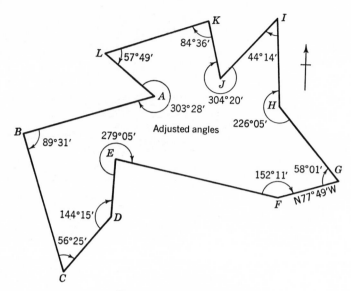

Fig. Ans. Problem 5-2

Sta.	Field		Cor., minutes	Adjusted	Course	Bearings
A	303°	30′	−2	303°28′	FG	N77°49′E
B	89	33	−2	89 31	GH	N44°10 W
C	56	27	−2	56 25	HI	N 1 55 E
D	144	17	−2	144 15	IJ	S46 09 W
E	279	07	−2	279 05	JK	N 9 31 W
F	152	13	−2	152 11	KL	S75 05 W
G	58	03	−2	58 01	LA	S47 06 E
H	226	07	−2	226 05	AB	S76 22 W
I	44	16	−2	44 14	BC	S14 07 E
J	304	22	−2	304 20	CD	N42 18 E
K	84	38	−2	84 36	DE	N 6 33 E
L	57	51	−2	57 49	EF	S74 22 E
	1796°	264′		1796°240′		
	+4	−240				
	1800°	24′				
	−1800					
Error		+24′				

5-4.

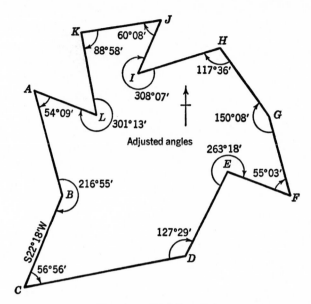

Fig. Ans. Problem 5-4

Sta.	Field		Cor., minutes	Adjusted		Course	Bearing
A	54°	08′	+1	54°	09′	BC	S22°18′W
B	216	54	+1	216	55	CD	N79 14 E
C	56	55	+1	56	56	DE	N26 43 E
D	127	28	+1	127	29	EF	S69 59 E
E	263	17	+1	263	18	FG	N14 56 W
F	55	02	+1	55	03	GH	N44 48 W
G	150	07	+1	150	08	HI	S72 48 W
H	117	35	+1	117	36	IJ	N20 55 E
I	308	06	+1	308	07	JK	S81 03 W
J	60	07	+1	60	08	KL	S 9 59 E
K	88	57	+1	88	58	LA	N68 46 W
L	301	12	+1	301	13	AB	S14 37 E
	1795°	288′		1795°300′			
	+4	−240′					
	1799°	48′					
	1800	00					
Error		−12′					

5-6.

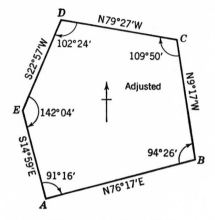

Fig. Ans. Problem 5-6

Sta.	Bearing	Unadjusted		Corrections		Adjusted	
		Lat.	Dep.	L	D	N	E
A						868.59	461.57
	N76°17′E	+131.39	+538.29	+.02	+.14	+131.41	+538.43
B						1000.00	1000.00
	N 9 17 W	+419.74	− 68.61	+.01	+.10	+419.75	− 68.51
C						1419.75	931.49
	N79 27 W	+ 78.01	−418.85	+.01	+.10	+ 78.02	−418.75
D						1497.77	512.74
	S22 57 W	−317.95	−134.64	+.01	+.09	−317.94	−134.55
E						1179.83	378.19
	S14 59 E	−311.25	+ 83.30	+.01	+.08	−311.24	+ 83.38
A						868.59	461.57
		−0.06	−0.51	+.06	+.51		

$$\frac{0.51}{2073} = 1:4100$$

5-8.

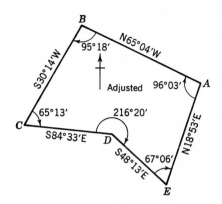

Fig. Ans. Problem 5-8

Sta.	Bearing	Unadjusted		Corrections		Adjusted	
		Lat.	Dep.	L	D	N	E
A						1425.55	1145.65
	N65°04′W	+236.19	−508.05	− .08	+ .08	+236.11	−507.97
B						1661.66	637.68
	S30 14 W	−418.32	−243.79	− .07	+ .07	−418.39	−243.72
C						1243.27	393.96
	S84 33 E	− 35.66	+373.72	− .05	+ .05	− 35.71	+373.77
D						1207.56	767.73
	S48 13 E	−207.52	+232.23	− .04	+ .04	−207.56	+232.27
E						1000.00	1000.00
	N18 53 E	+425.62	+145.58	− .07	+ .07	+425.55	+145.65
A						1425.55	1145.65
		+ 0.31	− 0.31	−0.31	+0.31		

$$\frac{0.44}{2181} = 1:5000$$

5-10.

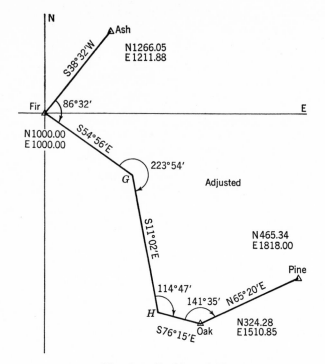

Fig. Ans. Problem 5-10

Sta.	Bearing	Unadjusted		Corrections		Adjusted	
		Lat.	Dep.	L	D	N	E
Ash							
	S38°32′W						
Fir		1000.00	1000.00			1000.00	1000.00
	S54 56 E	−199.45	+284.14	− .13	.00	−199.58	+284.14
G						800.42	1284.14
	S11 02 E	−441.51	+ 86.09	− .17	+ .01	−441.68	+ 86.10
H						358.74	1370.24
	S76 15 E	− 34.41	+140.61	− .05	.00	− 34.46	+140.61
Oak						324.28	1510.85
	N65 20 W						
Pine							
		324.63	1510.84	−0.35	+0.01		
	Oak	−324.28	−1510.85				
		+ 0.35	− 0.01				

$$\frac{0.35}{942} = 1:2700$$

5-12.

Angles and sides				Logs; Log sines	Cor. logs	Final logs	Final sides
CA	375.42			2.574517	−21	2.574496	375.40
B-CA	70°08′ 53″	+2″	55″	−9.973394			
C-AB	48 06 25	+2	27	2.601123 +9.871806			
A B				2.472929	−21	2.472908	297.10
A-BC	61 44 36	+2	38	+9.944897			
BC				2.546020	−21	2.545999	351.56
	179°58′114″	+6″	120″				
E-BC	82°36′ 08″	+1″	09″	−9.996370			
B-EC	48 31 21	+1	22	2.549650 +9.874609			
EC				2.424259	−21	2.424238	265.61
C-BE	48 52 28	+1	29	+9.876953			
BE				2.426603	−21	2.426582	267.04
	179°59′ 57″	+3″	60″				
D-EB	52°04′ 07″	−1″	06″	−9.896936			
E-BD	49 33 46	−1	45	2.529667 +9.881450			
BD				2.411117	−21	2.411096	257.69
B-DE	78 22 10	−1	09	+9.990990			
DE				2.520657	−21	2.520636	331.62
	180°00′ 03″	−3″	60″				

	Logs
DE computed	2.520657
DE measured, 331.60	2.520615
Error in logs	+42
Correction log	−21

CHAPTER 6

6-2.

Sta.	+	H.I.	−	Rod	Elev.	
B.M. 2	9.605	61.768			52.163	
T.P. 1	4.572	58.919	7.421		54.347	
T.P. 2	7.241	60.425	5.735		53.184	
B.M. 7	3.645	55.144	8.926		51.499	
T.P. 3	7.283	53.785	8.642		46.502	*Error*
B.M. 2			1.634		52.151	−0.012
	32.346		32.358			−0.012

6-4.

Sta.	+	H.I.	−	Rod	Elev.	
B.M. 4	2.643	85.427			82.784	
T.P. 1	4.771	82.596	7.602		77.825	
B.M. 9	6.039	78.801	9.834		72.762	
T.P. 2	7.541	82.693	3.649		75.152	
T.P. 3	10.455	87.422	5.726		76.967	*Error*
B.M. 4			4.652		82.770	−0.014
	31.449		31.463			−0.014

6-6.

Sta.	+	H.I.	−	Rod	Elev.	
B.M. 6	7.111	81.749			74.638	
T.P. 1	4.787	83.477	3.059		78.690	
T.P. 2	5.487	86.352	2.612		80.865	
B.M. 11	4.816	81.525	9.643		76.709	
T.P. 3	6.035	77.576	9.984		71.541	*Error*
B.M. 6			2.958		74.618	−0.020
	28.236		28.256			−0.020

6-8.

Sta.	+	H.I.	−	Rod	Elev.	
B.M.	8.562	75.990			67.428	
T.P. 1	9.714	81.627	4.077		71.913	
T.P. 2	4.758	83.991	2.394		79.233	
B.M. 3	2.625	74.971	11.645		72.346	
T.P. 4	8.481	76.697	6.755		68.216	*Error*
B.M.			9.262		67.435	+0.007
	34.140		34.133			+0.007

6-10.

Sta.	+	H.I.	−	Rod	Elev.	
B.M.	7.546	46.918			39.372	
T.P. 1	6.851	50.127	3.642		43.276	
T.P. 2	8.221	52.928	5.420		44.707	
T.P. 3	10.687	57.602	6.013		46.915	
T.P. 4	1.632	50.488	8.746		48.856	*Error*
B.M.			11.103		39.385	+0.013
	34.937		34.924			+0.013

6-12.

Sta.	+	H.I.	−	Rod	Elev.	
B.M.	1.732	93.378			91.646	
T.P. 1	4.231	88.333	9.276		84.102	
T.P. 2	5.524	85.630	8.227		80.106	
T.P. 3	9.063	90.074	4.619		81.011	
T.P. 4	8.362	92.189	6.247		83.827	*Error*
B.M.			0.559		91.630	−0.016
	28.912		28.928			−0.016

6-14.

Sta.	+	H.I.	−	Rod	Elev.	
B.M. 14	4.674	40.466			35.792	
0 + 0				7.1	33.4	
1 + 0				10.7	29.8	
2 + 0				12.3	28.2	
3 + 0				7.8	32.7	
T.P. 1	8.149	41.773	6.842		33.624	
4 + 0				4.0	37.8	
5 + 0				2.7	39.1	
T.P. 2	9.677	41.836	9.614		32.159	*Check*
6 + 0				6.8	35.0	35.792
7 + 0				9.6	32.2	+22.500
8 + 0				6.6	35.2	58.292
9 + 0				5.8	36.0	−23.623
B.M. 15			7.167		34.669	34.669
	22.500		23.623		B.M. 15	34.680
					Error	− 0.011

See also Fig. Ans. Problem 6-14.

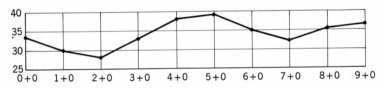

Part of Fig. Ans. Problem 6-14

6-16.

Sta.	+	H.I.	−	Rod	Elev.	
B.M. 16	5.628	70.866			65.238	
0 + 0				12.8	58.1	
1 + 0				8.1	62.8	
2 + 0				4.7	66.2	
3 + 0				8.4	62.5	
T.P. 1	3.754	70.583	4.037		66.829	
4 + 0				12.8	57.8	
5 + 0				10.6	60.0	
T.P. 2	6.281	68.743	8.121		62.462	*Check*
6 + 0				6.5	62.2	65.238
7 + 0				6.0	62.7	+15.663
8 + 0				5.4	63.3	80.901
9 + 0				11.6	57.1	−17.379
B.M. 17			5.221		63.522	63.522
	15.663		17.379		B.M. 17	63.534
					Error	− 0.012

See also Fig. Ans. Problem 6-16.

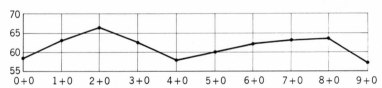

Part of Fig. Ans. Problem 6-16

6-18. See Fig. Ans. Problem 6-18.

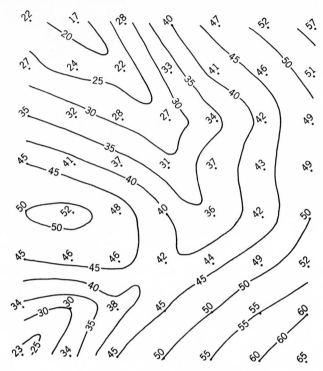

Fig. Ans. Problem 6-18

6-20. See Fig. Ans. Problem 6-20.

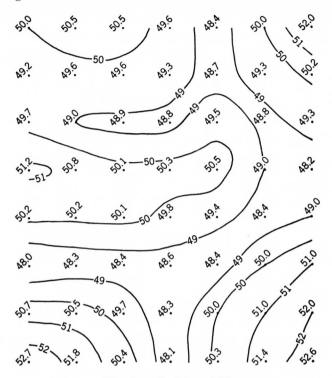

Fig. Ans. Problem 6-20

CHAPTER 7

7-2. See Fig. Ans. Problem 7-2.

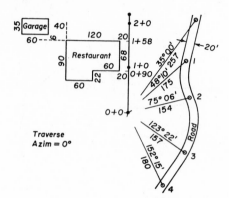

Fig. Ans. Problem 7-2

7-4. See Fig. Ans. Problem 7-4.

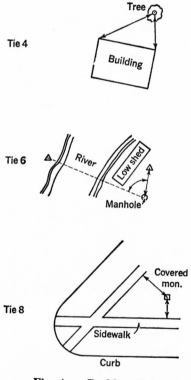

Fig. Ans. Problem 7-4

7-6. See Fig. Ans. Problem 7-6.

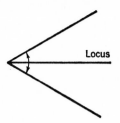

Fig. Ans. Problem 7-6

7-8. 1. Two distances should be measured at right angles. 2. A distance with an angle should be measured along the side of the angle. 3. Two angles from the control should add to 90°.

CHAPTER 8

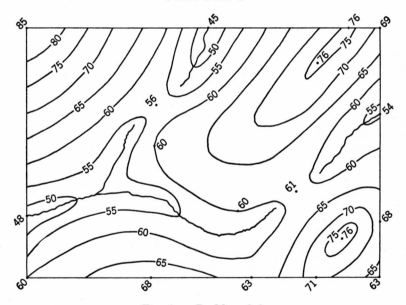

Fig. Ans. Problem 8-2

8-4.a See also Fig. Ans. Problem 8-4*a* and 8-4*b*.

Course	S	VA	H	V	Cor.	Elev. V
AB	4.03	$-3°31'$				92.07
	4.03	$-3°33'$				
	4.03	$-3°32'$	401	-24.78	$+0.04$	-24.74
BC	3.82	$+2°14'$				67.33
	3.82	$+2°14'$				
	3.82	$+2°14'$	381	$+14.86$	$+0.04$	$+14.90$
CD	3.11	$-1°55'$				82.23
	3.13	$-1°53'$				
	3.12	$-1°54'$	312	-10.33	$+0.03$	-10.30
DA	2.70	$+4°18'$				71.93
	2.68	$+4°18'$				2
	2.69	$+4°18'$	267	$+20.12$	$+0.03$	$+20.14$
			1361	-0.13	$+0.13$	92.07

$$\text{Estimated error } 0.02 \sqrt{13.61} = 0.07$$

$$\text{Cor.} = \frac{0.13}{13.61} \times \text{Col. } H$$

		PLANT SITE #2						Ch Smith	Date
								TT Jones	Windy
Sta.	S	Azim	V∡	H	V		Elev.	Rod Kole	
⊼A	h.i.		4.73				92.07		
D	2.68	87° 30'	$-4°18'$						
B	4.03	176° 58'	$-3°31'$	4 01					
①	1.48	155° 30'	$-3°27'$	147	-8.9		83.2		Bld. 50
Mon.BM	1.46	311° 32'	$+3°07'$	146	$+7.93$		100.00	assumed	① 145
⊼ B	h.i.		4.59				67.33		
A	4.03	356°58'	$+3°33'$						
C	3.82	70° 42'	$+2°14'$	381					
Stream	1.47	245°20'	$-1°41'$	147	-4.3		63.0		
Mon.	2.08	219° 50'	$-0°35'$	208	-2.1		65.2		
Stream	1.33	108°15'	$-0°10'$	133	-0.4		66.9		
⊼C	h.i.		4.82				82.23		
B	3.82	250°42'	$-2°14'$						
D	3.11	338°45'	$-1°55$	312					
Saddle	1.46	291° 25'	$-3°57'$	145	-10.0		72.2		
Mon.	3.02	159°40'	$+0°06'$	302	$+0.5$		82.7		
Boundary	1.21	121°50'	$+3°43'$	120	$+7.9$		90.1		
⊼D	h.i.		4.61				71.93		
C	3.13	158°45'	$+1°53'$						
A	2.70	267°28'	$+4°18'$	267					
②	1.19	249°20'	$+5°08'$	118	$+10.6$		82.5		②
Stream	1.43	55° 05'	$-3°08'$	142	-7.8		64.1		Bld.
Mon.	2.31	69° 20'	$-0°28'$	231	-1.9		70.0		

Fig. Ans. Problem 8-4*a*

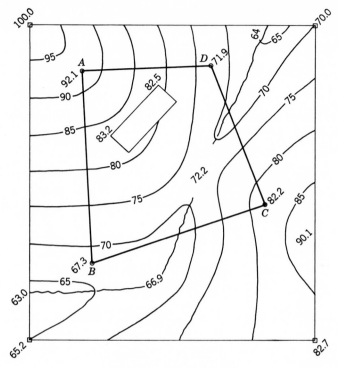

Fig. Ans. Problem 8-4*b*

CHAPTER 9

9-2. 0.017 ft

9-4.

$$\frac{21.23}{5.868} = +3.6179$$

.203 × 3.6179 = 0.7344
.50 × 3.6179 = 1.8090
.165 × 3.6179 = 0.5970

Sta.	Grade	Grade used	Sta.	Grade	Grade used
6 + 29.7	51.26	51.26	+ 50	62.8484	62.85
+ 50	51.9944	51.99	10 + 0	64.6574	64.66
7 + 0	53.8034	53.80	+ 50	66.4664	66.47
+ 50	55.6124	55.61	11 + 0	68.2754	68.28
8 + 0	57.4214	57.42	+ 50	70.0844	70.08
+ 50	59.2304	59.23	12 + 0	71.8934	71.89
9 + 0	61.0394	61.04	+ 16.5	72.4904	72.49

9-6.

3'6¼''	6'3''
4'9⅛''	7'9¾''
9'2¾''	2'11¼''
10'1⅞''	5'0¾''
8'8⅝''	6'8''

9-8.

2.56	4.51
1.84	7.24
3.64	5.40
6.29	8.61
9.71	10.44

9-10.

Sta.	Grade	Elev. mark	Cut or fill	Write on mark
0 + 0	47.28	46.17	F 1.11	F 1'1⅜''
0 + 50	45.28	41.62	F 3.66	F 3'7⅞''
1 + 0	43.28	45.10	C 1.82	C 1'9⅞''
1 + 50	41.28	40.83	F 0.45	F 0'5⅜''
2 + 0	39.28	36.15	F 3.13	F 3'1½''
2 + 50	37.28	42.14	C 4.86	C 4'10⅜''
3 + 0	35.28	34.75	F 0.53	F 0'6⅜''
3 + 50	33.28	35.29	C 2.01	C 2'0⅛''
4 + 0	31.28	32.67	C 1.39	C 1'4⅝''
4 + 50	29.28	33.48	C 4.20	C 4'2⅜''

9-12.

Sta.	+	H.I.	−	Rod	Elev.	Grade	Mark	Grade rod compt.		Rod on ground
B.M.		81.29						81.29		
0 + 0				0.97	80.32	80.32	G	80.32		1.4
								0.97	81.29	
+50				0.97	80.32	81.32	F 1′0″		81.32	1.0
								81.29	−0.03	
1 + 0				0.47	80.82	82.32	F 1′6″	82.32		0.6
								−1.03	81.29	
+50				1.47	79.82	83.32	F 3′6″		83.32	1.7
								81.29	−2.03	
2 + 0				1.47	79.82	84.32	F 4′6″	84.32		1.9
								−3.03		
T.P.		92.42							92.42	
+50				2.60	89.82	85.32	C 4′6″		85.32	2.8
								92.42	7.10	
3 + 0				3.10	89.32	86.32	C 3′0″	86.32		3.2
								6.10	92.42	
+50				4.10	88.32	87.32	C 1′0″		87.32	4.2
								92.42	5.10	
4 + 0				6.60	85.82	88.32	F 2′6″	88.32		6.7
								4.10	92.42	
+50				7.60	84.82	89.32	F 4′6″		89.32	7.8
								3.10		

CHAPTER 10

10-2. See Fig. Ans. Problem 10-2.

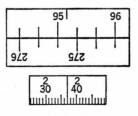

Fig. Ans. Problem 10-2

10-4. See Fig. Ans. Problem 10-4.

Fig. Ans. Problem 10-4

10-6.   0°00′00″
      30 01 40
      60 03 20
      90 05 00
     120 06 40
     150 08 20

10-8.   0°00′
      60 10
     120 20
     180 30
     240 40
     300 50

CHAPTER 11

11-2.

Sta.	Chord	Defl.	Curve data
+ 50			
+26.94 PT	26.94	33°09′15″	$R = 400'$
50 + 0	49.97	31 13 30	
			$\triangle = 66°18′24″$
+ 50		27 38 45	
			$\triangle/2 = 33°09′12″$
49 + 0		24 03 45	
			$= 33°09.2'$
+ 50		20 29 00	
+25.32 PI			$T = 261.29$
48 + 0		16 54 00	
			$L = 462.91$
+ 50		13 19 15	
47 + 0		9 44 15	
+ 50	49.97	6 09 30	
46 + 0	35.96	2 35 00	
+64.03 PC		0	
+ 50			

```
tan 33°09.2' = 0.65322
         R =        400
         T = 261.28800

   Plus  PI = 4825.32
   Less   T =  261.29
   Plus PC = 4564.03
   Add    L =  462.91
   Plus PT = 5026.94

Plus 1st sta. = 4600.00
Less plus PC =  4564.03
                  35.97
                1719
                 323.73
                 359.7
                25179
                3597
                61832.43

        154.58.11 = 2°34.58'
   4.00)618.32.43
```

```
△60°   1.04720
  6°   0.10472
 10'   0.00291
  8'   0.00233
 20"   0.00010
  4"   0.00002
       1.15728
         ×400
     = 462.91200

        1719
          50
       85950
      214.88 = 3°34.88'
  4.00)859.50

       26.94
       1719
      242.46
      269.4
     18858
     2694
     46309.86
     115.77.46 = 1°55.77'
  4.00)463.09.86
```

```
 46   2°34.58'   2°35'00"
      3 34.88
+50   6 09.46    6 09 30
      3 34.88
 47   9 44.34    9 44 15
      3 34.88
+50  13 19.22   13 19 15
      3 34.88
 48  16 54.10   16 54 00
      3 34.88
+50  20 28.98   20 29 00
      3 34.88
 49  24 03.86   24 03 45
      3 34.88
+50  27 38.74   27 38 45
      3 34.88
 50  31 13.62   31 13 30
      1 55.77
 PT  33°09.39'  33 09 15
```

11-4.

Sta.	Chord	Defl.	Curve data
+ 50			$R = 600'$
+49.48 PT	49.45	21°17'15"	
30 + 0	49.97	18 55 30	
			$\triangle = 42°34'28"$
+ 50		16 32 15	
			$\triangle/2 = 21°17'14"$
29 + 0		14 09 00	
			$= 21°17.23'$
+ 50		11 45 45	
+37.42 PI			$T = 233.78$
28 + 0		9 23 30	
			$L = 445.84$
+ 50		6 59 15	
27 + 0	49.97	4 36 00	
+ 50	46.33	2 12 45	
PC + 03.64		0	
26 + 0			

11-6.

$$\begin{array}{lrr}
\text{Plus PI} & = & 1287.93 \\
\text{Less } T_1 & = & -407.14 \\
\hline
\text{Plus PC} & = & 880.79 \\
\text{Add } L_1 & = & +326.26 \\
\hline
\text{Plus PCC} & = & 1207.05 \\
\text{Add } L_2 & = & +379.49 \\
\hline
\text{Plus PT} & = & 1586.54 \\
T_2 & = & 545.52
\end{array}$$

11-18.

$$\Delta_1 = 80°35'14'' \qquad \Delta_2 = 61°20'24''$$

$$\begin{array}{lrr}
\text{Plus PC} & = & 1532.71 \\
\text{Add } L_1 & = & +172.75 \\
\hline
\text{Plus PRC} & = & 1705.46 \\
\text{Add } L_2 & = & +131.49 \\
\hline
\text{Plus PT} & = & 1836.95
\end{array}$$

CHAPTER 12

12-2.

Station		x	x^2	$C = -0.2278x^2$	Tangent elev.	Curve elev.
PVC	$12 + 50$	0	0	0	68.64	68.64
	$13 + 00$	1	1	-0.23	70.34	70.11
	$13 + 50$	2	4	-0.91	72.04	71.13
	$14 + 00$	3	9	-2.05	73.74	71.69
	$14 + 50$	4	16	-3.64	75.44	71.80
	$15 + 00$	5	25	-5.70	77.14	71.44
	$15 + 50$	6	36	-8.20	78.84	70.64
	$16 + 00$	7	49	-11.16	80.54	69.38
	$16 + 50$	8	64	-14.58	82.24	67.66
PVT	$17 + 00$	9	81	-18.45	83.94	65.49

12-4.

Station		x	x^2	$C = 0.2250x^2$	Tangent elev.	Curve elev.
PVC	$7 + 50$	0	0	0	62.71	62.71
	$8 + 00$	1	1	$+0.22$	61.11	61.33
	$8 + 50$	2	4	$+0.90$	59.51	60.41
	$9 + 00$	3	9	$+2.02$	57.91	59.93
	$9 + 50$	4	16	$+3.60$	56.31	59.91
	$10 + 00$	5	25	$+5.62$	54.71	60.33
	$10 + 50$	6	36	$+8.10$	53.11	61.21
	$11 + 00$	7	49	$+11.02$	51.51	62.53
	$11 + 50$	8	64	$+14.40$	49.91	64.31
	$12 + 00$	9	81	$+18.22$	48.31	66.53
PVT	$12 + 50$	10	100	$+22.50$	46.71	69.21

Lowest point:

$$a = g_1 \frac{L}{g_1 - g_2} = -0.032 \frac{500}{-0.032 - 0.058} = 177.78'$$

$$C = \frac{a^2}{L} H = \left(\frac{177.78}{500}\right)^2 22.50 = 2.85'$$

Station	*Elevation*
PVC $= \quad 7 + 50.00$	PVC $= \quad 62.71$
$a = +1 \quad 77.78$	$(-0.032)177.78 = \quad -5.69$
Low point $= \quad 9 + 27.78$	Tangent elev. $= \quad 57.02$
	$C = \quad +2.85$
	Low point $= \quad 59.87$

CHAPTER 13

13-2. Same. Move more.

13-6. About 82.5

13-10. About 86.5

13-4. Opposite. Move less.

13-8. About 53

13-12. About 62

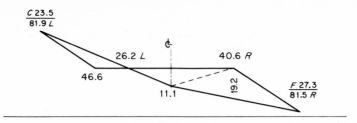

Fig. Ans. Problem 13-14.

13-14

	Cut		*Fill*
	$20.4 \times 23.5 = 479$		$66.8 \times 11.1 =$ 741
			$19.2 \times 81.5 = \underline{1565}$
			2306

$$50 \times {}^{479}\!/_{27} = 887 \text{ cu yd} \qquad 50 \times {}^{2306}\!/_{27} = 4270 \text{ cu yd}$$

CHAPTER 14

USE OF TABLES

14-2. 0.48107

14-6. 0.57780

14-10. 25°14′07″

14-14. 12°14′45″

14-18. 2.379215

14-22. 0.470234

14-26. 5531.70

14-30. −2.90802

14-34. 9.738261 − 10

14-38. 9.839204 − 10

14-42. 24°58′27″   155°01′33″

14-46. 18°10′27″

14-50. 8.429111 − 10

14-54. 2.084223

14-58. 89°24′28″

14-4. 0.31902

14-8. 0.24176

14-12. 22°52′16″

14-16. 28°41′24″

14-20. 8.794586 − 10

14-24. 1.562085

14-28. 0.00268001

14-32. 446.905

14-36. 9.6486440 − 10

14-40. 9.495253 − 10

14-44. 78°52′02″

14-48. 31°27′58″

14-52. 8.429032 − 10

14-56. 1°23′23″   178°36′37″

14-60. 88°38′17″

RIGHT TRIANGLES

M = by machine; L = by logarithms

14-62. M: opp. = 294.76 ft   adj. = 322.96 ft $B = 47°36′48″$
 L: 294.76   322.96   47 36 48

14-64. M: hyp. = 393.23 ft   adj. = 187.75 ft $B = 28°31′13″$
 L: 393.23   187.76   28 31 13

14-66. M: hyp. = 441.45 ft   opp. = 258.05 ft $B = 54°13′43″$
 L: 441.45   258.05   54 13 43

14-68. M: $A = 66°55′11″$ $B = 23°04′49″$ adj. = 169.71 ft
 L: 66 55 12   23 04 48   169.70

14-70. M: $A = 55°00′38″$ $B = 34°59′22″$ opp. = 386.40 ft
 L: 55 00 36   34 59 24   386.40

14-72. M: $A = 53°18′15″$ $B = 36°41′45″$ hyp. = 459.54 ft
 L: 53 18 14   36 41 46   459.54

OBLIQUE TRIANGLES

M = by machine; L = by logarithms

14-74. M: $C = 71°21′22″$ $b = 168.04$ ft $c = 282.82$ ft
 L: 71 21 22   168.04   282.82

14-76. M: $C = 67°15′44″$ $a = 320.89$ ft $b = 387.77$ ft
 L: 67 15 44   320.88   387.77

14-78. M: $B = $ 73°20′30″ $C = 81°35′14″$ $c = 533.93$ ft
 or   106 39 30   48 16 14   402.80
 L: 73 20 37   81 35 07   533.93
 or   106 39 23   48 16 21   402.82

14-80. M: $B = 49°23′22″$ $C = 71°19′15″$ $c = 497.10$ ft
 L: 49 23 29   71 19 08   497.08

14-82. M: $B = 64°09′12″$ $C = 60°08′13″$ $a = 391.82$ ft
 L: 64 09 14   60 08 11   391.83

14-84. M: $B = 93°00'26''$ $C = 19°54'53''$ $a = 438.79$ ft
  L: 93 00 26   19 54 53   438.80
14-86. M: $A = 32°13'45''$ $B = 58°00'28''$ $C = 89°45'47''$
  L: 32 13 45   58 00 28   89 45 47
14-88. M: $A = 113°15'48''$ $B = 23°15'28''$ $C = 43°28'44''$
  L: 113 15 48   23 15 28   43 28 44

APPENDIX A

A-2.

v_i	v_i^2
0.21	0.0441
0.45	0.2025
0.30	0.0900
0.27	0.0729
0.26	0.0676
1.49	0.4771
$\div 5 = 0.30$	0.0954

Expected error $= 0.30 \sqrt{12}$ $= \pm 1.04$

$\sigma = \sqrt{0.0954} \sqrt{12} = \pm 1.07$

$E_{90} = (1.645)(1.07)$ $= \pm 1.76$

Design accuracy $= 1.76/12000$ $= \frac{1}{6818}$

A-4.

v_i	v_i^2
0.009	0.000081
0.019	0.000361
0.007	0.000049
0.018	0.000324
0.007	0.000049
0.060	0.000864
$\div 5 = 0.012$	0.000173

Expected error $= 0.012 \sqrt{12}$ $= 0.042$

$\sigma = \sqrt{0.000173} \sqrt{12}$ $= 0.046$

$E_{90} = (1.645)(0.046)$ $= 0.076$

Design accuracy $= 0.076 \sqrt{5280/12000} = 0.050$

APPENDIX B

B-2.

$$S = \frac{0.024}{3.4} = 0.007$$

Standardized length $= 100.0060$ ft

$$-C_s = +\frac{(0.024)^2(100)^3}{24(10)^2} = +0.2400$$

$$+C_p = +\frac{100(20-10)}{28,000,000(0.007)} = +0.0051$$

$$+C_s = -\frac{(0.024)^2(100)^a}{24(20)^2} = -0.0600$$

$$= \overline{100.1911}$$

Call it 100.191 ft

APPENDIX C

C-2. $\frac{40}{60} \times \frac{D}{2}$. One-third of the distance from C to A.

C-4. 3.557

C-6. The bubble would not remain centered when the telescope was turned in azimuth.

C-8. Balance the horizontal lengths of the plus and minus sights.

APPENDIX D

D-2.

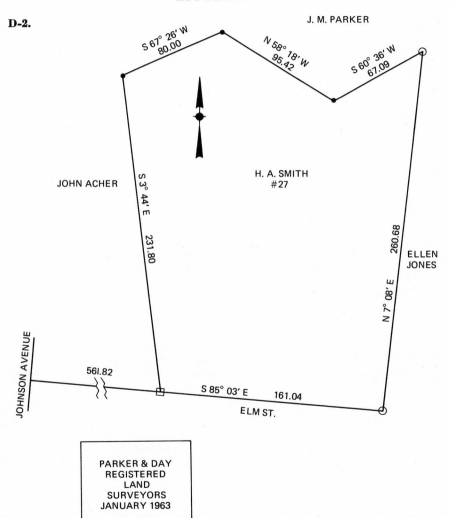

Index